ELECTROCHEMISTRY OF POROUS MATERIALS

ELECTROCHEMISTRY OF POROUS MATERIALS

Second Edition

Antonio Doménech-Carbó

CRC Press is an imprint of the
Taylor & Francis Group, an **informa** business

Second edition published [2021]
by CRC Press
4 Park Square, Milton Park, Abingdon, Oxon, OX14 4RN

and by CRC Press
6000 Broken Sound Parkway NW, Suite 300, Boca Raton, FL 33487-2742

First issued in paperback 2022

[First edition published by CRC Press 2009]
CRC Press is an imprint of Informa UK Limited

Publisher's Note
The publisher has gone to great lengths to ensure the quality of this reprint but points out that some imperfections in the original copies may be apparent.

British Library Cataloguing-in-Publication Data
A catalogue record for this book is available from the British Library

Library of Congress Cataloging-in-Publication Data
Names: Doménech-Carbó, Antonio, author.
Title: Electrochemistry of porous materials / Antonio Doménech Carbó.
Description: Second edition. | Boca Raton : CRC Press, 2021. | Includes bibliographical references and index.
Identifiers: LCCN 2020055151 (print) | LCCN 2020055152 (ebook) | ISBN 9780367366506 (hardback) | ISBN 9780429351624 (ebook)
Subjects: LCSH: Porous materials--Electric properties. | Electrochemistry.
Classification: LCC QD575.P67 C37 2021 (print) | LCC QD575.P67 (ebook) | DDC 620.1/16--dc23
LC record available at https://lccn.loc.gov/2020055151
LC ebook record available at https://lccn.loc.gov/2020055152

ISBN 13: 978-1-03-247426-7 (pbk)
ISBN 13: 978-0-367-36650-6 (hbk)
ISBN 13: 978-0-429-35162-4 (ebk)

DOI: 10.1201/9780429351624

Typeset in Times
by MPS Limited, Dehradun

Dedication

Als meus pares

Contents

Foreword

It is a remarkable feature of modern electrochemistry that research is directed, on one side, to a deeper understanding of the very fundamentals of the elementary steps of charge transfer *at* interfaces and charge propagation *in* phases (i.e. the simplest electrochemical systems) and, on the other, to the investigation of increasingly complex systems (e.g., multi-phase systems in which sometimes several electrochemically active centers are present) where charge propagation may proceed on complex pathways, etc. where electrocatalysis may be involved and electrochemically-initiated chemical conversions take place. These complex systems are not only of applied significance (e.g., in batteries and fuel cells), but they also prompt new developments for the understanding of fundamental processes. Now, the time is ripe for such complicated systems to be studied with all the modern techniques, including the most advanced spectroscopic and microscopic methods. The author of this book—equipped for the undertaking as he has published extensively about it—attempted to survey a specific but large area of modern electrochemical research. Porous materials are very complex with respect to possible electrochemical reactions; the author covers materials from nanopores to micropores, and he treats all these under the aspect of insertion electrochemistry, as electron and ion transfer processes proceed together. The range of different compounds and materials is impressive, and it is very rewarding for the reader to see a presentation of that great variety in one volume. This is a unique book where, for the first time, a comprehensive treatment of the electrochemical features of porous materials is given. Because of the great technological importance of these materials, the book will be welcomed by the electrochemical community and I also have the confidence that it will give an impetus to the theoreticians who may see—in one glance—what interesting and intriguing systems the experimentalist has already studied, along with theoretical questions derived from these investigations.

Fritz Scholz
University of Greifswald, Germany

Preface

Over the last decades, research on porous materials has increased considerably, with applications for sensing, gas storage, catalysis, energy transformation, and storage—among others. The term porous materials designates a wide variety of substances, from clay minerals and silicates to metal oxides, metal-organic frameworks, or even thin films and membranes. Porous metals and carbons can also be included within such systems.

Electrochemistry plays an important role in both research and application of porous materials via electroanalysis, electrosynthesis, sensing, fuel cells, capacitors, electro-optical devices, etc.

The purpose of the current text is to provide an approach to the electrochemistry of porous materials that combines the presentation of a generalized theoretical model with a description of redox processes for different porous materials and a view of their electrochemical applications.

Because of the considerable variety of materials classified as porous, the text will be limited to several groups: porous silicates and aluminosilicates, porous metal oxides and related compounds, porous polyoxometalates, metal-organic frameworks, porous carbons, carbon nanotubes, and several hybrid materials. All these can be viewed as relatively homogeneous from the electrochemical point of view. Metal and metal oxide nanoparticles, 'organic metals', fullerenes, and dendrimers—also regarded as nanostructured materials, displaying distinctive electrochemical features—will not be treated here for obvious reasons of brevity, although their appearance in hybrid materials as modifiers for microporous materials will be discussed.

This book is devoted to conjointly present the advances in electrochemistry of nanostructured materials. More specifically, the text presents the foundations and applications of the electrochemistry of microporous materials with the incorporation of recent developments in applied fields (fuel cells, supercapacitors, etc.) and fundamental research (fractal scaling, photo-electrocatalysis, magnetoelectrochemistry, etc.). The book attempts to make electrochemistry accessible to researchers and graduate students working on the chemistry of materials, but also approximate porous materials chemistry to electrochemists. To provide a reasonable volume of literature, citations are limited to fundamental articles. When possible, textbooks and review articles have been cited or, alternatively, recent articles covering wide citations of previous literature have been used to facilitate access to a more extensive literature for the reader interested in monographic topics.

The book includes part of research performed in collaboration with Elisa Llopis, María José Sabater, Mercedes Alvaro, Pilar Navarro, María Teresa Doménech, Antonio Cervilla, Javier Alarcón, Avelino Corma, Hermenegildo García, and their coworkers who have kindly provided materials for texts and figures. I gratefully acknowledge the accurate revision made by Milivoj Lovrić with respect to theoretical aspects. I would like to express my appreciation and thanks to Fritz Scholz for his friendship and accurate revision of the overall manuscript and its valuable comments, criticisms, and suggestions. Finally, I would like to thank my family for their continuous support, attention, and patience.

Preface to the 2nd edition

During the last ten years, research on the electrochemistry of porous materials has experienced considerable growth. This growth has been oriented in three main lines: the development of theoretical models on the electrochemical behavior of porous electrodes, the incorporation of new porous materials, and the expansion of the applications of such materials to new fields.

This edition attempts to reflect this growth through the incorporation of four new chapters and the reorganization of the contents in the entire text. Recent developments in sensing, energy production and storage, degradation of pollutants, desalination, and drug release have been introduced in the text accompanied by a description of advanced theoretical approaches. To facilitate the revision of the sources, a list of direct citations has been added to the end of each chapter, complemented by a list of additional literature at the end of the text. The author hopes that this new version of the book can offer the reader a comprehensive view of electrochemistry—its origins, recent developments, challenges, and future issues.

Contributors

Antonio Doménech-Carbó (Valencia, Spain, 1953) is a Professor at the Department of Analytical Chemistry, University of Valencia (PhD from 1989). His research is focused on solid-state electrochemistry with particular emphasis on the study of porous materials and the development of electroanalytical methods for archaeometry, conservation, and restoration. He has supervised eight PhD projects and directed several national R+D projects. For his research in cultural heritage, he received the 'Demetrio Ribes' award from the Regional Government of Valencia in 2006 and 2019. He is the author of more than 250 articles, including one IUPAC's technical report, and several books; among them, *Electrochemical Methods in Archeometry, Conservation and Restoration* (Springer, 2009), and *Electrochemistry of immobilized particles and droplets* (2nd ed. Springer, 2014). Currently, he is a member of the editorial board of *ChemTexts* (Springer), topical editor of *Journal of Solid State Electrochemistry* (Springer) and *Periodico di Mineralogía* (La Sapienza University). He is a reviewer at the European Research Council and national research agencies of Argentina, Brazil, Chile, Croatia, Czech Republic, Flanders, France, Kazakhstan, Poland, Romania, Spain, Swiss, and The Netherlands, and a referee of more than 150 indexed journals.

List of Abbreviations

General Concepts:

AFC	Alkaline fuel cell
CA	Chronoamperometry
CDI	Capacitive deionization
CECE	Electrochemical pathway consisting of chemical reaction-electron transfer-chemical reaction-electron transfer successive steps
COF	Covalent organic frameworks
CPE	Constant phase element
CV	Cyclic voltammetry
DDS	Drug delivery system
DFT	Density functional theory
DMFC	Direct methanol fuel cell
EAFM	Electrochemical atomic force microscopy
ECEC	Electrochemical pathway consisting of electron transfer-chemical reaction-electron transfer-chemical reaction successive steps
EDL	Electrochemical double-layer capacitor
EIS	Electrochemical impedance spectroscopy
EPA	Environmental Protection Agency
EQCMB	Electrochemical quartz crystal microbalance
FESEM	Field emission scanning electron microscopy
FET	Field-effect transistor
FIA	Flux injection analysis
FTIR	Fourier-Transform infrared spectroscopy
FTO	Fluor doped tin oxide
GCE	Glassy carbon electrode
HER	Hydrogen evolution reaction
HOMO	High occupied molecular orbital
HPLC	High-performance liquid chromatography
ICB	Ionic current blockade
ITO	Indium doped tin oxide
ISE	Ion-selective electrode
IUPAC	International Union of Pure and Applied Chemistry
LUMO	Low unoccupied molecular orbital
MCDI	Membrane capacitive deionization
MCE	Mineralization current efficiency
MCFC	Molten carbonate fuel cell
MIP	Molecularly imprinted polymer
MOF	Metal-organic framework
MSU	Mesoporous molecular sieve
MWCNTs	Multi-walled carbon nanotubes
NHE	Normal hydrogen electrode
NiMH	Nickel metal-hydride battery
NP	Nanoparticle
OCP	Open circuit potential
OER	Oxygen evolution reaction
OMC	Ordered mesoporous carbon

ORR	Oxygen reduction reaction
OSC	Organic solar cell
PAFC	Phosphoric acid fuel cell
PEC	Photoelectrocatalysis
PEFC	Polymer electrolyte fuel cell
RDE	Rotating disk electrode
SCE	Saturated calomel electrode
SCEM	Scanning electrochemical microscopy
SEM	Scanning electron microscopy
RHE	Reversible hydrogen electrode
SHE	Standard hydrogen electrode
SOFC	Solid oxide fuel cell
SWV	Square wave voltammetry
TEM	Transmission electron microscopy
UPD	Underpotential deposition
UV	Ultraviolet radiation
VIMP	Voltammetry of immobilized particles
XRD	X-ray diffraction
ZCP	Point of zero charge

Chemicals:

ACNTs	Activated carbon nanotubes
BTC	Benzene tricarboxylic acid
CT	Catechol
CNTs	Carbon nanotubes
DHI	Dehydroindigo
DOBPDC	4,4'-Dioxidobiphenyl-3,3'-dicarboxylate
FeTPP	Fe(II)-tetraphenylporphyrinato complex
Fc	Ferrocene
GO	Graphene oxide
GR	Graphene
GRO	Graphite oxide
Hb	Hemoglobin
HQ	Hydroquinone
IND	Indigo
LAO	Lithium aluminum oxide ($LiAlO_2$)
LDH	Layered double hydroxide
LND	Leucoindigo
LPS	Li10SnP2S12
MB	Maya blue
MSU	Mesoporous molecular sieve
MWCNTs	Multiple-walled carbon nanotubes
NASICON	Na superionic conductor
OMD	Octahedral molecular sieve
PBS	Phosphate buffer saline
PC	Propylene carbonate
PDDA	poly(diallyldimethylammonium)
PEDOT	poly(3,4-ethylene dioxythiophene)
PEDOT-PSS	poly(3,4-ethylene dioxythiophene) poly-(styrenesulfonate)
P(EO)	Poly(ethylene oxide)

PIM Polymer of intrinsic (micro)porosity
POM Polyoxometalate
PTE Polyethylene terephthalate
PTFE Pentafluoroethylene
rGO Reduced graphene oxide
RS Resorcinol
SWCNTs Single-walled carbon nanotubes
TPY 2,4,6-triphenyl pyrylium ion
YSZ Yttria-stabilized zirconia

1 Porous Materials and Electrochemistry

1.1 POROUS MATERIALS, CONCEPT, AND CLASSIFICATIONS

Porous materials have attracted considerable attention in the last decades because of their wide variety of scientific and technological applications. In its most generalized meaning, the term *pore* designs a limited space or cavity in a (at least apparently) continuous material. Porous materials comprise inorganic compounds, such as aluminosilicates, and biological membranes and tissues. According to the International Union of Pure and Applied Chemistry (IUPAC), pores are classified into three categories: micropores (size less than 2 nm), mesopores (size between 2 and 50 nm), and macropores (size larger than 50 nm) [1].

Porous materials discussed at the International Conference on Materials for Advanced Technologies in 2005 included clay minerals, silicates, aluminosilicates, organosilicon, metals, silicon, metal oxides, carbons and carbon nanotubes, polymers and coordination polymers or metal-organic frameworks (MOFs), metal and metal oxide nanoparticles, thin films, membranes, and monoliths [2].

Fundamental and applied research dealing with novel porous materials is addressed to improve template-synthesis strategies, chemical modification of porous materials via molecular chemistry, construction of nanostructures of metals and metal oxides with controlled interior nanospace, and reticular design of MOFs with pore sizes ranging from the micropore to the mesopore scales, among others. Porous materials are useful for sensing, catalysis, shape- and size-selective absorption and adsorption of reagents, gas storage, electrode materials, etc.

Because of the considerable variety of materials to be categorized as porous, several classifications can be proposed. Thus, based on the distribution of pores within the material, we can distinguish between regular and irregular porous materials; according to the size distribution of pores, one can separate between uniformly-sized and non-uniformly-sized porous materials.

From a structural point of view, porous materials can be the result of building blocks following an order of construction that can be extended from centimeter to the nanometer levels. Porous materials can range from highly ordered crystalline materials, such as aluminosilicates or metal-organic frameworks, to amorphous sol-gel compounds, polymers, and fibers. This text will focus on materials that have porous structures; ion-insertion solids that have no micro- or mesoporous structures will not be treated here. To present a systematic approach from the 'electrochemical' point of view, porous materials will be divided here into:

- Porous metals;
- Porous silicates and aluminosilicates;
- Porous metal oxides and related compounds (including pillared oxides, laminar hydroxides, and polyoxometalates);
- Porous sulfides, nitrides, and phosphides;
- Metal-organic frameworks (MOFs);
- Porous carbons, nanotubes (CNTs), graphene and its derivatives, and fullerenes; and
- Porous organic polymers and hybrid materials.

This list—although not exhaustive of the entire range of porous materials—attempts to cover those that can be described in terms of extended porous structures and whose electrochemistry has been extensively studied. In the last years, additionally, there has been growing interest in the preparation of nanostructures of metal and metal oxides with controlled interior nanospace, while a variety of nanoscopic porogen such as dendrimers, cross-linked and core-corona nanoparticles, hybrid copolymers, and cage supramolecules are currently under intensive research [3]. Several of such nanostructured systems will be treated along with the text, although the study *in extenso* of their electrochemistry should be treated elsewhere.

The most relevant characteristic of porous materials is the disposal of a high effective surface/volume relationship, usually expressed in terms of their specific surface area (area per mass unit) that can be determined from nitrogen adsorption/desorption data. Different methods are available to determine the specific surface area (BET, Langmuir, and Kaganer), micropore volume (*t*-plot, α_s, and Dubinin-Astakhov), and mesopore diameter (Barrett-Joyner-Halenda). Table 1.1 summarizes the values of specific surface area reported for selected porous materials. A functional classification of porous materials (see Figure 1.1) can be made based on their degree of long-range order (influencing molecular sieving capacity) and their intermolecular bond strengths (influencing thermal and/or chemical stability) [4].

TABLE 1.1
Typical Values for the Specific Surface Area of Selected Porous Materials

Material	Specific surface area ($m^2\ g^{-1}$)
Zeolite X	700
SBA-15	650
MCM-41	850
Activated carbon	2,000
Nanocubes MOF-5	3,500

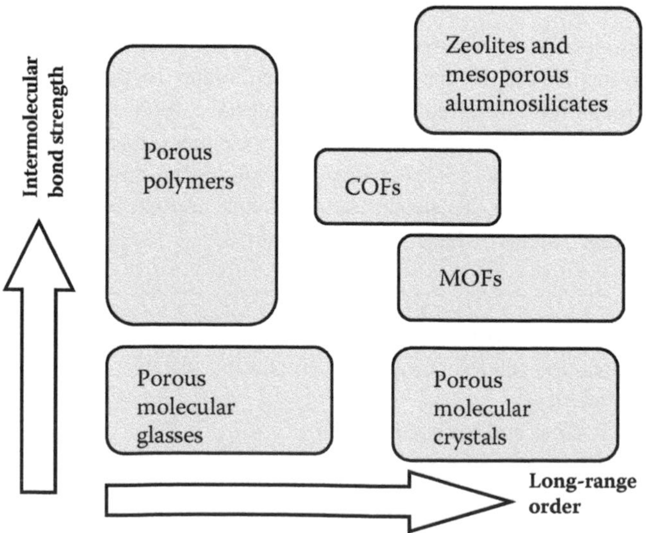

FIGURE 1.1 A Model of Functional Classification of Porous Solids.

1.2 MIXED POROUS MATERIALS

Porous material chemistry involves a variety of systems, which will generically be termed here as mixed systems, from the combination of different structural moieties resulting in significant modifications of the properties of the pristine porous materials. In this group, we can include quite varied materials:

- Composites constructed by the addition of a binder to porous materials and, eventually, other components, forming mixtures for definite applications. This kind of system is frequently used to prepare composite electrodes.
- Functionalized materials prepared by attachment of functional groups to a porous matrix.
- Materials with encapsulated species where molecular guests are entrapped into cavities of the porous host material.
- Doped materials where a structural component of the material becomes partially substituted by a dopant species, or when external species ingress in the original material as an interstitial ion. The term doping is applied to yttria-doped zirconia used for potentiometric determination of O_2, but also to describe the incorporation of Li^+ in polymers and nanostructured carbons.
- Intercalation materials where different nanostructured components are attached to the porous matrix. This is the case with metal and metal oxide nanoparticles generated into zeolites and mesoporous silicates, or organic polymers intercalated between laminar hydroxides.

From several applications, it is convenient to describe much of the above systems as resulting from the modification of the parent porous materials by a second component. In this sense, one can separate network modification, network building, and network functionalization processes. Network modification exists when the final structure of the parent material is modified because of its combination with the second component, resulting in the formation of a new system of links. Network building occurs when the material is formed by assembling the units of both components. Finally, functionalization involves the attachment of selected molecular groups to the host porous material without modification of its structure.

1.3 ELECTROCHEMISTRY AND POROUS MATERIALS

All the aforementioned materials, despite their variety of physico-chemical and structural properties, can be studied by electrochemical methods and can be treated as materials for electrochemical applications. In most cases, porous materials can be synthesized, modified, or functionalized by electrochemical methods. Intersection of electrochemistry with porous materials science can be associated to:

- Electroanalytical methods to gain compositional and structural information on porous materials.
- Electrosynthetic routes to prepare or modify porous materials.
- The design and performance of electrocatalysts for synthesis and sensing.
- The characterization of photochemical and magnetochemical properties.
- The design and performance of electrochemical, electro-optical, etc. sensors.
- The design and performance of porous materials as electrode materials, fuel cells, etc.
- The design and performance of capacitors, electro-optical devices, solar cells, etc.
- Electrosynthetic applications and pollutant degradation systems.
- The design and operation of desalination arrangements.
- The applications for drug release.

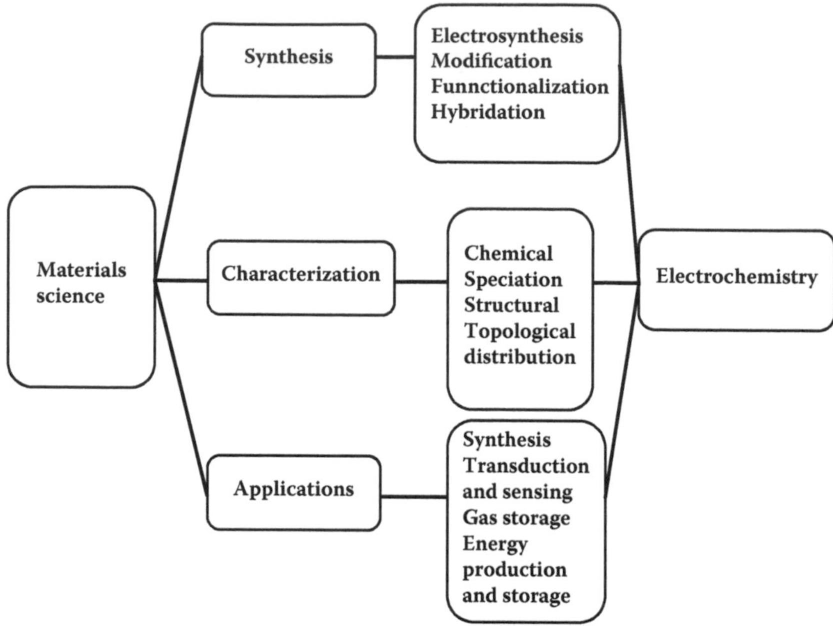

FIGURE 1.2 Schematics for the Relationships Between Electrochemistry and Porous Materials Science.

Relations between electrochemical items and materials science can be grouped into three main aspects, as schematized in Figure 1.2. First, it should be noted that electrosynthetic methods allow the preparation of a variety of materials—from porous oxide layers in metal anodes to metal-organic frameworks (MOFs) [5], layered double hydroxides (LDHs) [6], and porous carbons [7]. Further, porous materials can be modified, functionalized, or hybridized (*vide infra*) through electrochemically-assisted procedures, resulting in the preparation of novel materials.

Electrochemical methods can also be used to obtain analytical information on porous materials. Voltammetric methods and related techniques have been used to gain information on reaction mechanisms for species in the solution phase, whereas impedance techniques have been extensively used in corrosion and metal surface studies. Over the last decades, the scope of available methods has been increased by the development of the voltammetry of microparticles by Scholz et al. [8,9]. This methodology, conceived as the recording of the voltammetric response of a solid material mechanically transferred to the surface of an inert electrode, provides information on the chemical composition, mineralogical composition, and speciation of solids [10,11].

Electrochemical applications of porous materials cover principal issues. These include transduction (electro-optical, magneto-optical devices) and sensing, gas production and storage, electrosynthesis at an industrial scale, and pollutant degradation as well as energy production and storage, including high-performance dielectric materials for advanced integrated circuits in the microelectronics industry. In the analytical domain, porous materials can be used in electroanalytical techniques (potentiometry, amperometry) to determine a wide variety of analytes, from gas composition to pollutants or bioanalytes, with applications for tissue engineering, DNA sequencing, cell-makers, and medical diagnosis. Of course, porous materials also find application in batteries, capacitors and supercapacitors, and fuel cells, among others.

1.4 SYNTHESIS OF POROUS MATERIALS

Although traditional synthetic methods can be used to prepare a variety of porous materials, the development of template synthesis strategies has prompted an explosive-like growth of synthetic methods. Template synthesis roughly involves two steps: the use of a structure-directing reagent that facilitates the porous material to adopt the desired structure followed by the template release. Three main types of templates—soft, hard, and complex—can be used [1]:

Soft templates can be removed by heat treatment. These are usually molecules and molecular associations such as amines, thermolabile organic polymers, and surfactants. Additionally, vesicles, ionic liquids, self-assembled colloidal crystals, and air bubbles have been used for soft-templating synthesis.

Hard templates require acid or basic attack for their release. Zeolites and mesoporous silica, used as templates for porous carbon preparation, [11] can be taken as examples. Complex templates combine soft and hard template techniques. This methodology is used to synthesize hierarchically bi-modal and tri-modal meso-macroporous materials with interconnected pore channels combining a surfactant template with a colloidal crystal template [12]. In parallel, sol-gel technologies have contributed to a significant growth of synthetic procedures for the preparation of all kinds of materials [13].

In recent times, there was much attention to the preparation of films of hybrid materials. Here, the composition (homogeneous, heterogeneous), structure (monolayer, multilayer), thickness, and texture (roughness) can notably influence the resulting optical and electrical properties of the system. Layer-by-layer (LbL) preparation involves the sequential deposition of oppositely charged building blocks modulated by their interaction with counterions.

A plethora of synthetic routes, however, is currently being developed. These include Ostwald ripening to build hollow anatase spheres and Au-TiO_2 nanocomposites, laser ablation, and spray pyrolysis, among others. Interestingly, porous materials can act as templates for synthesizing other porous materials. This is the case of metal-organic frameworks [14] and organo-modified layered double hydroxides [3] for porous carbon synthesis. Techniques for thin film deposition include vacuum thermal evaporation and organized assembly. Additionally, electrosynthetic methods can be applied to prepare or modify porous materials. Within an extensive list of procedures, one can mention:

a. The preparation of porous oxide films by the anodization of metal electrodes via electrosynthesis of metal-organic frameworks, electrosynthesis of porous carbons and nanotubes, or electropolymerization, forming porous polymers.
b. Electrochemical modification of porous materials through electrochemical doping via ion insertion in materials for lithium batteries, electropolymerization of polymers attached to porous substrates, nanoparticle electrogeneration, and attachment to porous materials.

1.5 MATERIAL-MODIFIED ELECTRODES

Roughly, electrochemical methods consist of recording the signal response of an electrode, which is immersed into an electrolyte solution, under the application of an electrical excitation signal. The potential of this electrode—the working electrode—is controlled with respect to a reference electrode also immersed in the electrolyte. In solution electrochemistry, electroactive species are in the liquid electrolyte, although eventually, the formation of gas and/or solid phases can occur during electrochemical experiments. In solid-state electrochemistry, the interest is focused on solid materials deposited (or forming) the electrode, in contact with a liquid—or eventually, solid—electrolyte.

A significant part of solid-state electrochemistry is concentrated in the attachment of solid materials to the surface of a basal, inert electrode. This process will be labeled as electrode modification in the following. Several methods have been proposed for electrode modification with porous materials:

a. Direct deposition from suspensions: in this procedure, a drop of a suspension of the solid in a volatile liquid is placed on the surface of the basal electrode and allow the solvent to evaporate [15].

b. Fixation/coverage into a polymer coating: preparing a suspension of the solids in a solution of the polymer in a volatile solvent and allowing the solvent to evaporate [16]. As a result, a coating of the solid particles embedded into the polymer coating is deposited onto a basal electrode. Alternatively, a microparticulate deposit obtained from evaporation of a suspension of the studied solid in a volatile solvent is covered by a polymer solution, then evaporating the solvent [17].
c. Attachment to carbon paste electrodes and formation of material/carbon/polymer composites: the powdered material is mixed with a paste made of graphite powder and a binder. This is usually a nonconducting, electrochemically silent, and viscous liquid (nujol oil, paraffin oil), but electrolyte binders such as aqueous H_2SO_4 solutions have also been used. Rigid electrodes can be prepared from mixtures of the material, graphite powder, a monomer, and a cross-linking agent, followed by radical-initiated copolymerization [18].
d. Formation of material/conductive powder mixtures or pressed graphite-material pellets: this method involves powdering and mixing with graphite powder and pressing the powder mixture into electrode grids, as commonly done in the battery industry. The pressed mixture can be attached to a graphite electrode and immersed into a suitable electrolyte or, eventually, dry films of pressed pellets can be placed between planar electrodes.
e. Co-electrodeposition with conducting polymers: from a material-monomer slurry submitted to electropolymerization conditions.
f. Mechanical transference: this method is based on the transference by abrasion of few micrograms (or nanograms, if necessary) of solid particles of the sample to the surface of an inert electrode, typically paraffin-impregnated graphite electrodes (PIGEs) [8–10].
g. Adsorptive and covalent link to electrode surfaces: particles of porous materials can be adsorptively or covalently bound to electrode surfaces by intermediate groups that can connect the basal conducting electrode and the porous particles. For instance, the use of silanes enables covalent binding for the covalent attachment of bifunctional silane to a single dense layer of zeolite Y to a SnO_2 electrode [19]. Adsorption can be facilitated by pendant groups, typically thiols of high affinity with gold surfaces. The use of thiol-alkoxysilanes has been applied to attach aluminosilicate materials to gold electrodes, combining the thiol affinity for gold with the ease functionalization of aluminosilicates with alkoxysilanes.
h. Layer and multilayer preparation methods: under the above designation, a variety of methods recently developed to prepare material-modified electrodes can be included, namely spin coating and formation of Langmuir-Blodgett films accompanied by continuous film synthesis on electrodes [20], self-assembled monolayer formation [21], layer-by-layer deposition [22], salinization, charge modification and seeding of the surface prior to the hydrothermal crystallization of the porous material, and layer-by-layer assembly by ionic linkages mediated by multilayers of oppositely charged electrolytes, among others. Hydrothermal crystallization on conductive substrates involves previous treatment of the basal electrode; for instance, zeolite-modified electrodes on glassy carbon electrode previously treated with a polycationic macromolecule to ensure durable binding of the negatively-charged zeolite seeds [23].

1.6 ELECTRODE-MODIFIED MATERIALS

Porous materials can be electrochemically synthesized and/or electrochemically modified by using electrolysis methodologies. Apart from the synthesis of, for instance, metal-organic frameworks [5] or carbons [24], porous materials can be electrochemically modified in several ways.

One of the most intensively investigated possibilities is that resulting in the attachment of nanometric units to porous, electrochemically silent frameworks. This is the case of metal and metal oxide nanoparticles anchored to micro- and mesoporous aluminosilicates prepared by electrolyzing dispersions of Pd(II)- and/or Cu(II)-exchanged zeolites in appropriate electrolytes.

The application of reductive potentials leads to the formation of metal and/or metal oxide nanoparticles in the zeolite framework. With appropriate control of the synthetic conditions, metal nanoparticles can be predominantly confined to particular sites (e.g., super cages in zeolites) in the porous framework [25]. Zeolite-supported Pt or RuO_2 nanoparticles act as electron transfer mediators rather than controlling heterogeneous electron-transfer surfaces, and improves Faradaic efficiency in electrolytic processes even in low ionic strength solutions [26].

Metal nanoparticles housed in zeolites and aluminosilicates can be regarded as arrays of microelectrodes placed in a solid electrolyte with shape and size selectivity. Remarkably, the chemical and electrochemical reactivity of metal nanoparticles differ from those displayed by bulk metals and are modulated by the high ionic strength environment and shape and size restrictions imposed by the host framework. In the other extreme of existing possibilities, polymeric structures can be part of the porous materials from electropolymerization procedures as is the case of polyanilines incorporated into microporous materials. The electrochemistry of these kinds of materials, which will be termed—*sensu lato*—hybrid materials, will be discussed in Chapter 11.

Another interesting and widely studied case is the formation of porous metal oxides by anodization of metals. Here, the electrolytic procedure yields a thin layer of porous materials applicable in catalysis, anti-corrosion, batteries, and other applications. Such materials will be treated in Chapter 7.

1.7 GENERAL ELECTROCHEMICAL CONSIDERATIONS

First, we can divide the electrochemical processes into two main groups: 1) electrochemical processes (termed as Faradaic) where charge transfer occurs between an electronic conductor (electrode) and an ionic conductor (electrolyte); and 2) processes that occur at the electrode/electrolyte interface without net charge transfer [27]. A variety of electrochemical techniques can be applied to obtain information on the composition and structure of microporous materials. A set of techniques—mainly cyclic voltammetry (CV), chronoamperometry (CA), chronopotentiometry, and coulometry—are focused on Faradaic processes. A second group of techniques—those that require impedance measurements, particularly focused in electrochemical impedance spectroscopy (EIS)—mainly involves non-Faradaic ones. This brief enumeration, however, does not exhaust the scope of available techniques because other extended methods, such as differential pulse and square wave voltammetry, electrochemical quartz microbalance (EQCM), or electrochemical atomic force microscopy (EAFM) can be used to characterize microporous solids. Apart from this, electrochemical techniques can be combined with other experimental procedures so that coupling with UV-Vis spectrometry, Fourier-transform infrared spectroscopy (FTIR), X-ray diffraction (XRD), and others is possible.

In a wide sense, electrochemical phenomena involve at least two types of processes: (a) electron transfer processes through a two-dimensional boundary (interface) separating the electrode (metal-type conductor) and the electrolyte (ionic conductor), and (b) charge transport in the involved media (the electrolyte and the electrodes and external circuit). In the study of such phenomena, one can distinguish between electronics, focused on the heterogeneous electrode/electrolyte charge transfer process, and ionics devoted to the study of ionically conducting liquid or solid phases [28].

About porous materials, it should be noted that restricted ionic conductivity is a general property of porous materials that can vary significantly depending on doping, type and concentration of defects, and temperature. Interestingly, several porous materials, like hydrated aluminosilicates, can behave like electrolyte-like conductors in liquid and solid ionic conductors when dry.

The classical model for describing the electrode-liquid electrolyte junction considers a highly structured region close to the electrode surface, the double-layer with dipole-oriented solvent molecules, and a double layer of charge-separated ions that creates a capacitive effect. At a larger distance from the electrode surface, there is a less structured region—the diffuse layer—that finally reduces to the randomly organized bulk electrolyte solution. The earlier formulation according to Helmholtz distinguishes between the inner (Helmholtz) layer, which comprises all species that are specifically adsorbed on the electrode surface, and the outer (Helmholtz) layer, comprising all ions that are closest to the electrode surface but are not specifically adsorbed [29]. As far as the area and geometry of the electrode surface influence the double-layer capacitance, porous materials with large effective surface areas can yield significant capacitance effects that will be further commented on.

When a difference of potential is established between the electrode and the electrolyte, there are several coupled processes occurring in the electrode/electrolyte region (the interphase)—a process of charge transfer through the electrode/electrolyte interface (two-dimensional region of contact) and concomitant charge transport processes in the electrolyte and the electrode, in particular involving ion restructuring in the double layer zone. As a result, the current flowing when a potential positive or negative relative to the potential where the interface is not charged (potential of zero charge of the system) can be described in terms of the sum of a Faradaic current, associated with the electron transfer process across the interface, and a capacitive (or double-layer charging) current associated to ion restructuring in the vicinity of the electrode surface.

Let us first consider an 'ordinary' electrochemical process consisting of the reduction (or oxidation) of a given electroactive species in solution at an inert electrode. Because the flow of Faradaic current is a direct expression of the rate of the electron transfer reaction at the electrode/electrolyte interface, the rate of mass transport of the electroactive species from the bulk solution to the electrode surface decisively influences the magnitude of the Faradaic current. Mass transport can occur via diffusion (whose driving forces are concentration gradients), convection (driven by momentum gradients), and migration of charged species (driven by electric fields). Convection phenomena appear when the solution is stirred (or undergoes unwanted room vibrations) or submitted to gas bubbling. To suppress the effect of ionic migration, an electrochemically silent supporting electrolyte in sufficiently high concentration is generally used in voltammetric experiments.

The process of electron transfer in the electrode/electrolyte interface can be treated as a heterogeneous reaction influenced by mass transport through the electrolyte. The description of the overall electrochemical process involves charge conservation (electroneutrality condition) and Fick's diffusion laws. In the simplest case, an electroactive species in solution approaches the electrode surface where interchange n electrons. This process results in the formation of an oxidized or reduced species also in solution. Under planar, semi-infinite diffusion conditions (*vide infra*), the Faradaic current, I, is proportional to the gradient of concentration of the electroactive species at the electrode surface. This can be expressed as:

$$I = -nFAD\left(\frac{\partial c}{\partial x}\right)_{x=0} \tag{1.1}$$

where x represents the distance from the electrode, c the concentration of electroactive species at x, D its diffusion coefficient, A the electrode area, and n the number of transferred electrons per mol of electroactive species.

The rate at which electron transfer takes place across that interface is described in terms of a heterogeneous electron-transfer rate constant. The kinetics can be described by means of the Butler-Volmer equation:

$$I = -nFAk^{\text{o}}\left[c^{\text{o}}{}_{\text{red}}\exp\left[\frac{nF(1-\alpha)(E-E^{\text{o}\prime})}{RT}\right] - c^{\text{o}}{}_{\text{ox}}\exp\left[-\frac{\alpha nF(E-E^{\text{o}\prime})}{RT}\right]\right] \tag{1.2}$$

In this equation, c°_{ox} and c°_{red} represent, respectively, the surface concentrations of the oxidized and reduced forms of the electroactive species; k°, the standard rate constant for the heterogeneous electron transfer process at the standard potential (cm/s^{-1}); and α, the symmetry factor, a parameter that characterizes the symmetry of energy barrier that must be surpassed during charge transfer. In Eq. (1.2), E represents the applied potential, and $E^{\circ\prime}$ thc formal electrode potential, usually close to the standard electrode potential. The difference $E - E^{\circ\prime}$ represents the overvoltage, a measure of the extra energy imparted to the electrode beyond the equilibrium potential for the reaction. Notice that the Butler-Volmer equation reduces the Nernst equation when the current is equal to zero (i.e. under equilibrium conditions) and when the reaction is extremely fast (i.e. when k° tends to ∞). The latter is the condition of electrochemical reversibility [30,31].

Now, let us consider an idealized experiment where the current initially passing is recorded when a given potential E is applied to a solution containing the oxidized form of an electroactive species existing in solution in concentration c_{ox}. Then, one can assume that, initially, $c^\circ_{ox} = c_{ox}$ and $c^\circ_{red} = 0$ so that, according to Eq. (1.2), the application of a given potential E is much lower (in absolute value) than $E^{\circ\prime}$, the current will be practically null. If the applied potential approaches $E^{\circ\prime}$, the current will be larger. Then, plotting recorded I values vs. the applied potential, one will obtain an exponential increase of I on E. This situation is roughly termed as Nernstian control. Under ideal, quasi-static conditions, when the applied potential is clearly larger in absolute value than $E^{\circ\prime}$, the current value becomes quite similar tending to a limiting value. Under these last conditions, the evolution of the system is controlled by the transport of electroactive species towards the electrode surface.

It should be noted that the overall electrochemical process can involve coupled chemical reactions in the solution phase, gas evolution, and/or deposition of solids, and/or formation/release of adsorbates onto/from the electrode surface so that electrochemical processes can, in general, be regarded as multi-step reaction processes. As far as electrochemical responses are strongly conditioned, not only by the kinetics of the interfacial electron transfer process but also by the kinetics of coupled chemical processes, electrochemical methods are able to yield mechanistic information of interest in a wide variety of fields.

1.8 DIFFUSIVE ASPECTS

Oxidation or reduction of electroactive species at an electrode surface produces a depletion of its concentration in the diffusion layer, generating a concentration gradient between the interface and the bulk solution, which is the driving force for net diffusion of electroactive molecules from the bulk of the solution. In general, it is possible to assume that the concentration gradient is confined to a limited region near the electrode surface, the diffusion layer, of thickness δ, so that at $x > \delta$, the concentration of electroactive species remains constant (and equal to the bulk concentration). In the following, it will be assumed that electrochemical experiments were conducted in conditions where no complications due to convection and migration effects appear. In short, this means that experiments are performed under quiescent, non-stirred solutions in the presence of a supporting electrolyte in sufficiently high concentration.

For disk-type electrodes usually with a radius of 0.1–1.0 cm^2, the thickness of the diffusion layer that is depleted of reactant is much smaller than the electrode size so that mass transport can be described in terms of planar diffusion of the electroactive species from the bulk solution to the electrode surface as schematized in Figure 1.3a. This situation corresponds to the so-called semi-infinite planar diffusion conditions.

The simplest electrochemical experiment involves stepping the potential from an initial value far from $E^{\circ\prime}$, where no electrode reaction occurs, to one where the electrochemical process proceeds at a diffusion-controlled rate. The corresponding current/time record is the chronoamperometric curve. The thickness of the diffusion layer can be estimated as $(Dt)^{1/2}$ for a time electrolysis t and usually ranges between 0.01 and 0.1 mm [28]. For an electrochemically reversible n-electron transfer process

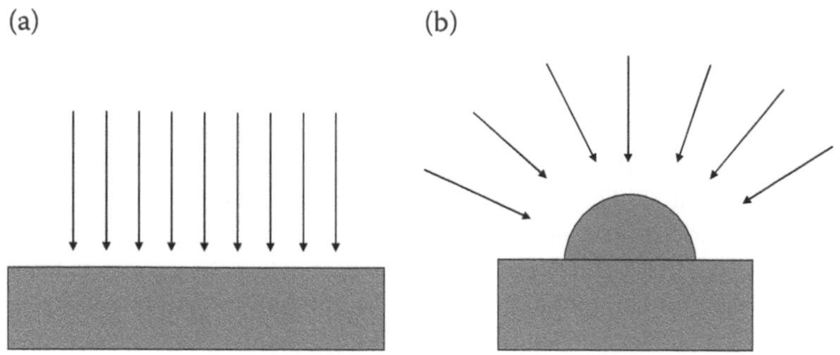

FIGURE 1.3 Schematics for (a) Linear Diffusion and (b) Radial Diffusion to Electrodes.

of an electroactive species whose concentration is c, in the absence of parallel chemical reactions, the variation of the Faradaic current with time is then given by the Cottrell equation:

$$I = \frac{nFAcD^{1/2}}{\pi^{1/2}t^{1/2}} \tag{1.3}$$

It should be noted, however, that at short times in the experimentally recorded curves deviations due to double-layer charging can appear, whereas at log times, convection can cause deviations from the expected response.

For microelectrodes, of typical of 5–10 μm, radial hemispherical diffusion conditions need to be considered, as schematized in Figure 1.3b [32]. In the case of a spherical electrode of radius r, the chronoamperometric curve is described by:

$$I = \frac{nFADc}{r} + \frac{nFAcD^{1/2}}{\pi^{1/2}t^{1/2}} \tag{1.4}$$

At sufficiently short times, the second term of the above equation dominates over the first, so that the current/time response approximates that described by Eq. (1.3). At long times, the second, Cottrell-type term decays to the point where its contribution to the overall current is negligible and then the current tends to a constant, steady-state value where the rate of electrolysis equals the rate at which molecules diffuse to the electrode surface [33].

At porous electrodes, diffusion will be conditioned by the electrode geometry and pore size distribution, so that, under several conditions, semi-infinite diffusion holds, but under some others, the porous electrode can be treated as an array of microelectrodes [25]. Detailed modeling of these situations will be described in Chapter 2.

1.9 VOLTAMMETRY AND RELATED TECHNIQUES

As previously noted, electrochemical methods are based on recording of the response of an electrode—in contact with an electrolyte—to an electrical excitation signal. Depending on the characteristics of the excitation potential signal applied to the working electrode and the measured signal response, one can distinguish different electrochemical techniques. Voltammetry consists of the recording of current (I) which flows between the working electrode and an auxiliary electrode vs. potential (E) which is applied between a working electrode and a reference electrode. In conventional three-electrode arrangements, a potentiostat controls the potential so that the current flows almost exclusively between the working electrode and the auxiliary electrode while a small, practically negligible current is passing through the reference electrode.

In linear potential scan (LSV) and cyclic (CV) voltammetry, a potential varying linearly with time is applied between an initial potential, E_i, usually at a value where no electrode processes occur, and a final potential (LSV) or cycled between two extreme (or switching) potential values at a given potential scan rate v (usually expressed in mV s^{-1}). In other techniques such as normal and differential pulse voltammetry or square wave voltammetry (SWV), the excitation signal incorporates potential pulses to a linear or staircase potential/time variation.

In a typical CV experiment, the potential scan is initiated at the open circuit potential and directed in a positive or negative direction. For a reversible process, when the potential approaches the formal potential of the involved couple, the current increases rapidly while the concentration of the electroactive species in the vicinity of the electrode is depleted. As a result, a maximum current is obtained, defining a voltammetric (cathodic or anodic). Notice that the linear sweep voltammetric and the CV peak appear at a certain voltage fraction past the formal potential. From which, the current slowly decreases. In the subsequent cathodic scan, a similar cathodic peak is measured, defining a cathodic peak potential, E_{pc}, and a cathodic peak current, I_{pc}. Then, the current reaches a maximum and subsequently decays so that ca. 150–200 mV after the voltammetric peak, the current becomes diffusion-controlled. The general expression for the current in the case of a reversible n-electron transfer is:

$$I = nFAcD^{1/2}\left(\frac{nF}{RT}\right)^{1/2} v^{1/2}\Psi(E - E^{o\prime}) \tag{1.5}$$

where $\Psi(E - E^{o\prime})$ represents a tabulated function of the difference between the applied potential and the formal electrode potential of the redox couple [34]. Notice that, in contrast with the idealized experiment described at the end of Section 1.7, a potential varying constantly with time is applied. This means that the oxidized form of the electroactive species is continuously reduced through the experiment and its concentration near the electrode is depleted, whereas the concentration of the reduced form increases. This determines that, for a certain potential, the measured current reaches a maximum, then subsequently decreases. In the reverse scan, the oxidized (or reduced) species electrochemically generated—which remains in the vicinity of the electrode surface (because the diffusion of products into the bulk of solution is slow)—is reduced (or oxidized) to the parent reactant following a similar scheme. As a result, CVs for reversible electron transfer processes involving two forms (oxidized and reduced) of an electroactive species in the solution phase consists of two peaks, cathodic and anodic, at potentials $E_{pc\text{ and }}E_{pa}$, whose separation is related to the number of transferred electrons, n. In the case of electrochemically irreversible or quasi-reversible electrode systems, the current/potential profile also depends on the kinetics of the electron transfer process, and also—if existing—on the kinetics of coupled chemical reactions, adsorptions, etc. [34]. Figure 1.4 shows a typical CV for a 1.0 mM aqueous solution of $K_4Fe(CN)_6$ in 0.10 M KCl. This is an example of essentially reversible behavior corresponding to the one-electron oxidation of $Fe(CN)_6{}^{4-}$ to $Fe(CN)_6{}^{3-}$.

For a reversible process involving species in solution, the peak potentials are independent on potential scan rate v, and the absolute value of the peak potential separation, $|E_{pa} - E_{pc}|$ approaches to $59/|n$ (mV at 298 K), while the half sum of such potentials can, in principle, be equal to the formal electrode potential of the couple. Under the above conditions, the peak current, I_p, is given by the Randles-Ševčik equation [29–31]:

$$I_p = 0.446nFAc\sqrt{\frac{nFvD}{RT}} \tag{1.6}$$

The peak current is then proportional to the concentration of the electroactive species and the square root of the potential scan rate. This last condition applies in general for diffusion-controlled

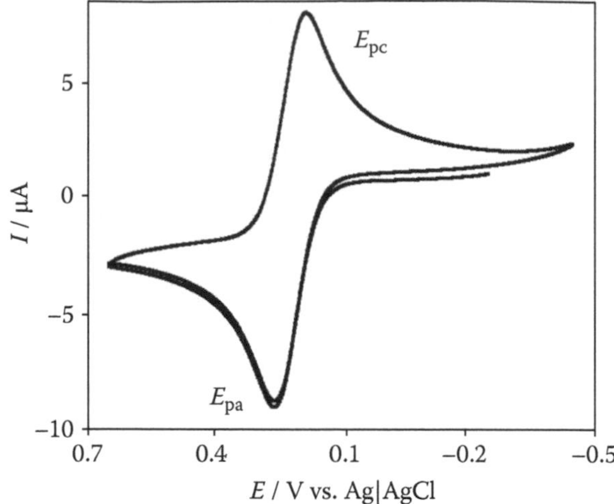

FIGURE 1.4 CV at Pt Electrode for a 1.0 mM Solution of $K_4Fe(CN)_6$ in 0.10 M KCl. Aqueous Solution. Potential Scan Rate 50 mV s^{-1}.

processes, regardless of their degree of reversibility. The extreme case is a fully irreversible electrochemical process; here, the peak potential varies linearly with the logarithm of scan rate:

$$E_p = E^{o\prime} - \frac{RT}{\alpha n_a F}\left[0.780 - \ln k^o + \ln\sqrt{\frac{\alpha n_a F D v}{RT}}\right] \tag{1.7}$$

In this equation, n_a denotes the number of electrons transferred in the rate-determining step. These processes are characterized by the absence of acathodic (or anodic) peak in the reverse scan. It is important to realize that this feature does not necessarily imply irreversibility of the electron transfer process; in most cases, this is accompanied by coupled chemical reactions resulting in significant modifications of the voltammetric pattern relative to the reversible case represented in Figure 1.4. A paradigmatic example is the oxidation of dopamine in an aqueous phosphate buffer illustrated in Figure 1.5. The process initiates by a two-electron, two-proton oxidation yielding the oxidation of the catechol unit to *o*-quinone followed by a cyclization reaction yielding leucoaminechrome. This species is more easily oxidized than the parent catecholamine and can experience a further oxidation to aminechrome, competing with a disproportionation reaction [35]. As a result, the initial oxidation wave has no cathodic counterpart and the voltammogram shows reduction peaks at more cathodic potentials for the reduction of leucoaminechrome and aminechrome.

A case of particular interest is when the electroactive species is confined to the electrode surface where it reaches a surface concentration Γ. In the reversible case, symmetric, bell-shaped current/potential curves described by [31]

$$I = \frac{n^2F^2 v A\Gamma_{ox}(b_{ox}/b_{red})\exp[nF(E - E^{o\prime})/RT]}{RT\{1 + (b_{ox}/b_{red})\exp[nF(E - E^{o\prime})/RT]\}^2} \tag{1.8}$$

may be obtained again for reversible behavior. Here, b_i (i = ox, red) is equal to $\Gamma_i \exp(-\Delta G^o{}_i/RT)$, ΔG^o being the standard free energy for surface attachment. The peak current is then given by:

$$I_p = \frac{n^2F^2 v A\Gamma}{4RT} \tag{1.9}$$

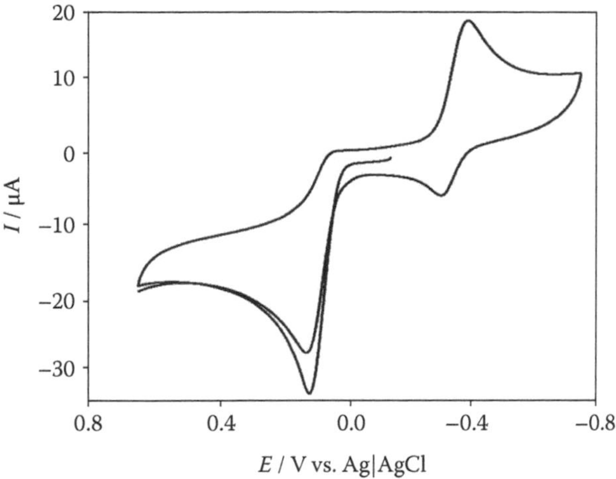

FIGURE 1.5 CV at Glassy Carbon Electrode for a 2.0 mM Solution of Dopamine in 0.10 M Potassium Phosphate Buffer at pH 7.0. Potential Scan Rate 50 mV s^{-1}.

Now, the peak current becomes proportional to the potential scan rate. It should be noted that Eqs. (1.8) and (1.9) are formally analogous to those obtained for species in solution diffusing in a restricted space, under the so-called thin layer conditions (by contraposition to unrestricted space diffusion, thick layer conditions).

Laviron [36] studied the voltammetric response of electroactive species confined to the electrode surface. Interestingly, interactions between species in the adsorbed layer may lead to peak splitting, a situation relevant to the electrochemistry of solids. The expressions for the cathodic and anodic peak potentials in the case of a small concentration of surface-confined species are:

$$E_{pc} = E^{o\prime} + \frac{RT}{\alpha n_a F} \ln\left[\alpha F/RT\right] \tag{1.10}$$

$$E_{pa} = E^{o\prime} + \frac{RT}{\alpha n_a F} \ln\left[(1 - \alpha)F/RT\right] \tag{1.11}$$

The electrochemical rate constant, k_s, is given by the equation:

$$\ln k_s = \alpha \ln(1 - \alpha) + (1 - \alpha)\ln \alpha - \ln(RT/n_a F v) - \alpha(1 - \alpha)n_a F \Delta E_p/RT \tag{1.12}$$

Here, α represents the electron-transfer coefficient, k_s the apparent charge-transfer rate constant, v the potential scan rate, and ΔE_p denotes the peak potential separation ($=E_{pa} - E_{pc}$).

Pulse voltammetric techniques are of interest because of their reluctance to charging effects. Their application is made difficult by the influence of pulse width in the shape of voltammetric curves. For SWV under usual conditions, the net current flowing during the anodic and cathodic half-cycles, I_{dif}, can be approached by [37]:

$$I_{dif} = C\frac{n^2F^2AD^{1/2}cE_{SW}f^{1/2}}{RT\pi^{1/2}}\frac{\exp(nF(E - E^{o\prime})/RT)}{[1 + \exp(nF(E - E^{o\prime})/RT)]^2} \tag{1.13}$$

f being the square wave frequency, E_{SW} the square wave amplitude (typically between 5 and 50 mV), C a numerical constant, and the other symbols have their customary meaning.

Obtaining information on the composition, structure, etc. of solid materials using voltammetric and related techniques can be performed by (a) recording the response of the material attached to an inert electrode and immersed into a suitable electrolyte, or (b) recording the modification of the response of an electroactive probe in the electrolyte solution in contact with the material-modified electrode. Additionally, (c) the electrochemical response of such systems under the application of optical or magnetic inputs can also be used.

In the first case, the voltammetric response can be associated with reductive/oxidative dissolution processes and topotactic or epistatic solid-to-solid transformations, eventually confined to thin surface layers of the parent material [10]. In the second case, among other possibilities, the solid can act as a preconcentrating system to enhance the signal of the electroactive probe in solution, but also as a catalyst to this process.

In the case of porous materials incorporating intercalated or entrapped electroactive species, the response of such species will be significantly conditioned by electrolyte ions because, as will be treated in Chapter 2, charge conservation imposes severe constraints for possible charge transfer processes. This aspect is also relevant for doping of nanostructured carbons and conducting polymers, treated in Chapters 7 and 8, respectively.

1.10 RESISTIVE AND CAPACITIVE EFFECTS

It is well known that experimental CVs for species in the solution phase frequently diverge from theoretical ones for n-electron reversible couples. The divergence can be caused by a variety of factors—deviations from reversibility, occurrence of coupled chemical reactions and/or surface effects, and resistive and capacitive effects [38]. These will be briefly treated here because of their potential significance when microheterogenous deposits or, more or less homogeneous coatings of microporous materials, cover the electrode surface.

The electrochemical cell can be treated as an electric circuit with a certain resistance R and a certain capacity C. Let us consider an experiment where a potential scan initiated at a potential E_{start} and conducted with a potential scan rate v is applied. The applied potential E satisfies the relationship [31]

$$E = E_{start} + vt = R(dq/dt) + q/C \tag{1.14}$$

where q represents the charge passed at a time t, and R and C represent the resistance and the capacity of the system. The above equation leads to the following expression for the current I at a time t:

$$I = vC + (E_{start}/R - vC)\exp(-t/RC) \tag{1.15}$$

This current does not include the contribution of the Faradaic and double-layer phenomena (i.e. it can be considered as the background current in the voltammetric experiment). Accordingly, the current/potential background will be :

$$I = vC + (E_{start}/R - vC)\exp[-(E - E_{start})/RC] \tag{1.16}$$

The combination of capacitive plus resistive effects results in an enhancement of the background currents in both the positive- and negative-directed scans, ultimately producing square voltammograms that will be treated as charge storage devices (Chapter 14). Here, the purely capacitive current is proportional to the potential scan rate. The repetition of the same box-shaped profile upon repetitive cycling the potential scan communicates the repeatability of the charge/discharge process, an essential property of systems devoted to charge storage.

The consideration of resistive and capacitive effects is important in voltammetric measurements because they also influence the Faradaic response so that the peak is flattened and decreased and shifted towards more negative (cathodic peak) or more positive (anodic peak) potentials [38]. These effects determine that the peak potential separation is essentially reversible systems different from the theoretical value of $59/n$ mV at 298 K. Then, peak potential separations larger than $59/n$ mV can result from the combination of deviations from reversibility and the appearance of uncompensated ohmic and capacitive effects in the electrochemical cell. Then, the ohmic drop can be mistaken for decelerated electrode kinetics [39]. This is reflected in the influence of the type and concentration of supporting electrolyte in the voltammetric profile, which becomes broader by decreasing this concentration. Figure 1.6 depicts an illustrative example of the variation of the voltammetric response with the type of supporting electrolyte [40].

Equation (1.12) predicts that the capacitive plus resistive current is proportional to v. Since, in the case of diffusion-controlled processes the peak current will vary with $v^{1/2}$, one can expect that the capacitive plus resistive effects will decrease on decreasing potential scan rate. Accordingly, the cathodic-to-anodic peak potential separation ΔE_p (= $|E_{pc} - E_{pc}|$) increases on increasing v but tends to the value in the absence of resistive effects when v tends to zero. The corresponding variation with the potential scan rate of peak potentials for the $Fe(CN)_6^{3-}/Fe(CN)_6^{4-}$ couple at zeolite Y-modified glassy carbon electrode is depicted in Figure 1.7.

To separate kinetic and resistive effects, one can perform experiments at variable scan rates and at different concentrations of electroactive species. As a result, the peak potential separation increases on increasing v and the concentration of the depolarizer. An estimate of the uncompensated

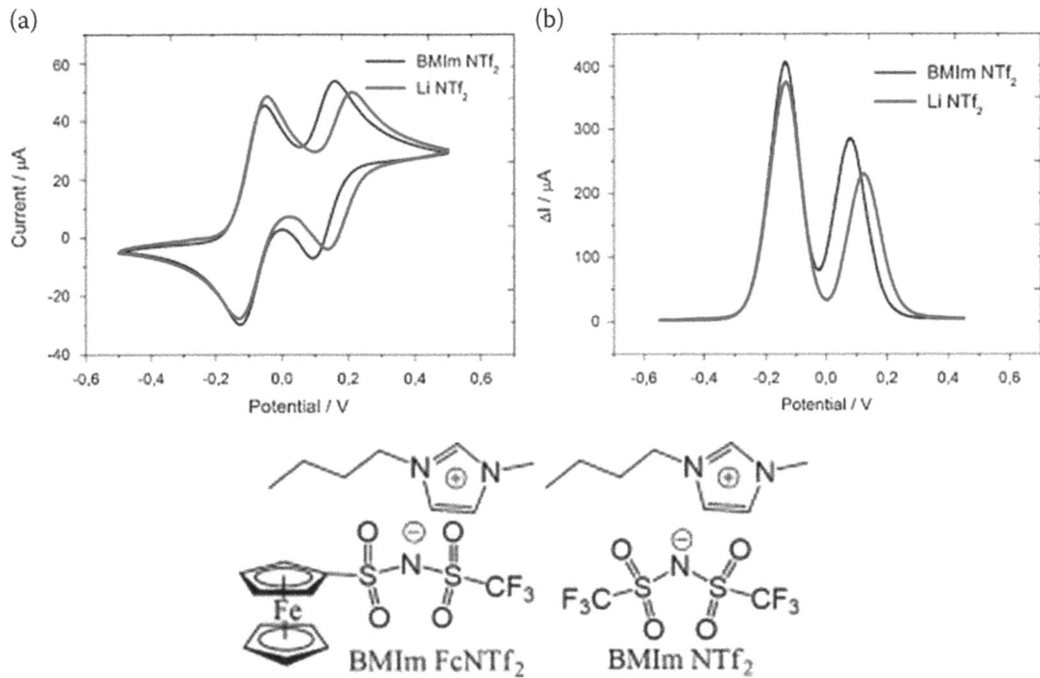

FIGURE 1.6 CVs of 10 mM Fc and 10 mM 1-Butyl-3-Methylimidazolium Ferrocenylsulfonyl-(Trifluoromethylsulfonyl)-Imide (BMIm $FcNTf_2$) in CH_3CN. Supporting Electrolytes 0.75 M 1-Butyl-3-Methylimidazolium Bis(Trifluoromethylsulfonyl)-Imide (BMIm NTf_2, black line) and Lithium Bis (Trifluoromethylsulfonyl)-Imide ($LiNTf_2$, red line). Potential Scan Rate 100 mV s^{-1}. Reproduced from Ref. [40] (Gélinas and Rochefort, *Electrochim. Acta.* 2015, 162: 36–44), with Permission.

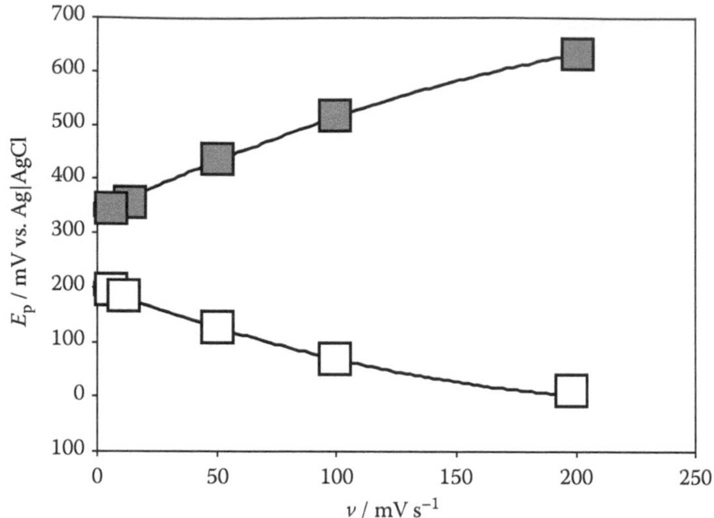

FIGURE 1.7 Variation with the Potential Scan Rate of Peak Potentials for the $Fe(CN)_6^{3-}$ /$Fe(CN)_6^{4-}$ Couple Recorded from CVs in a 2.5 mM Solution of $K_4Fe(CN)_6$ in 0.15 M $NaClO_4$ at a Zeolite Y-modified Glassy Carbon Electrode.

resistance can be made from the slope of the peak potential separation vs. peak current plot for different analyte concentrations at a given potential scan rate using the relationship [41]:

$$\Delta E_p = (\Delta E_p)_{kin} + 2I_pR \quad (1.17)$$

In this equation, ΔE_p represents the measured cathodic-to-anodic peak potential separation and $(\Delta E_p)_{kin}$ represents the value determined as the ordinate at the origin in the ΔE_p vs. I_p plot for different concentrations of electroactive species. The value for $(\Delta E_p)_{kin}$ can directly be related to kinetic parameters for the interfacial electron transfer reaction [34]. The slope of the above representation allows the calculation of the uncompensated ohmic resistance in the cell. It should be considered, however, that cathodic-to-anodic peak potential separation can also be increased as a result of coupled chemical reactions.

1.11 ELECTROCHEMICAL IMPEDANCE SPECTROSCOPY

Application to a time-dependent potential to an electrochemical cell, in general, gives rise to the appearance of a phase difference between the applied potential and the current response because diffusion, electron transfer, etc. processes yield an impedance effect similar to that typically observed in alternating current circuits. Electrochemical impedance spectroscopy is a technique based on the measurement—under steady-state conditions—of the complex impedance of the electrochemical cell as a function of the frequency f (or angular frequency $\omega = 2\pi f$) of an imposed sinusoidal input of small amplitude. This situation can be represented in terms of a complex formulation where all involved quantities can have one real and one imaginary component. A vectorial formulation is usually used to represent impedances. The common circuit elements—resistors, capacitors, and inductances—can be described as impedances of magnitude Z satisfying:

$$Z = Z_{real} + jZ_{imag} \quad (j = \sqrt{-1}) \quad (1.18)$$

Considering a conventional electrical circuit submitted to an alternating potential input of angular frequency ω, the impedance for a resistor is $Z = R$, R being the resistance of the resistor). For a capacitor of capacitance C, the impedance is $Z = -j/C\omega$, while for an inductance L, the impedance is $Z = jL\omega$. For an idealized alternating current circuit containing a resistor R, the phase angle φ is zero, whereas for purely capacitive and purely inductive circuits, the phase angle would be –90° and 90°, respectively.

Electrochemical impedance spectroscopy (EIS) produces a set of impedance values at different frequencies constituting the impedance spectrum. Typically, EIS experiments are conducted from mHz to kHz, so that available information covers a wide range of time scales [42]. It is pertinent to note that impedance is only defined for a stationary system fulfilling the constraints of the linear systems theory (LST). This means that the measuring system must be time-invariant during the time of acquisition of the EIS data. This condition can be assessed by applying Kramers-Kronig (K-K) transforms whose application determines whether the impedance data are influenced by time-dependent phenomena [43].

The electrochemical cells can be represented by an equivalent circuit formed by an association of impedances that pass current with the same amplitude and phase angle that the real cell under a given potential input. After acquiring the impedance data at different frequencies, they are represented usually as the so-called Nyquist and Bode plots. The Nyquist diagram consists of a representation of (minus) the imaginary component of total impedance vs. the real component of this quantity ($-Z_{imag}$ vs. Z_{real}), whereas the Bode plots correspond to the representation of the logarithm of modulus of total impedance, $\log|Z_{tot}|$, and (minus) the phase angle, $-\varphi$, vs. logarithm of frequency.

To properly describe the action of an alternating potential input on electrochemical cells, one can consider—in principle—at least two coupled interface processes influencing the impedance of the system: the electron transfer process across the electrolyte/electrode interface and the double layer effect. The equivalent circuit typically used to represent this situation is the Randles circuit, constituted by a solution resistance R_u, a charge transfer resistance R_{ct} (often called electron transfer or polarization resistance with a meaning different to that used in other contexts), and a double-layer capacitance C_{dl}. This circuit is schematized in Figure 1.8, where R_u represents the uncompensated resistance of the electrolyte and other possible ohmic resistances and R_{ct} represents the ohmic drop that can be associated with the electron-transfer process through the electrode/electrolyte interface. The double-layer effect, which roughly consists of charge separation in the electrode/electrolyte boundary as a result of charge migration, can be assimilated to a capacitor of capacitance C_{dl}.

The impedance equations for the Randless circuit are:

$$Z_{real} = R_s + \frac{R_{ct}}{1 + R_{ct}^2 C_{dl}^2 \omega^2} \tag{1.19}$$

$$Z_{imag} = -\frac{jR_{ct}^2 C_{dl} \omega}{1 + R_{ct}^2 C_{dl}^2 \omega^2} \tag{1.20}$$

The variation of the total impedance with the angular frequency (in double logarithmic scale), termed the Bode plot, is shown in Figure 1.9a. Interestingly, at the high-frequency limiting ($\omega \to \infty$), the total impedance approaches R_u, while at the low-frequency limiting ($\omega \to 0$), Z approaches $R_u + R_{ct}$. For the Randles circuit, the phase angle varies with ω, as depicted in Figure 1.9b, the maximum φ value being lower than 90°. At intermediate frequencies, there is a linear variation of the total impedance (absolute value) with the angular frequency ω (or the frequency υ), so that the slope of the corresponding linear representation is $\partial(Z)/\partial(\log\upsilon) = -1$ while the ordinate at the origin equals to $1/C_{dl}$. The Nyquist plot (Figure 1.10) provides a circumference arc with a maximum of imaginary impedance located at $\omega_{max} = 1/R_{ct}C_{dl}$.

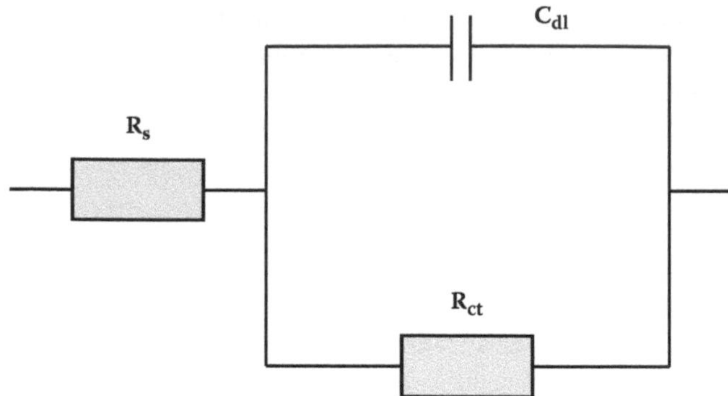

FIGURE 1.8 Schematics for the Randles Circuit.

The charge transfer resistance can, in principle, be used to evaluate the apparent heterogeneous rate constant, k_{app}, for the electron transfer process using the relationship:

$$k_{app} = \frac{RT}{F^2 c R_{ct}} \tag{1.21}$$

To properly describe the impedance response of many processes, additional impedance elements have been introduced. The constant phase elements can be represented by means of the equation:

$$Z(\omega) = Q_o (j\omega)^{-g} \tag{1.22}$$

where Q_o is a constant and β is a constant exponent. For $g = 0$, the CPE reduces to a resistance $R = Z_o$, whereas for $g = 1$, the CPE reduces to a capacitance. The case $g = ½$ corresponds to the

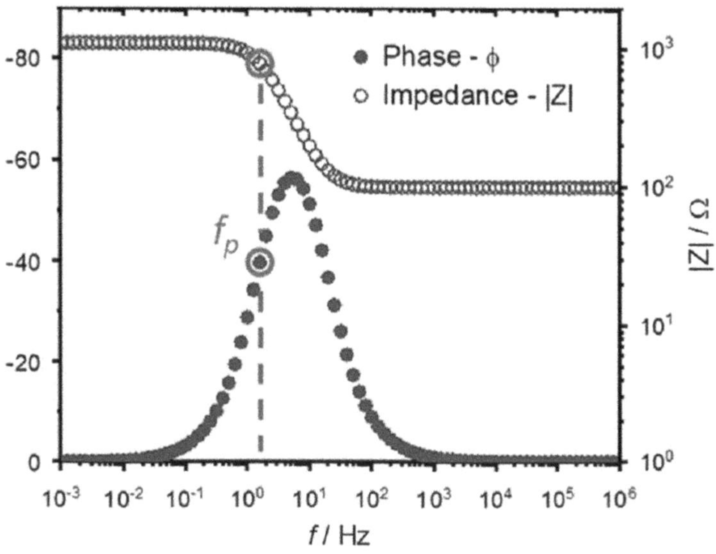

FIGURE 1.9 Bode Plots Representing Total Impedance and Phase Angle vs. Frequency Variations for a Randles Circuit Taking $R_s = 100\ \Omega$, $R_{ct} = 1000\ \Omega$, $C_{dl} = 100\ \mu F$. Reproduced from Ref. [44] (Bredar et al. *ACS Appl. Energ. Mater.* 2020, 3: 66–98), with Permission.

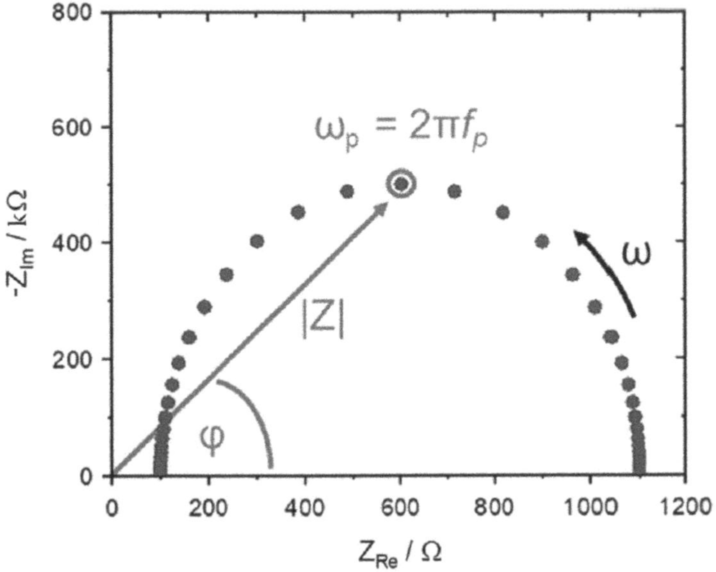

FIGURE 1.10 Nyquist Plot for a Randles Circuit Taking R_s = 100 Ω, R_{ct} = 1000 Ω, C_{dl} = 100 μF. Reproduced from Ref. [44] (Bredar et al. *ACS Appl. Energ. Mater.* 2020, 3: 66–98), with Permission.

Warburg impedance, which represents a kind of resistance to mass transfer. The impedance of the Warburg element for the diffusion of species in solution is:

$$Z_W = \frac{\sigma_W \sqrt{2}}{(j\omega)^{1/2}} \tag{1.23}$$

where

$$\sigma_W = \frac{RT}{n^2F^2A\sqrt{2}}\left(\frac{1}{c_{ox}D_{ox}^{1/2}} + \frac{1}{c_{rd}D_{rd}^{1/2}}\right) \tag{1.24}$$

c_{ox}, c_{rd} being the concentration of the oxidized and reduced forms of the electroactive species and D_{ox}, D_{rd}, their corresponding diffusion coefficients.

One of the most remarkable aspects of this technique is that, in a unique experiment, it provides information in phenomena occurring at considerably different time scales because frequencies between mHz (or even μHz) to kHz (or even MHz) are available. The critical frequency, f_c, for an electrochemical process is defined as the frequency at maximum of the impedance arc associated with the relaxation process. The relaxation time, τ, for this process can be calculated from the value of the critical frequency as $\tau = 1/2\pi f_c$. This technique is of particular interest in studying electrode surfaces and has become an essential tool in examining electrode materials.

1.12 OTHER TECHNIQUES

Two additional electrochemical techniques—electrochemical quartz crystal microbalance (EQCM), and scanning electrochemical microscopy (SECM)—can be mentioned. EQCM is based on the piezoelectric properties of quartz and permits the evaluation of mass changes occurring on an electrode supported over a quartz lamina.

Mass changes are determined from the measurement of the change in the oscillation frequency of the quartz crystal [45], given by the Sauerbrey equation:

$$\Delta f = -\frac{2\pi f_o^{\,2}\Delta m}{(\mu\rho)^{1/2}} \tag{1.25}$$

where Δf represents the frequency change, Δm is the mass change per unit area, f_o is the fundamental frequency of the quartz resonator, μ is the shear modulus, and ρ is the density of quartz. Frequency changes can be monitored simultaneously to the charge passed during electrochemical cycles, Δq, so that, for the deposition or dissolution of a compound of molar mass M, one can write:

$$\Delta m = \frac{M\Delta q}{nF} \tag{1.26}$$

Accordingly, M can be determined from the slope of the Δm vs. Δq representation [46]. Adsorption/desorption of ions and solvent molecules, however, can complicate data interpretation.

EQCM provides *in situ* measurement of mass changes accompanying electrochemical processes—adsorption, electrodeposition, electrocrystallization, electrodissolution, intercalation, electrochromism, etc. Studies dealing with, for instance, manganese oxides [47] and fullerenes [48], illustrate the capabilities of such technique in the context of the electrochemistry of porous materials.

Scanning electrochemical microscopy (SECM) is a technique of surface analysis based on the measurement of the current flowing through an ultramicroelectrode tip (diameter between 1 and 25 μm), submitted to potential inputs whose position, with respect to the surface of the system to be studied, remains constant or is changed. The tip electrode is associated with a three-electrode cell with a bipotentiostatic server. There are several modes of operation of SECM. The simplest approach involves the disposal of a solution of a redox probe (typically, $Fe(CN)_6^{3-}/Fe(CN)_6^{4-}$) in contact with the surface under analysis. The reference and auxiliary electrodes are immersed into that solution, as well as the tip electrode, which is placed at a short distance from the surface. In turn, a layer of the studied material is deposited over a base electrode forming a substrate electrode. The potentials of the tip (E_{tip}) and the substrate (E_s) electrodes can be varied while the tip position can be scanned to cover a given micrometric area. Then, if E_{tip} is fixed to a value at which the redox probe is electrochemically reduced or oxidized under diffusion-controlled conditions, the tip current varies depending on the distance (topography) to the surface and the conductivity of the same. In short, the tip can be regarded as a force sensor, thus providing topographic images of the studied surface, or as an electrochemical sensor, yielding an 'electrochemical' image of the studied surface at a nanometric scale [49].

REFERENCES

[1]. Rouquerol, J.; Avnir, D.; Fairbridge, C.W.; Everett, D.H.; Haynes, J.M.; Pernicone, N.; Ramsay, J.D.F.; Sing, K.S.W.; Unger, K.K. 1994. Recommendations for the characterization of porous solids (IUPAC Technical Report). *Pure and Applied Chemistry*. 66: 1739–1758.

[2]. Zhao, X.S. 2006. Novel porous materials for emerging applications. *Journal of Materials Chemistry*. 16: 623–625.

[3]. Leroux, F.; Raymundo-Piñero, E.; Fedelec, J.-M.; Béguin, F. 2006. Textural and electrochemical properties of carbon replica obtained from styryl organo-modified layered double hydroxide. *Journal of Materials Chemistry*. 16: 2074–2081.

[4]. Slater, A.G.; Cooper, A.I. 2015. Function-led design of new porous materials. *Science*. 348, aaa8075.

[5]. Mueller, U.; Schubert, M.; Teich, F.; Puetter, H.; Schierle-Arndt, K.; Pastré, J. 2006. Metal-organic frameworks-prospective industrial applications. *Journal of Materials Chemistry*. 16: 626–636.

[6]. Yarger, M.S.; Steinmiller, E.M.P.; Choi, K.-S. 2008. Electrochemical synthesis of Zn-Al layered double hydroxide (LDH) films. *Inorganic Chemistry*. 47: 5859–5865.
[7]. Noked, M.; Soffer, A.; Aurbach, D. 2011. The electrochemistry of activated carbonaceous materials: past, present, and future. *Journal of Solid State Electrochemistry*. 15: 1563–1578.
[8]. Scholz, F.; Nitschke, L.; Henrion, G. 1989. A new procedure for fast electrochemical analysis of solid materials. *Naturwissenschaften*. 76: 71–72.
[9]. Scholz, F.; Nitschke, L.; Henrion G.; Damaschun, F. 1989. A new technique to study the electrochemistry of minerals. *Naturwissenschaften*. 76: 167–168.
[10]. Scholz, F.; Schröder, U.; Gulabowski, R.; Doménech-Carbó, A. 2014. *Electrochemistry of Immobilized Particles and Droplets*, 2nd edit. Springer, Berlin.
[11]. Yang, Q.-H.; Xu, W.; Tomita, A.; Kyotani, T. 2005. The template synthesis of double coaxial carbon nanotubes with nitrogen-doped and boron-doped multiwalls. *Journal of the American Chemical Society*. 127: 8956–8957.
[12]. Yuan, Z.-Y.; Su, B.-L. 2004. An inherent macroperiodic assembly or an artificial beauty? *Studies in Surface Science and Catalysis*. 154B: 1525–1531.
[13]. Wright, J.D.; Sommerdijk, N. 2000. *Sol-Gel Materials: Their Chemistry and Biological Properties*. Taylor & Francis, London.
[14]. Liu, B.; Shioyama, H.; Akita, T.; Xu, Q. 2008. Metal-organic frameworks as a template for porous carbon synthesis. *Journal of the American Chemical Society*. 130: 5390–5391.
[15]. Li, H.-Y.; Anson, F.C. 1985. Electrochemical behavior of cationic complexes incorporated in clay coatings on graphite electrodes. *Journal of Electroanalytical Chemistry*. 184: 411–417.
[16]. Ghosh, P.K.; Mau, A.W.-H.; Bard, A.J. 1984. Clay-modified electrodes: Part II. Electrocatalysis at bis (2,2′-bipyridyl) (4,4′-dicarboxy-2,2′-bipyridyl)Ru(II)-dispersed ruthenium dioxide—hectorite layers. *Journal of Electroanalytical Chemistry*. 169: 315–317.
[17]. Calzaferri, G.; Lanz, M.; Li, J.-w. 1995. Methyl viologen-zeolite electrodes: intrazeolite charge transfer. *Chemical Communications*. 1313–1314.
[18]. Shaw B.R.; Kreasy, K.E. 1988. Carbon composite electrodes containing alumina, layered doublé hydroxides, and zeolites. *Journal of Electroanalytical Chemistry*. 243: 209–217.
[19]. Li, Z.; Lai, C.; Mallouk, T.E. 1989. Self-assembling trimolecular redox chains at zeolite Y modified electrodes. *Inorganic Chemistry*. 28: 178– 182.
[20]. Kornik, S.; Baker, M.D. 2002. Nanoporous zeolite film electrodes. *Chemical Communications*. 1700–1701.
[21]. Jiang, Y.-X.; Si, D.; Chen, S.-P.; Sun, S.-G. 2006. Self-assembly film of zeolite Y nanocrystals loading palladium on an Au electrode for electrochemical applications. *Electroanalysis*. 18: 1173–1178.
[22]. Zhang, Y.; Chen, F.; Shan, W.; Zhuang, J.; Dong, A.; Cai, W.; Tang, Y. 2003. Fabrication of ultrathin nanozeolite film modified electrodes and their electrochemical behavior. *Microporous and Mesoporous Materials*. 65: 277–285.
[23]. Walcarius, A.; Ganesan, V.; Larlus, O.; Valtchev, V. 2004. Low temperature synthesis of zeolite films on glassy carbon: towards designing molecularly selective electrochemical devices. *Electroanalysis*. 16: 1550–1554.
[24]. Kavan, L. 1997. Electrochemical carbon. *Chemical Reviews*. 97: 3061–3082.
[25]. Rolison, D.R. 1994. The intersection of electrochemistry with zeolite science, in *Advanced Zeolite Science and Applications*, Jansen, J.C.; Stöcker, M.; Karge, H.G.; Weitkamp, J. Eds. *Studies in Surface Science and Catalysis*. 85, 543–587.
[26]. Rolison, D.R.; Stemple, J.Z. 1993. Electrified microheterogeneous catalysis in low ionic strength media. *Chemical Communications*. 25–27.
[27]. Moked, M.; Soffer, A.; Aurbach, D. 2011. The electrochemistry of activated carbonaceous materials: past, present, and future. *Journal of Solid State Electrochemistry*. 15: 1563–1578.
[28]. Bockris, J.O.M.; Reddy, A.K.N. 1977. *Modern Electrochemistry*. Plenum Press, New York.
[29]. Bard, A.J.; Inzelt, G.; Scholz, F. Eds. 2008. *Electrochemical Dictionary*. Springer, Berlin-Heidelberg.
[30]. Oldham, K.B.; Myland, J.C. 1994. *Fundamentals of Electrochemical Science*. Academic Press, San Diego.
[31]. Bard, A.J.; Faulkner, L.R. 2001. *Electrochemical Methods*, 2nd edit. John Wiley & Sons, New York.
[32]. Štulík, K.; Amatore, C.; Holub, K.; Mareček, V.; Kutner, W. 2000. Microelectrodes. Definitions, characterization and applications (IUPAC Technical Report). *Pure and Aopplied Chemistry*. 72: 1483–1492.
[33]. Forster, R.J. 1994. Microelectrodes: New dimensions in electrochemistry. *Chemical Society Reviews*. 289–297.

[34]. Nicholson, R.S.; Shain, I. 1964. Theory of stationary electrode polarography. *Analytical Chemistry*. 36: 706–723.

[35]. Amatore, C.; Savéant, J.-M. 1978. Do ECE mechanisms occur in conditions where they could be characterized by electrochemical kinetic techniques? *Journal of Electroanalytical Chemistry*. 86: 227–232.

[36]. Laviron, E. 1979. General expression of the linear potential sweep voltammogram in the case of diffusionless electrochemical systems. *Journal of Electroanalytical Chemistry*. 101: 19–28.

[37]. Ramaley, L.; Krause, M.S. 1969. Theory of square wave voltammetry. *Analytical Chemistry*. 41: 1362–1365.

[38]. Nicholson, R.S. 1965. Some examples of the numerical solution of nonlinear integral equations. *Analytical Chemistry*. 37: 667–671.

[39]. Krulic, D.; Fatouros, N. J. 2011. Peak heights and peak widths at half-height in square wave voltammetry without and with ohmic potential drop for reversible and irreversible systems . *Journal of Electroanalytical Chemistry*. 652 : 26–31.

[40]. Gélinas, B.; Rochefort, D. 2015. Synthesis and characterization of an electroactive ionic liquid based on the ferrocenylsulfonyl(trifluoromethylsulfonyl)imide anion. *Electrochimica Acta*. 162: 36–44.

[41]. DuVall, S.; McCreery, R.L. 1999. Control of catechol and hydroquinone electron transfer kinetics on native and modified glassy carbon electrodes. *Analytical Chemistry*. 71: 45594–45602.

[42]. Retter, H.; Lohse, H. 2005. Electrochemical impedance spectroscopy, in Scholz, F. Ed. *Electroanalytical Methods*. Springer, Berlin.

[43]. Macdonald, D.D.; Sikora, A.; Engelhardt, G. 1998. Characterizing electrochemical systems in the frequency domain. *Electrochimica Acta*. 43: 87–107.

[44]. Bredar, A.R.C.; Chown, A.L.; Burton, A.R.; Farnum, B.H. 2020. Electrochemical impedance spectroscopy of metal oxide electrodes for energy applications. *ACS Applied Energy Materials*. 3: 66–98.

[45]. Chen, L.C. 2008. *Electrochemistry for Biomedical Researchers*. National Taiwan University Press, Taipei.

[46]. Uchida, H.; Ikeda, N.; Watanabe, M. 1997. Electrochemical quartz crystal microbalance study of copper adatoms on gold electrodes Part II. Further discussion on the specific adsorption of anions from solutions of perchloric and sulfuric acid. *Journal of Electroanalytical Chemistry*. 424: 5–12.

[47]. Wu, B.L.; Lincot, D.; Vedel, J.; Yu, L.T. 1997. Voltammetric and electrogravimetric study of manganese dioxide thin film electrodes. Part 1. Electrodeposited films. *Journal of Electroanalytical Chemistry*. 420: 159–165.

[48]. Bond, A.M.; Miao, W.; Raston, C.L. 2000. Identification of processes that occur after reduction and dissolution of C_{60} adhered to gold, glassy carbon, and platinum electrodes placed in acetonitrile (electrolyte) solution. *Journal of Physical Chemistry B*. 104: 2320–2329.

[49]. Bard, A.J.; Mirkin, M.V. Eds. 2001. *Scanning Electrochemical Microscopy*. Marcel Dekker, New York-Basel.

2 Electrochemical Processes at Porous Electrodes

2.1 INTRODUCTION

The electrochemistry of porous electrodes is an expanding area that exploits the capabilities of conducting electrodes for sensing and energy production and storage. The key point is the possibility to modulate the electrochemical response of a given electroactive species in solution by varying the composition of the material forming the porous electrode and the geometrical characteristics of the pores. In the current chapter, we study the theoretical approaches for these systems, assuming that the porous electrode is non-electroactive (i.e. does not experience chemical transformations throughout electrochemical turnovers).

This electrochemistry was studied early by de Levie [1,2] and has received significant theoretical inputs over the last decade. In the following, it will be assumed that the microporous material forms a layer onto an inert nonporous conducting block and that this working electrode will be accompanied by auxiliary and reference electrodes in contact with the electrolyte solution containing electroactive species and the supporting electrolyte. In principle, there are two different situations to be treated: when the porous material is electron-conducting and when it is insulating. Remarkably, there is a possibility to build a variety of porous nanoarchitectures upon metal deposition onto different structure-directing agents. Figure 2.1 illustrates the influence of pore distribution in the voltammetric response showing the cyclic voltammograms recorded at a porous gold electrode and a bimodally porous Au structure formed using an opal template in contact with 1 M H_2SO_4 [3].

2.2 POROUS ELECTRODES: IMPEDANCE ANALYSIS

Let us consider a system constituted by N cylindrical pores of length l_p and radius r_p as schematized in Figure 2.2 [4]. The impedance of the system should incorporate the contributions of the electrolyte filling the pores and the walls plus additional terms associated with the solution resistance and the barrier layer schematized in Figure 2.3 [5].

The classical model of de Levie assumes that only the pore side walls are conducting and that the potential gradient exists in the axial direction only. Then, the ohmic resistance of an individual pore can be expressed as:

$$R_{por} = \frac{\rho l_p}{\pi r_p{}^2} \tag{2.1}$$

Replacing the capacitors by constant phase elements (Eq. (1.22)) to account for deviations from ideal capacitive behavior, one obtains

$$Z_{por} = \sqrt{\frac{R_{por}}{C_{dl}\,(j\omega)^{1-n}}}\,\coth\left(\sqrt{R_{por}\,C_{dl}\,(j\omega)^{1-n}}\right) \tag{2.2}$$

As a consequence of the coupling of the solution resistance and double layer impedance for the signal penetration depth, which is smaller than the pore length, the circuit can be approached, at low frequencies, by a R_{por}-C_{dl} connection in series so that

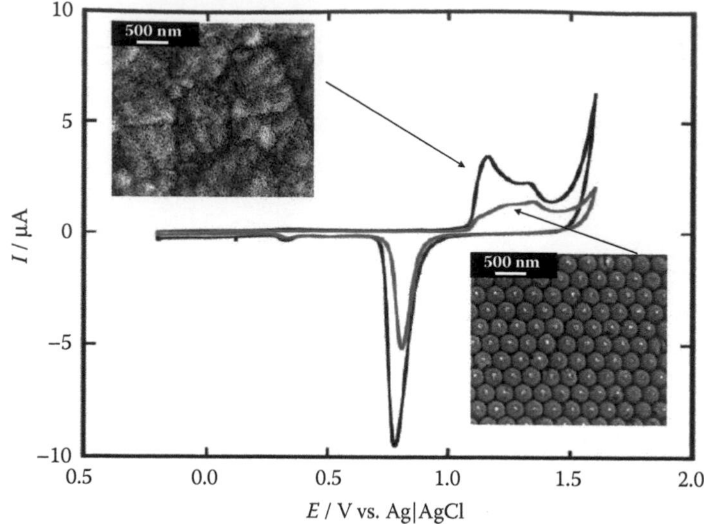

FIGURE 2.1 SEM Images of on a Porous Gold Film (Upper Image) and Porous Gold Opal (Lower Image) Electrodes and the Respective Cyclic Voltammograms Collected in 1 M H_2SO_4. Potential Scan Rate 50 mV s^{-1}. Reproduced from Ref. [3] (Chae et al. *Appl. Mater. Interfaces*. 2012, 4: 3973–3979), with Permission.

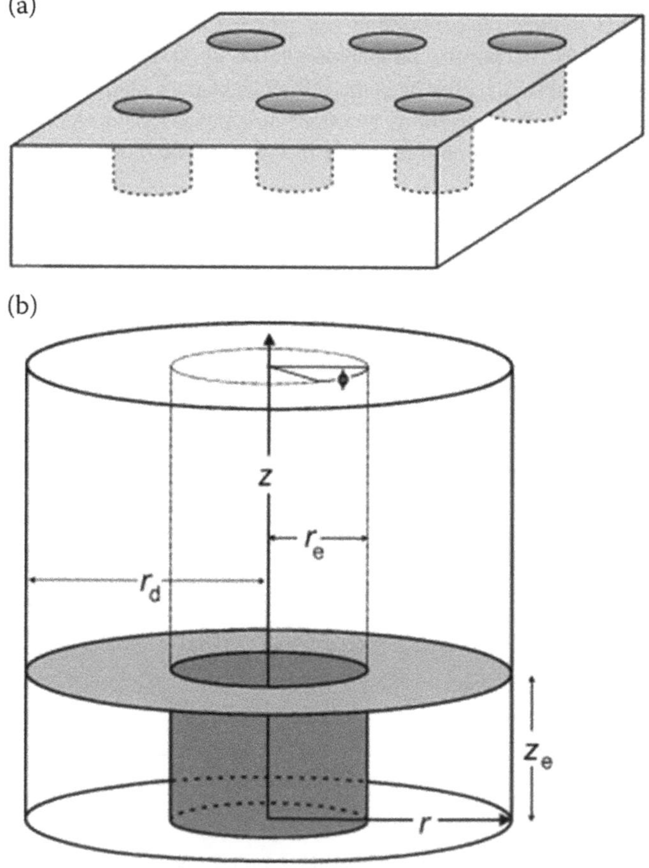

FIGURE 2.2 Schematics of a Porous Electrode and Representative Pore Parameters. Reproduced from Ref. [4] (Ward and Compton, *J. Electroanal. Chem.* 2014, 724: 43–47), with Permission.

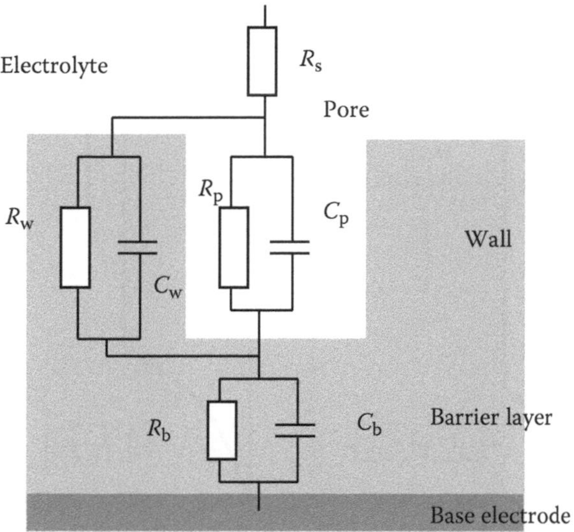

FIGURE 2.3 Equivalent Circuit Incorporating the Contributions of Pore Walls Usable to Model the Impedance Behavior of Porous Electrodes.

$$Z_{por} = R_s + R_{por} + \frac{1}{j\omega C_{dl}} \tag{2.3}$$

where R_s is the solution resistance. In this idealized case, the complex plane plots display a vertical capacitance straight line intersecting the real axis at $Z_{real} = R_{por} + R_s$.

At high frequencies, Eq. (2.2) approximates to

$$Z_{por} = \sqrt{\frac{R_{por}}{C_{dl}\,(j\omega)^{1-n}}} \tag{2.4}$$

This corresponds to a 45° line in the Nyquist plot. Introducing the charge transfer resistance instead of a straight line at 45°, a semicircle is obtained [6]. Figure 2.4a depicts a schematic representation of the theoretical Nyquist plots for porous electrodes according to the previous considerations.

A more realistic approach, however, considers that the pores are embedded in a layer of conducting material [7,8]. Then, the capacitance of the top flat layer must be incorporated as $Z_{flat} = 1/j\omega C_{flat}$ in parallel to the pore impedance (see Figure 2.4b) so that the total impedance of the circuit will be

$$Z_{tot} = R_s + \frac{1}{j\omega C_{flat} + \frac{1}{Z_{por}}} \tag{2.5}$$

As a result, the high-frequency region tends to a capacitive loop in the Nyquist plot (see Figure 2.4b).

2.3 VOLTAMMETRY AT POROUS ELECTRODES

The electrochemistry at porous electrodes has been recently theoretically discussed in a series of works by Compton et al. [4,9,10]. Let us consider a conducting solid block penetrated by a

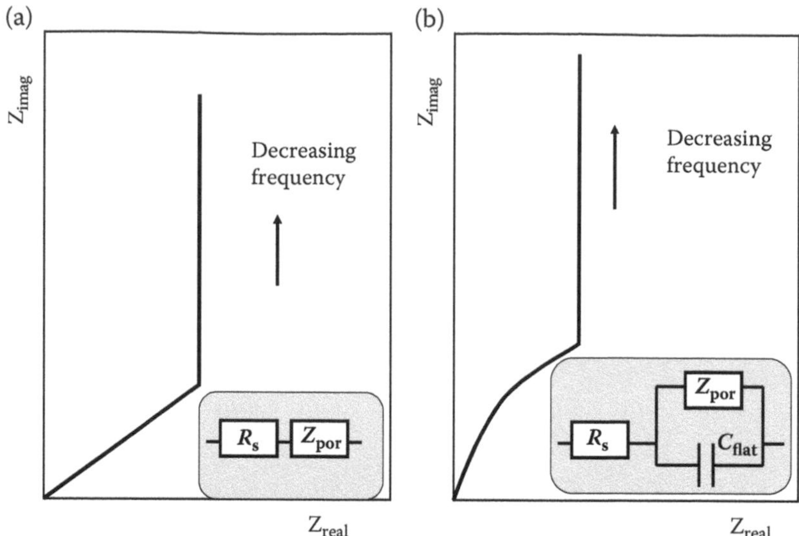

FIGURE 2.4 Equivalent Circuits and Theoretical Nyquist Plots for a Porous Electrode (a) without and (b) with Account of the Contribution of the Flat Surface of the Film Embedding the Pores.

regularly distributed array of cylindrical pores, as represented in Figure 2.2. The exposed upper surface of the block, of geometric area A and height z_e, contains a series of cylindrical pores with radius r_e, whose interior surfaces (bottom and walls) are electron-conducting. It is assumed that the block is immersed in an electrolyte solution that fills the interior of the pores and contains an electroactive species A, which can undergo a reversible one-electron reduction process: $A + e^- \rightarrow B$.

The electrode surface can be treated as an array of identical unit cells, each with a pore at its center having a diffusion domain (i.e. the region where there is a significant gradient of concentration of electroactive species, see ref. [11]) of radius r_d. The voltammetric behavior of the system depends on the above parameters and potential scan rate v so that peak current varies between two extremes. In the limit of low scan rate, $v \rightarrow 0$, the solution can fill the pores so that the voltammogram looks like a typical diffusion-controlled reversible process at macroelectrodes, the peak current I_p being described by the reversible Randles–Ševčik equation (Eq. (1.6)). In the limit of high scan rate, $v \rightarrow \infty$, there is no time for fresh solution species to diffuse from the bulk solution to the pores. Then, the peak current is limited by the volume of solution that permeates the porous surface at the start of the scan and the voltammogram is equivalent to that seen in the voltammetry of surface-confined species. The corresponding expression for peak current,

$$I_p = \frac{n^2F^2vV_p}{4RT} \tag{2.6}$$

where V_p denotes the pore volume, is formally analog to Eq. (1.9). The crossover between these behaviors occurs for a threshold scan rate given by

$$v = \frac{3.183DRT}{z_e^2 F \left(\frac{r_e}{r_d}\right)^4} \tag{2.7}$$

where D is the diffusion coefficient of the electroactive species A.

If the reduction of A is a fully irreversible process at low scan rates, and when adjacent pores are sufficiently close together that diffusion field is perpendicular to the plane of the bulk electrode surface, the peak potential varies with the potential scan rate as

$$E_p = E^{o\prime} - \frac{RT}{\alpha F}\left[0.780 - \ln k^{o}\Psi + \ln\sqrt{\frac{\alpha F D v}{RT}}\right] \tag{2.8}$$

This equation equals that for planar nonporous electrodes (Eq. (1.7)) except by the coefficient Ψ. This coefficient represents the ratio of the electroactive surface area to the geometric surface area, which can be expressed as:

$$\Psi = 1 + 2\left(\frac{r_e}{r_d}\right)^2 \frac{z_e}{r_e} \tag{2.9}$$

where the $(r_e/r_d)^2$ ratio represents the porosity of the system. It is worth noting that this situation parallels that of planar electrodes when covered by a dense array of nanoparticles. Then, the effective area is larger than the geometrical area of the base electrode so that there is an apparent electrocatalytic effect [4]. This is an important factor to be accounted for when treating electrochemical sensing (Chapter 12).

At high scan rates, the peak potential tends to vary with v according to the expression analog to thin layer voltammetry

$$E_p = E^{o\prime} + \frac{RT}{\alpha F}\ln\left(\frac{ARTk^{o}}{\alpha F V_p v}\right) \tag{2.10}$$

In these conditions, there is no apparent catalytic enhancement of the current.

2.4 CONFINEMENT AND DIFFUSION IN PORES

As discussed in Section 2.2, charge transport behavior at porous electrodes is dominated by the transition from semi-infinite to thin-layer diffusion. This is superimposed to effects associated with confinement at the nanoscale, due to the altered structure of solvents when they are confined to nano-sized pores. To distinguish transport effects arising from nano-confinement from simple Fickian diffusion, Compton et al. modeled the voltammetry at a ring-tube model [12]. This system consists of an insulating cylindrical pore of infinite length and radius r_e containing a conducting band of height z_e. The pore is filled with the electrolyte solution with an electroactive species that can undergo a reversible one-electron oxidation. This formalism represents the case of conducting pores by taking $z_e \to \infty$ and the case of insulating pores by assuming $z_e \to 0$. As schematized in Figure 2.5, four limiting cases can be discerned. First, the case when $r_e \to 0$ and $z_e \to 0$ corresponds to a capillary pore with a thin annular band, while the case $r_e \to 0$ and $z_e \to \infty$ corresponds to an entirely conducting capillary pore. For large pores, the curvature of the band becomes negligible on the scale of the diffusion layer so that the case $z_e \to 0$ is equivalent to a ring electrode and the case $z_e \to \infty$ is equivalent to a macroelectrode.

Since the voltammetric response depends not only on the dimensions above but also on the value of the diffusion coefficient of the electroactive species and the time scale of the experiment (i.e. the potential scan rate v), the different cases can be expressed in terms of two parameters

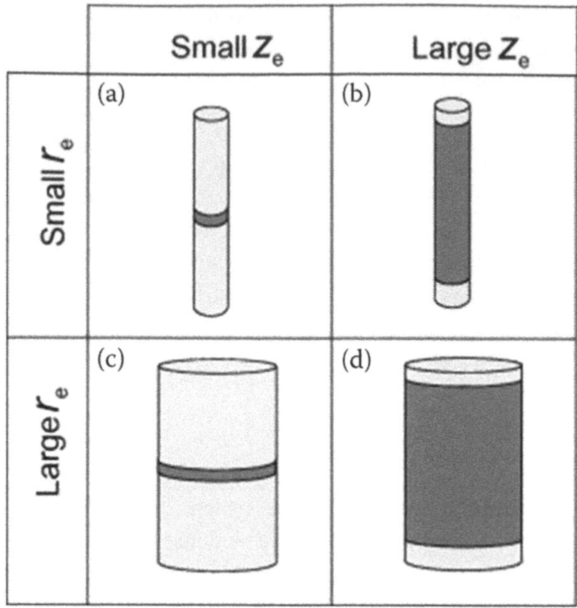

FIGURE 2.5 Scheme of the Four Limiting Cases of the Model of a Cylindrical Insulating Pore of Radius r_e Containing a Conducting Band of Height z_e. Darker Areas are Conducting While Lighter Areas are Insulating. Reproduced from Ref. [12] (Ward et al. *J. Electroanal. Chem.* 2013, 695: 15–24), with Permission.

$$\sigma_r = \frac{Fr_e^2}{DRT}v; \qquad \sigma_z = \frac{Fz_e^2}{DRT}v \tag{2.11}$$

When $\sigma_z \to \infty$, $\sigma_r \to \infty$, the system behaves like a planar macroelectrode with peak current given by the Randles-Ševčik equation (Eq. 1.6) (i.e. peak current proportional to $v^{1/2}$). When $\sigma_z \to \infty$, $\sigma_r \to 0$, the behavior of thin layer macroelectrodes is reproduced, the peak current being

$$I_p = \frac{F^2 \pi r_e^2 z_e c}{4RT} v \tag{2.12}$$

where c is the concentration of the electroactive species and v the potential scan rate. In turn when the limit $\sigma_z \to 0$, $\sigma_r \to 0$, the peak current becomes

$$I_p = 0.446(2\pi r_e^2) Fc\sqrt{\frac{FD}{RT}v} \tag{2.13}$$

Under these micro-planar conditions, the volume of electroactive solution near the electrode is rapidly exhausted, and the peak current is limited by the rate at which fresh material is transported to the electroactive surface. In the case of $\sigma_z \to 0$, $\sigma_r \to \infty$, corresponds to the so-called micro-band. Here, the curvature of the band is negligible compared to the scale of the diffusion layer and the voltammetric response approximates that of a flat band electrode. Figure 2.6 compares a series of theoretical voltammograms for a conducting porous electrode ($z_e \to \infty$) represented in terms of normalized currents and potentials for different values of the parameter σ_r taken from Ward et al. [12]. This figure illustrates the pass from a symmetrical peak at low σ_r values, characterizing the thin-layer behavior, to infinite diffusion asymmetric peaks on increasing σ_r.

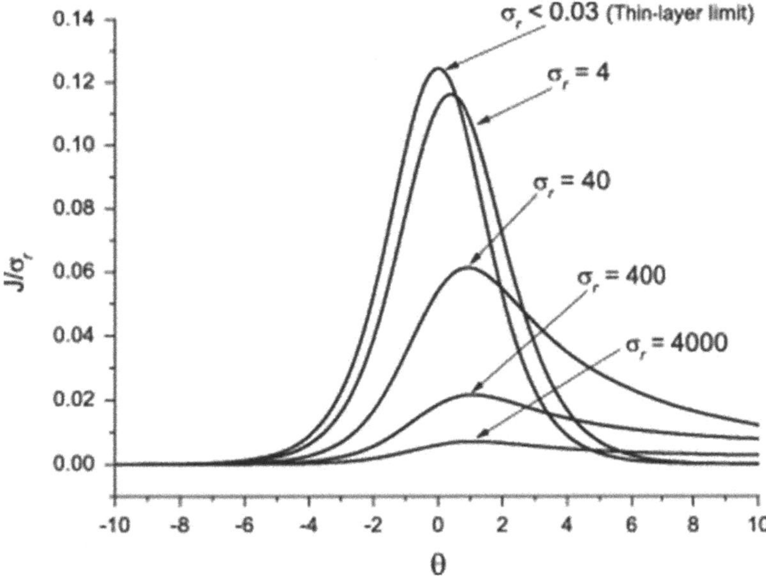

FIGURE 2.6 Theoretical Voltammograms at a Conducting Porous Electrode for Different Values of the Parameter σ_r in Terms of Normalized Currents and Potentials. Reproduced from Ref. [12] (Ward et al. *J. Electroanal. Chem.* 2013, 695: 15–24), with Permission.

2.5 INSULATING POROUS FILMS

A situation of particular interest corresponds to the previously described case of a conducting plane electrode covered by many insulating cylindrical bars aligned perpendicularly on the supporting electrode, as schematized in Figure 2.7 [10]. Electrochemical reactions involving species in the electrolyte intercalated between the cylinders are assumed to occur solely at the supporting electrode. In principle, since the exposed area of the substrate electrode is lower than that of the same electrode in the absence of insulating bars, the peak current for the oxidation or reduction of a species in solution will be diminished. In most cases of interest, however, the electroactive species can experience adsorption and desorption processes at the bar surfaces. Then, the porous electrode acts as a preconcentrating device, increasing the effective concentration of the electroactive species. The study from Chan et al. [10] emphasizes the importance of adsorption resulting in a significant enhancement of the currents relative to the bare substrate electrode. Figure 2.8 shows a series of theoretical voltammograms for the oxidation or reduction of a species that is adsorbed, according to Langmuir isotherm, to a set of insulating bars covering a conducting substrate. K_{ads} represents the ratio between the adsorption and desorption rates for different values of the equilibrium adsorption constant. The peak current increases as K_{ads} increases, thus determining — in the case of analytical determinations — a notable sensitivity increase. It must be underlined, however, that this increase can be compensated by the partial blocking of the substrate electrode surface exerted by the porous layer.

2.6 FRACTAL SURFACES

Many systems exhibit fractal geometries, characterized by structures that look the same on all length scales [13]. Fractal structures can be characterized by a scaling law where the number of discrete units is proportional to the d_F-power of the size of those units, d_F being the fractal dimension of the structure. Diffusion in fractal media and diffusion towards, from, and through interfaces are of interest to electrochemical processes involving porous materials. Since the pioneering works by

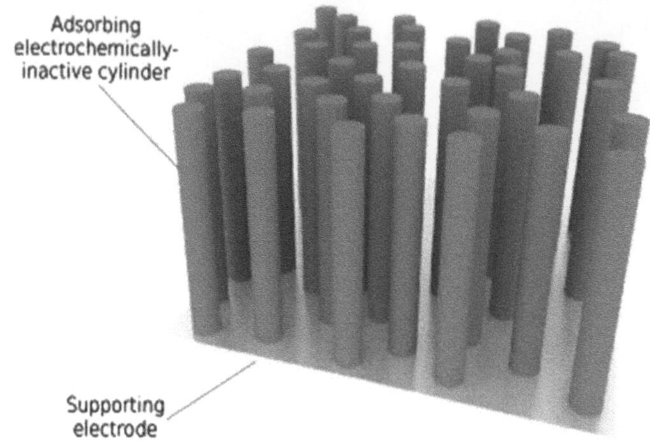

FIGURE 2.7 Scheme for a Set of Cylindrical Insulating Bars Covering a Plane Conducting Electrode. Reproduced from Ref. [10] (Chan et al. *J. Electroanal. Chem.* 2017, 801: 135–140), with Permission.

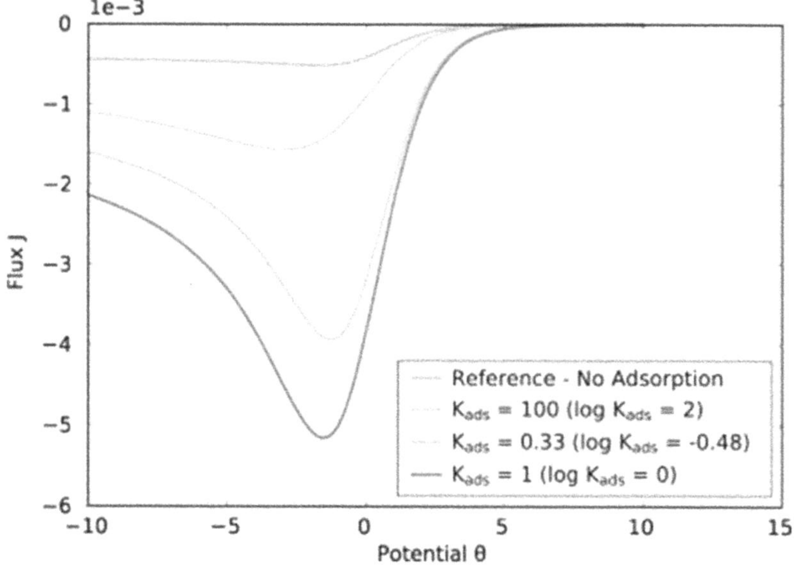

FIGURE 2.8 Theoretical Voltammograms for the Oxidation or Reduction of a Species Which is Adsorbed, According to Langmuir Isotherm, to a Set of Insulating Bars Covering a Conducting Substrate, for Different Values of the Ratio Between the Adsorption and Desorption Rates. Reproduced from Ref. [10] (Chan et al. *J. Electroanal. Chem.* 2017, 801: 135–140), with Permission.

Nyikos and Pajkossy [14], fractal geometry has provided a powerful tool to treat problems of mass transport towards and from interfaces between the electrode and the electrolyte.

As previously indicated, when mass transport is purely controlled by the reaction-diffusion in the electrode or electrolyte, the electrochemical responses at the rough interface obey the generalized form of the Cottrell equation which can be expressed as [15,16]:

$$I = \frac{nFAcD^{1/2}}{\pi^{1/2} t^{\left(\frac{d_F - 1}{2}\right)}} \tag{2.14}$$

The above equation predicts a linear dependence of I on $t^{-\alpha}$ ($\alpha = (d_F - 1)/2$) which can be tested from experimental CAs. For a cyclic voltammetric experiment at a potential scan rate v, involving oxidized and reduced species in solution, the Randles–Ševčik equation (Eq. (1.6)) can be generalized to

$$I_p = \frac{0.2518(nF)^{3/2}A^{d_F/2}K^{2-d_F}cD^{3-d_F}\Gamma}{(RT)^{1/2}}\left(\frac{d_F - 1}{2}\right)v^{\left(\frac{d_F-1}{2}\right)} \quad (2.15)$$

Here, Γ is the gamma function, K is a constant related to the fractal dimension of the interface which, in principle, is taken as a unit, and A is the macroscopic area of the electrode. This equation predicts that the peak current in voltammetric experiments will be proportional to $v^{(d_F-1)/2}$ rather than $v^{1/2}$ (Eq. (1.6)), a relationship that can easily be tested. Interestingly, under such conditions, the diffusion layer length acts as a yardstick length to probe the fractal topography of the electrode surface [17] and can be estimated as

$$\delta = \frac{nFAcD}{I_p} \quad (2.16)$$

In the above equations, d_F values should be comprised between 2 and 3, respectively corresponding to a planar and a volumic electrode surfaces. Following Andrieux and Audebert [16], both chronoamperometry and cyclic voltammetric experiments would be suitable to determine the fractal dimension d_F of a volumic modified electrode. In the first case (CA), plots of ln I vs. ln t from CA data should provide straight lines of slope $-(d_F - 1)/2$ (Eq. (2.14)). In CV, successive experiments using different potential scan rates, v must be performed so that plots of I_p vs. v from CV experiments yield straight lines of slope α.

Analogously, the generalized Warburg equation, representative of the response of constant phase elements in EIS experiments, becomes [15]

$$Z(\omega) = \frac{1}{nFAD^{1/2}}\left(\frac{dE}{dc}\right)(j\omega)^{-\alpha} \quad (2.17)$$

where E represents the applied potential and c the concentration of the electroactive species. The thickness of the diffusion layer can now be determined from the frequencies where the CPE operates, f_{CPE}, as [18]

$$\delta = \sqrt{\frac{D}{f_{CPE}\,g}} \quad (2.18)$$

The corresponding interval of frequencies can be obtained from the $\tan^{-1}$ $[d(-Z_{image})/dZ_{real}]$ vs. f plots. Figure 2.9 shows AFM images determined from different V_2O_5-porous film electrodes and Figure 2.10 shows the corresponding representation for V_2O_5 film electrodes with various pore structures in 1 M $LiClO_4$/propylene carbonate solution [19]. Studies dealing with diffusion phenomena at fractal interfaces for diffusion-controlled and non-diffusion-controlled electron transfer processes are of interest in modeling hydrogen transport through hydride-forming electrode, Li^+ intercalation in $Li_{1-x}CoO_2$ electrodes, and flow patterns in microfluidic devices, among others [20].

2.7 THE PROBLEM OF THE OXIDATION STATE

The determination of the redox state of an electroactive species is a frequent problem in electrochemistry [21]. This problem is of particular significance in the case of porous solids where the

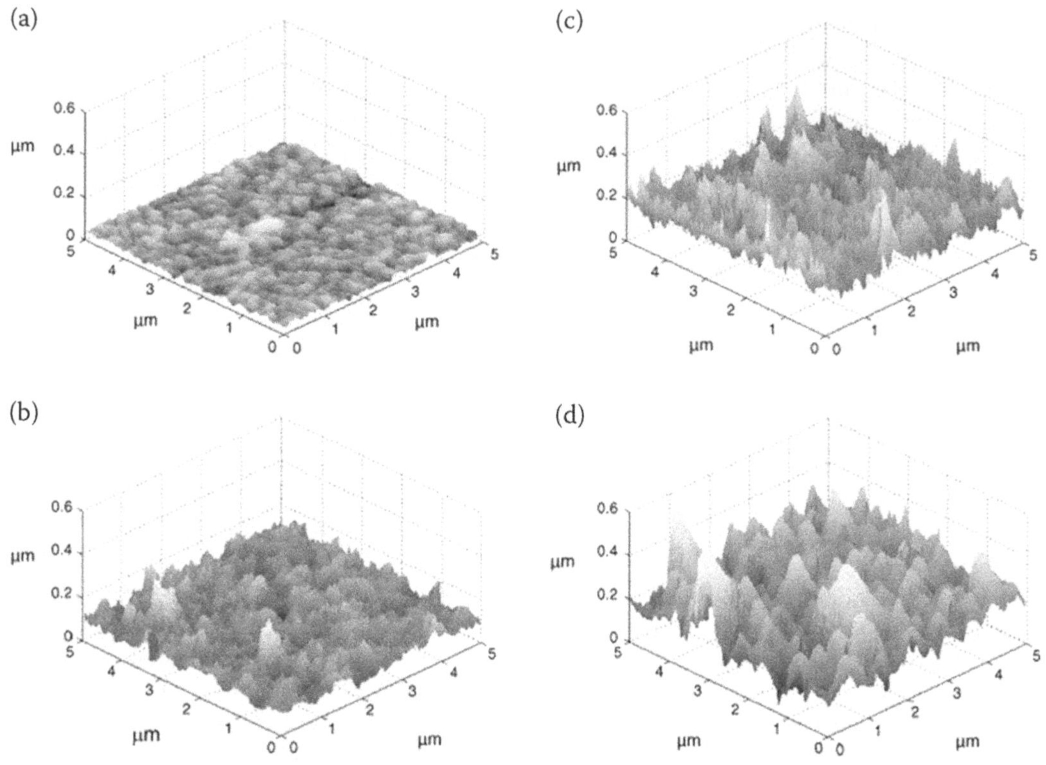

FIGURE 2.9 Three-Dimensional AFM Images Determined from Different V_2O_5-Porous Film Electrodes. Reproduced from Ref. [19] (Jung and Pyun. *Electrochim. Acta.* 2006, 51: 2646–2655), with Permission.

attachment of electroactive guests may produce redox tuning to some extent [22]. In the solution phase, CVs of an electrochemically reversible couple are of identical shape following the first cycles, and their shape is the same for the oxidized and reduced forms as well as on the proportion of both forms when they coexist in the solution.

The discrimination between the extreme cases (i.e. solution of A^{ox} or A^{rd}) can be made by measuring the open circuit potential of the solution or by recording the initial currents upon starting the potential once on the positive and again on the negative limit. A straightforward and sensitive discriminating method of application was proposed by Scholz and Hermes [23] in CV experiments in the solution phase. The method exploits the shift along the current axis of the entire voltammogram depending on the redox state of the depolarizer and involves the measurement of the currents at the lower ($I_{\lambda c}$) and upper ($I_{\lambda a}$) limits of the potential range. Simulations indicated that the difference $\Delta I = | I_{\lambda a} | - |I_{\lambda c}|$ varied linearly on the molar ratio of the reduced (or oxidized) electroactive species of the couple provided that: (i) the potential limits are chosen symmetric to the formal potential, (ii) the potential range is relatively large (at least eight times the peak separation), and (iii) repetitive voltammetry (fifth scan is recommended in order to avoid transient phenomena) is used [23].

The presence of large background currents due to uncompensated resistive and capacitive effects can distort voltammetric curves. A second procedure derives from the analysis of currents in CVs initiated in anodic (or cathodic) direction at the midpeak potential of the redox couple [24]. Qualitatively, we can use two diagnostic criteria to define the oxidation state, as illustrated in Figure 2.11, where CVs at Pt electrode of aqueous solutions of (a) $Fe(CN)_6^{3-}$ and (b) $Fe(CN)_6^{4-}$ initiated at the midpeak potential in the positive direction, are depicted. The first criterion is determined by the initial current recorded in the first scan. If the solution contains the oxidized form,

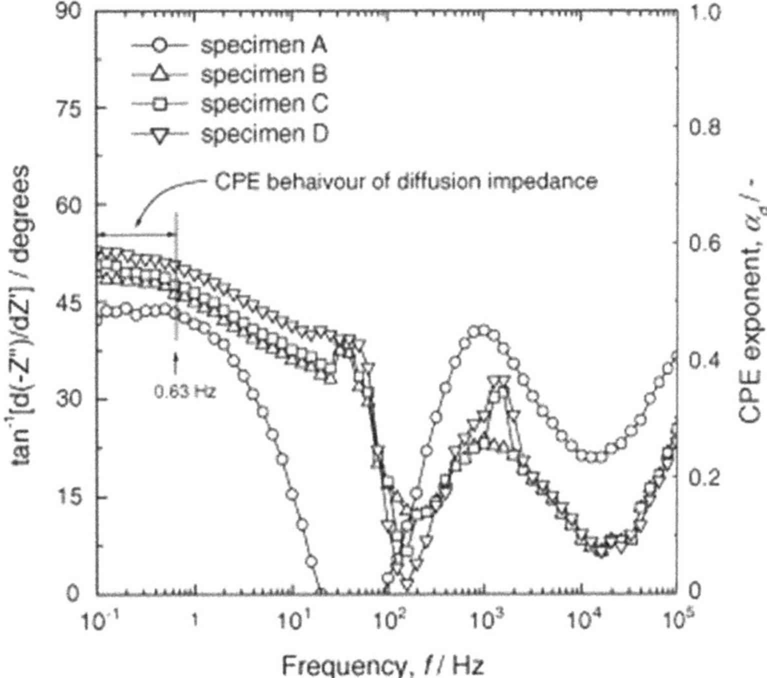

FIGURE 2.10 Plots of $\tan^{-1}$ [d($-Z_{imag}$)/dZ_{real}] vs. f for V_2O_5 Film Electrodes with Various Pore Structure in 1 M $LiClO_4$/Propylene Carbonate Solution at an Electrode Potential of 3.1 V vs. Li^+/Li. Reproduced from Ref. [19] (Jung and Pyun. *Electrochim. Acta.* 2006, 51: 2646–2655), with Permission.

the initial current is slightly cathodic (Figure 2.11a); if the solution contains the reduced form, the initial current is slightly anodic (Figure 2.11b).

The second criterion, which can be extended to quantify the proportion of both species in mixed solutions, is based on the determination of the ratio between the anodic peak current in the first (i_{pa}^{I}) and second (i_{pa}^{II}) scans in initial anodic scan voltammograms. As can also be seen in Figure 2.11, these currents differ when the oxidized form exists in solution, and equal when it's in the reduced form. Figure 2.12 shows the variation of i_{pa}^{I}/i_{pa}^{II} on the molar fraction of oxidized form, α_{ox}, for solutions of $K_4Fe(CN)_6$ plus $K_3Fe(CN)_6$ in 1.0 mM total concentration from CVs such as in Figure 2.11 [25]. The peak currents were measured from the initial current in the first anodic scan and fit well to linear variations of i_{pa}^{I}/i_{pa}^{II} on α_{ox} varying with the potential scan rate. These methods, although can compensate to some extent ohmic and capacitive effects, are conditioned by the frequent difficulty in defining a proper baseline for current measurements. Among other strategies, minimizing capacitive effects using SWV has been reported [25].

2.8 ELECTROCHEMISTRY AT NANOPORES

There is an increasing interest in the electrochemistry of a variety of nanopore systems. These include membranes dividing two electrolyte solutions, detection of transport to/from nanoelectrodes lying at the bottom of nanopores, and electrophoretic and electroosmotic transport in nanochannels [26] as well as applications in microfluidic sensing [27]. Single nanopores in membranes have been made by milling nanometer-size holes in a variety of substrates (carbon nanotubes, graphene, SiO_2, epoxy membranes, etc.). Biological nanopores comprise a lipid bilayer with pore-containing proteins such as α-hemolysin [28]. This permits the spatial resolution of

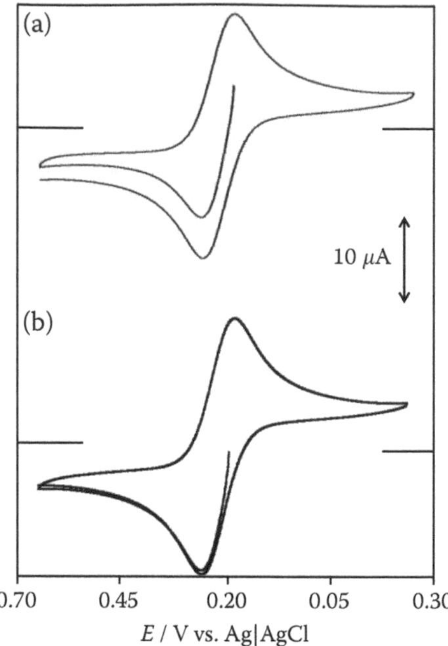

FIGURE 2.11 CVs at Pt Electrode of (a) 1.0 mM $K_3[Fe(CN)_6]$, (b) 1.0 mM $K_4[Fe(CN)_6]$ Solutions in 0.10 M KCl. Potential Scan Initiated at the Midpeak Potential in the Positive Direction; Potential Scan Rate 50 mV s^{-1}.

nucleotides and the differentiation between single and double-stranded DNA and DNA modifications.

These systems have been used (Coulter counting) as molecular counting devices schematized in Figure 2.13 [30]. The two chambers containing an electrolyte solution are divided by a membrane with a single channel. Applying an appropriate potential difference across the membrane, an ionic current is recorded. If the solution contains particles of an appropriate size and charge, they will enter the channel and reduce the ion current displaying a series of current pulses. The height of the pulse, Δi_C, is related to particle size, and the pulse width, Δt, corresponds to the particle transit time. This process is termed ionic current blockade (ICB). Figure 2.14 shows the histogram of dsDNA blockage in an ICB experiment at 1 M KCl solution at pH 7.2 with 0.01% Triton-X100 [31]. The histogram has been fitted with the sum of two Gaussians at "short" and "long" blockage duration times.

Corresponding data can provide information about the size, charge, and concentration of the particles. When the nanopore is too small to permit the passage of the blocking species, a "capture and release" experiment can be made. Here, the target particle occludes the pore entrance, attenuating the ionic flow. Then, the particle is released by reversing the applied potential bias [32].

Nanopore electrodes can also be fabricated. The procedure involves pore generation in a glass shroud by sealing a platinum nanowire (whose edge consists of a conical tip) followed by electrochemically metal etching. This leaves a glass-walled pore whose dimensions are approximately equal to that of the original nanoelectrode [33]. The voltammetric behavior of nanopore electrodes using usual redox probes (ferrocene, hexacyanoferrate ions) at long time scales gives steady-state limiting currents such as microelectrodes, representative of radial diffusion. These currents decrease for increasing pore depth. When this quantity is more than 50-fold larger than the pore diameter, the currents become almost constant, depending on the cone angle and nanopore orifice diameter. In short-time voltammetric experiments, the behavior is close to that expected for linear diffusion within the pore cone while at intermediate times, the CV response approaches that for a thin layer cell [26].

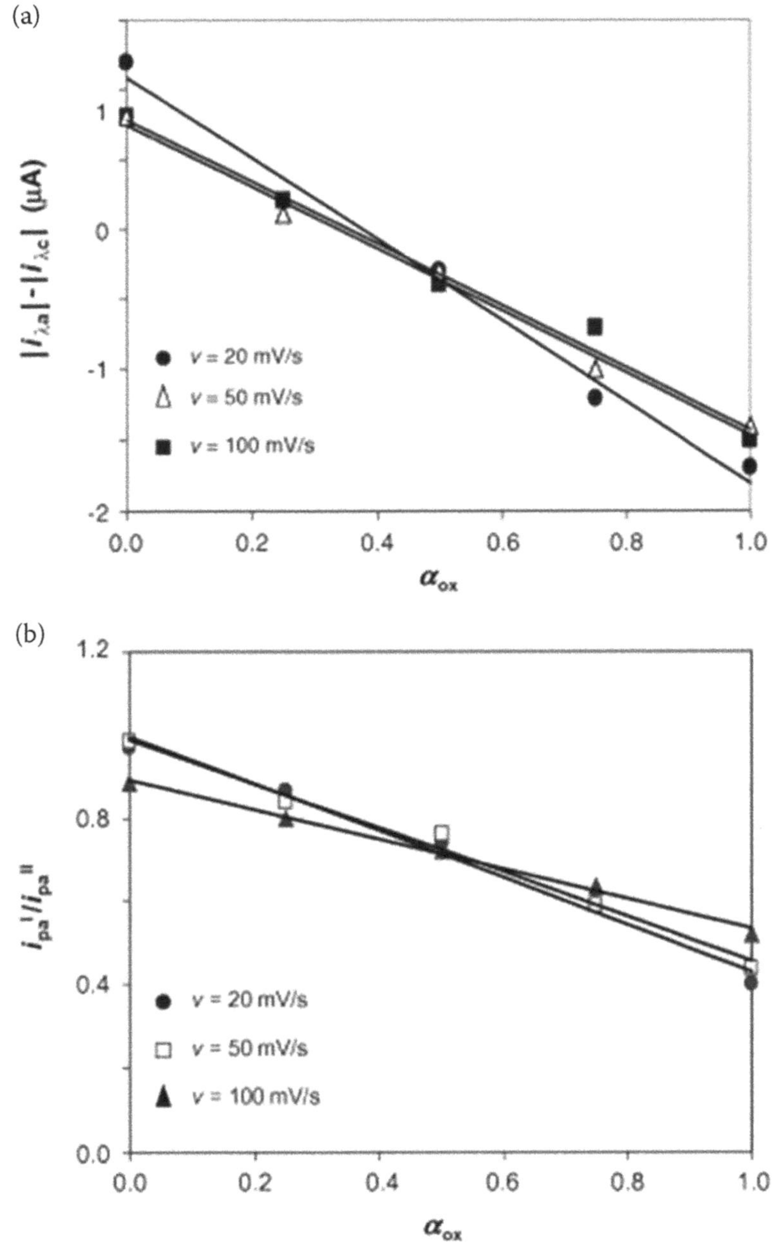

FIGURE 2.12 Plots of (a) ΔI vs. α_{ox}, and (b) i_{pa}^{I}/i_{pa}^{II} vs. α_{ox} for Different $K_4Fe(CN)_6$ Plus $K_3Fe(CN)_6$ Mixtures (Total Concentration 1.0 mM) in 0.10 M $NaClO_4$ from CVs Such as in Figure 2.11. Reproduced from Ref. [25] (Doménech-Carbó et al. *J. Electroanal. Chem.* 2012, 684: 13–19), with Permission.

The use of nanopore membranes and nanopore electrodes offers interesting possibilities for analytical purposes as far as the detection limits may be lowered to single molecules. Their use in real samples, however, is conditioned by the selectivity that can be attained resulting from the pore size. Since measurements involve the pore blockage, identification of a given analyte may be possible from the mean duration of the ionic current blockage/pore residency times. Accordingly, pore functionalization should promote selectivity (other than size selectivity associated with the pore diameter) directly related to the specific interactions of the visiting species

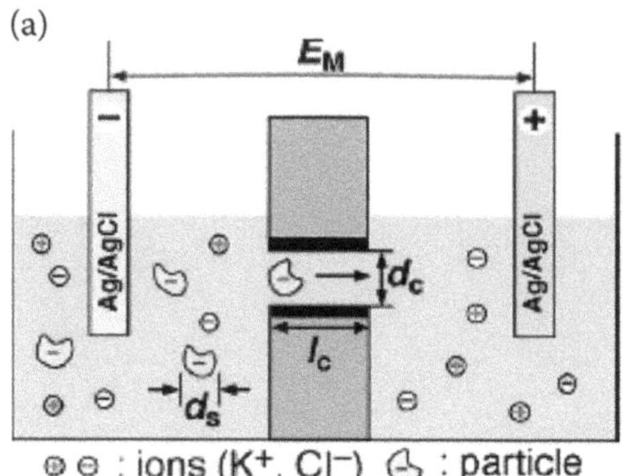

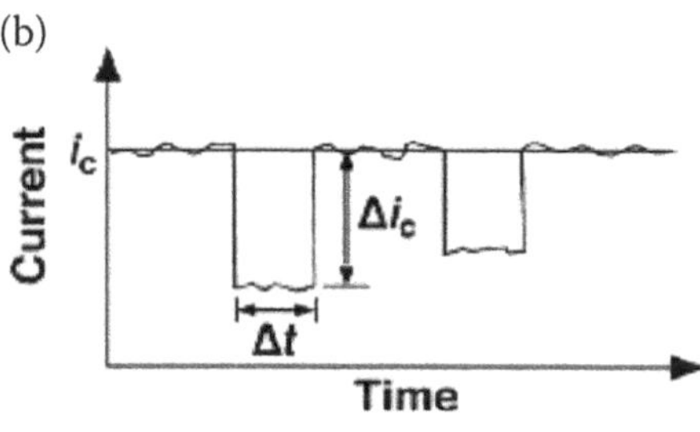

FIGURE 2.13 (a) Scheme for Electric Field Driven Coulter Counting and (b) the Corresponding Current/Time Profiles. Reproduced from Ref. [29] (Ito et al., *Acc. Chem. Res.* 2004, 37: 937–945), with Permission.

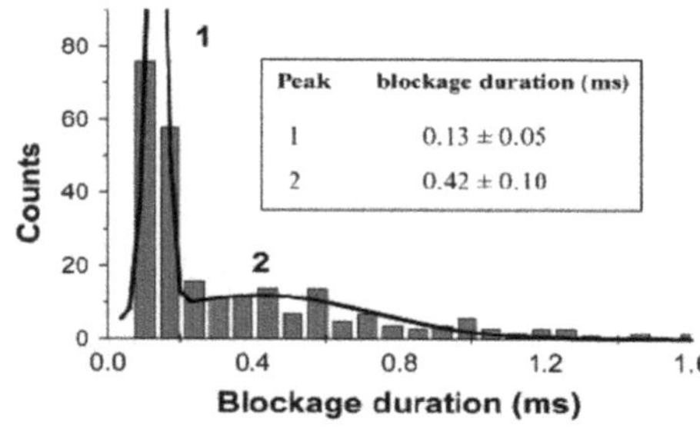

FIGURE 2.14 Histograms of dsDNA Blockage Durations in 1 M KCl, pH 7.2, 0.01% Triton-X100. Solution on the Small-Pore Side of the Membrane Contained 284 bp and 4.1 kb dsDNA Fragments. The Histogram Has Been Fitted with the Sum of Two Gaussians. Reproduced from Ref. [31] (Mara et al. *Nano Letters*. 2004, 4: 497–501), with Permission.

in the nanopore [27]. For nanopore diameters larger than 100 nm, functionalized nanopores (with amino-organosilanes, spiropyrans) display ion-selective responses which, in the case of spiropyrans, can be modulated by light, acting as electrochemical gates [34].

REFERENCES

[1]. de Levie, R. 1963. On porous electrodes in electrolyte solutions: I. Capacitance effects. *Electrochimica Acta*. 8: 751–814.

[2]. de Levie, R.; Vogt, A. 1990. On the electrochemical response of rough electrodes: Part II. The transient response in the presence of slow faradaic processes. *Journal of Electroanalytical Chemistry*. 281: 23–28.

[3]. Chae, W.-S.; Van Gough, D.; Ham, S.-K.; Robinson, D.B.; Braun, P.V. 2012. Effect of ordered intermediate porosity on ion transport in hierarchically nanoporous electrodes. *Applied Materials & Interfaces*. 4: 3973–3979.

[4]. Ward, K.R.; Compton, R.G. 2014. Quantifying the apparent 'catalytic' effect of porous electrode surfaces. *Journal of Electroanalytical Chemistry*. 724: 43–47.

[5]. Poljacek, S.M.; Risovic, D.; Cigula, T.; Gojo, M. 2012. Application of electrochemical impedance spectroscopy in characterization of structural changes of printing plates. *Journal of Solid State Electrochemistry*. 16: 1077–1089.

[6]. Tremblay, M.-L.; Martin, M.H.; Lebouin, C.; Lasia, A.; Guay, D. 2010. Determination of the real surface area of powdered materials in cavity microelectrodes by electrochemical impedance spectroscopy. *Electrochimica Acta*. 55: 6283–6291.
[7]. Lasia, A. 1995. Impedance of porous electrodes. *Journal of Electroanalytical Chemistry*. 397: 27–33.
[8]. Pajkossy, T.; Kolb, D.M. 2008. Anion-adsorption-related frequency-dependent double layer capacitance of the platinum-group metals in the double layer region. *Electrochimica Acta*. 53: 7403–7409.
[9]. Barnes, E.O.; Chen, X.; Li, P.; Compton, R.G. 2014. Voltammetry at porous electrodes: A theoretical study. *Journal of Electroanalytical Chemistry*. 720–721: 92–100.
[10]. Chan, H.T.H.; Kätelhön, E.; Compton, R.G. 2017. Voltammetry at electrodes decorated with an insulating porous film: Understanding the effects of adsorption. *Journal of Electroanalytical Chemistry*. 801: 135–140.
[11]. Amatore, C.; Savéant, J.M.; Tessier, D. 1983. Charge transfer at partially blocked surfaces. A model for the case of microscopic active and inactive sites. *Journal of Electroanalytical Chemistry*. 147: 39–51.
[12]. Ward, K.R.; Gara, M.; Xiong, L.; Lawrence, N.S.; Hartshorne, R.S.; Compton, R.G. 2013. Thin-layer vs. semi-infinite diffusion in cylindrical pores: A basis for delineating Fickian transport to identify nano-confinement effects in voltammetry. *Journal of Electroanalytical Chemistry*. 695: 15–24.
[13]. Mandelbrot, B. 1982. *The Fractal Geometry of Nature*. Freeman, San Francisco.
[14]. Nyikos, L.; Pajkossy, T. 1986. Diffusion to fractal surfaces. *Electrochimica Acta*. 31: 1347–1350.
[15]. Dassas, Y.; Duby, P. 1995. Diffusion toward fractal interfaces. Potentiostatic, galvanostatic, and linear sweep voltammetric techniques. *Journal of the Electrochemical Society*. 142: 4175–4180.
[16]. Andrieux, C.P.; Audebert, P. 2001. Electron transfer through a modified electrode with a fractal structure: Cyclic voltammetry and chronoamperometry responses. *Journal of Physical Chemistry B*. 105: 444–448.
[17]. Pajkossy, T. 1991. Electrochemistry at fractal surfaces. *Journal of Electroanalytical Chemistry*. 300: 1–11.
[18]. Go, J.-Y.; Pyun, S.-I. 2004. A study on lithium transport through fractal $Li_{1-\delta}$ CoO_2 film electrode by analysis of current transient based upon fractal theory. *Electrochimica Acta*. 49: 2551–2562.
[19]. Jung, K.-N.; Pyun,S.-I. 2006. Effect of por estructure on anomalous behaviour of the lithium intercalation into porous V_2O_5 film electrode using fractal geometry concept. *Electrochimica Acta*. 51: 2646–2655.
[20]. Go, J.-Y.; Pyun, S.-I. 2007. A review of anomalous diffusion phenomena at fractal interface for diffusion-controlled and non-diffusion-controlled transfer processes. *Journal of Solid State Electrochemistry*. 11: 323–334.
[21]. Galus, Z. 1970. *Fundamentals of Electrochemical Analysis*, 2nd edit. Ellis Horwood, New York, 1970.
[22]. Sabater, M.J.; Corma, A.; Doménech-Carbó, A.; Fornés, V.; García, H. 1997. Chiral salen manganese complex encapsulated with zeolite Y: A heterogeneous enantioselective catalyst for the epoxidation of alkenes. *Journal of the Chemical Society Chemical Communications*. 1285–1286.
[23]. Scholz, F.; Hermes, M. 1999. The determination of the redox state of a dissolved depolariser by cyclic voltammetry in the case of electrochemically reversible systems. *Electrochemistry Communications*. 1: 345–348. See corrigendum (2000) in *Electrochem. Commun*. 2: 814.
[24]. Doménech-Carbó, A.; Sánchez-Ramos, S.; Doménech-Carbó, M.T.; Gimeno-Adelantado, J.V.; Bosch-Reig, F.; Yusá-Marco, D.J.; Saurí-Peris, M.C. 2002. Electrochemical determination of the Fe(III)/Fe(II) ratio in archaeological ceramic materials using carbon paste and composite electrodes. *Electroanalysis*. 14: 685–696.
[25]. Doménech-Carbó, A.; Martini, M.; de Carvalho, L.M.; Doménech-Carbó, M.T. 2012. Square wave voltammetric determination of the redox state of a reversibly oxidized/reduced depolarizer in solution and in solid state. *Journal of Electroanalytical Chemistry*. 684: 13–19.
[26]. Murray, R.W. 2008. Nanoelectrochemistry: Metal nanoparticles, nanoelectrodes, and nanopores. *Chemical Reviews*. 108: 2688–2720.
[27]. Batchelor-McAuley, C.; Dickinson, E.J.F.; Rees, N.V.; Toghill, K.G.; Compton, R.G. 2012. New electrochemical methods. *Analytical Chemistry*. 84: 669–684.

[28]. White, R.J.; White, H.S. 2007. Influence of electrophoresis waveforms in determining stochastic nanoparticle capture rates and detection sensitivity. *Analytical Chemistry*. 79: 6334–6349.
[29]. Ito, T.; Sun, L.; Henriquez, R.R.; Crooks, R.M. 2004. A carbon nanotube-based coulter nanoparticle counter. *Accounts of Chemical Research*. 37: 937–945.
[30]. Ma, L.; Cockroft, S.L. 2010. Biological nanopores for single-molecule biophysics. *ChemBioChem*. 11: 25–34.
[31]. Mara, A.; Siwy, Z.; Trautmann, C.; Wan, J.; Kamme, F. 2004. An asymmetric polymer nanopore for single molecule detection. *Nano Letters*. 4: 497–501.
[32]. Zhang, B.; Zhang, Y.; White, H.S. 2006. Steady-state response of the nanopore electrode. *Analytical Chemistry*. 78: 477–483.
[33]. Zhang, Y.; Zhang, B.; White, H.S. 2006. Electrochemistry of nanopore electrodes in low ionic strength solutions. *Journal of Physical Chemistry B*. 110: 1768–1774.
[34]. Wang, G.; Bohaty, A.K.; Zharov, I.; White, H.S. 2006. Photon gated transport at the glass nanopore electrode. *Journal of the American Chemical Society*. 128: 13553–13558.

3 Electrochemical Processes at Ion-Permeable Solids

3.1 INTRODUCTION

In this chapter, theoretical approaches to describe the behavior of electrochemically active, ion-permeable porous solids will be treated. Electroactive solids can be divided into insulating solids containing immobile redox-active centers, and solids where the entire porous material is electrochemically responsive. There is, however, a variety of combinations of electronic and ionic conductivity in porous solids offering different electrochemical responses. In this chapter, this electrochemistry will be seen in the context of the electrochemical analysis of solids [1,2] (i.e. considering the electrochemical processes affecting the porous material in contact with a suitable electrolyte). Here, the microporous material will be attached to an inert electrode, forming the working electrode while the reference and auxiliary electrodes complete the usual three-electrode cell.

Depending on the distribution of electroactive centers, we can distinguish two extreme situations: first, when the material itself is electroactive, such as several transition metal oxides or metal-organic frameworks (MOFs); second, when the electroactive centers are isolated within a nonconducting matrix, such as in zeolites incorporating electroactive cations. Remarkably, in the first case, the advance of the electrochemical reaction through the material can involve significant structural modifications and eventual phase changes. In the second case, there is a variety of possible electroactive units, the most frequent being simple ions or neutral molecules encapsulated into the pores, functional groups anchored to the microporous support, or metal nanoparticles or nanoclusters (polyoxometalates, silicates, etc.). In these cases, the overall electrochemical process involves the propagation of the redox reaction across a non-conducting porous solid via electron hopping between immobile redox centers. Charge conservation demands that this process be coupled with ion transport across the porous material. Eventually, such charge transport processes can be restricted to the surface of the material particles.

In general, the electrochemistry of microporous materials involves charge transfer in a (at least) three-phase system constituted by the material, a basal, electron-conducting electrode (usually a metal or graphite, but also other materials like boron-doped diamond or indium-doped tin oxide (ITO)), and a liquid electrolyte. Depending on the spatial distribution of the corresponding interfaces, we can consider three separate situations, schematically depicted in Figure 3.1. In the first case (Figure 3.1a), a discontinuous set of microparticles of the porous material are deposited on the electrode surface, remaining in contact with the electrolyte. In the second case (Figure 3.1b), a continuous layer of the microporous material is interposed between the basal electrode and the electrolyte. Another frequent arrangement is obtained when the particles of the microporous material are embedded into a conducting matrix like a conducting polymer or a carbon paste, forming a composite. Finally, the electrolyte can be embedded within the microporous material that is sandwiched between two metal electrodes (Figure 3.1c). The first two arrangements are those usually taken to characterize the electrochemistry of solids while the last situation directly concerns batteries and will separately be treated.

3.2 GENERAL APPROACH

Let us consider the case of a particle of microporous solid deposited on an inert electrode immersed into a suitable electrolyte. It will be assumed that the material is not a metallic conductor but incorporates immobile redox centers so that electron transport and ion transport are allowed through

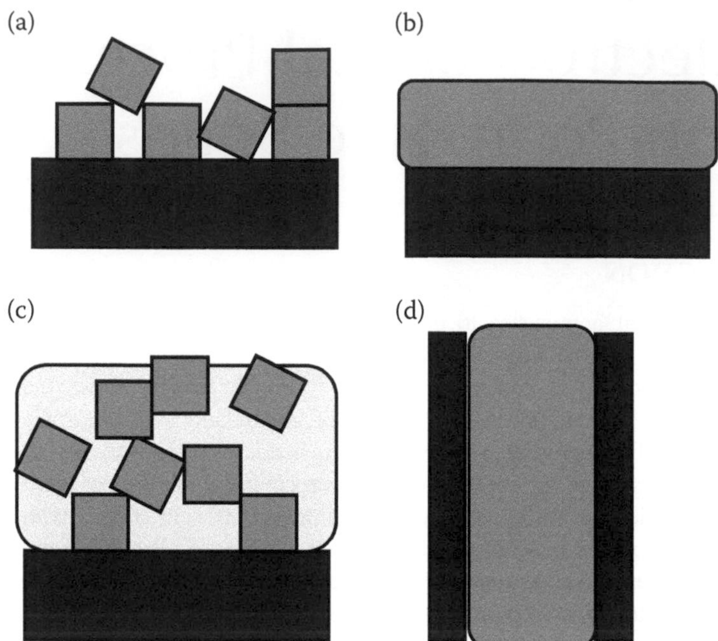

FIGURE 3.1 Schematic Representation of Four Possible Configurations for Studying the Electrochemistry of Microporous Materials. (a) Discontinuous deposit of microparticles; (b) continuous layer; (c) material embedded into a composite matrix; (d) material sandwiched between two electrodes.

the solid via electron hopping between redox-active centers and ion diffusion across the micropores of the material. This situation can be described based on theoretical studies dealing with the electrochemistry of redox polymers [3,4] and the formulation of Lovrić, Scholz, Odham, and their coworkers [5–8] on the voltammetry of redox conductive microparticles. Although several aspects of the electrochemistry of porous materials can be approached by the concepts and methods developed for redox polymers, it should be noted that, in such materials, a mixed-valent, self-exchange-based electron hopping mechanism occurs because the flexibility and ability for segmental motion of polymer chains makes it possible to approach redox centers, thus facilitating electron hopping. Porous materials, however, cannot be generally treated as organic polymers capable of segmental motion, so electron transfer between immobile redox centers attached to the porous matrix cannot be physically equated to polymeric motion-assisted occurring at redox polymer-modified electrodes [9].

In the presence of a suitable monovalent electrolyte, MX, it will be assumed that each redox-active unit of the microporous material can be reduced (or oxidized) to an equally immobilized form. Charge conservation implies that the reduction process involves the ingress of electrolyte cations into the microporous system (denoted in the following by { }):

$$\{\mathrm{Ox}\}_{\mathrm{solid}} + n\mathrm{M}^{+}_{\mathrm{sol}} + n\mathrm{e}^{-} \rightarrow \{\mathrm{Rd}^{n-}\cdots n\mathrm{M}^{+}\}_{\mathrm{solid}} \tag{3.1}$$

Equivalently, the oxidation of immobile redox centers in the material could involve the concomitant entrance of electrolyte anions into the microporous material:

$$\{\mathrm{Rd}\}_{\mathrm{solid}} + n\mathrm{X}^{-}_{\mathrm{sol}} \rightarrow \{\mathrm{Ox}^{n+}\cdots n\mathrm{X}^{-}\}_{\mathrm{solid}} + n\mathrm{e}^{-} \tag{3.2}$$

In CV experiments under conditions of reversibility, the inverse of the processes represented by Eqs. (3.1) and (3.2) will occur.

It is convenient to note that in most of the studied systems, such as in zeolites, the parent microporous solid frequently incorporates insertion ions (Li^+, Na^+, NH_4^+, etc.). As a result, the oxidation (reduction) process can occur via coupled issue of electrons and cations (anions) from the material to the electrolyte solution, i.e. there are different reductive

$$\{Ox^{m+}\cdots mA^-\}_{solid} + nM^+_{sol} + ne^- \rightarrow \{Rd^{(m-n)+}\cdots mA^-\cdots nM^+\}_{solid} \tag{3.3}$$

$$\{Ox^{m+}\cdots mA^-\}_{solid} + ne^- \rightarrow \{Rd^{(m-n)+}\cdots (m-n)A^-\}_{solid} + nA^-_{sol} \tag{3.4}$$

and oxidative pathways:

$$\{Rd^{m-}\cdots mQ^+\}_{solid} + nX^-_{sol} \rightarrow \{Ox^{(m-n)-}\cdots mQ^+\cdots nX^-\}_{solid} + ne^- \tag{3.5}$$

$$\{Rd^{m-}\cdots mQ^+\}_{solid} \rightarrow \{Ox^{(m-n)-}\cdots (m-n)Q^+\}_{solid} + nQ^+ + ne^- \tag{3.6}$$

From a thermodynamic point of view, the variation of standard Gibbs free energy associated with the electron transfer process represented by (3.1), ΔG°_{ec}, can be related to the variation of such thermodynamic quantity for the electron transfer process for species in the solution phase ΔG°_{es}, and for the transfer of the oxidized ΔG°_{tox}, and reduced ΔG°_{tred} forms of the electroactive species and the electrolyte cations ΔG°_M, from the solution phase to the porous solid. In the following, two extra-thermodynamic assumptions will be made: (i) there is no accumulation of net charge in the solid complex/electrolyte boundary, and (ii) the structure of the solid and the ion binding to the solid are not affected by the solvent. The Born-Haber-type cycle corresponding to the above equations is shown in Figure 3.2, taken for simplicity a neutral oxidized form as a parent species.

The relation between these Gibbs free energies in this figure is:

$$\Delta G^\circ_{EC} = \Delta G^\circ_{ES} + \Delta G^\circ_{tred} - \Delta G^\circ_{tox} + n\Delta G^\circ_M \tag{3.7}$$

Here, ΔG°_{EC} denotes the Gibbs free energy of electron transfer associated with the electrochemical conversion of the oxidized into the reduced form. ΔG°_{ES} corresponds to the electrochemical process in the solution phase, ΔG°_{tox} and ΔG°_{trd} are the free energies associated with the trapping

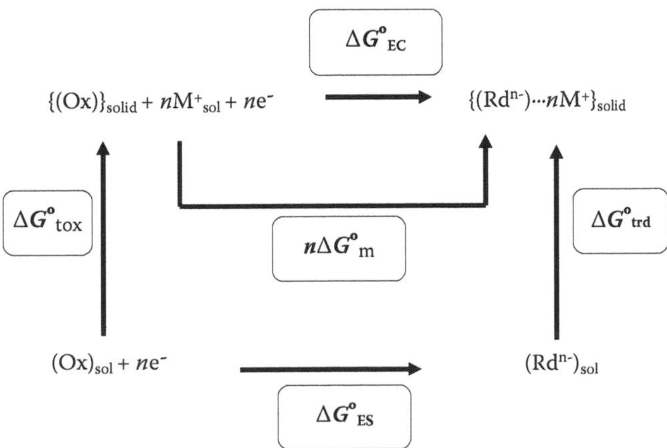

FIGURE 3.2 Thermochemical Cycle for the Electrochemical Reduction of a Guest Species Attached to a Porous Solid.

of the electroactive guest species into the microporous solid, and ΔG°_{M} is the free energy of trapping of the charge-balancing cation in the solid. This equation clearly reveals that in determining the spontaneity of the redox process, not only the stability of the oxidized and reduced forms of the electroactive species but also that of the charge-balancing cations in the porous solid relative to their stability in the solution phase, are significant.

3.3 INSERTION ELECTROCHEMICAL ELECTRODES

The above systems correspond to insertion electrochemical electrodes where there is simultaneous electron transfer and ion transfer occurring across different interfaces as schematized in Figure 3.3 [10]. This scheme is valid for microparticulate deposits and droplets immobilized onto solid electrodes [1]; formally, it permits the separation of the contribution of ion and electron transfer:

$$\Delta G^{\circ}_{EC} = \Delta G^{\circ}_{\text{electron transfer}} + \Delta G^{\circ}_{\text{ion transfer}} \tag{3.8}$$

In the following, we will consider a porous solid attached to an inert base electrode that experiences a reversible n-electron reduction involving the ingress of n monovalent electrolyte cations, as described by (3.1). This may result, as schematized in Figure 3.4, in the formation of segregated layers [11] or mixed crystals [12] of the parent oxidized solid and the generated M^{+}-permeated reduced form. This process can formally be described as the sum of the processes:

$$\{Ox\}_{solid} + ne^{-} \rightarrow \{Rd^{n-}\}_{solid} \tag{3.9}$$

$$\{Rd^{n-}\}_{solid} + nM^{+}_{sol} \rightarrow \{Rd^{n-}\cdots nM^{+}\}_{solid} \tag{3.10}$$

The Gibbs free energy of the overall process ((3.1)), ΔG°_{EC}, can be calculated, in the case of electrochemical reversibility, from the standard potential of the couple E°_{EC} as $-nFE^{\circ}_{EC}$. This can be related with the voltammetric midpeak potential, E_{mp}, by means of the relationship:

$$E_{mp} = E^{\circ}_{EC} + \frac{RT}{nF}\ln\frac{a_{\{Ox\}}}{a_{\{Rdn^{-}\cdots nM^{+}\}}a_{M^{+}_{sol}}} \tag{3.11}$$

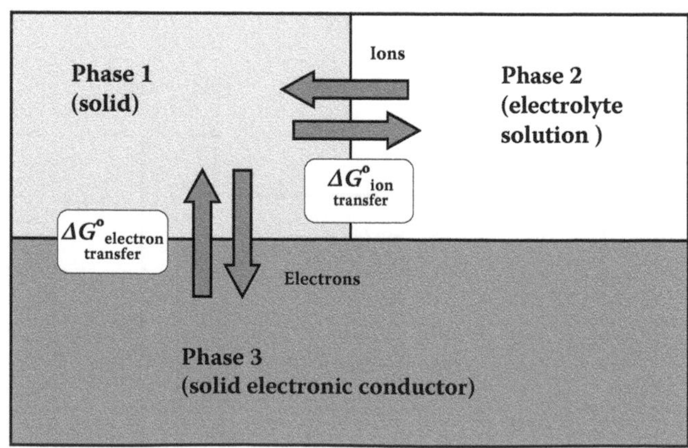

FIGURE 3.3 Schematic Representation of the Three-Phase System Forming Insertion Electrochemical Electrodes.

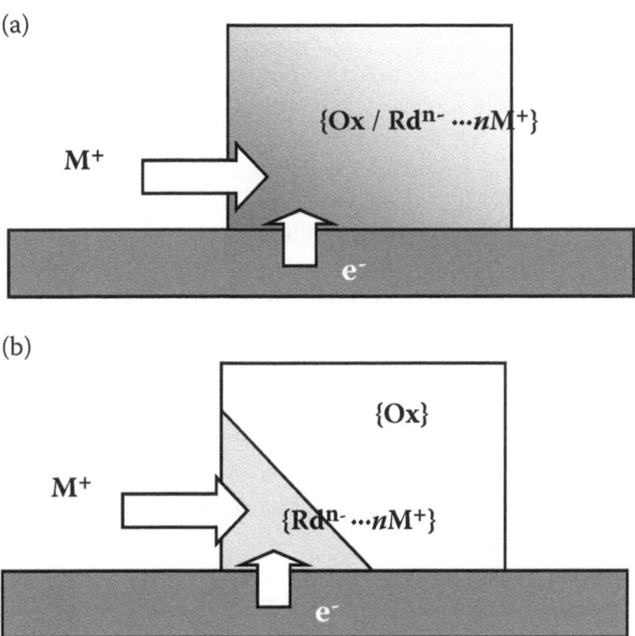

FIGURE 3.4 Schematic Representation of the Electrochemical Reduction of a Cation-Permeable Porous Solid Forming (a) Mixed Crystals and (b) Segregated Layers.

The electrochemical process described by (3.1) will produce a layer of M^+-permeated, reduced material. Then, the process described by (3.10) will correspond to the equilibrium between the reduced thin layer of the porous solid and the electrolyte. The Gibbs free energy change associated with this process can be estimated from the corresponding standard potential $E°_{CM}$ as $-nFE_{CM}$. This can be estimated from the measurement of the open circuit potential E_{OCP}, as [13]:

$$E_{OCP} = E°_{CM} + \frac{RT}{nF} \ln \frac{a_{\{Rd^{n-}\}} a_{M^{+}sol}}{a_{\{Rd^{n-} \cdots nM^{+}\}}} \tag{3.12}$$

The measured E_{OCP} contains the potential of the reference electrode and the potential drops at the electrode/oxidized solid and oxidized solid/reduced solid layer interfaces. These potentials remain constant in all measurements and can be neglected. In turn, the ratio can be considered as essentially constant under several conditions favored by the application of a conditioning potential at the voltammetric midpeak potential [14]. Then, the electrolyte-independent standard potential for the electron transfer process described by (3.9), $E°_{ET}$, can be calculated as:

$$E°_{ET} = E°_{EC} - E°_{CM} = E_{OCP} - E_{mp} \tag{3.13}$$

This scheme has also been tested for anion transfer on different solids, revealing consistent differences between the midpeak and open circuit potentials independent of the electrolyte [15]. Then, the Gibbs free energy of electron transfer can be calculated as:

$$\Delta G°_{ET} = nF\,(E_{OCP} - E_{mp}) \tag{3.14}$$

3.4 ION TRANSPORT

Let us consider that a cylinder of an electroactive porous solid is pressed into the surface of a conducting matrix (graphite or graphite-based composites) in a way that only a circular surface of the cylinder is in contact with an electrolyte solution, as depicted in Figure 3.5 [12,16].

It will be assumed that the solid material experiences a reversible reduction process involving the ingress of electrolyte cations, M^+, as described by (3.1). Then, the cations will diffuse through the electrolyte/solid circular interface along the longitudinal axis of the cylinder. Assuming that the diffusion layer in the porous material is shorter than its length, cation diffusion can be described in terms of the Fick equation under planar, semi-infinite boundary conditions. When $x \le 0$ (i.e. in the electrolyte solution):

$$\frac{\partial c}{\partial t} = D_{M^+sol}\left(\frac{\partial^2 c}{\partial x^2}\right) \tag{3.15}$$

where D_{M^+sol} represents the diffusion coefficient of the cations in the electrolyte. When $x \ge 0$ (i.e. within the cylinder of porous material):

$$\frac{\partial c}{\partial t} = D_{M^+solid}\left(\frac{\partial^2 c}{\partial x^2}\right) \tag{3.16}$$

Now, D_{M^+solid} is the diffusion coefficient of the cations through the porous solid. Based on the separation between the electronic and ionic contributions made in the precedent section, Lovrić et al. [16] obtained that the voltammetric response is dependent on the mass transfer parameter $g = \rho(D_{M^+solid}/D_{M^+sol})^{1/2}/[M^+]_{bulk}$, ρ being the molar density of the solid and $[M^+]_{bulk}$ is the concentration of M^+ in the bulk of the electrolyte. For $g < 0.1$ (in general, for large concentrations of

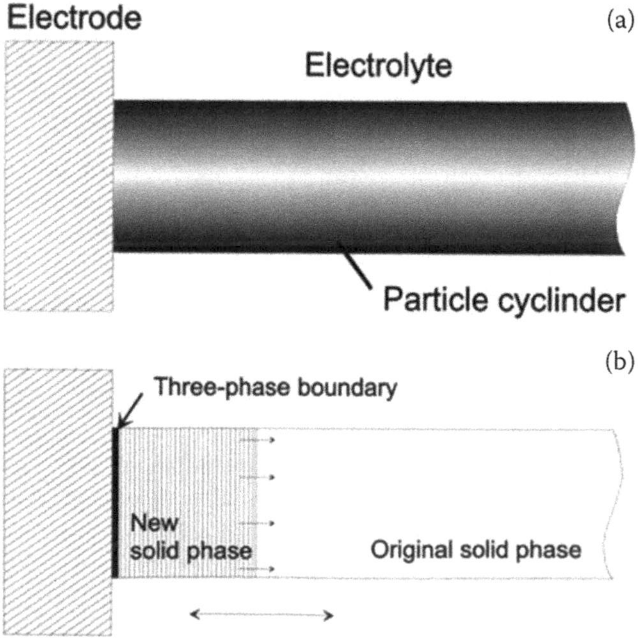

FIGURE 3.5 Schematics for Ion Insertion in a Cylinder of a Porous Material Pressed into a Graphite Matrix. Reproduced from Ref. [12] (Hermes et al. *J. Electroanal. Chem.* 2001, 501: 193–204), with Permission.

M^+ in the electrolyte), the voltammogram approaches an n-electron reversible couple involving species in solution. The voltammetric midpeak potential can be expressed as a function of the concentration (activity) of charge-balancing cations, M^+, in the electrolyte:

$$E_{mp} = E^{\circ} + (RT/nF)\ln K + (RT/F)\ln [M^+]_{bulk} \tag{3.17}$$

where K is the equilibrium constant for the ion transfer process described by (3.10). In the opposite case, when $g > 10$, the midpeak potential becomes independent on the concentration of M^+ in the solution:

$$E_{mp} = E^{\circ} + (RT/nF)\ln K + (RT/2nF)\ln\left(\frac{D_{M^+\ solid}}{D_{M^+\ sol}}\right) + (RT/nF)\ln \rho \tag{3.18}$$

Figure 3.6 depicts the experimental cyclic voltammograms recorded for microparticulate deposits of two bisferrocenyl-functionalized pseudopeptides in contact with 0.10 M K_2HPO_4 aqueous solution [14]. The electrochemistry of these compounds involves the oxidation of the ferrocene units coupled with anion insertion into the solid exhibiting significant selectivity for hydrogen phosphate and dihydrogen phosphate anions relative to other common inorganic anions. The voltammograms were recorded in electrolyte concentration high enough to promote the response described by (3.17). In the following, it will be assumed that the concentration of electrolyte counterions in a solution is sufficiently high to ensure that ion transport phenomena occurring in the electrolyte can be neglected.

3.5 MIXED PHASES AND MISCIBILITY GAPS

The above treatment of the cation-assisted electrochemical reduction of a solid {Ox} to an $\{Rd^{n-} \cdots n M^+\}$ form assumed that a mixed phase of variable composition, $\{(Ox)_x \cdots (Rd^{n-} \cdots . nM^+)_{1-x}\}$ was

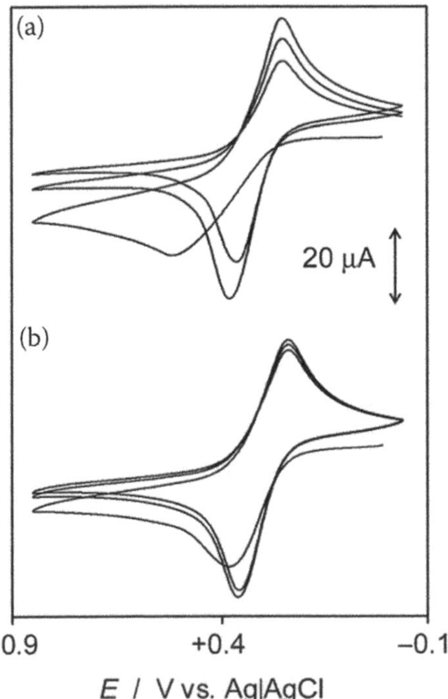

FIGURE 3.6 Cyclic Voltammograms Recorded for Microparticulate Deposits on Glassy Carbon Electrode of Bisferrocenyl-Functionalized Pseudopeptides (a) Et-Val-Fc_2 and (b) But-Val-Fc_2, in Contact with 0.10 M K_2HPO_4 Aqueous Solution. Potential Scan Rate 50 mV s^{-1}. Reproduced from Ref. [14] (Pandey et al. *RSC Adv.* 2016, 6: 35257–35266), with Permission.

generated. However, this process can also produce the segregation of such oxidized and reduced forms in different phases. This situation was studied by Lovrić, Scholz, et al. [17] considering that equilibrium is established between the solid forms and the species existing in the solution phase, as schematically depicted in Figure 3.2, and assuming that the activities in solid-state are proportional to the molar fractions of the respective components. Then, the activities of the species in solid-state can be related to the activities of the corresponding species in solution as:

$$\ln \frac{a_{\{Ox\}} a_{M^+ \text{ solution}}}{a_{\{Rd^{n-} \cdots nM^+\}}} = \frac{nF}{RT}(E^{o}_{sol} - E^{o}_{ss}) + \ln \frac{a_{Ox \text{ solution}}}{a_{Rd^{n-} \text{ solution}}} \tag{3.19}$$

where E^{o}_{sol} and E^{o}_{ss} are, respectively, the standard potentials for the reduction processes in solution phase and in solid-state.

Interestingly, this electrochemistry opens the opportunity for electrochemical doping, as illustrated in Figure 3.7, where the CV response of Prussian blue in contact with 0.1 M KNO_3 (a) without and (b) with addition of Cd^{2+} is depicted [12].

The pair of peaks at ca. 0.2 V vs. Ag/AgCl corresponds to the electrochemical reduction of Prussian blue ($Fe_4[Fe(CN)_6]_3$ or $FeK[Fe(CN)_6]$, depending on the preparation procedure), a process that can be formulated as:

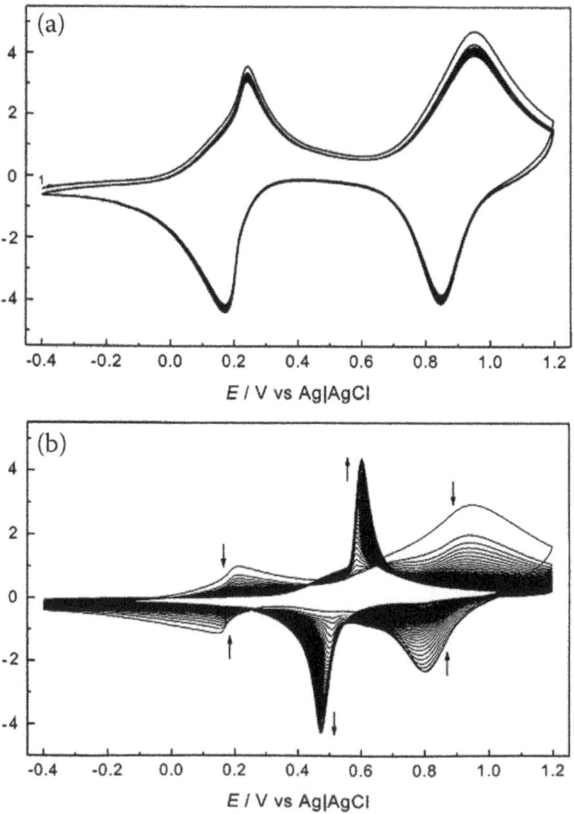

FIGURE 3.7 Repetitive CV of Prussian Blue Microparticles in Contact with 0.1 M KNO_3 (a) Without and (b) with Addition of Cd^{2+}. Potential Scan Rate 100 mV s^{-1}. Reproduced from Ref. [12] (Hermes et al. *J. Electroanal. Chem.* 2001, 501: 193–204), with Permission.

$$\{FeK[Fe(CN)_6]\}_{solid} + K^+_{aq} + e^- \rightarrow \{FeK_2\{Fe(CN)_6\}\}_{solid} \quad (3.20)$$

The process at ca. 0.9 V can be described as the oxidation of Prussian blue:

$$\{FeK[Fe(CN)_6]\}_{solid} \rightarrow \{Fe\{Fe(CN)_6\}\}_{solid} + K^+_{aq} + e^- \quad (3.21)$$

In the presence of Cd^{2+}, both the reduction and oxidation of Prussian blue processes occur with the entrance of such ions into the lattice, giving rise to a solid species reduced/oxidized at ca. 0.5 V. This process can be described as:

$$\{CdK[Fe(CN)_6]\}_{solid} + K^+_{aq} + e^- \rightarrow \{CdK_2\{Fe(CN)_6\}\}_{solid} \quad (3.22)$$

resulting from the reductive/oxidative substitutions of Fe^{3+} ions by Cd^{2+} in the Prussian blue lattice forming a new compound that is segregated as a new phase [12].

An interesting situation arises when the oxidized and reduced forms are completely immiscible. This does not allow an equilibrium description, but the intermediate case, where there are immiscibility gaps between the oxidized and reduced solid forms, can be treated similarly as mixed phases. Here, the essential idea is that {Ox} is gradually reduced, and the activity of the reduced form, $\{Rd^{n^-} \cdots n\,M^+\}$, will increase until a maximum value given by

$$Z_{rd/ox} = \frac{m_{\{Rd^{n^-} \cdots nM^+\}\,lim}}{m_{\{Ox\}} + m_{\{Rd^{n^-} \cdots nM^+\}\,lim}} \quad (3.23)$$

that expresses the maximum solubility of $\{Rd^{n^-} \cdots nM^+\}$ in {Ox} in terms of the numbers of moles (m) of the oxidized and reduced forms. When the reduction process advances sufficiently, a massive ingress of M^+ ions into the solid phase occurs and the system reaches the maximum solubility of {Ox} in $\{Rd^{n-} \cdots n\,M^+\}$, given by:

$$Z_{ox/rd} = \frac{m_{\{Ox\}\,lim}}{m_{\{Ox\}\,lim} + m_{\{Rd^{n^-} \cdots nM^+\}}} \quad (3.24)$$

Accordingly, the solid system evolves from pure {Ox} to a solid solution of variable composition until the maximum concentration of reduced form in {Ox}, $\{(Ox)_{x'} \cdots (Rd^{n-} \cdots. nM^+)_{1-x'}\}$ is reached. Then, there exists a miscibility gap until the mixed crystal reaches a composition $\{(Ox)_{x''} \cdots (Rd^{n-} \cdots. nM^+)_{1-x''}\}$ (logically, $x' \ll x''$) when the reduction to a segregated phase $\{(Rd^{n-} \cdots. nM^+)\}$ takes place. The first transition (saturation of mixed crystals in {Ox} phase) corresponds to a critical potential of the saturated mixed crystal given by:

$$E_{C,1} = E^{\circ\prime} + \frac{RT}{nF}\ln\left(\frac{1 - Z_{rd/ox}}{Z_{rd/ox}}\right) \quad (3.25)$$

Then, the composition of the crystal remains constant and independent on the electrode potential until it equals a second value,

$$E_{C,2} = E^{\circ\prime} + \frac{RT}{nF}\ln\left(\frac{Z_{ox/rd}}{1 - Z_{ox/rd}}\right) \quad (3.26)$$

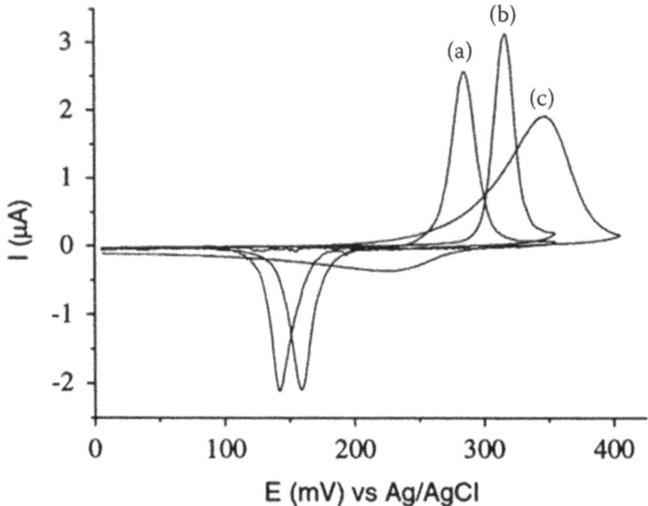

FIGURE 3.8 Dependence of Voltammograms for the $TTF^{0/+}$ Process on Nitrate Concentration When TTF is Adhered to a Platinum Electrode (A = 49 μm^2) Which is Immersed in (a) 0.1, (b) 0.05, and (c) 0.01 M Aqueous KNO_3 Electrolyte. Voltammograms Were Recorded Without iR_u Compensation at a Scan Rate of 25 mV s^{-1}. Reproduced from Ref. [18] (Wooster et al. *Anal. Chem.* 2003, 75: 586–592), with Permission.

Here, the composition of the system changes abruptly again. These features result, according to the indicated modeling, in voltammograms where the cathodic and anodic peaks are separated by a large potential hiatus. An example of a CV for a reversible redox reaction with miscibility gaps is shown in Figure 3.8 and Figure 3.9 [18]. Here, the CVs for tetrathiafulvalene particles attached to Pt electrode in contact with different KNO_3 aqueous solutions are depicted [18]. The voltammograms show the

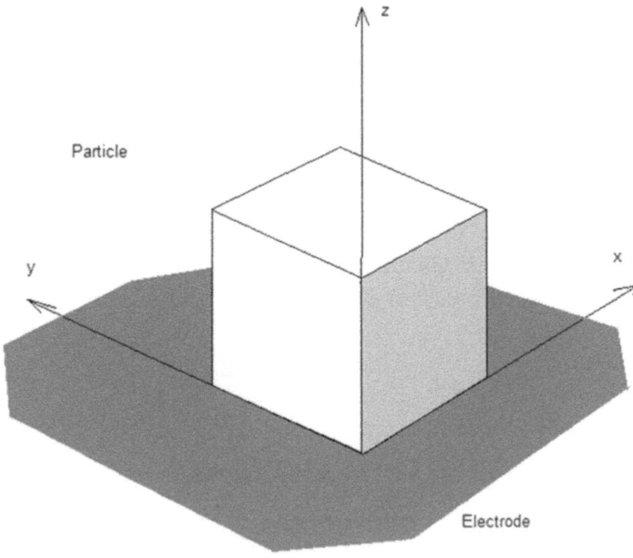

FIGURE 3.9 Coordinate System for the Idealized Representation of a Particle of a Microporous Material Containing Redox-Active Centers Deposited on an Electrode in Contact with a Suitable Electrolyte. the Positive Semi-axes x and y are Located in the Crystal/Electrode Interface and the Positive Semi-axis z Lies with the Crystal/Electrolyte Interface.

cathodic and anodic peaks corresponding to the $TTF^{0/+}$ interconversion with a significant separation in the potential axis. Notice the variation of the voltammetric response with the concentration of the supporting electrolyte. It is pertinent to note, however, that deviations from reversibility as well as kinetic constraints due to oversaturation and nucleation phenomena can also influence the voltammetric response.

3.6 MICROPARTICLES ELECTROCHEMISTRY

Let us consider the case of a cubic particle of microporous solid material attached to a plane electrode put in contact with an electrolyte containing monovalent cations, as depicted in Figure 3.8. Again, it will be assumed that upon application of a reductive potential input, a redox reaction as described by (3.1) occurs. Then, the progress of the reaction across the microporous material involves a series of electron-exchange plus ion-exchange reactions between immobile redox centers that can be represented by means of:

$$\{\mathrm{Red}\cdots \mathrm{M}^{+}\}_{\mathrm{site\ A}} + \{\mathrm{Ox}\}_{\mathrm{site\ B}} \rightarrow \{\mathrm{Red}\cdots \mathrm{M}^{+}\}_{\mathrm{site\ B}} + \{\mathrm{Ox}\}_{\mathrm{site\ A}} \qquad (3.27)$$

According to Schröder et al. [8], it can be assumed that the redox reaction initiates at the three-phase particle/electrode/electrolyte boundary, further expanding through the crystal. Fick's law can be expressed as:

$$\frac{\partial[\mathrm{Rd}^{n^-}\}\cdots n\mathrm{M}^{+}]}{\partial t} = D_z\left(\frac{\partial^2[\mathrm{Rd}^{n^-}\}\cdots n\mathrm{M}^{+}]}{\partial x^2}\right) + D_y\left(\frac{\partial^2[\mathrm{Rd}^{n^-}\}\cdots n\mathrm{M}^{+}]}{\partial y^2}\right) + D_z\left(\frac{\partial^2[\mathrm{Rd}^{n^-}\}\cdots n\mathrm{M}^{+}]}{\partial z^2}\right) \qquad (3.28)$$

where D_x, D_y, D_z represent the charge diffusion coefficients along the x, y, and z directions. These diffusion coefficients will depend, in general, on the orientation of the particles of the microporous material. This formulation assumes that both electrons and ions are exchanged simultaneously and that no charge separation effects occur. For simplicity, it will be considered the idealized case of a two-dimensional particle located in the x-z plane, assuming that (i) the concentrations of the oxidized and reduced centers at the three-phase junction are thermodynamically equilibrated and (ii) the flux of the electrons proceeds only in the z-direction while the flux of the cations proceeds only in the x-direction. Then, finite difference simulations lead to predict the chronoamperometric behavior of the system when a constant potential sufficiently cathodic to promote diffusion-controlled conditions, as:

a. If charge diffusion is significantly slower so that the distance of charge transport L $(= 2(Dt)^{1/2})$ is clearly lower than the size of the cuboid particle, the electrochemical response will be equivalent to that recorded when reactants freely diffuse from an infinite volume of solution to the electrode. This situation—often termed as thick-layer behavior—corresponds to semi-infinite boundary conditions, and Cottrell-type behavior is observed, for instance, in cyclic voltammetry and chronoamperometry. The resulting (current intensity)/(time) (I/t) chronoamperometric current for a deposit of N cuboid crystals of finite size can be expressed as [8,19]:

$$I = nNFc\left[u\left(\left(\frac{\Delta x D_{\mathrm{e}}^{1/2} + \Delta y D_{\mathrm{M}}^{1/2}}{2\pi^{1/2}t^{1/2}}\right) + (D_{\mathrm{e}}D_{\mathrm{M}})^{1/2}\right) - 4D_{\mathrm{M}}(2D_{\mathrm{e}}t)^{1/2}\right] \qquad (3.29)$$

In this equation, u represents the length of the three-phase junction (i.e. the perimeter of the electrode/crystal interface) and Δx and Δz represent the size of the discrete boxes where the crystal is divided for numerical simulation procedures. It is assumed that the diffusion coefficient in the x-direction, D_x, equals the diffusion coefficient of electrolyte cations, D_M, and the diffusion coefficient in the z-direction, D_z, equals that of electrons, D_e. The above equation contains a Cottrell-type term, characterized by $I \propto t^{-1/2}$, accompanied by a time-independent term and a third term for which $I \propto t^{1/2}$. The time-independent term can be associated with the restrictions imposed by the finite character of the crystals, while the third term can be described as an edge effect, resulting from the overlap of the cation diffusion near the corners of the crystal and its influence on the entire diffusion process.

Under these conditions, the iso-concentration lines approach straight lines whose slope is a function of the ratio between D_e and D_M [8]. Plots of the product $it^{1/2}$ vs. t yields characteristic curves with a maximum at a transition time t^* given by:

$$t_{max} = \frac{p^2}{128D_M} \tag{3.30}$$

This transition time describes the point at which the transition from the three-dimensional diffusion to the planar diffusion conditions occurs. This magnitude is of interest because it relates diffusion and crystal size parameters.

Experimental data for zeolite-associated species agreed with that model, as illustrated for 2,4,6-triphenyl pyrylium ion (= PY^+) immobilized in zeolite Y immersed into 0.10 M Et_4NClO_4/MeCN. The voltammogram of these zeolite-modified electrodes display a well-defined cathodic peak at –0.26 V vs. Ag/AgCl (Figure 3.10a) corresponding to the process (notice that, for simplicity, the zeolite system is represented as no charge-balanced):

$$\{PY^+@Y\}_{solid} + Et_4N^+_{sol} + e^- \rightarrow \{PY@Y \cdots Et_4N^+\}_{solid} \tag{3.31}$$

The chronoamperograms recorded applying a constant potential of –0.35 V (Figure 3.10b) display a well-defined maximum in the $It^{1/2}$ vs. t plot at $t = 15$ ms. Taking a mean perimeter for the crystal/electrode junction of 500 nm, estimated from TEM examination of deposits, one obtains $D_M = 1.3 \times 10^{-9}$ cm^2s^{-1}.

b. At relatively long experimentation times, semi-infinite diffusion does not hold, and the predicted behavior depends on the values of the diffusion coefficients for electrons and cations relative to the crystal dimensions. Roughly, when the diffusion of cations is fast compared to the diffusion of the electrons, the cations spread along the electrode/crystal interface into the bulk of the crystal so that the oxidized redox centers along this interface are exhausted. Now, electron diffusion becomes rate-determining and the orientation of the equiconcentration lines becomes increasingly parallel to the electrode surface so that the systems tend to reach a two-dimensional diffusion. Then, the chronoamperometric current becomes [8]:

$$I = \frac{2nFNAD_e c}{H} \sum_{j=1}^{\infty} \exp\left(\frac{-(2j-1)^2\pi^2 D_e t}{4H^2}\right) \tag{3.32}$$

where H denotes the crystal height. Here, the transition time for crystals of length L and width B is given by:

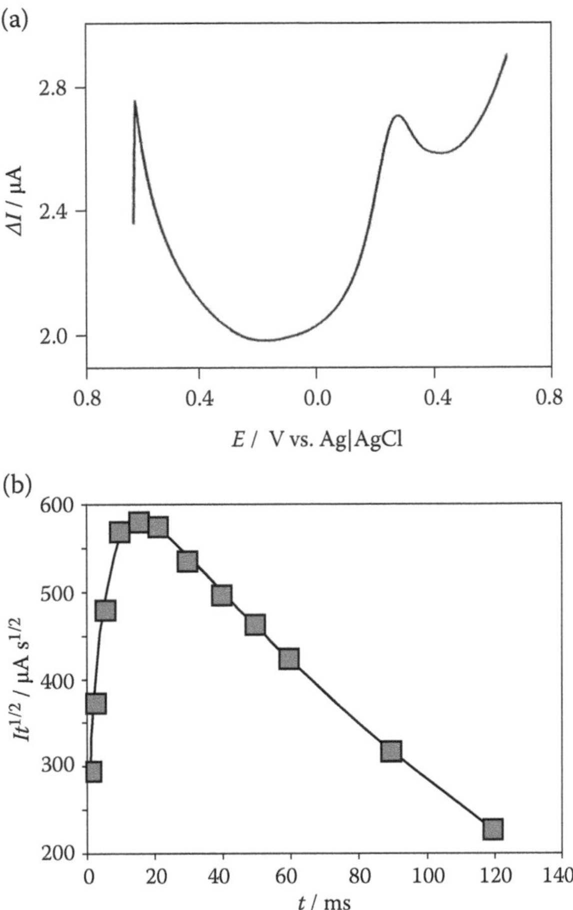

FIGURE 3.10 Electrochemistry of Microparticulate Deposit of 2,4,6-Triphenylpyrylium Ion Attached to Zeolite Y Deposited on Paraffin-Impregnated Graphite Electrode in Contact with 0.10 M Et_4NClO_4/MeCN. (a) SWV Recorded by Scanning the Potential in the Negative Direction; Potential Step Increment 4 mV; Square Wave Amplitude 25 mV; Frequency 5 Hz. (b) Plots of $It^{1/2}$ vs. t from Chronoamperometric Data at an Applied Potential –0.35 V vs. AgCl/Ag.

$$t^* = \frac{L^2B^2}{4.45(L^2 + B^2)D_M} \tag{3.33}$$

When electron diffusion is clearly faster than cation diffusion, the reaction zone initially spreads along the z-axis and reaches the top surface of the crystal. Now, the cation diffusion in the *x-direction* becomes rate-determining and the transition time is given by $t^* = H^2/1.1D_e$. The chronoamperometric long-time curves for crystals where $B \ll L$ can be approached by:

$$I = \frac{NnFLHcD_M}{B}\exp\left(\frac{-\pi^2 D_M t}{B^2}\right) \tag{3.34}$$

Interestingly, this model predicts that there is redox conductivity even if one of the diffusion coefficients equals zero (i.e. when one of the charge transport processes is hindered through the solid). If $D_e = 0$, the reaction propagates along the particle/electrode interface; while if $D_M = 0$, the reaction layer is confined to the lateral sides of the particle. Comparable situations can be obtained when $D_M \gg D_e$ or $D_e \gg D_M$, respectively. The last appears to be the case for proton-assisted electron transfer processes involving organic molecules in contact with aqueous buffers. Here, charge transfer is ensured by proton hopping between immobile redox centers via chemical bond breaking and re-forming rather than cation insertion into the organic lattice. As a result, one can assume that $D_e \gg D_M$, so the electroactive region is confined to a narrow layer in the lateral sides of the crystals.

It is pertinent to underline that the foregoing set of equations have formal similarities with those derived for the chronoamperometry of redox polymers. For instance, the finite diffusion current-time relationship derived by Daum et al. [20] for redox polymers forming a lamina of thickness δ and surface A containing immobile redox centers in concentration c is:

$$I = \frac{nFAcD^{1/2}}{\pi^{1/2}t^{1/2}}\left[\sum_{k=0}^{\infty}(-1)^k[\exp(-k^2\delta^2/Dt) - \exp(-(k+1)^2\delta^2/Dt]\right] \tag{3.35}$$

This equation contains both Cottrell-type (I proportional to $t^{1/2}$) thick layer and thin layer terms.

Recently, González-Meza et al. [21] modeled the cyclic voltammetry of ion-insertion solids comparing different approaches. Analytical solutions are obtained when the activities of the solid phase are assumed to be constant using spherical coordinates. The essential idea is that the potential can be expressed as a function of the molar fractions of the oxidized and reduced forms of the solid (see also (3.11)):

$$E = E^\circ + \frac{RT}{nF}\ln\left(\frac{\gamma_{\{Ox\}}\gamma_{M^+\ sol}}{\gamma_{\{Rd^{n-}\}\cdots nM^+\}}}\right) + \frac{RT}{nF}\ln\left(\frac{\alpha_{\{Ox\}}}{\alpha_{\{Rd^{n-}\}\cdots nM^+\}}}\right) + \frac{RT}{nF}\ln[M^+] \tag{3.36}$$

where α denotes the molar fraction and γ the activity coefficient of the respective species. In the case of planar coordinates, numerical solutions lead to the following expressions for the peak current and the peak potential:

$$I_p = 0.6105nFAc_{M^+}\left(\frac{nFD_{M^+}}{RT}v\right)^{1/2} \tag{3.37}$$

$$E_p = E^{o\prime}{}_{app} + \frac{RT}{nF}\ln[M^+] - 0.8540\frac{RT}{nF} \tag{3.38}$$

where $D_M{}^+$ represents the diffusion coefficient of the electrolyte cations within the solid and $c_M{}^+$, the concentration of such ions in the electrolyte bulk and an apparent formal potential resulting from the standard potential and the activity coefficient terms in (3.36), $E^{o\prime}{}_{app}$, is introduced. In the case of spherical diffusion, the analytical solution for the peak current incorporates an additional term:

$$I_p = 0.6105nFAc_{M^+}\left(\frac{nFD_{M^+}}{RT}v\right)^{1/2} + \frac{nFAc_{M^+}D_{M^+}}{r_o} \tag{3.39}$$

in which r_o represents the radius of a spherical electrode. Equations (3.37) and (3.39) are analog to the Randles-Ševčik equation (Eq. (1.6)) for reactants in the solution phase.

3.7 DETERMINATION OF THERMOCHEMICAL PARAMETERS OF INDIVIDUAL IONS

One interesting issue of the electrochemistry of ion insertion solids is the possibility of its application to determine "absolute" values of thermochemical properties of individual ions. These properties—in particular, solvation free energies—cannot be obtained from conventional thermochemical experiments. For reasons of charge conservation, such experiments can only yield the sum of solvation free energies of a pair of oppositely charged ions. Then, the attribution of individual free energies of solvation requires extra-thermodynamic assumptions [22]. The introduction of three-phase electrochemistry, however, has prompted the determination of free energies of solvation for individual ions. A first group of methods involves liquid-liquid interfaces [1], but a second group uses porous solids [10].

Let us consider the thermochemical cycle depicted in Figure 3.11, corresponding to the cation-assisted reduction of a porous solid when both the oxidized and the reduced forms of the same are insoluble in the solvent. It is assumed that a microparticulate deposit of the oxidized form of the solid is attached to a graphite electrode and put in contact with a given electrolyte containing a monovalent cation, M^+. The Gibbs free energy of the electrochemical reduction of the complex, ΔG°_{EC}, is related to the corresponding quantity for the reduction process in the gas phase, ΔG°_{EG}, the reticular free energy of the oxidized and reduced solid forms, $\Delta G^\circ_{ret}(Ox)$, $\Delta G^\circ_{ret}(Rd)$, the free energy of the electron transfer from the vacuum (gas phase) to the base electrode, Σ_e, and the free energy of solvation of the cation, $\Delta G^\circ_{solv}(M^+)$. These Gibbs free energies can be related as:

$$\Delta G^\circ_{EC} = \Delta G^\circ_{EG} + \Delta G^\circ_{ret}(Rd) - \Delta G^\circ_{ret}(Ox) - n\Delta G^\circ_{solv}(M^+) - n\Sigma_e \tag{3.40}$$

This expression applies for two solvents, A and B, so that the free energy of ion transfer from one solvent to another, $\Delta G^\circ_{A\rightarrow B\ transfer}(M^+)$, can be calculated from the values of the free energies of electron transfer in the respective solvents, $\Delta G^\circ_{EC\ solvent\ A}$, $\Delta G^\circ_{EC\ solvent\ B}$, as:

$$\Delta G^\circ_{A\rightarrow B\ transfer}(M^+) = (\Delta G^\circ_{EC\ solventA} - \Delta G^\circ_{EC\ solventB})/n \tag{3.41}$$

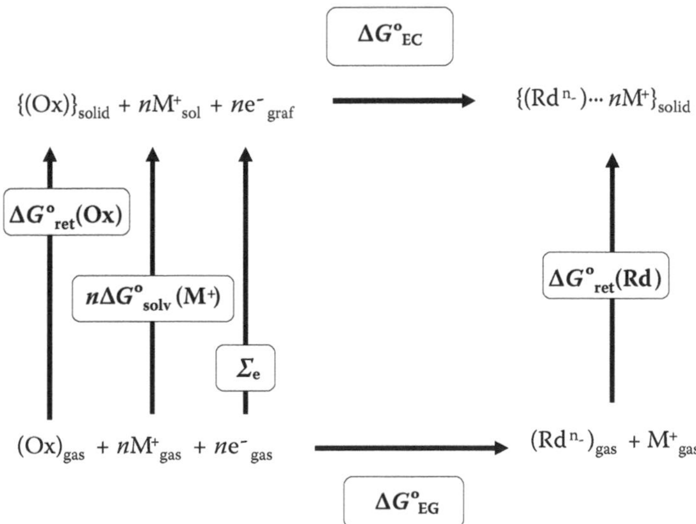

FIGURE 3.11 Thermochemical Cycle Used to Describe the Cation-Assisted Electrochemical Reduction of a Porous Solid Whose Oxidized and Reduced Forms are Insoluble in the Solvent Attached to a Base Graphite Electrode.

These free energies can easily be calculated from the electrode potentials, in turn, determined from the voltammetric midpeak potentials, $E_{mp\ solventA}$ and $E_{mp\ solventB}$, as,

$$\Delta G^{\circ}_{A\rightarrow B\ transfer}(M^{+}) = F\,(E_{mp\ solvent\ B} - E_{mp\ solvent\ A}) \tag{3.42}$$

provided that several extra-thermodynamic conditions are accomplished: (i) the electrochemical reduction of the porous solid is reversible in contact with both solvents; (ii) the reference electrode potentials for the two solvents can be related to each other; (iii) there is no accumulation of net charge in the solid complex/electrolyte boundary; and (iv) the structure of the solid forms is not affected by the solvent. Condition (ii) involves a general problem in electrochemistry—the absence of a solvent-independent scale of potentials because of the existence of unknown liquid junction potentials [23]. Currently, this problem is solved by assuming that the potential of the ferrocenium/ferrocene couple (or some other ferrocene derivatives) is solvent-independent [24,25]. The described methodology has been applied to calculate the free energy of ion transfer from water to several organic solvents using Prussian blue reduction for cations [26], and alkynyl-based Au(I) dinuclear and Au(I)-Cu(I) cluster complexes containing ferrocenyl groups for anions [27,28]. Table 3.1 summarizes the free energies of ion transfer between water and MeCN estimated for several inorganic ions using ion-insertion solids.

Just the electrochemistry of porous solids provides another way—involving extra-thermodynamic assumptions—to face the above problem. As described in Section 3.3, there is a possibility to define and evaluate the potential E°_{ET} corresponding to the process described by (3.9) as solvent-independent. Apart from the measure of open-circuit potentials and voltammetric midpeak potentials, E°_{ET} can be estimated by combining voltammetric and chronoamperometric data [29] with satisfactory application to both cation- and anion-insertion processes. The general problem, however, is the translation of electrode potentials to an absolute scale [30,31].

3.8 REFINEMENTS

The description of electrochemical processes of porous solids made in precedent sections can be refined by introducing additional considerations. First, the studied exchange reactions, like those represented by (3.19), can be described in terms of a second-order reaction kinetics [32] so that the apparent diffusion coefficient, D_{app}, measured in electrochemical experiments (e.g., chronoamperometry) diffusion-controlled conditions, can be related to the "true" diffusion coefficient, D, by the relationship:

TABLE 3.1
Free Energy of Ion Transfer from Water to MeOH and MeCN Calculated from Voltammetric Data for the Reduction of Microparticulate Deposits of Prussian Blue [a] and the Oxidation of Alkynyldiphosphine Dinuclear Au(I) Complexes and Heterometallic Au(I)_Cu(I) Cluster Complexes Containing Ferrocenyl Units [b] [26–28]

Ion	$\Delta G^{\circ}_{water\rightarrow MeOH}$ kJ mol^{-1}	$\Delta G^{\circ}_{water\rightarrow MeCN}$ kJ mol^{-1}
Li^{+a}	5.2 ± 0.4	9.1 ± 0.6
Na^{+a}	8.7 ± 0.4	8.6 ± 0.8
K^{+a}	11.5 ± 0.6	7.1 ± 0.8
F^{-b}	26 ± 2	65 ± 5
Cl^{-b}	13.5 ± 1.2	36 ± 2
Br^{-b}	11.5 ± 1.5	36 ± 2

$$D_{\text{app}} = D + \frac{k_e \delta^2 c}{s} \tag{3.43}$$

where $s = 2$, 4, and 6, for one-, two-, and three-dimensional diffusion, respectively. In the above equation, k_e is the second-order rate constant, and c and δ are the concentration and the distance between the centers of the chemically equivalent species involved in the electron exchange, respectively.

The aforementioned picture of diffusion via electron hopping between localized centers and ion motion across the pore and channel system of the material does not specify whether the barriers for electron hopping or those for ion motion control the rate of charge transport. If the barrier is expressed in terms of a thermally activated diffusion, the following equation holds:

$$cD^{1/2} = cD_o^{1/2} \exp(-E_a/RT) \tag{3.44}$$

E_a is the activation energy for the diffusion process.

Another possibility is to consider the transport of electrolyte ions through the electrolyte/solid interface. This involves the desolvation of the ion, its anchorage to the solid surface, and the cross of the interface. Assuming that these processes are fast, they can be treated, following White et al. [33], as a partitioning equilibrium. This can be expressed as the partition equilibrium, K_{pt}, equal to the ratio between the thermochemical activities of the mobile ion in the solid and solution phases. This lead—for experiments such as described in Section 3.4,—to a chronoamperometric curve given by [29]:

$$I = \frac{2nFNAc_{M^+\text{ solution}} K_{pt} D^{1/2}_{M^+\text{ solid}}}{(g + K_{pt})(\pi t)^{1/2}} \sum_{j=0}^{\infty} \left(\frac{1 - K_{pt}/g}{1 + K_{pt}/g} \right) \exp\left(\frac{-(j + 1/2)^2 \delta^2}{D_{M^+\text{ solid}} t} \right) \tag{3.45}$$

where g, as before, represents the square root of the $D_{M^+\text{ solution}}/D_{M^+\text{ solid}}$ ratio. At short times, it is possible to replace $D_{M^+\text{ solid}}$ and the effective diffusion coefficient given by:

$$D^{\text{eff}}_{M^+\text{ solid}} = \frac{D_{M^+\text{ solid}}}{1 + Kc_{Ox}[1 + K_{bd} K_{pt} c_{M^+\text{ solution}}]} \tag{3.46}$$

K_{bd} being the binding constant equal to $a_{\{Rdn- \cdots nM+\}}/a_{\{Ox\}}a_{\{M+\text{ solution}\}}$, assuming that the equilibrium is rapidly established at all points of the porous solid.

3.9 OVERVIEW

The electrochemistry of porous materials forming microparticulate deposits attached to inert electrodes in contact with suitable electrolytes constitutes involves a variety of thermochemical and kinetic considerations. The precedent treatments correspond to ideal cases but experimental data can deviate from the expectancies due to several factors—deviations from reversibility of the electrochemical process, continuum deviation when the number of immobile redox centers is not too large to ensure the continuum approximation where the basis of Fick's laws remains operative, discontinuities, and inhomogeneities in the microporous material and/or in the distribution of redox centers throughout the material, uncompensated capacitive and resistive effects, presence of ion transport control due to a strong disequilibrium between the concentration of cations in the electrolyte and the material because ion diffusion constants vary over the gradients of ion activity coefficients, solvent content and electrostatic ion-size effects, and the appearance of concentration polarization associated to large changes in the ion concentration in the porous material.

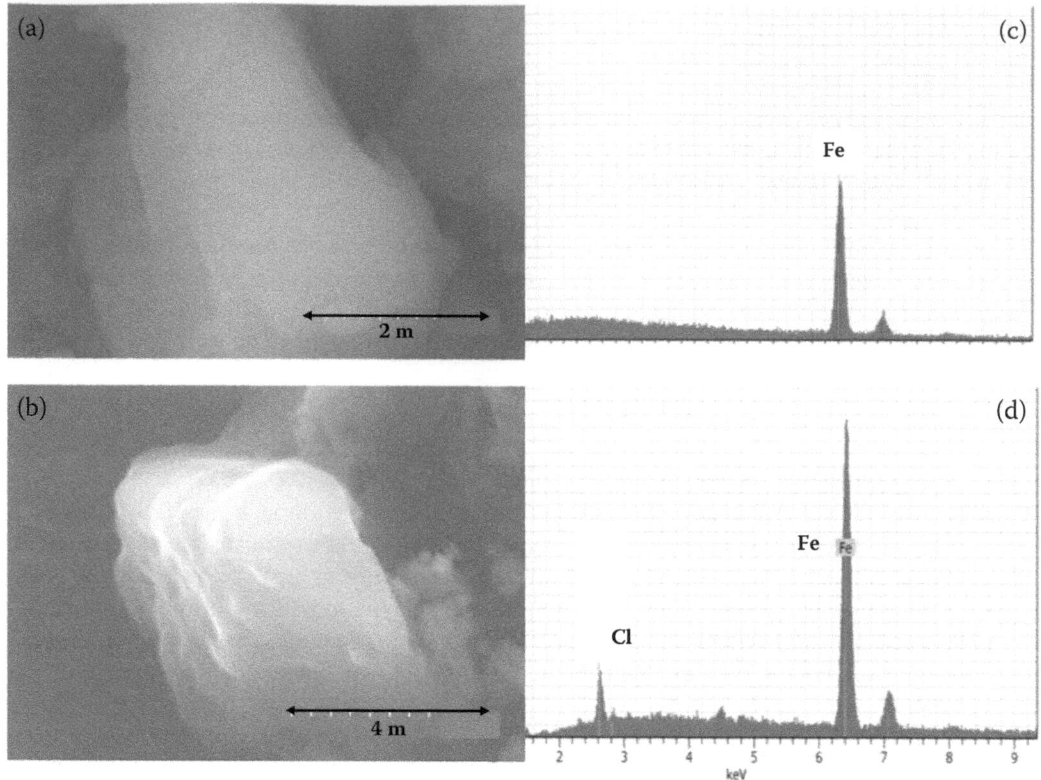

FIGURE 3.12 (a,b) FESEM Images and EDX Analysis of Microparticulate Deposits of the Bisferrocenyl-Functionalized Pseudopeptide But-Val-Fc_2 on a Graphite Plate Immersed into an Aqueous 0.10 M $NaClO_4$ Solution. (a,c) Before, and (b,d) After Application of a Potential Step at +0.65 V During 10 min. Adapted from Ref. [14] (Pandey et al. *RSC Adv.* 2016, 6: 35257–35266), with Permission.

It is also interesting to recognize the importance of the structural features occurring at solid materials at a scale that we call a mesoscopic, intermediate between the nanoscopic and microscopic scales, consisting, among others, in crystalline defects, dislocations, cracks, etc., whose influence in charge transport can be of importance. Figure 3.12 shows the FESEM images of individual crystals recorded in a microparticulate deposit of the bisferrocenyl-functionalized pseudopeptide But-Val-Fc_2 attached to a graphite plate in contact with an aqueous 0.10 M $NaClO_4$ solution, before and after application of a constant potential step at 0.50 V for 10 min. This process involves the oxidation of the pendant ferrocene units of the compound [14]. The bar-shaped crystals exhibit smooth surfaces that, after the oxidative step, become slightly corrugated and superficially delaminated. The accompanying EDX spectra in Figure 3.12 clearly show the presence of Cl in the crystals after the anion-assisted oxidative treatment, evidencing the occurrence of ion insertion. The features suggest that the structural integrity of the crystal is maintained as expected for a topotactic process, but ion permeation occurs through preferential channels involving dislocations or other crystalline defects at a scale larger than vacancies, interstitial atoms, or impurities acting at the nanoscopic scale. Related results were reported for water-insoluble alkynyl-triphosphine tetranuclear Au(I) complexes containing ferrocenyl motifs [15]. This means that electrochemical ion insertion processes can acquire significant structural complexity. In view of the capabilities of this type of process regarding thermochemical and synthetic (doping) fields, future research should explore the structural and kinetic constraints influencing ion insertion electrochemistry.

REFERENCES

[1]. Scholz, F.; Schröder, U.; Gulabowski, R.; Doménech-Carbó, A. 2014. *Electrochemistry of Immobilized Particles and Droplets*, 2nd edit. Springer, Berlin.

[2]. Doménech-Carbó, A.; Labuda, J.; Scholz, F. 2013. Electroanalytical chemistry for the analysis of solids: Characterization and classification (IUPAC Technical Report). *Pure and Applied Chemistry*. 85: 609–631.

[3]. Andrieux, C.P.; Savéant, J.-M. 1980. Electron transfer through redox polymer films. *Journal of Electroanalytical Chemistry*. 11: 377–381.

[4]. Laviron, E. 1980. A multilayer model for the study of space-distributed redox modified electrodes: Part I. Description and discussion of the model. *Journal of Electroanaytical Chemistry*. 112: 1–9.

[5]. Lovrić, M.; Scholz, F. 1997. A model for the propagation of a redox reaction through microcrystals. *Journal of Solid State Electrochemistry*. 1: 108–113.

[6]. Oldham, K.B. 1998. Voltammetry at a three-phase junction. *Journal of Solid State Electrochemistry*. 2: 367–377.

[7]. Lovrić, M.; Scholz, F. 1999. A model for the coupled transport of ions and electrons in redox conductive microcrystals. *Journal of Solid State Electrochemistry*. 3: 172–175.

[8]. Schröder, U.; Oldham, K.B.; Myland, J.C.; Mahon, P.J.; Scholz, F. 2000. Modelling of solid state voltammetry of immobilized microcrystals assuming an initiation of the electrochemical reaction at a three-phase junction. *Journal of Solid State Electrochemistry*. 4: 314–324.

[9]. Rolison, D.R. 1994. The intersection of electrochemistry with zeolite science, in *Advanced Zeolite Science and Applications*, Jansen, J.C.; Stöcker, M.; Karge, H.G.; Weitkamp, J. Eds. *Studies in Surface Science and Catalysis*. 85: 543–587.

[10]. Scholz, F.; Doménech-Carbó, A. 2019. The thermodynamics of insertion electrochemical electrodes – A team play of electrons and ions across two separate interfaces. *Angewandte Chemie International Edition*. 58: 3279–3284.

[11]. Leidner, C.P. ; Denisevich, P.; Willman, K.W.; Murray, R.W. 1984. Charge trapping reactions in bilayer electrodes . *Journal of Electroanalytical Chemistry*. 164: 63–78.

[12]. Hermes, M.; Lovrić, M.; Hartl, M.; Retter, U.; Scholz, F. 2001. On the electrochemically driven formation of bilayered systems of solid Prussian-blue-type metal hexacyanoferrates: A model for Prussian blue cadmium hexacyanoferrate supported by finite difference simulations. *Journal of Electroanalytical Chemistry*. 501: 193–204.

[13]. Cisternas, R.; Kahlert, H.; Wulff, H.; Scholz, F. 2015. The electrode response of a tungsten bronze electrode differ in potentiometry and voltammetry and give access to the individual contributions of electron and proton transfer. *Electrochemistry Communications*. 56: 34–37.

[14]. Pandey, M.D.; Martí-Centelles, V.; Burguete, M.I.; Montoya, N.; Luis, S.V.; García-España, E.; Doménech-Carbó, A. 2016. Bisferrocenyl-functionalized pseudopeptides: Access to separated ionic and electronic contributions for electrochemical anion sensing. *RSC Advances*. 6: 35257–35266.

[15]. Doménech-Carbó, A.; Koshevoy, I.O.; Montoya, N. 2016. Separation of the ionic and electronic contributions to the overall thermodynamics of the insertion electrochemistry of some solid Au(I) complexes. *Journal of Solid State Electrochemistry*. 20: 673–681.

[16]. Lovrić, M.; Hermes, M.; Scholz, F. 1998. The effect of the electrolyte concentration in the solution on the voltammetric response of insertion electrodes. *Journal of Solid State Electrochemistry*. 2: 401–404.

[17]. Scholz, F.; Lovrić, M.; Stojek, Z. 1997. The role of redox mixed phases $\{ox_x(C_n\ red)_{1-x}\}$in solid state electrochemical reactions and the effect of miscibility gaps in voltammetry. *Journal of Solid State Electrochemistry*. 1: 134–142.

[18]. Wooster, T.J.; Bond, A.M.; Honeychurch, M.J. 2003. An analogy of an ion-selective electrode sensor based on the voltammetry of microcrystals of tetracyanoquinodimethane or tetrathiafulvalene adhered to an electrode surface. *Analytical Chemistry*. 75: 586–592.

[19]. Doménech-Carbó, A. 2004. A model for solid-state voltammetry of zeolite-associated species. *Journal of Physical Chemistry B*. 108: 20471–20478.

[20]. Daum, P.; Lenhard, J.R.; Rolison, D.; Murray, R.W. 1980. Diffusional charge transport through ultrathin films of radiofrequency plasma polymerized vinylferrocene at low temperature. *Journal of the American Chemical Society*. 102: 4649–4653.

[21]. Gozález-Meza, O.A.; Larios-Durán, E.R.; Gutiérewz-Becerra, A.; Casillas, N.; Escalante, J.I.; Bárcena-Soto, M. 2019. Development of a Randles-Ševčík-like equation to predict the peak current of cyclic voltammetry for solid metal hexacyanoferrates. *Journal of Solid State Electrochemistry*. 23: 3123–3133.

[22]. Tissandier, M.D.; Cowen, K.A.; Feng, W.Y.; Gundlach, E.; Cohen, M.H.; Earhart, A.D.; Coe, J.V.

1998. Free energy of transfer of hydrated ion clusters from water to an immiscible organic solvent. *Journal of Physical Chemistry A*. 102: 7787–7794.

[23]. Izutsu, K. 2011. Liquid junction potentials between electrolyte solutions in different solvents. *Analytical Sciences*. 27: 685–694.

[24]. Gritzner, G. 1990. Polarographic half-wave potentials of cations in nonaqueous solvents. *Pure and Applied Chemistry*. 62: 1839–1858.

[25]. Gritzner, G. 1998. Single-ion transfer properties: A measure of ion solvation in solvents and solvent mixtures. *Electrochimica Acta*. 44: 73–83.

[26]. Doménech-Carbó, A.; Montoya, N.; Scholz, F. 2011. Estimation of individual Gibbs energies of cation transfer employing the insertion electrochemistry of solid Prussian blue. *Journal of Electroanalytical Chemistry*. 657: 117–122.

[27]. Doménech-Carbó, A.; Koshevoy, I.O.; Montoya, N.; Pakkanen, T.A. 2011. Estimation of free energies of anion transfer from solid-state electrochemistry of alkynyl-based Au(I) dinuclear and Au(I)-Cu(I) cluster complexes containing ferrocenyl groups. *Electrochemistry Communications*. 13: 96–98.

[28]. Doménech-Carbó, A.; Koshevoy, I.O.; Montoya, N.; Karttunen, A.J.; Pakkanen, T.A. 2011. Determination of individual Gibbs energies of anion transfer and excess Gibbs energies using an electrochemical method based on insertion electrochemistry of solid compounds. *Journal of Chemical & Engineering Data*. 56: 4577–4586.

[29]. Doménech-Carbó, A. 2012. Solvent-independent electrode potentials of solids undergoing insertion electrochemical reactions: Part I. Theory. *Journal of Physical Chemistry C*. 116: 25977–25983.

[30]. Trasatti, S. 1986. The absolute electrode potential: An explanatory note. *Pure and Applied Chemistry*. 58: 955–966.

[31]. Trasatti, S. 1990. The "absolute" electrode potential – the end of the history. *Electrochimica Acta*. 35: 269–271.

[32]. Botár, L.; Ruff, I. 1986. Effect of exchange reaction on transport processes: Fick's second law for diffusion on lattice points. *Chemical Physics Letters*. 126: 348–351.

[33]. White, H.S.; Leddy, J.; Bard, A.J. 1982. Polymer films on electrodes. 8. Investigation of charge-transport mechanisms in Nafion polymer modified electrodes. *Journal of the American Chemical Society*. 104: 4811–4817.

4 Electrocatalysis

4.1 INTRODUCTION

Electrocatalytic processes have received considerable attention in the last decades because of their application in synthesis and sensing. The term catalysis is used to describe the modification in the reaction rate of a given chemical reaction as an effect of the addition of a catalyst species. Two essential conditions must be accomplished by catalytic processes: the thermodynamics of the reaction is unaltered, and the catalyst stays unchanged. Additionally, a common demand for catalysis is that it must involve small concentrations of the catalyst. In the most general view, the rate of the reaction can either be increased (positive catalysis) or decreased (negative catalysis), although obviously, positive catalysis is preferentially desired.

The term electrocatalysis will be used in the following to design electrochemical processes involving the oxidation or reduction of a substrate species, Z, whose reaction rate is varied in the presence of a given catalytic species. The effect of electrocatalysis is an increase of the standard rate constant of the electrode reaction, resulting in a shift of the electrode reaction to a lower overpotential at a given current density and a current increase. The Faradaic current resulting from the occurrence of a catalytic electrode mechanism is called catalytic current. For a positive electrocatalysis, the current obtained in the presence of the catalyst must exceed the sum of the currents obtained for the catalyst and the substrate—separately—under selected experimental conditions [1]. Three possible situations can be discerned:

- The catalyst and the substrate are in the same phase, usually dissolved in the bulk solution (homogeneous catalysis).
- The catalyst or the substrate is immobilized (via fixation, functionalization, adsorption, etc.) at the electrode surface (heterogeneous catalysis).
- The catalyst is electrochemically generated at the electrode/electrolyte interface.

Homogeneous electrocatalysis in the solution phase can be described by a regeneration mechanism, where the catalyst reacts with the product of the electrode reaction involving the substrate to regenerate the initial electroactive species. This can be represented by the reaction sequence:

$$\mathrm{Ox} + e^- \rightarrow \mathrm{Rd} \tag{4.1}$$

$$\mathrm{Rd} + \mathrm{Z} \rightarrow \mathrm{Ox} + \mathrm{P} \tag{4.2}$$

The reaction described by Eq. (4.1) must have a formal potential that is larger (more positive) than the formal potential of the P/Z redox system. This system, however, is so irreversible that the direct reduction

$$\mathrm{Z} + e^- \rightarrow \mathrm{P} \tag{4.3}$$

does not occur at the potentials where Ox is reversibly reduced to Rd. In this context, the regeneration of the catalyst (Ox) leads to much steeper concentration profiles of the catalyst in the diffusion-reaction layer (i.e. to a steeper concentration gradient that (see Chapter 1) means larger current). The resulting effect is that, in the presence of Z, the currents recorded in voltammetric experiments for the process (4.1) become enhanced.

The catalytic mechanism in the solution phase represented by the above equations involves a reaction operating under conditions of pseudo-first-order kinetics [2]. Then, the shape of the cyclic voltammograms depends on the parameter k_fRT/nFv, k_f being the rate constant for reaction (4.2) and v the potential scan rate. For low values of this parameter (i.e. at relatively high scan rates), the catalytic reaction has no effect on the CV response, and a profile equivalent to a single n-electron transfer process is approached. For high k_fRT/nFv values (i.e. at low scan rates), the current tends to a limiting value,

$$I_{\mathrm{lim}} = nFAc\sqrt{Dk_f\, c_{\mathrm{cat}}} \tag{4.4}$$

so that s-shaped voltammetric curves that can be described by the equation

$$I \approx \frac{nFAc\,(Dk_f)^{1/2}}{1 + \exp[(nF/RT)(E - E^{\circ\prime})]} \tag{4.5}$$

are obtained. Figure 4.1 shows an example of how the voltammetric profile varies with the potential scan rate illustrated by the electrochemical Fenton reaction. Here, the reduction of H_2O_2 to OH^- is catalyzed by Fe(II) generated electrochemically by reduction of the parent Fe(III) in aqueous solution (charges omitted for brevity):

$$\mathrm{Fe(III)_{aq}} + e^- \rightarrow \mathrm{Fe(II)_{aq}} \tag{4.6}$$

$$2\mathrm{Fe(II)_{aq}} + \mathrm{H_2O_2} \rightarrow 2\mathrm{Fe(III)_{aq}} + 2\mathrm{OH^-_{aq}} \tag{4.7}$$

There are electrocatalytic systems where the catalyst is immobilized at the electrode surface. Typical examples involve the use of redox polymers or adsorbed species so that the substrate

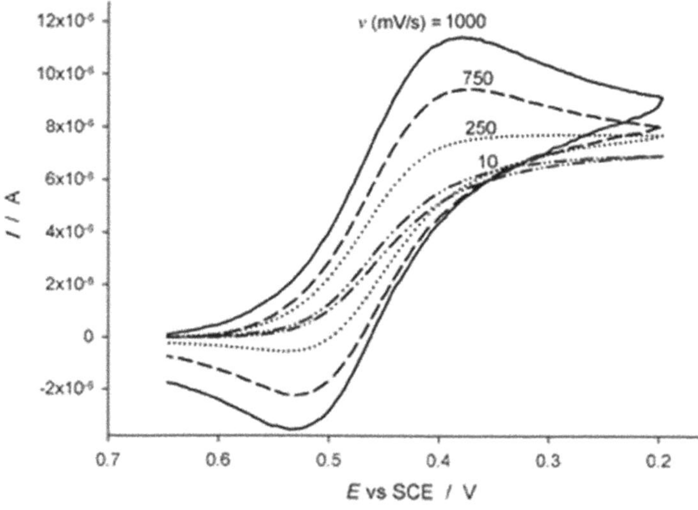

FIGURE 4.1 Experimental Cyclic Voltammograms Corresponding to the Reduction of 1 mM Fe(III) plus 0.5 M KCl and 0.02 M HCl Aqueous Solution at a Gold Macroelectrode (A = 2.1 mm^2) at Different Scan Rates in the Presence of 50 mM H_2O_2. Reproduced from Ref. [3] (Molina et al. *PhysChem.* 2011, 13: 14694–14704), with Permission.

interchanges electrons with the catalyst that is electrochemically regenerated [4]. In this chapter, the electrocatalytic effects associated with porous materials will be discussed regarding two more frequent situations. The first one corresponds to cases where the porous material acts as a catalyst for a given electrochemical process involving a substrate in the solution phase; the second corresponds to the case where the species in the solution phase acts as a catalyst for an electrochemical process involving electroactive species in the porous material. Both situations can be applied to isolated electroactive centers that become attached to an electrochemically silent porous network, as in zeolites with encapsulated redox-active guests, and cases where the entire porous material, which eventually can have semiconducting character and/or incorporate electroactive centers, possesses electrocatalytic activity, as occurring in transition metal oxides and related materials.

4.2 HETEROGENEOUS ELECTROCATALYSIS

Heterogeneous electrocatalysis proceeds via electron exchange between the species in solution and the catalyst forming a layer onto the electrode surface. Figure 4.2a shows a scheme for this catalytic pathway that can be enhanced by incorporating a second redox mediator in the solution phase (Figure 4.2b).

Rotating disk voltammetry is a technique widely used for studying catalytic processes. For a reversible n-electron transfer process controlled by mass transport in solution, the limiting current, I_{lim} (μA), recorded in a linear potential scan voltammogram, varies with the rotation rate, ω (s^{-1}), following the Levich equation [5]:

$$I_{\text{lim}} = 0.620nFAcD^{2/3}\upsilon^{-1/6}\omega^{1/2} \tag{4.8}$$

where D denotes the diffusion coefficient of the electroactive species, c its bulk concentration, υ the kinematic viscosity (cm^2s^{-1}) of the solution, A the electrode area, and the other symbols with their customary meaning. This equation predicts that plots of I_{lim}, vs. $\omega^{1/2}$ should be a straight line intersecting the origin. Deviations from the expected behavior can be attributed to deviations from ideality and kinetic complications. Following Andrieux et al. [6] for a mediated catalytic process, the limiting current should satisfy:

(a)

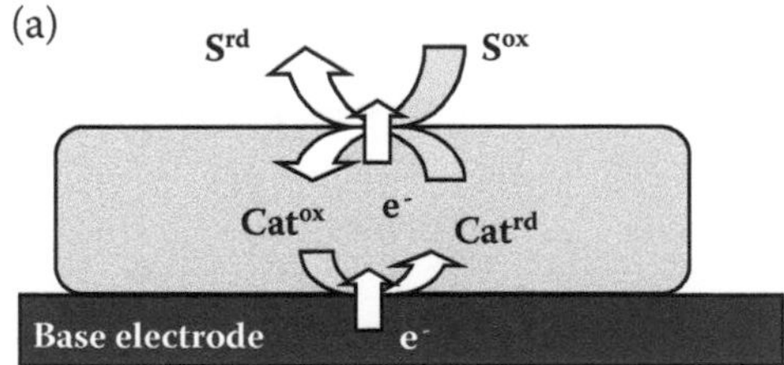

(b)

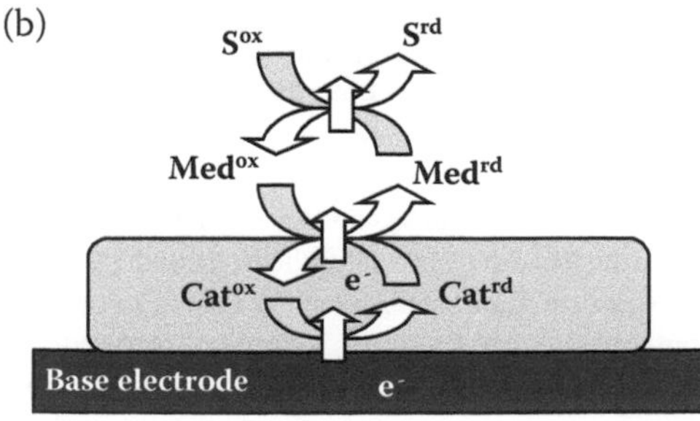

FIGURE 4.2 Schematics for the Electrocatalysis Performed with an Immobilized Catalyst (a) and Incorporating a Redox Mediator in Solution (b).

$$1/I_{\lim} = 1/i_{\mathrm{k}} + 1/nFAck\Gamma[1 - (\delta/FAc_{\mathrm{cat}}D_{\mathrm{e}})] \tag{4.9}$$

where the kinetic current, i_{k}, representing the current in the absence of mass transfer effect, is introduced. In the above equation, Γ denotes the surface concentration of the catalyst in the film, k the second-order rate constant for the catalytic reaction, δ the film thickness, D_{e} the diffusion coefficient for electrons in the film, and c_{cat} the total volume concentration of the catalyst within the film. When the catalytic reaction becomes dominant, Eq. (4.9) reduces to the Koutecky-Levich equation [5]:

$$1/I_{\lim} = 1/nFAck\Gamma + 1/0.620nFAcD^{2/3}\upsilon^{-1/6}\omega^{1/2} \tag{4.10}$$

and:

$$I_{\mathrm{k}} = nFAck\Gamma \tag{4.11}$$

Figure 4.3 shows CVs of nanoparticles in N-doped porous carbon polyhedrons in Ar-(black) and O_2-(red) saturated 0.1 M KOH (Figure 4.3a) and the corresponding LSVs in rotating disk

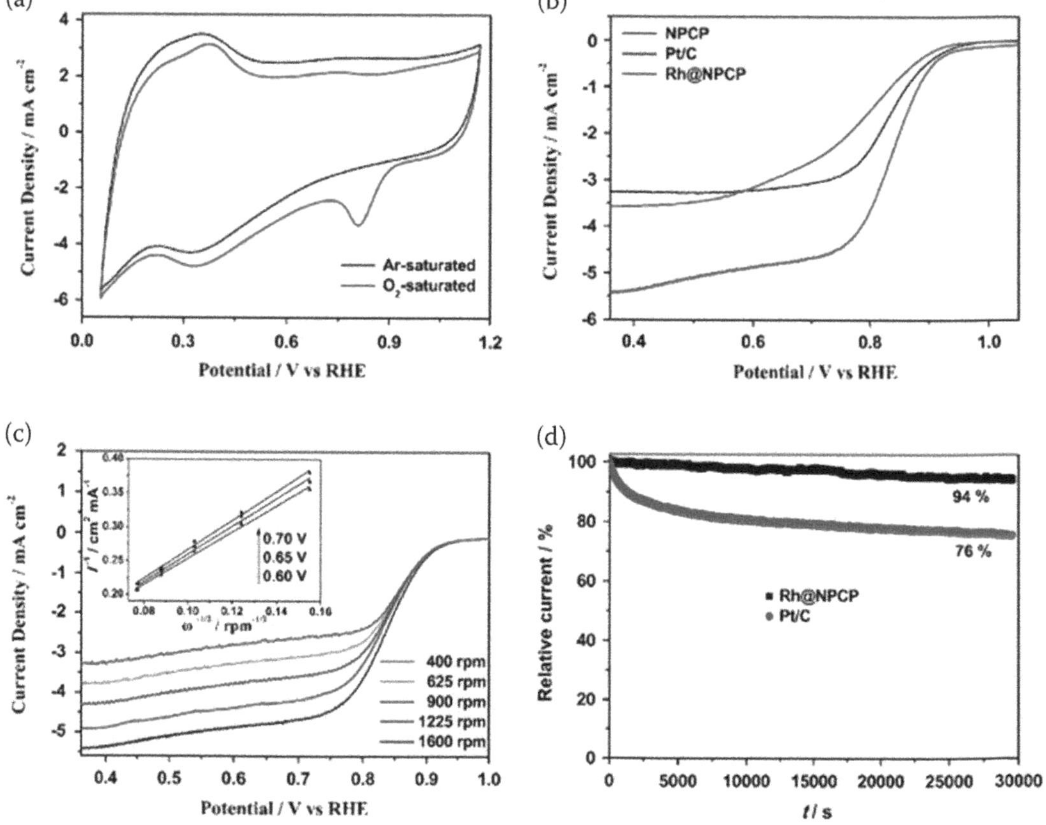

FIGURE 4.3 (a) CVs of Nanoparticles in N-Doped Porous Carbon Polyhedrons in Ar- (Black) and O_2-(Red) Saturated 0.1 M KOH; (b) LSV Curves of the Studied Composite, the Carbon Polyhedrons and Pt/C in O_2 -Saturated 0.1 M KOH. Potential Scan Rate 10 mV s^{-1}, Rotation Rate 1600 rpm; (c) LSV Curves at Different Rotation Rates, Inset Shows the Corresponding Koutecky-Levich Plots; (d) Chronoamperometric Curves of the Studied Composite and Pt/C at 0.70 V and a Rotation Speed of 1600 rpm. From Ref. [7] (Huang et al. *Electrochim. Acta.* 2019, 326: artic. 134982), with Permission.

voltammetry [7]. There is an electrocatalytic effect on ORR. The slope of Koutecky-Levich plots agrees with that expected for four-electron pathway:

$$O_2 + 2H_2O + 4e^- \rightarrow 4OH^- \tag{4.12}$$

Rather than a peroxide-based (also called serial or indirect) reduction pathway

$$O_2 + H_2O + 2e^- \rightarrow HO_2^- + OH^- \tag{4.13}$$

$$HO_2^- + H_2O + 2e^- \rightarrow 3OH^- \tag{4.14}$$

whose rate-determining step is a two-electron process.

Figure 4.4 shows Koutecky-Levich plots for the oxidation of methanol at Pt oxide [8].

As was discussed in the precedent chapter, conducting porous electrodes produces an apparent catalytic effect due to the enhancement of the effective surface area relative to the geometrical area of the base electrode. In these circumstances, there is no redox mediation such as schematically depicted in Figure 4.2 and, therefore, there is no "true" electrocatalysis.

In most cases, however, there is a second catalytic pattern resulting in the enhancement of currents and the decrease of the overpotential of kinetically hindered electrochemical processes. This involves the interaction of the substrate species in solution with the so-called active sites of the solid catalyst. A classic example is the hydrogen evolution reaction (HER) at Pt and Au electrodes. Here, different types of processes can be discerned—platinum oxidation, double layer region with no Faradaic processes, hydrogen underpotential deposition (UPD), and hydrogen evolution reaction (HER). Figure 4.5 depicts the typical CV recorded on a polycrystalline Pt electrode supported on carbon fiber in an acidic solution where the double layer region is marked [9].

This is a multi-step process initiated by the reduction of a hydronium ion on an active site (labeled as Pt*) of the catalyst surface (Volmer step),

$$H_3O^+_{aq} + Pt^* + e^- \rightarrow H\cdot_{ads} + H_2O \tag{4.15}$$

yielding an H atom adsorbed onto the metal surface. This process characterizes the UPD region and presumably involves cation adsorption and OH desorption [10,11]. The next step is the evolution of molecular H_2, either through a new proton/electron transfer (Heyrovsky step),

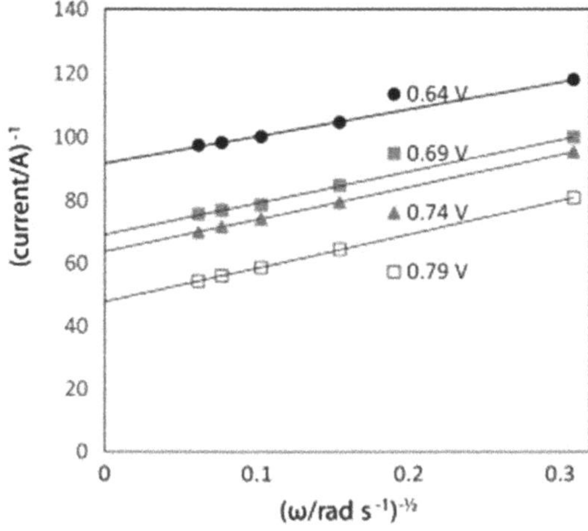

FIGURE 4.4 Steady-State Koutecky-Levich Plots for Oxidation of 0.1 M Methanol in 1 M H_2SO_4 at a GC/PtRu Black (ca. 8 mg cm^{-2}) Electrode. Reproduced from Ref. [8] (Sayadi and Pickup, *Electrochim. Acta.* 2016, 199: 12–17), with Permission.

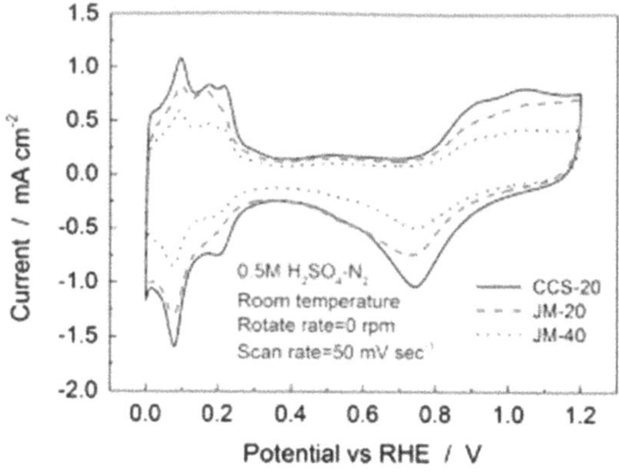

FIGURE 4.5 CVs of Different Pt Catalysts in 0.5 M H_2SO_4; Potential Scan Rate 50 mV s^{-1}. Reproduced from Ref. [9] (Song et al. *Electrochim. Acta.* 2015, 177: 174–180), with Permission.

$$H\cdot_{ads} + H_3O^+_{aq} + e^- \rightarrow H_2 + H_2O \tag{4.16}$$

or through the recombination of two adsorbed hydrogen atoms (Tafel step),

$$2H\cdot_{ads} \rightarrow H_2 \tag{4.17}$$

In summary, there are two possible catalytic pathways for HER in both acid and alkaline media initiated by the Volmer step and ended by the Tafel or Heyrovsky mechanisms. Broadly, these can be distinguished by the value of the Tafel slope obtained from the plots of the potential vs. log (current density). The ideal Tafel slopes for the Volmer, Heyrovsky, and Tafel step are of 120, 40, and 30 mV decade^{-1}, respectively.

In the past decade, studies on HER at single crystal surfaces have underlined the crucial importance of defect sites and crystal orientation relative to adsorption/desorption properties. Given their high specific surface area and the possibility of modulating their adsorption properties, there is considerable effort to develop nano- or microporous materials to be used in large-scale H_2 production, overcoming the difficulty in the use of Pt-based materials due to its scarcity and excessive cost. Among others, tungsten and molybdenum carbides and sulfides and bi- and polymetallic porous materials are under extensive study.

The study of catalysts for HER, as well as for the oxygen evolution reaction (OER), is generally made from LSV and CV at low potential scan rates (usually below 5 mV s^{-1}), optionally using rotating disk electrode arrangements (RDE) to ensure quasi-static conditions. Figure 4.6 compares the LSVs recorded at different porous electrocatalysts deposited onto GCE for OER [12].

The onset potential, E_{onset}, is defined as the intersection with the potential axis of the prolongation of the linear-like portion of the current/potential curve. Usually, it is computed as the onset overpotential relative to the standard potential of the involved couple, η_{onset}. In the case of HER, the objective of the catalysts is to shift this potential to less negative values (less positive values in the case of OER). Similar considerations apply for a second widely used parameter, the overpotential relative to the standard potential of the couple for which the current density equals 10 mA cm^{-2}, $\eta_{j\,=\,10}$.

Another important parameter used to compare different electrocatalysts is the Tafel slope (or the inverse Tafel slope), determined from the linear representation of ln I vs. E in the foot of the voltammetric rising curve. Here, the following relationship (Tafel equation) applies:

$$\ln j \approx \ln j_o + \frac{\alpha n F}{RT}\eta \tag{4.18}$$

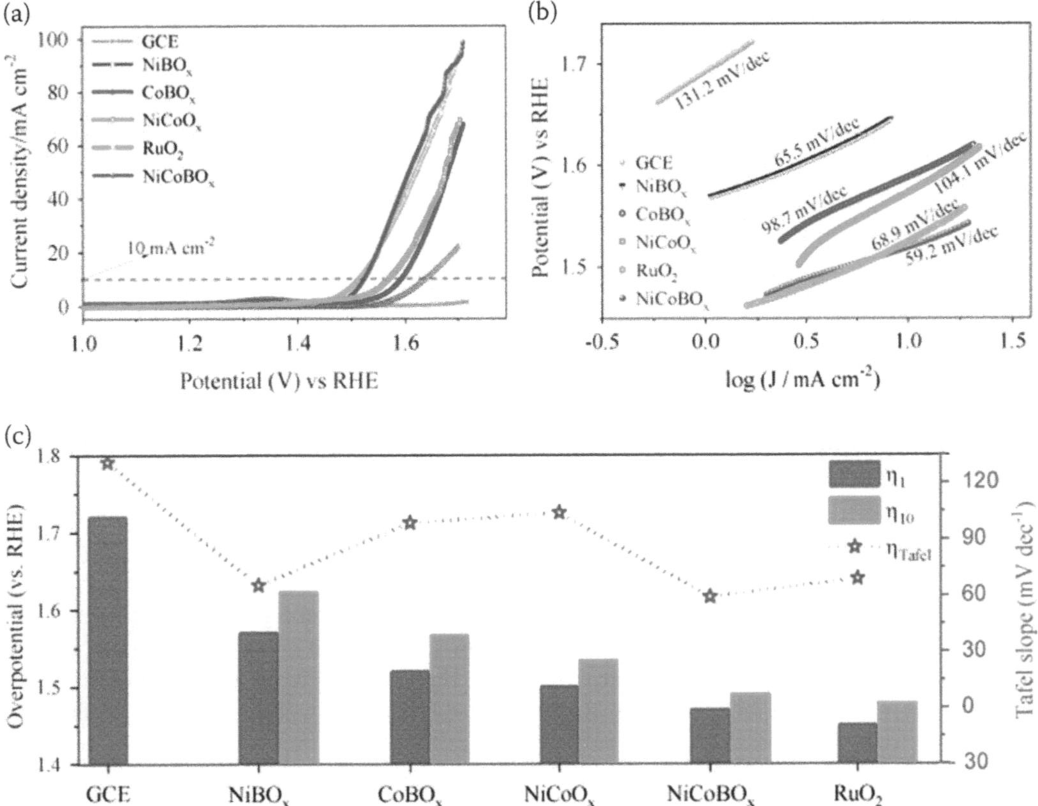

FIGURE 4.6 (a) Polarization Curves, (b) Corresponding Tafel Plots, and (c) Performance Comparison of Bare GCE, $NiBO_x$, $CoBO_x$, $NiCoO_x$, $NiCoBO_x$, RuO_2 Electrocatalysts in 1 M KOH. From Ref. [12] (Ramadoss et al. *Electrochimica Acta*. 321: artic. 134666), with Permission.

where j is the current density at an applied potential overpotential η relative to the standard potential of the couple, and j_o is the exchange current density, often used as another parameter representative of the catalyst efficiency. Usually, the term Tafel slope, representative of the kinetics of the process, is applied to the $2.303RT/\alpha nF$ ratio expressed in mV decade^{-1} (or simply, mV).

4.3 STRUCTURAL AND ELECTRONIC EFFECTS

Heterogeneous electrocatalytic systems possess considerable complexity. The variety of factors influencing such processes can be divided into two major effects: structural and electronic [13]. The essential idea is that the activity of a heterogeneous catalyst occurs through active sites, the concept introduced by Taylor in 1925 [14]. The advances experienced by surface science in the last decades have notably increased the modeling of surfaces and the energetic aspects of the different structural arrangements. Studies on single-crystal electrochemistry have contributed significantly to this field [15]. These studies suggest that many electrocatalytic processes at metal surfaces can be grouped into [16]:

a. Catalytic processes that take place preferentially on steps and defect sites within (111) terraces. Breaking/forming of O–H or C–H bonds (oxidation of methanol or CO) are typically involved.

b. Electrocatalytic processes that take place on (100) terrace sites preferentially without steps and defects. These reactions involve the breaking/forming of C–O, C–C, N–O, N–N, or O–O bonds.
c. Processes with no clear preference for a particular site or crystal surface are typically influenced by adsorption effects.

Figure 4.7 depicts the unit stereographic triangle of face-centered cubic single-crystal surfaces with the corresponding surface atomic arrangements [13,16,17]. The oxidation of CO falls within the first type of electrocatalytic processes. Here, the rate-determining step can be represented as:

$$H_2O + \{ * \} \rightarrow \cdot OH_{ads} + H^+_{aq} + e^- \tag{4.19}$$

{*} being the active site. The reaction involves the adsorption of CO (i.e. CO + {*} → CO_{ads}). Then, the adsorbed hydroxyl reacts with adsorbed CO via a Langmuir–Hinshelwood reaction to yield CO_2 [18]:

$$CO_{ads} + \cdot OH_{ads} \rightarrow CO_2 + 2\{ * \} + H^+_{aq} + e^- \tag{4.20}$$

The reduction of nitrite at stepped Pt surfaces is illustrative of the second group of electrocatalytic processes. Figure 4.8 shows the LSV for the reduction of 2 mM nitrite in 0.1 M NaOH on (a) Pt[*n* (100)-(111)] single-crystal electrodes. Here, two peaks corresponding to the reduction of nitrite to ammonia and nitrogen appear, respectively, the second reaction being exclusive of Pt(100) facet [13].

$$NO_2^- + 5H_2O + 6e^- \rightarrow NH_3 + 7OH^- \tag{4.21}$$

$$2NO_2^- + 4H_2O + 6e^- \rightarrow N_2 + 8OH^- \tag{4.22}$$

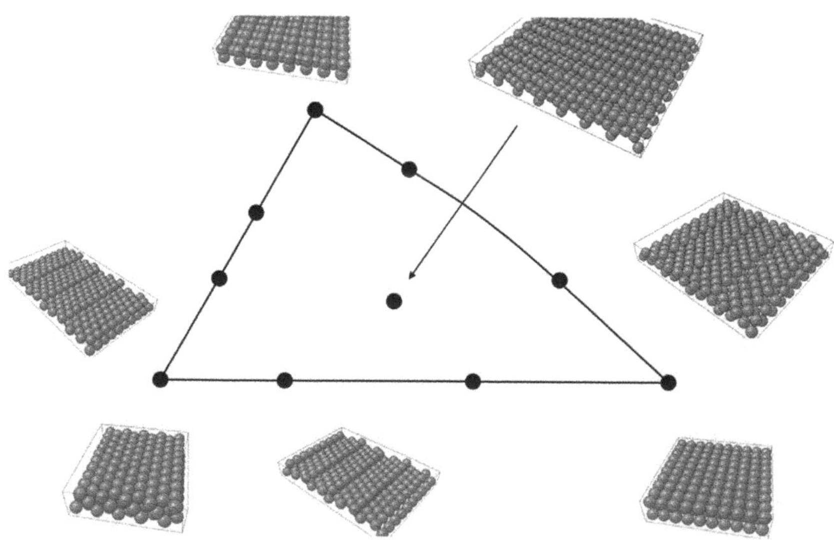

FIGURE 4.7 This Figure Depicts the Unit Stereographic Triangle of Face-Centered Cubic Single-Crystal Surfaces with the Corresponding Surface Atomic Arrangements. Reproduced from Ref. [13] (Bandarenka and Koper, *J. Catal.* 2013, 308: 11–24), with Permission.

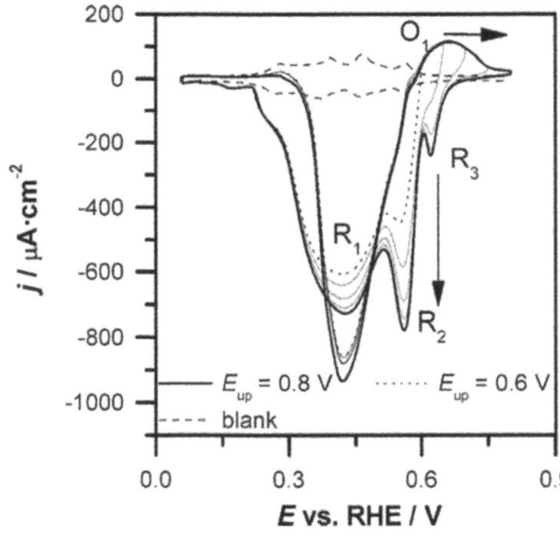

FIGURE 4.8 LSV for the Reduction of 2 mM Nitrite in 0.1 M NaOH on (a) Pt[*n*(100)-(111)] Single-Crystal Electrodes. Potential Scan Rate 50 mV s^{-1}. From Ref. [18] (Duca et al. *J. Am. Chem. Soc.* 2011, 133: 10928–10939), with Permission.

Within the third group of electrocatalytic processes, the ORR process on platinum in perchloric and sulfuric acid solutions, which has been proven to be sensitive to electrolyte anions, is one of the relevant examples [19].

Of course, the above considerations can be extended from metals to a variety of compounds—metal oxides in particular. In principle, the catalytic ability of a surface, with respect to a given electrochemical process, depends on the composition of the material. Two related demands must be fulfilled corresponding to the historical Sabatier principle [20]: the surface should be able to activate the reactants but also not bind the reaction intermediates to prevent poisoning of active sites. The binding or adsorption energies of the intermediates to the surface, in turn depending on the electronic structure of the surface, provide a quantitative approach for this aspect.

Recent advances in computational quantum chemistry permit estimate binding energies for different surfaces accounting for the electronic structure of the material and the structural properties of the surface. These are reflected usually in the so-called volcano plots whose introduction for HER electrocatalysis was pioneered by Trasatti [21]. Volcano plots draw the exchange current density—representative of the reaction rate—vs. the binding energy between the active-site and the intermediate species. Calculation of binding/adsorption energies by density functional theory calculations [22] has significantly expanded the use of volcano plots that can be applied to different electrocatalytic systems. Roughly, if the catalytic process involves a two-step reaction passing through an adsorbed intermediate, the adsorption energy should be minimal. If it is high, the adsorption and desorption are slow, limiting the overall reaction rate in both cases.

In the case of HER electrocatalysis at metal (M) surfaces, M-H binding energies were taken [21], while in the case of the electrocatalysis of OER at metal oxide surfaces, the M-OH energy is computed. Figure 4.9 shows an example of volcano plots for ORR reaction at nonaqueous Li-air batteries [23]. Here, the electrocatalytic activity of the ORR at GCE and different metal electrodes is compared with the binding energy of O_2 to the different surfaces.

It is important to recognize the high complexity that heterogeneous electrocatalytic systems can reach. In particular, the superposition of different processes, such as the reduction of CO_2 and HER [12] or the evolution of chlorine from chloride solutions at RuO_2, surfaces [24]. Two of the basic pathways that have been proposed for chloride to chlorine oxidation that parallel those for HER reaction are:

Volmer-Tafel process:

$$2\{ * \} + 2Cl^-_{aq} \rightarrow 2Cl_{ads} + 2e^- \tag{4.23}$$

$$2Cl_{ads} \rightarrow 2\{ * \} + Cl_2 \tag{4.24}$$

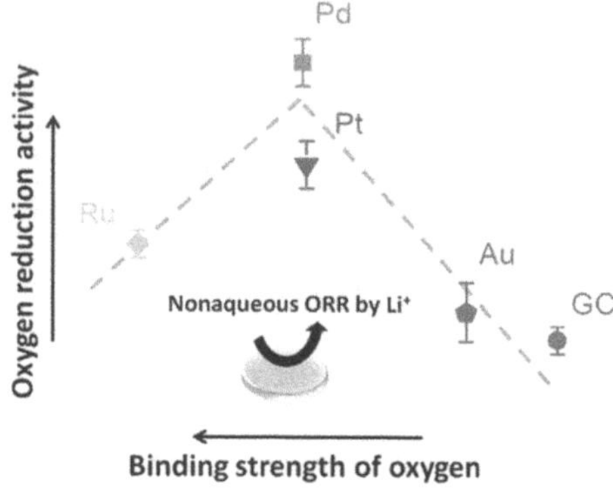

FIGURE 4.9 Volcano Plot for ORR in Nonaqueous Li^+-Containing Electrolytes at GCE and Different Metals. From Ref. [23] (Lu et al. *J. Am. Chem. Soc.* 2011, 133: 19048–19051), with Permission.

Volmer-Heyrovsky process:

$$\{ * \} + Cl^-_{aq} \rightarrow Cl_{ads} + e^- \tag{4.25}$$

$$Cl_{ads} + Cl^-_{aq} \rightarrow Cl_2 + \{ * \} + e^- \tag{4.26}$$

The electrochemical process is superimposed to water oxidation so that DFT calculations suggest that Cl_2 formation occur through ClO and $Cl(O_2)$ adsorbed onto different surface sites, one of the possible pathways being:

$$O_{ads} + 2Cl^-_{aq} \rightarrow ClO_{ads} + Cl^-_{aq} + e^- \rightarrow Cl_2 + O_{ads} + 2e^- \tag{4.27}$$

4.4 ELECTROCATALYSIS AT MICROHETEROGENEOUS DEPOSITS OF POROUS MATERIALS

This is a situation analog to the catalysis at redox polymers, where it is assumed that a microheterogeneous layer is permeated by the substrate. It will be assumed that there is a formation of substrate-catalyst adducts experiencing electron exchange and that the cycle ends with the electrochemical reduction of the catalyst at the base electrode/catalyst interface [25].

$$S^{ox} + \{Cat^{red}\} \overset{K_M}{\rightarrow} S^{ox} \cdots \{Cat^{red}\} \overset{k_c}{\rightarrow} S^{red} + \{Cat^{ox}\} \tag{4.28}$$

$$\{Cat^{ox}\} + e^- \overset{k_e}{\rightarrow} \{Cat^{red}\} \tag{4.29}$$

In these and in subsequent equations, charge-compensating metal ions in the solid will be omitted for brevity. It will be assumed that the catalyst is homogeneously distributed in a thin film of the surface of a basal electrode (see Figure 4.2). If the substrate can diffuse through the catalyst layer under steady-state conditions, the equation describing the transport and kinetics in the layer will be:

$$D_S \frac{d^2[S]}{dx^2} - \frac{k_c c_{cat} [S]}{K_M + [S]} = 0 \tag{4.30}$$

where D_S represents the diffusion coefficient of the substrate through the layer, [S] is its concentration at a distance x from the electrode, and c_{cat} is the concentration of catalyst. Following Lyons et al. [25], it will be assumed that Michaelis-Menten (or Langmuir-Hinshelwood) kinetics applies, k_c being the catalytic rate constant (s^{-1}) and K_M the Michaelis constant (mol cm^{-3}). Neglecting concentration polarization in the solution, we consider several cases. First, if the concentration of the substrate can be taken as uniform throughout the catalyst layer, a situation approached by thin films, and assuming that there is no concentration polarization of the substrate in the film, the flux of substrate Φ_S (mol cm^{-2} s^{-1}) is given by:

$$\Phi_S = \frac{k_c c_{cat} L c_S}{K_M + K c_S} \tag{4.31}$$

In this equation, L represents the thickness of the catalyst layer, c_S the concentration of substrate in the bulk of the solution, and K the partition coefficient that can be approached for unity in porous materials. When the concentration of substrate in the film is not uniform, four extreme cases can be considered depending on the film thickness and substrate concentration. For low substrate concentrations and thin films, the flux of substrate is described by the equation:

$$\Phi_S = \frac{k_{cat} c_{cat} L c_S}{K_M} \tag{4.32}$$

Whereas for low substrate concentrations and thick films, the flux approaches:

$$\Phi_S = c_S \sqrt{\frac{k_{cat} c_{cat} D_S}{K_M}} \tag{4.33}$$

Here, the reaction kinetics is more rapid than the diffusive transport of the substrate and there is a large concentration gradient of substrate across the film.

For high substrate concentrations, the catalyst is saturated by the substrate. Then, the flux becomes independent of substrate concentration and the rate-determining step is the decomposition of the intermediate substrate catalyst adduct, described by the rate constant k_c. Here, the flux of substrate satisfies

$$\frac{1}{\Phi_S} = \frac{k_M}{K_c L K c_{cat} c_S} + \frac{1}{k_c L c_{cat}} \tag{4.34}$$

This equation corresponds to the Lineweaver-Burk kinetic equation and can be transformed into an equation giving the dependence of kinetic currents, I_k, on the concentration of substrate in the solution bulk. The values of I_k at different concentrations of substrate can be determined from the limiting, steady-state currents, I_{lim}, obtained, for instance, as plateau currents in rotating disk voltammetry using Eq. (4.10). For the case of thin films with a surface concentration of catalytic centers Γ_{cat} (mol cm^{-2}) over an electrode of area A, one can write:

$$\frac{1}{I_k} = \frac{K_M}{nFAk_c \Gamma_{cat} c_S} + \frac{1}{nFAk_c \Gamma_{cat}} \tag{4.35}$$

Accordingly, plots of $1/I_k$ vs. $1/c_S$ should provide straight lines whose slope and ordinate at the origin enable calculation of K_M and k_c, provided that Γ_{cat} is known. The above expression can be rewritten to give the Eadie-Hofstee plot, which arises from the following equation:

$$\frac{I_L}{c_S} = \frac{nFAk_c\Gamma_{cat}}{K_M} + \frac{I_L}{K_M} \tag{4.36}$$

This treatment is largely focused on the characterization of sensors based on microparticulate materials embedded into polymer matrices.

Electrochemical data recorded under no steady-state conditions can also be used to study electrocatalytic processes involving porous materials. In cases where the catalytic system can be approached by homogeneous electrocatalysis in the solution phase, variation of cyclic voltammetric profiles with potential scan rate [2] and/or square wave voltammetric responses with square wave frequency can be used. This situation can be taken for highly porous materials where substrate transport, as well as charge-balancing ion transport, is allowed. On first examination, the catalytic process can be approached as described by Eqs. (4.1) and (4.2). The catalytic current is then expressed by:

$$I_{cat}/I_d = (kc_{cat}t)^{1/2}\{(kc_{cat}t)^{1/2}\,\mathrm{erf}[(kc_{cat}t)^{1/2}] + (kc_{cat}t)^{-1/2}\exp(-kc_{cat}t)\} \tag{4.37}$$

i_d being the current for the unaltered, diffusion-controlled, single electron transfer process and erf the error function. When the term $kc_{cat}t$ exceeds 2, the error function (erf) reduces to 1, and the above equation can be approached as:

$$I_{cat}/I_d \approx (\pi kc_{cat}t)^{1/2} \tag{4.38}$$

In several cases when the electrocatalytic pathway involves adduct formation between immobilized units of the catalyst and the mobile substrate permeating the porous solid, the mechanism can be described in terms of a charge transfer process preceded by a relatively slow chemical reaction. Here, the chronoamperometric current can be approached by the expression developed by Koutecky and Brdicka [5]:

$$I = nFAcD^{1/2}(Kk_f)^{1/2}\exp[Kk_ft][\mathrm{erf}(Kk_ft)^{1/2}] \tag{4.39}$$

where k_f represents the rate constant of the direct preceding (adduct formation) reaction and K, the equilibrium constant for this process. The case of the charge transfer process preceded by a first-order bulk-surface reaction was described by Guidelli [26]. Here, the parent electroinactive species is transformed into an electroactive species both through a homogeneous chemical reaction taking place within a thin solution layer adjacent to the electrode (with rate constant, k_f) and through a heterogeneous chemical reaction catalyzed by the electrode surface (with rate constant k_{ads}). The chronoamperometric current becomes:

$$\frac{I}{I_d - I} = t^{1/2}\frac{(3/5)[k_fK\Gamma + (k_{ads}D/K)^{1/2}t^{1/2}]}{(12Dt/7\pi)^{1/2} + K\Gamma} \tag{4.40}$$

In this equation, i_d represents the diffusion limiting current given by $I_d = (12D/7\pi)^{1/2}nFAct^{1/2}$.

4.5 ELECTROCATALYSIS AT ION-PERMEABLE SOLIDS

Microparticulate deposits of nonconducting porous solids mechanically transferred to the surface of inert electrode—dispersed in carbon paste or composite electrodes or embedded into a conducting polymeric matrix further fixed on the electrode surface—can also exert important electrocatalytic effects. A first example can be the reduction of Cr(VI) with silicomolybdic acid assisted by proton insertion. The overall process can be summarized as [27]:

$$\begin{aligned}&\{H_{10}SiMo_6^{VI}Mo_6^{V}O_{40}\}_{solid} + Cr_2O_7^{2-}{}_{aq} + 8H^+_{aq} \rightarrow \\ &\{H_4SiMo_{12}^{VI}O_{40}\}_{solid} + 2Cr^{3+}_{aq} + 7H_2O\end{aligned} \tag{4.41}$$

As for thin films of redox polymers, there are three elements of porous electrode behavior crucial to its performance in electrocatalysis: the transport of a solution reactant to the catalytic sites within the porous system, the transport of electrolyte charge-balancing ions, and the electron transport across the solid—a process responsible for the regeneration of the initial oxidation state of the catalyst. There are several possibilities depending on the conductivity of the porous support, its pore size distribution, the presence and mobility of electrolyte ions and/or mobile redox mediators, and the type of attachment of catalytic centers to the support and its distribution. In the following, electrocatalysis at microheterogeneous deposits of insolating materials on inert electrodes will be described, assuming that the microparticulate deposit that is in contact with a given electrolyte solution is distributed along a microheterogeneous layer covering the surface of a basal, metal-type conducting electrode. In the most direct electrocatalytic case, the solid incorporates electroactive centers, Cat, producing an electrocatalytic effect on the electrochemical oxidation/reduction of a given substrate, S, in solution. In classical, electrocatalysis via redox films, there is a charge transfer process based on the diffusion of substrate molecules through the solution, interchange of electron with the immobile redox centers in the film, and subsequent electrochemical regeneration of the mediator, as schematically depicted in Figure 4.10.

According to the modeling of the electrochemistry of ion-insertion solids, electron transfer processes in these systems involve an initial reaction at the three-phase electrolyte/particle/electrode junction further extended via transport of electrons and charge-balancing ions in

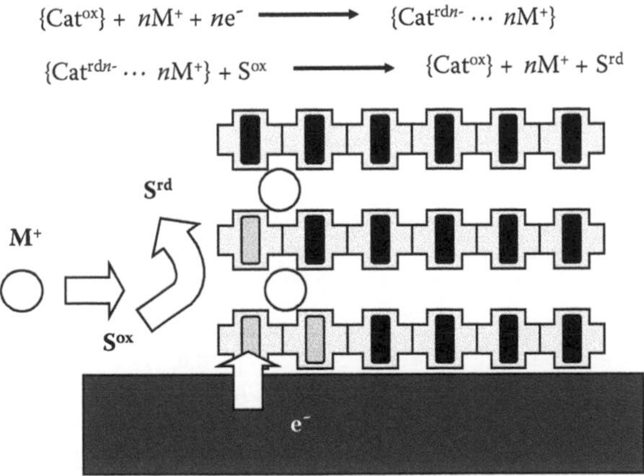

FIGURE 4.10 Diagram for Possible Electrocatalytic Processes Involving Immobilized Catalyst Species into a Porous Solid Acting on a Substrate in Solution Phase.

perpendicular directions. Here, we consider that the solid catalyst can undergo a reduction process of the type

$$\{Cat^{ox}\}_{solid} + nM^{+}_{sol} + ne^{-} \rightarrow \{Cat^{redn-} \cdots nM^{+}\}_{solid} \tag{4.42}$$

involving the entrance of electrolyte cations into the porous solid. Then, the substrate in the solution phase is reduced via heterogeneous redox reaction with Cat^{red} (substrate charges omitted for simplicity):

$$S^{ox}_{sol} + \{Cat^{redn-} \cdots nM^{+}\}_{solid} \rightarrow \{Cat^{ox}\}_{solid} + S^{red}_{sol} + nM^{+}_{sol} \tag{4.43}$$

Another classical example of electrocatalysis by ion-permeable solids is the catalysis of H_2O_2 reduction at metal hexacyanoferrates. The standard potential of the H_2O_2/H_2O couple is 1.56 V, higher than those of the O_2/H_2O_2 (1.01 V) and O_2/H_2O (0.45) couples, but it is kinetically hindered, so electrocatalysis is needed to reduce H_2O_2. Figure 4.11 compares the CVs at unmodified GCE and Prussian blue-modified GCE, forming two different composites with multi-walled carbon nanotubes (MWCNTs) and chitosan-reduced graphene oxide (rGO) [28]. The signal at ca. 0.15 V corresponds to the electrochemical reduction of Prussian blue ($Fe_4[Fe(CN)_6]_3$ or $FeK[Fe(CN)_6]$, depending on the preparation procedure), a process that can be formulated as [29,30] (see also Chapter 3):

$$\{KFe[Fe(CN)_6]\}_{solid} + K^{+}_{aq} + e^{-} \rightarrow \{K_2Fe\{Fe(CN)_6\}\}_{solid} \tag{4.44}$$

The electrocatalytic process presumably occurs via dissociative adsorption of H_2O_2,

$$H_2O_2 \rightarrow 2{\cdot}OH_{ads} \tag{4.45}$$

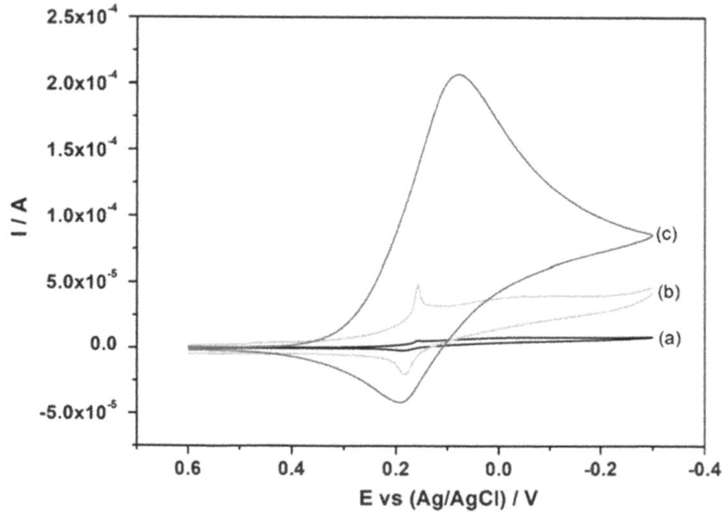

FIGURE 4.11 Voltammograms of GCE Unmodified (Black Line) and Modified with Prussian Blue Plus MWCNTs (Red Line), and Prussian Blue Plus Chitosan/Reduced Graphene Oxide (Blue Line) Composites in 5 mM H_2O_2 Solution in 0.1 M KCl at pH 2.7. Potential Scan Rate 50 mV s^{-1}. Reproduced from Ref. [28] (Yang et al. *Electrochim. Acta.* 2012, 81: 37–43), with Permission.

occurring, according to the theoretical scheme revised in Chapter 3, in the vicinity of the particle/base electrode/electrolyte three-phase boundary. This process will be followed by reaction with Fe (II) centers of Prussian blue:

$$\{FeK_2\{Fe(CN)_6\}\}_{solid} + \cdot OH_{ads} \rightarrow \{FeK[Fe(CN)_6]\}_{solid} + K^+_{aq} + OH^-_{aq} \quad (4.46)$$

Depending on the size of electrolyte ions and substrate molecules with respect to the size of the pores and channels of the solid, one can consider that (i) both substrate molecules and electrolyte counterions can move across the solid, and the catalytic process can occur in the entire bulk of the (porous) solid; (ii) only electrolyte counterions can diffuse within the solid or only substrate molecules can move across the solid; or (iii) neither substrate molecules nor electrolyte ions can diffuse in the solid—here, the catalytic process is confined to the surface of the solid particles.

Accordingly, there is a variety of possible catalytic situations. The inverse to that can also be conceived (i.e. that the substrate is immobilized, and the catalyst dissolved). Now, the kinetically constrained electrochemical reduction of immobilized species $\{S^{ox}\}$ can be represented as:

$$\{S^{ox}\}_{solid} + nM^+_{sol} + ne^- \rightarrow \{S^{red\,n-} \cdots nM^+\}_{solid} \quad (4.47)$$

Then, the catalytic process is initiated by the reduction process (charges omitted for simplicity)

$$Cat^{ox}_{sol} + ne^- \rightarrow Cat^{rd}_{sol} \quad (4.48)$$

followed by the catalytic reaction:

$$\{S^{ox}\}_{solid} + Cat^{red}_{sol} + nM^+_{sol} \rightarrow Cat^{ox}_{sol} + \{S^{red\,n-} \cdots nM^+\}_{solid} \quad (4.49)$$

Another conceivable possibility is that the electrocatalysis can proceed via adduct formation between immobilized sites and species in solution and subsequent solid-state reduction of that adduct species. This situation can be represented as:

$$S^{ox}_{sol} + \{Cat^{red\,n-} \cdots nM^+\}_{solid} \rightarrow \{S^{ox} \cdots Cat^{red\,n-} \cdots nM^+\}_{solid} \quad (4.50)$$

$$\{S^{ox} \cdots Cat^{red\,n-} \cdots nM^+\}_{solid} \rightarrow \{S^{red} \cdots Cat^{ox} \cdots nM^+\}_{solid} \quad (4.51)$$

$$\{S^{red} \cdots Cat^{ox} \cdots nM^+\}_{solid} \rightarrow S^{red}_{sol} + \{Cat^{ox}\}_{solid} + nM^+_{sol} \quad (4.52)$$

Here, the catalyst is electrochemically regenerated by means of a process similar to that described by Eq. (4.48). The possibility of different electrocatalytic pathways can be illustrated by the oxidation of 1,4-dihydro benzoquinone, H_2Q, in aqueous media at nicotinamide adenine nucleotide coenzyme (NADH) encapsulated within SBA-15 and MCM-14 mesoporous aluminosilicates, respectively, NADH@MCM and NADH@SBA [31]. Here, a molecule with biocatalytic activity, NADH, that displays a rich electrochemistry [32], was attached to two different mesoporous aluminosilicates. These aluminosilicates have a common ordered structure based on hexagonal channels with a complementary set of disordered micropores that provide connectivity between the channels through the silica, although differing in their pore size and aluminum content (Si/Al molar ratio equal to 13 for MCM-41 and ∞ for SBA-15). Figure 4.12 compares the CV response of a 2.5 mM solution of H_2Q in phosphate buffer (pH 7.0) at unmodified and NADH@MCM- and NADH@SBA-modified glassy carbon electrodes.

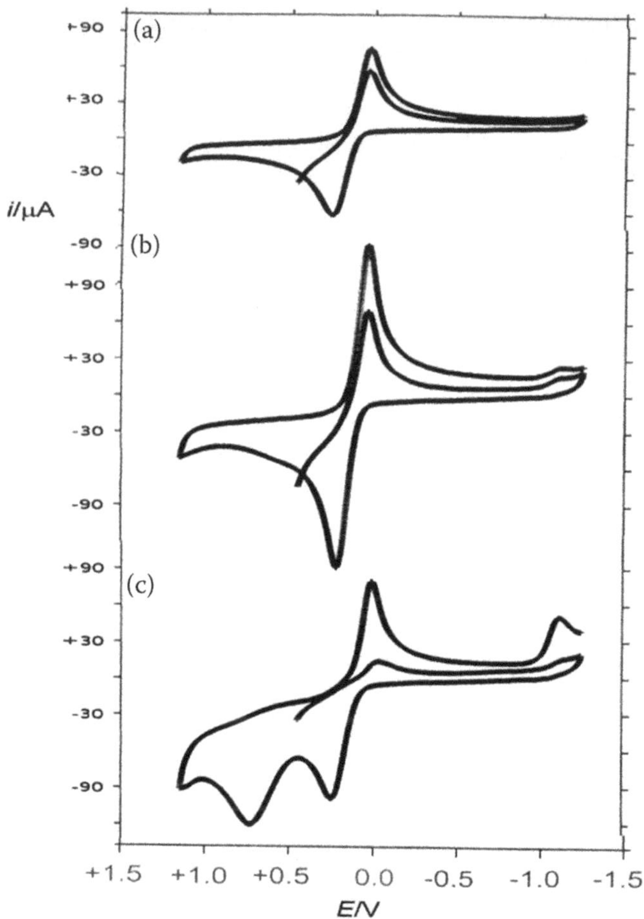

FIGURE 4.12 CVs at a 2.5 mM H_2Q Solution in 0.50 M Phosphate Buffer, pH = 7 at (a) Unmodified Glassy Carbon Electrode; (b) NADH@MCM-, and (c) NADH@SBA-Modified Electrodes. Potential Scan Rate 50 mV s^{-1}. Reproduced from Ref. [31] (Doménech-Carbó et al. *Electrochim. Acta.* 2006, 51: 4897–4908), with Permission.

At unmodified electrodes, the hydroquinone/quinone redox couple exhibits well-defined peaks at potentials of 0.22 (anodic) and 0.04 V (cathodic), as depicted in Figure 4.12a. This response corresponds to a relatively complicated mechanism involving proton-assisted electron transfer processes also mediated by solution electron transfer reactions [33,34]. In contact with phosphate buffer, NADH@MCM- and NADH@SBA-modified electrodes display a weak, apparently irreversible oxidation peak at ca. 0.74 V. This process corresponds to the electrochemical oxidation of NADH to β-nicotinamide adenine dinucleotide, NAD^+, which—following literature [35]—is electrochemically reduced to NADH at potentials ca. −1.0 (small cathodic peak in Figure 4.12b).

At NADH@MCM-modified electrodes, the voltammetric profile remained essentially unchanged, while H_2Q-centered peak currents were significantly enhanced (Figure 4.12b). At NADH@SBA, however, the response of H_2Q solutions consist of two anodic peaks at 0.22 V and 0.74 V (Figure 4.12c), both being significantly increased with respect to peak recorded for NADH@SBA in the absence of H_2Q and in H_2Q solutions at unmodified electrodes. This response can be rationalized assuming that two different electrocatalytic pathways are operative. In the case of NADH@MCM, a reaction/regeneration mechanism can be proposed. The catalytic process is initiated by the electrochemical oxidation of hydroquinone to quinone, Q, in solution phase:

$$H_2Q_{aq} \rightarrow Q_{aq} + 2H^+_{aq} + 2e^- \tag{4.53}$$

The resulting quinone diffuses across the mesoporous system and reacts with the NADH catalyst. As a result, the direct reduction of the immobilized NADH substrate does not proceed electrochemically, but only via heterogeneous redox reaction with the oxidized form of the hydroquinone catalyst in solution, possibly,

$$\{NADH \cdots Na^+\}_{solid} + Q_{aq} \rightarrow \{NAD^+\}_{solid} + HQ^-_{aq} + Na^+_{aq} \tag{4.54}$$

$$HQ^-_{aq} + H^+_{aq} \rightarrow H_2Q_{aq} \tag{4.55}$$

Due to the anionic nature of NADH (β-nicotinamide adenine dinucleotide reduced disodium salt), its attachment requires the parallel attachment of sodium ions so that the materials can be represented as $\{NADH \cdots Na^+\}$. The hydroquinone catalyst is regenerated, completing the catalytic cycle resulting in significant current enhancement.

For NADH@SBA, a different mechanism can be proposed, assuming that H_2Q forms a surface-confined adduct (denoted in the following equations by the subscript s) with the substrate, following the reaction scheme summarized by Eq. (4.41). As a result, two successive substrate- and catalyst-centered electron transfer processes are observed:

$$\{NADH \cdots Na^+\}_{solid} + H_2Q_{aq} \rightarrow \{H_2Q - NADH \cdots Na^+\}_s \tag{4.56}$$

$$\{H_2Q - NADH \cdots Na^+\}_s \rightarrow \{Q - NADH \cdots Na^+\}_s + 2H^+_{aq} + 2e^- \tag{4.57}$$

$$\{Q - NADH \cdots Na^+\}_s \rightarrow \{Q - NAD^+\}_{solid} + Na^+_{aq} + H^+_{aq} + 2e^- \tag{4.58}$$

Charge conservation is ensured by the release of protons and/or sodium ions from the mesoporous system to the electrolyte solution. This mechanism is consistent with the observation of two peaks (see Figure 4.12c) at the potentials observed for the non-catalyzed oxidation of H_2Q and NADH in solution. Consistently, chronoamperometric data can be approached to those theoretically predicted for an electron transfer process preceded by a chemical reaction in the solution phase, as demanded by a process where adduct formation acts as a rate-controlling step.

The remarkable differences between the electrocatalytic responses observed for NADH@MCM and NADH@SBA can be attributed to the existence of at least two competing pathways represented, respectively, by Eqs. (4.45–4.47) (reaction/regeneration scheme) and Eqs. (4.48–4.50) (adduct formation), where H_2Q acts as a catalyst in solution and NADH as an immobilized substrate. Hydroquinone-NADH adduct formation is considerably slower at NADH@SBA than at NADH@MCM, a feature that would be associated with the different acidity/polarity of the lattices because MCM-41 contains both Bronsted (Al(OH)Si groups) and Lewis (extra framework Al^{3+} centers), whereas SBA-15 is a pure silica material; their pore sizes are also different (3.2 nm for MCM-41 and 6.0 nm for SBA-15). Considering that adduct formation requires a pre-organization of NADH units to adopt the most favorable conformation lying with H_2Q, larger cavities should favor conformational pre-organizing and, consequently, adduct formation—in agreement with the aforementioned observations—but a significant influence of polarity/acidity effects cannot be discarded.

4.6 OVERVIEW

There is a variety of applications of electrocatalysis in electrosynthesis, electrochemical sensing, and energy production. A significant part of research in electrocatalysis is focused on hydrogen evolution reaction (HER) due to the environmental-friendly character of electrochemical hydrogen production. Because the most efficient electrocatalysts are expensive and are terrestrial scarce precious metals (Pt, Ru), the development of alternative efficient catalysts receives continuous interest [36]. This extends to new catalysts for the oxygen reduction reaction and (ORR) and oxygen evolution reaction (OER), necessarily coupled with HER that display a sluggish kinetics at Pt, Ru, etc. electrodes [37].

Frequently, the catalyst is prepared as a layer coated on the surface of a base electrode, forming a thin (or an ultra-thin) film. For electrocatalytic applications, the thin-film electrochemical process can be influenced by (a) the density (or total number) of catalytically active sites, (b) the charge-transport rate between the active sites and the base electrode, (c) the diffusion rate of the active sites within the film, and (d) the diffusion rate of charge-balancing ions within the thin film [38]. Obviously, porous solids are of particular interest, having a high specific surface area, porosity, and high density of catalytic centers [39]. The isolation of catalytic centers within porous solids facilitates the substrate to adopt favorable conformations. Another interesting field is developing around the electrochemical promotion of catalytic reactions by solid electrolytes interfaced with noble metal electrodes [40,41].

Porous solids can be used not only as electrocatalytically active materials but also as supports of electrocatalytic centers. Additionally, a variety of composite materials can be prepared by combining porous materials with secondary supports like conductive polymers, carbon nanotubes (CNTs); graphene (and graphene oxide and reduced graphene oxide), in most cases, also have catalytic ability. Typical catalyst units are metal nanoparticles, quantum dots, and a variety of nanoarchitecture of metal oxides and metal sulfides, among others. These strategies lead to prepare bifunctional or multifunctional catalysts that we will see different examples of in succeeding chapters.

REFERENCES

[1]. Bard, A.J.; Inzelt, G.; Scholz, F. Eds. 2008. *Electrochemical Dictionary*. Springer, Berlin-Heidelberg.

[2]. Nicholson, R.S.; Shain, I. 1964. Theory of Stationary Electrode Polarography. *Analytical Chemistry*. 36: 706–723.

[3]. Molina, A.; González, J.; Laborda, E.; Wang, Y.; Compton, R.G. 2011. Catalytic mechanism in cyclic voltammetry at disc electrodes: An analytical solution. *Physical Chemistry Chemical Physics*. 13: 14694–14704.

[4]. Murray, R.W. 2008. Nanoelectrochemistry: Metal nanoparticles, nanoelectrodes, and nanopores. *Chemical Reviews*. 108: 2688–2720.

[5]. Bard, A.J.; Faulkner, L.R. 2001. *Electrochemical Methods,* 2nd edit. John Wiley & Sons, New York.

[6]. Andrieux, C.P.; Savéant, J.-M. 1982. Kinetics of electrochemical reactions mediated by redox polymer films: Reversible ion-exchange reactions. *Journal of Electroanalytical Chemistry*. 142: 1–30.

[7]. Huang, J.; Du, C.; Nie, J.; Zhou, H.; Zhang, X.; Chen, J. 2019. Encapsulated Rh nanoparticles in N-doped porous carbon polyhedrons derived from ZIF-8 for efficient HER and ORR electrocatalysis. *Electrochimica Acta*. 326: artic. 134982.

[8]. Sayadi, A.; Pickup, P.G. 2016. Evaluation of methanol catalysts by rotating disc voltammetry. *Electrochimica Acta*. 199: 12–17.

[9]. Song, J.; Li, G.; Qiao, J. 2015. Ultrafine porous carbon fiber and its supported platinum catalyst for enhancing performance of proton exchange membrane fuel cells. *Electrochimica Acta*. 177: 174–180.

[10]. van der Niet, M.J.T.C.; Garcia-Araez, N.; Hernández, J.; Feliu, J.M.; Koper, M.T.M. 2013. Water dissociation on well-defined platinum surfaces: The electrochemical perspective . *Catalysis Today*. 202: 105–113.

[11]. Chen, X.; McCrum, I.T.; Schwarz, K.A.; Janik, M.J.; Koper, M.T.M. 2017. Co-adsorption of cations as the cause of the apparent pH dependence of hydrogen adsorption on a stepped platinum single-crystal electrode . *Angewandte Chemie International Edition*. 56: 15025–15029.

[12]. Ramadoss, M.; Chen, Y.; Hu, Y.; Yang, D. 2019. Three-dimensional porous nanoarchitecture constructed by ultrathin $NiCoBO_x$ nanosheets as a highly efficient and durable electrocatalyst for oxygen evolution reaction. *Electrochimica Acta*. 321: artic. 134666.

[13]. Bandarenka, A.S.; Koper, M.T.M. 2013. Structural and electronic effects in heterogeneous electrocatalysis: Toward a rational design of electrocatalysts. *Journal of Catalysis*. 308: 11–24.
[14]. Taylor, H.S. 1925. A theory of the catalytic surface. *Proceedings of the Royal Society of London*. 108A: 105–111.
[15]. Vidal-Iglesias, F.J.; Solla-Gullon, J.; Montiel, V.; Feliu, J.M.; Aldaz, A. 2005. Ammonia selective oxidation on Pt(100) sites in alkaline medium . *Journal of Physical Chemistry B*. 109: 12914–12919.
[16]. Koper, M.T.M. 2011. Structure sensitivity and nanoscale effects in electrocatalysis. *Nanoscale*. 3: 2054–2073.
[17]. Tian, N.; Zhou, Z.-Y.; Sun, S.-G. 2008. Platinum metal catalysts of high-index surfaces: From single-crystal planes to electrochemically shape-controlled nanoparticles. *Journal of Physical Chemistry C*. 112: 19801–19817.
[18]. Duca, M.; Figueiredo, M.; Climent, V.; Rodriguez, P.; Feliu, J.M.; Koper, M.T.M. 2011. *Journal of the American Chemical Society*. 133: 10928–10939.
[19]. Kuzume, A.; Herrero, E.; Feliu, J.M. 2007. Oxygen reduction on stepped platinum surfaces in acídica media. *Journal of Electroanalytical Chemistry*. 599: 333–343.
[20]. Sabatier, P. 1911. Hydrogenations et deshydrogenations par catalyse. *Berichte der Deutschen Chemischen Gesellschaft*. 44: 1984–2001.
[21]. Trasatti, S. 1972. Work function, electronegativity, and electrochemical behaviour of metals: III. Electrolytic hydrogen evolution in acid solutions. *Journal of Electroanalytical Chemistry*. 39: 163–184.
[22]. Nørskov, J.K.; Rossmeisl, J.; Logadottir, A.; Lindqvist, L.; Kitchin, J.R.; Bligaard, T.; Jónsson, H. 2004.Origin of the overpotential for oxygen reduction at fuel cell cathode. *Journal of Physical Chemistry B*. 108: 17886–17892.
[23]. Lu, Y.-C.; Gasteiger, H.A.; Shao-Horn, Y. 2011. *Journal of the American Chemical Society*. 133: 19048–19051.
[24]. Hansen, H.A.; Man, I.C.; Studt, F.; Abild-Pedersen, F.; Bligaard, T.; Rossmeisl, J. 2010. Electrochemical chlorine evolution at rutile oxide (110) surfaces. *Physical Chemistry Chemical Physics*. 12: 283–290.
[25]. Lyons, M.E.G.; Lyons, C.H.; Michas, A.; Bartlett, P.N. 1992. Amperometric chemical sensors using microheterogeneous systems. *Analyst*. 117: 1271–1280.
[26]. Guidelli, R. 1971. Diffusion toward planar, spherical, and dropping electrodes at constant potential. *Journal of Electroanalytical Chemistry*. 33: 303–317.
[27]. Balamurugan, A.; Chen, S.-M. 2007. Silicomolybdate doped polypyrrole film modified glassy carbon electrode for electrocatalytic reduction of Cr(VI). *Journal of Solid State Electrochemistry*. 11: 1679–1687.
[28]. Yang, J.-H.; Myoung, M.; Hong, H.-G. 2012. Facile and controllable synthesis of Prussian blue on chitosan-functionalized graphene nanosheets for the electrochemical detection of hydrogen peroxide. *Electrochimica Acta*. 81: 37–43.
[29]. Dostal, A.; Meyer, B.; Scholz, F.; Schröder, U.; Bond, A.M.; Marken, F.; Shaw, S.J. 1995. Electrochemical study of microcrystalline solid prussian blue particles mechanically attached to graphite and gold electrodes: Electrochemically induced lattice reconstruction. *Journal of Physical Chemistry*. 99: 2096– 2103.
[30]. Dostal, A.; Kauschka, G.; Reddy, S.J.; Scholz, F. 1996. Lattice contractions and expansions which accompany the electrochemical conversion of Prussian blue and the reversible and irreversible insertion of rubidium and thallium ions. *Journal of Electroanalytical Chemistry*. 406: 155–163.
[31]. Doménech-Carbó, A.; García, H.; Marquet, J.; Bourdelande, J.L.; Herance, J.R. 2006. Modelling electrocatalysis of hydroquinone oxidation by nicotinamide dinucleotide coenzyme encapsulated within SBA-15 and MCM-41 mesoporous aluminosilicates. *Electrochimica Acta*. 51: 4897–4908.
[32]. Blaedel, W.J.; Jenkins, R.A. 1975. Electrochemical oxidation of reduced nicotinamide adenine dinucleotide. *Analytical Chemistry*. 47: 1337–1343.
[33]. Hillard, E.A.; Caxico de Abreu, F.; Melo Ferreira, D.C.; Jaouen, G.; Fonseca Goulart M.O.; Amatore, C. 2008. Electrochemical parameters and techniques in drug development, with an emphasis on quinones and related compounds. *Chemical Communications*. 2612–2628.
[34]. Costetin, C. 2008. Electrochemical approach to the mechanistic study of proton-coupled electron transfer. *Chemical Reviews*. 108: 2145–2179.
[35]. Moiroux, J.; Elving, P.J. 1979. Optimization of the analytical oxidation of dihydronicotinamide adenine dinucleotide at carbon and platinum electrodes. *Analytical Chemistry*. 51: 346–350.
[36]. Zheng, F.; Zhang, Z.; Zhang, C.; Chen, W. 2020. Advanced electrocatalysts based on metal–organic frameworks. *ACS Omega*. 5: 2495–2502.

[37]. Favaro, M.; Valero-Vidal, C.; Eichhom, J.; Toma, F.M.; Ross, P.N.; Yano, J.; Liu, Z.; Crumlin, E.J. 2017. Elucidating the alkaline oxygen evolution reaction mechanism on platinum. *Journal of Materials Chemistry A*. 5: 11634–11643.

[38]. Shi, M.; Anson, F.C. 1996. Rapid oxidation of $Ru(NH3)_6{}^{3+}$ by $Os(bpy)_3{}^{3+}$ within Nafion coatings on electrodes . *Langmuir*. 12: 2068–2075.

[39]. Ferey, G. 2008. Hybrid porous solids: Past, present, future. *Chemical Society Reviews*. 37: 191–214.

[40]. Toghan, A.; Rösken, L.M.; Imbihl, R. 2010. Origin of non-Faradayicity in electrochemical promotion of catalytic ethylene oxidation. *Physical Chemistry Chemical Physics*. 12: 9811–9815.

[41]. Fang, Y.; Li, X.; Li, F.; Lin, X.; Tian, M.; Long, X.; An, X.; Fu, Y.; Jin, J.; Ma, J. 2016. Self-assembly of cobalt-centered metal organic framework and multiwalled carbon nanotubes hybrids as a highly active and corrosion-resistant bifunctional oxygen catalyst. *Journal of Power Sources*. 326: 50–59.

5 Electrochemistry of Aluminosilicates

5.1 INTRODUCTION

Silicates are an important group of substances that have a skeleton based on tetrahedral SiO_4 units tailored to adopt a wide variety of structures. Their related aluminosilicates are those where a certain number of silicon atoms have been isomorphically substituted by aluminum ones. As a result, a defect of positive charge is produced in the lattice. Compensation of this excess of negative charge requires the formation of silanol groups and/or the attachment of cationic species to the aluminosilicate framework. SiO_4 and AlO_4 tetrahedral units act as a building unit in constructing a wide variety of structures, from porous (zeolites, sepiolite) to laminar silicates and aluminosilicates (montmorillonite, kaolinite). Since the preparation [1] of periodic mesoporous silica (MCM-41), a variety of silica-based mesoporous materials have been synthesized with a variable degree of order, stability, and Al/Si ratios. An important group of materials is constituted by different types of porous silica [2].

The ionic nature of the cationic attachment allows an easy ion exchange in such materials but, at the same time, rigid porous structures exert a remarkable size- and eventual charge-selectivity in incorporating guest ions. It should be noted, however, that adsorption of guest species onto the external surface of the crystals of such materials can occur in general. Laminar silicates can adsorb guest species in the interlaminar spaces, acting as flexible hosts with relatively size-selectivity. Porous silicates such as zeolites possess a system of pores, channels and cavities that provide opportunity for trapping different species. The dynamic size of species that can enter through the zeolite pores is given by their kinetic diameter, taken as the intermolecular distance of the closest approach for two molecules colliding with zero initial kinetic energy [3].

The first report using zeolites for electrochemical purposes, dated in 1939, was described by Marshall and was followed by works of Barrer and James in 1960 [4]. The concept of clay-modified electrodes was first developed by Gosh and Bard [5] and zeolite electrochemistry was intensively studied in the late 1980s and 1990s, with most research focused on the controversy about the intra- vs. extra-zeolitic nature of electrochemical processes [6]. Since then, research on zeolite (and micro and mesoporous silicate and aluminosilicate materials) electrochemistry has diverged into different application fields, mainly electrocatalysis and environmental remediation, and sensing (potentiometric, amperometric, and voltammetric), but also on fuel cells, batteries and capacitors, and water splitting [7].

5.2 ELECTROCHEMISTRY OF ZEOLITE-ASSOCIATED SPECIES

Zeolites are microporous crystalline aluminosilicates characterized by its molecular sieving performance, cation-exchange capacity, and catalytic ability. The chemical composition of zeolites can be represented by the general formula $M_{x/n}{}^{n+}(Al_xSi_yO_{2(x+y)})^{x-}\cdot zH_2O$, M being cationic species—generally alkaline or alkaline-earth metals, protons, ammonium ion, etc. The Al/Si molar ratio can vary from 1 (sodalite) to 0 (silicalites). As `previously noted, the isomorphic substitution of silicon atoms by aluminum ones produces a negative charge in the lattice that must be neutralized by such cationic species.

Remarkably, zeolites can be completely hydrated and dehydrated without damage to the crystalline network. The structure of zeolites is formally constructed by conjoining SiO_4 and AlO_4

tetrahedral building blocks so that a variety of ringed structures, generically labeled as Secondary Building Units (SBUs), can be formed. Figure 5.1 shows the structures of zeolites A and X, Y (faujasite). As previously noted, zeolites can restrict the size and shape of the molecules that enter, reside within, or exit the lattice. These structures of such zeolites comprise pore opening, cage, and channel structures; for instance, the pore structure of zeolite Y consists of almost spherical 13 Å cavities interconnected tetrahedrally through smaller apertures of 7.4 Å. Tables 5.1 and 5.2 show the characteristics of several types of zeolites used in electrochemical studies. Zeolites possess high porosity, with large surface areas, typically ranging from 400 to 650 m^2g^{-1} and pore volumes up to 0.1 cm^3g^{-1}.

For every aluminum atom in the lattice, a fixed negative charge results. This negative charge is counterbalanced by mobile cations, typically alkaline or alkaline-earth cations. In contact with aqueous solutions, these mobile cations can easily be exchanged for other cations (metal cations, cationic complexes) of appropriate dimensions, thus determining the remarkable ion-exchange properties of zeolites.

Zeolites are traditionally regarded as 'acid' materials. There are two types of acidic centers in zeolites. First, protons held in the zeolite forming bridging hydroxyl groups Si(OH)Al (Bronsted centers). A direct way to form such centers is to interchange alkaline ions (e.g., Na^+) by NH_4^+ ions and further calcinations of the material. As a result, NH_3 is released while Si(OH)Al groups are formed. Additionally, charge-balancing cations and extra-lattice Al^{3+} ions act as Lewis acidic centers.

The density of Al atoms in the pores directs the polarity of internal spaces, thus determining the hydrophobicity/hydrophilicity of the material. In general, the hydrophobic character of the zeolites

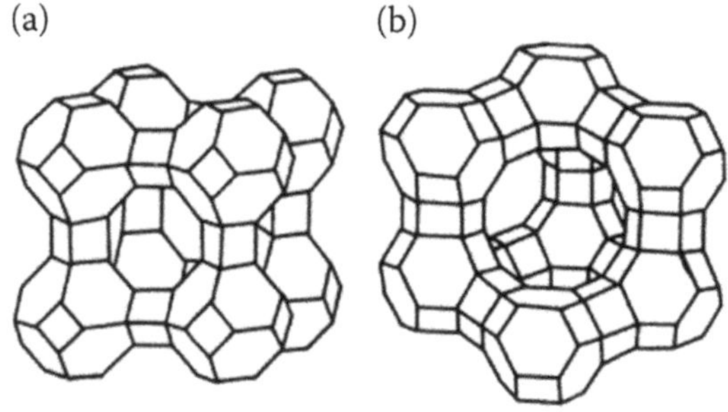

FIGURE 5.1 Three-Dimensional Representation of the Framework Structure of (a) Zeolite A, and, (b) X and Y.

TABLE 5.1
Characteristics of Several Zeolites

Type of pore	Pore size (Å)	Number of tetrahedral units	Channel directions	Examples
Middle	4.8 × 3.5 [010]	8	2	Ferrierite
	5.4 × 4.2 [001]	10		
Middle	5.6 × 5.3 [010]	10	2	ZSM-5
	5.1 × 5.5 [100]	10		
Large	6.5 × 7.0 [001]	12	1	Mordenite
	5.7 × 5.6 [001]	10		
Large	7.6 × 6.4 [100]	12	3	β-zeolite
	5.6 × 5.6 [001]	12		
Large	7.4 × 7.4 [111]	12	3	Y zeolite
Extra-large	8.2 × 8.1 [010]	14	1	UTD-1

TABLE 5.2
Physical and Chemical Properties of Zeolites

Zeolite type	Unit cell composition[a]	Si/Al molar ratio	Unit cell volume (nm^3)	Void volume ($cm^3\ cm^{-3}$)	Pore Openings (Å)	Supercage diameter (Å)	Kinetic diameter (nm)	t_c^c (°C)
A	$Na_{12}[(AlO_2)_{12}(SiO_2)_{12}]\cdot 27H_2O$	0.7–1.2	1.870	0.47	4.1	11.4	0.39	660
X	$Na_{86}[(AlO_2)_{86}(SiO_2)_{106}]\cdot 264H_2O$	1–1.5	15.36–15.67	0.50	7.4	13	0.81	660
Y	$Na_{56}[(AlO_2)_{56}(SiO_2)_{136}]\cdot 250H_2O$	>1.5–3	14.90–15.35	0.48	7.4	13	0.81	700
L	$K_9[(AlO_2)_9(SiO_2)_{27}]\cdot 22H_2O$	2.6–3.5	2.205	0.32	7.1		0.81	845
mord.	$Na_8[(AlO_2)_8(SiO_2)_{40}]\cdot 24H_2O$	4.7–5	2.794	0.28	6.5–7.0		0.39	

[a] Typical composition for the fully hydrated zeolite.
[b] Void volume/total volume.
Adapted from Refs. [3] (Breck, 1974) and [6] (Rolison and Bessel, 2000).

increases by decreasing the aluminum content. The coordinative environment of such aluminum centers influences notably their acidic properties, so additional Al^{3+} centers alters the charge density of the original acid site. Increasing the number of aluminum atoms in the second coordination sphere of a given acid site decreases the acidity of the material.

Interestingly, zeolites can also present basic character, associated with oxygen atoms in the lattice having a large density of negative charge [8]. The alkalinity of zeolites has been related to the aluminum content and the nature of charge-balancing cations. By the first token, alkalinity is promoted by high aluminum contents; by the second, highly polarizable cations increase the alkalinity of the zeolite with respect to less polarizable ones.

The capacity of zeolites to trap cationic species prompted the study of electroactive guest components attached to zeolite hosts. Shaw et al. [9] proposed three basic pathways to explain the electrochemistry of zeolites containing an electroactive species A in contact with a suitable electrolyte containing M^+ cations, able to enter the zeolite pore/channel system. Taking for simplicity the reduction of a monovalent cationic species A^+, these electrochemical pathways can be summarized as [4,10,11]:

a. Extrazeolitic mechanism, where the electroactive species is ion-exchanged being subsequently reduced/oxidized in the solution phase:

$$\{A^+\}_{zeolite} + M^+_{sol} \rightarrow A^+_{sol} + \{M^+\}_{zeolite} \tag{5.1}$$

$$A^+_{sol} + e^- \rightarrow A_{sol} \tag{5.2}$$

As in previous chapters, in these equations, sol denotes solvated species in an electrolyte solution and { } species attached to the zeolite framework.

b. The intrazeolitic mechanism involves the reduction of intrazeolite species coupled with the ingress of charge balancing electrolyte cations into the zeolite framework:

$$\{A^+\}_{zeolite} + M^+_{sol} + e^- \rightarrow \{A \cdots M^+\}_{zeolite} \tag{5.3}$$

c. In a boundary-mediated mechanism, the electron transfer process is confined to the external region of the zeolite crystals and propagates via electron-exchange reactions with oxidized species in the zeolite bulk:

$$\{A^+\}_{zeol,surf} + M^+_{solv} + e^- \rightarrow \{A \cdots M^+\}_{zeol,surf} \tag{5.4}$$

$$\{A \cdots M^+\}_{zeol,surf} + \{A^+\}_{zeolite} \rightarrow \{A^+\}_{zeol,surf} + \{A \cdots M^+\}_{zeolite} \tag{5.5}$$

This pathway can be combined with the presence of a redox mediator (for simplicity, taken as a monovalent cationic species B^+) existing in the electrolyte solution. This situation offers two possibilities: the first one is the external reduction of the mediator and its subsequent oxidation by the boundary-associated A^+ species:

$$B^+_{sol} + e^- \rightarrow B_{sol} \tag{5.6}$$

$$\{A^+\}_{zeol,surf} + B_{sol} + M^+_{sol} \rightarrow \{A \cdots M^+\}_{zeol,surf} + B^+_{sol} \tag{5.7}$$

The second possibility consists of the reduction of the redox mediator that is previously attached to the zeolite external region:

$$\{A^{+}\}_{zeolite} + B^{+}_{sol} \rightarrow A_{sol} + \{B^{+}\}_{zeol,surf} \tag{5.8}$$

$$\{B^{+}\}_{zeol,surf} + M^{+}_{sol} + e^{-} \rightarrow \{B \cdots M^{+}\}_{zeol,surf} \tag{5.9}$$

$$\{B \cdots M^{+}\}_{zeol,surf} + \{A^{+}\}_{zeolite} \rightarrow \{B^{+}\}_{zeol,surf} + \{A \cdots M^{+}\}_{zeolite} \tag{5.10}$$

The electrochemistry of zeolite-associated species was widely discussed in the 1990s, the debate being focused on the extrazeolite/intrazeolite character of electrochemical processes occurring at zeolite-modified electrodes (see additional literature). The more relevant empirical data can be summarized as: (a) voltammetric and coulometric data indicate that, under ordinary experimental conditions, only a small percentage (lower than 1–2%) of guest molecules is electroactive; (b) voltammetric responses displayed by zeolite probes containing electroactive guests in contact with size-excluded electrolyte countercations such as Hex_4N^+ and Bu_4N^+ are clearly lower than those of the same materials in contact with less size-excluded countercations like Et_4N^+ or Li^+; (c) in the presence of size-excluded electrolyte countercations, the voltammetric signals are rapidly exhausted under repetitive cycling of the potential, while in the presence of less size-excluded counteractions, voltammetric signals decay slowly under repetitive voltammetry; (d) voltammetry of zeolite-associated electroactive species in contact with size-excluded electrolyte counteractions produces single signals, while in the presence of less size-excluded cations, peak splitting frequently appears; and (e) electrocatalytic effects eventually exerted by zeolite-attached species on selected processes also become significantly dependent on the size of electrolyte cations [11].

Bessel and Rolison [12,13] concluded that the voltammetric response observed for zeolite-associated species reflects the behavior of extrazeolite ion-exchanged molecules and/or those located in the more external 'boundary' layer of the zeolite and proposed that different discernible responses can be attributed to electroactive species located in different sites in the zeolite matrix—the so-called topological redox isomers. Data on the electrochemistry of zeolite-associated transition metal complexes [14] and reactive organic intermediates [15] provided results in agreement with such ideas.

In this context, the electrochemistry of bulky species entrapped into the zeolite cages via 'ship-in-a-bottle' synthesis is particularly significant. From the seminal work of Lunsford et al. in the early 80s [16], 'ship-in-a-bottle' synthesis of metal complexes in the zeolite supercages, encapsulation of catalytically, optically, and/or electrochemically active species within micro- and mesoporous aluminosilicates has received considerable attention. Site isolation of individual guest molecules combined with shape and size restrictions imposed by the supercage steric limitations makes it possible to perform catalytic reactions combining the advantages of homogeneous and heterogeneous catalytic systems [17]. For instance, oxidation of water to O_2 by zeolite Y-encapsulated tris(2,2′-bipyridine)ruthenium(III), $Ru(bpy)_3^{3+}$, precludes multi-metal centered degradation reactions, resulting in ligand degradation and CO_2 evolution, typically observed in solution [18].

Figure 5.2 shows a representative example, corresponding to the voltammetry of triphenylamine (Ph_3N) encapsulated inside faujasite (KX) supercages [19]. In a MeCN solution, Ph_3N experiences a one-electron oxidation to the corresponding radical cation that dimerizes, so that the initial anodic signal is coupled with two overlapping cathodic peaks in the reverse scan associated with the reduction of the radical cation and the dimer. The initial voltammetric scans recorded at Ph_3N@KX–modified electrodes showed similar features but evolved after several cycles of potential to a stationary response formed by two almost reversible couples. These features can be

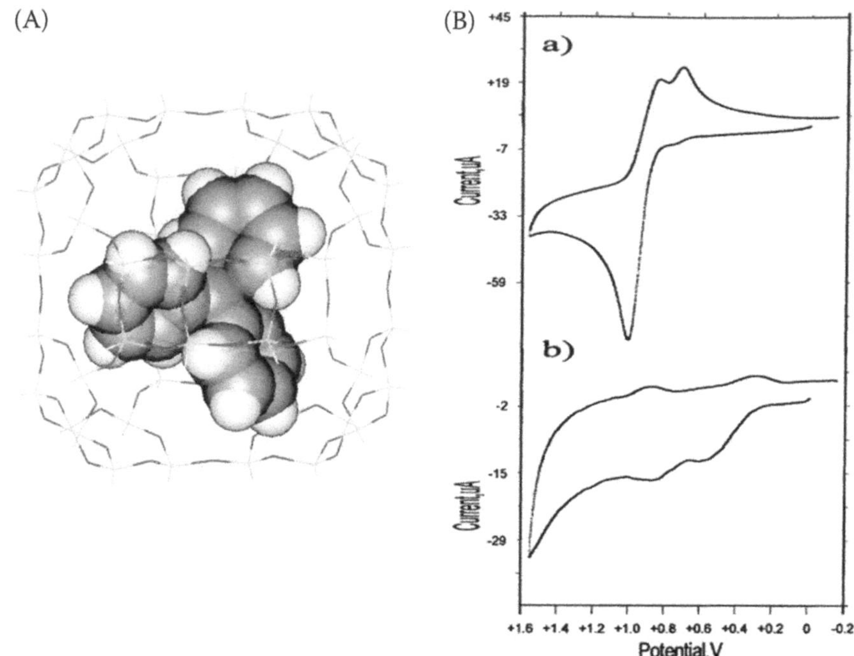

FIGURE 5.2 (A) Image of the Best Docking of Ph_3N Encapsulated Inside the Supercage of an Idealized All-Silica Faujasite (KX), and (B) CV of Ph_3N in 0.10 M $LiClO_4$ MeCN Solution; 1st (a) and 11th Cycle (b) Voltammograms Recorded at a Zeolite-Modified Electrode Containing $Ph_3N@PdCl_2$–KX in Contact with 0.10 M $LiClO_4$ MeCN Electrolyte. Reproduced from Ref. [19] (Doménech-Carbó et al. *Tetrahedron.* 2005, 61: 791–796), with Permission.

interpreted assuming that the initial voltammetric scans detect the electrochemical activity of external Ph_3N units. When these external molecules are exhausted, the voltammograms reflect the presence of zeolite-associated Ph_3N units presumably occupying different sites in the host framework.

Interestingly, anionic species can also be entrapped into zeolites, although studied cases are currently limited to a few inorganic [3,20] and organic [21] anions. The reason for this paucity is that zeolite frameworks produce a significant energy barrier for the ingress of anions due to the Coulostatic repulsion exerted by the zeolite framework. Since no mobile negatively charged species exists in zeolites, anchorage of anionic species requires the coupled ingress of cations to maintain charge neutrality in the solid. Accordingly, the encapsulation of anionic species into zeolites occurs as ionic pair ingress. The electrochemistry of Meisenheimer anions attached to zeolites can be described in terms of oxidation steps involving the issue of metal cations and reduction steps involving the ingress of electrolyte cations into the zeolite pore/channel system [22].

5.3 TOPOLOGICAL REDOX ISOMERS

In the context of photochemical studies, Turro and García-Garibay [23] introduced the idea that the guest molecules can occupy different positions in the zeolite framework. Bessel and Rolison [12] first applied it to the electrochemistry of zeolites, coining the term topological redox isomers in designing redox-active guest species associated with different sites in inorganic support. In fact, Li and Calzaferri [24] previously interpreted the electrochemistry of Ag^+-zeolites characterized in terms of the reduction of Ag^+ ions located in different sites of the zeolite framework to different conceivable Ag° species—'internal' clusters, externally adsorbed clusters, and 'external' metallic deposits. Accordingly, four basic types of topological redox isomers can be distinguished:

A. Located outside the zeolite, with no bonding or sorptive association;
B. Boundary isomer located to the external surface of zeolite particles;
C. Adsorbed into the voids of the zeolite but able to sample the global pore lattice topology; and
D. Size-included in the interior of supercages, cubo-octahedral cage, or channel intersections of the zeolite framework.

Figure 5.3 shows a schematic of these four basic types of topological redox isomers. It is pertinent to note, however, that the boundary-associated isomer could either be (i) adsorbed or ion-exchanged (but size-excluded) to the outer surface of the zeolite, (ii) adsorbed at the zeolite exterior as a result of insufficient desorption or displacement during purification of the zeolite, (iii) occupying truncated or partially broken zeolite supercages and/or occluded in defect sites at the zeolite boundary, or (iv) encapsulated in the first layer of complete supercages at the zeolite's outermost boundary.

Electrochemical methods provide information on the coexistence of different topological redox isomers. Discrimination between them can be obtained from differences in the electrolyte size influence and electrochemical mechanism that can be observed at zeolite-modified electrodes. The attachment of bulky cationic guests, unable to migrate within the zeolite framework, is of particular interest. As previously noted, zeolites can stabilize elusive cationic organic intermediates that tightly fit into the cavities of the aluminosilicate due to intense electrostatic fields experienced inside the pores, stabilizing cationic species, and by the blocking effect on the attack of nucleophilic reagents. In fact, several organic cations and radical cations become indefinitely persistent when they are encapsulated in appropriate zeolites [25].

The presence of different topological redox isomers is consistent with the aforementioned data on the electrochemistry of triphenylamine (Ph_3N) encapsulated inside faujasite and has been discussed for 2,4,6-triphenylthiopyrylium ion associated to zeolite Y ($TPY^+@Y$) based on the different voltammetric responses in electrolytes containing size-excluded Hex_4N^+ and Bu_4N^+ cations, and Et_4N^+ and Li^+ cations able to permeate the zeolite [26].

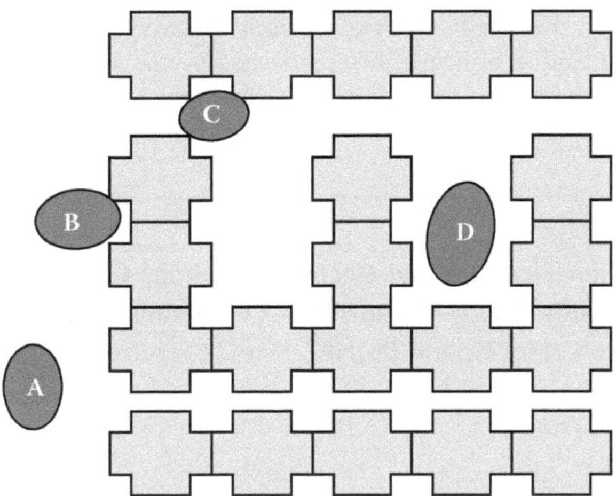

FIGURE 5.3 Schematic Illustration of the Four Basic Types of Topological Redox Isomers Existing in Zeolites. (A) Located Outside the Zeolite with No Bonding or Sorptive Association; (B) Boundary Isomer Located to the External Surface; (C) Adsorbed into the Voids of the Zeolite; (D) Size Included in the Interior of Supercages, Cubo-Octahedral Cage, or Channel Intersections of the Zeolite Framework.

It is worth noting that zeolite electrochemistry can be described based on the model described in Section 3.7. Using this, diffusion coefficients for electrons and electrolyte cations can be calculated from CA. Values for D_e and D_M are listed in Table 5.3 using data for graphite electrodes modified with Mn(salen)N_3- and 2,4,6-triphenylthiopyrylium-containing zeolite Y, immersed into Et_4NClO_4/MeCN, $LiClO_4$/MeCN and Bu_4NPF_6/MeCN electrolytes [14,16]. CA data for the first two electrolytes fit well with the theoretical model, whereas data obtained in the presence of Bu_4N^+ separate from theory, in agreement with the expected purely extrazeolitic nature of electrochemical processes in contact with that electrolyte cation. Accordingly, estimated D values contained in Table 5.3 for Bu_4N^+ represent merely the result of a forced fit of experimental data with theory.

It should be noted that the high site isolation of electroactive guest molecules into zeolite hosts suggests that electron hopping between zeolite-encapsulated species—demanded by the propagation of the redox reaction (see Chapter 3) through the material particles—should be highly hindered. Voltammetric data suggests that, although restricted to external zeolite regions, there is a place for a certain degree of redox reaction propagation across the boundary zone. Consistently, spectroscopic studies reveal that significant electronic interactions between adjacent entrapped molecules occur when assisted by small, mobile charge-balancing ions. Thus, studies on intrazeolitic photochemical charge transfer for methylviologen and $Ru(bpy)_3^{2+}$ indicate that an efficient electron transfer coupled with recombination occurs at high guest loadings [27].

5.4 SPECIES DISTRIBUTION

Modeling developed in Chapter 3 assumed that the guest electroactive molecules are uniformly distributed in the inorganic support. However, it appears reasonable to assume that molecules entrapped into the voids of microporous aluminosilicates will be not uniformly distributed in the bulk of the material. The determination of the distribution of guest molecules in the voids of the aluminosilicate host is an obvious target for elucidating the chemical and photochemical reactivity of the resulting materials [28]. In fact, for large loadings of $Ru(bpy)_3^{2+}$ in zeolite Y, the complex tends to accumulate toward the surface of the zeolite microcrystals instead of a random distribution in the aluminosilicate bulk [27], a situation schematically depicted in Figure 5.4 [29].

One can assume that the depth reached in each electrochemical experiment at porous aluminosilicate-modified can, in principle, be approached by the advance of the diffusion layer. This depth can be estimated, assuming isotropy, if the diffusion coefficient for the rate-determining charge transport is known. Then, at short times, only the electrochemical response of the guest

TABLE 5.3

Diffusion Coefficients for Electrons and Electrolyte Cations Calculated from CA Data for Graphite Electrodes Modified with Mn(salen)N_3 -Containing Zeolite Y and Immersed into Et_4NClO_4/MeCN, $LiClO_4$ /MeCN, and Bu_4NPF_6/MeCN Electrolytes (All in Conc. 0.10 M)

Electrolyte cation	Mn(salen)N_3@Y D_M ($cm^2\ s^{-1}$)	Mn(salen)N_3@Y D_e ($cm^2\ s^{-1}$)	TPY@Y D_M ($cm^2\ s^{-1}$)	TPY@Y D_e ($cm^2\ s^{-1}$)
Bu_4N^+	(1×10^{-12})	(2.3×10^{-12})	(1×10^{-12})	(1×10^{-12})
Et_4N^+	1.1×10^{-9}	3.5×10^{-11}	5.0×10^{-9}	2.0×10^{-11}
Li^+	7.8×10^{-10}	2.0×10^{-11}	4.6×10^{-9}	2.2×10^{-11}

In parenthesis uncertain values. Data taken from Refs. [14,26].

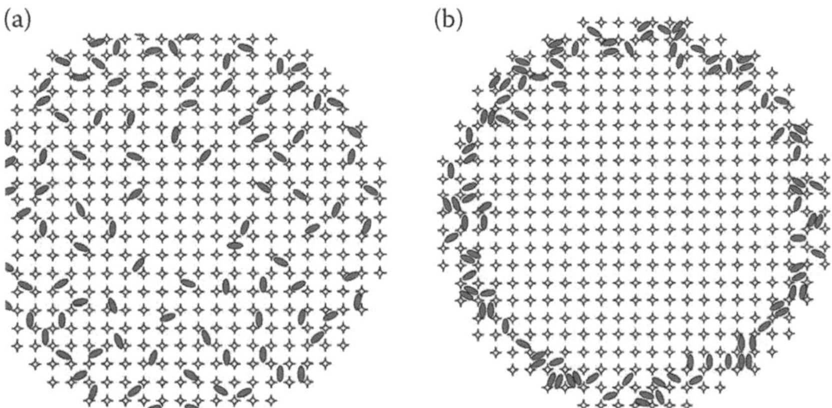

FIGURE 5.4 Schematic Representation of the Distribution of (a) Small, and (b) Bulky Guest Species into the Host Framework of a Microporous Aluminosilicate. Small Guest Can be Homogeneously Distributed Within the Host Framework While Large-Size Guests Can Only be Accommodated Near the Host Surface. Adapted from Ref. [29] (Hashimoto et al. *PhysChemChemPhys*. 2006, 8: 1451–1458.), with Permission.

molecules externally adsorbed or located in the more external layers of the crystals will be electroactive. At longer times, the observed response will reflect the contribution of molecules located at increased depth in the host crystals.

Then, the quotient between the concentrations of accessed molecules at a given depth and the total electrochemically-accessible molecules in the solid can be estimated from CVs taking a characteristic time, t_D, for the advance of the diffusion layer as a function of the potential scan rate as $t_D = \Delta E/v$, ΔE being the potential range covered in the experiment. Accordingly, the depth, x, can be estimated as $(Dt_D)^{1/2}$. Since the peak current can be taken as representative of the average concentration of electroactive species in the accessed zones of zeolite grains, the quotient between the concentration at a given depth and that of the surface can be approached by the quotient between the peak current at a given frequency and one at low frequencies (or extrapolated at frequency zero).

A slightly more elaborated approximation can be derived from Eq. (3.29), corresponding to the theoretical CA curve for the ion-insertion electrochemistry of a microparticulate deposit of a porous material containing immobile redox centers. This equation can be abbreviated as:

$$I = \frac{A}{t^{1/2}} + B - Ct^{1/2} \tag{5.11}$$

where A, B, and C are adjustable parameters depending on the crystal size and the diffusion coefficients of electrons and ions in the porous material. Rewriting this equation as $It^{1/2} = A + Bt^{1/2} - Ct$ and differentiating leads to express the slope of the $It^{1/2}$ vs. t curve as [30]:

$$\frac{\Delta(It^{1/2})}{\Delta t} = \frac{B}{2t^{1/2}} - C \tag{5.12}$$

This equation predicts that at relatively long times—assuming a uniform distribution of guest species in the porous framework—the slope of the $It^{1/2}$ vs. t curve will tend to a constant value, –C. Constructing the subsequent $\Delta(It^{1/2})/\Delta t$ vs. t curves, one can estimate the ratio between the concentrations at a depth x, $c(x)$ and the concentration in the external region, $c(0)$, as the ratio between the $\Delta(It^{1/2})/\Delta t$ values at a time t and that extrapolated at $t \to 0$.

$$\frac{c(x)}{c(0)} = \frac{\left[\frac{\Delta (It^{1/2})}{\Delta t}\right]}{\left[\frac{\Delta (It^{1/2})}{\Delta t}\right]_{t \to 0}} \tag{5.13}$$

Naturally, the concentration profiles are strongly dependent on the value adopted for the diffusion coefficient.

A more elaborated view can be obtained assuming, as before, mono-dimensional, relatively slow diffusion so that the size of the grains of porous material is large compared to the size of the diffusion layer, and that semi-infinite boundary conditions for planar diffusion apply. The model is of application to several possible situations when (a) electroactive molecules diffuse out of the pore-channel system of the solid; (b) charge-balancing counterions diffuse into the pores with fast electron hopping between immobile redox centers; (c) charge balancing ions diffuse out of the pores with fast electron hopping, and (d) slow electron shop between immobile electroactive centers accompanied by fast ion diffusion [31]. Cases (b) and (d) can be described in terms of diffusion coefficients corresponding to the processes of electron hopping between immobile redox centers (D_e) and ion transfer through the pores (D_M). In the following, it will be assumed that there is a unique rate-determining process so that the overall reaction rate can be expressed in terms of an averaged diffusion coefficient D, similar to several approaches to the electrochemistry of redox polymers.

Let us consider that the solid is constituted by a set of n laminas of thickness δ_j, so that the depth reached by the diffusion layer in a given voltammetric experiment varies between $x = 0$ and x_n. Let $c_n(x)$ the average concentration of immobile redox centers in the region between $x = 0$ and the nth layer, and let $c_{boundary}$ be the corresponding concentration at $x = 0$. Under the aforementioned semi-infinite diffusion conditions, one can assume that the ratio between the peak current and the square root of the potential scan rate, $I/v^{1/2}$, is proportional to the average concentration of redox centers in the porous solid. Then [31]:

$$\frac{I_n(x)}{v^{1/2}} = \frac{\sum_{j=1}^{n} g c_j(x)\delta_j}{\sum_{j=1}^{n}\delta_j} = \frac{\sum_{j=1}^{n} g c_j(x)\delta_j}{x_n} \tag{5.14}$$

$$\frac{I_{n+1}(x)}{v^{1/2}} = \frac{\sum_{j=1}^{n+1} g c_j(x)\delta_j}{\sum_{j=1}^{n+1}\delta_j} = \frac{\sum_{j=1}^{n+1} g c_j(x)\delta_j}{x_{n+1}} \tag{5.15}$$

In these equations, g represents an electrochemical constant depending on the experimental conditions (electrolyte, electrode area, etc.) but independent of the concentration of electroactive centers and the potential scan rate.

Experimental data for several studied systems (see Figure 5.5) indicate that the $I/v^{1/2}$ ratio tends to a limiting value at high scan rates, $(I/v^{1/2})_{lim}$, representative of the concentration of the external boundary region (i.e. $(I/v^{1/2})_{lim} = gc(0)$). Accordingly, the average concentration $c_k(x)$ of the k-layer, of thickness δ_k ($= x_{n+1} - x_n$) (i.e. located between the nth and $n + 1$)th layers) relative to the boundary concentration can be estimated as [31]

$$\frac{c_k(x)}{c(0)} = \left(\frac{I_{n+1}(x)}{I_{lim}} - \frac{I_n(x)}{I_{lim}}\right)\left(\frac{x_{n+1}}{\delta_k}\right) + \frac{I_n(x)}{I_{lim}} \tag{5.16}$$

using the experimental I values and estimating x_n and $x_n + 1$ as:

$$x = \left(\frac{2D\Delta E}{v}\right)^{1/2} \tag{5.17}$$

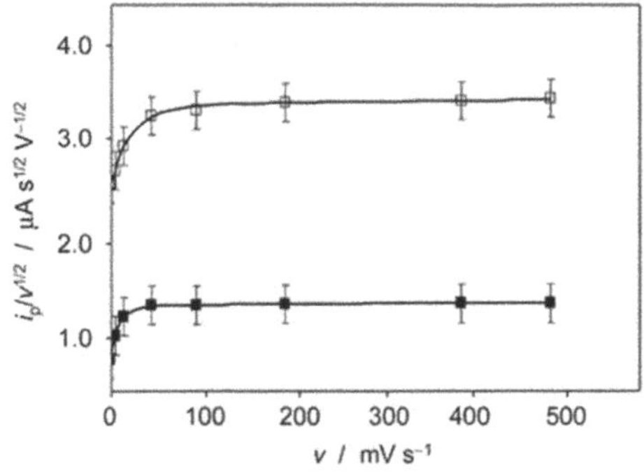

FIGURE 5.5 Variation of $I/v^{1/2}$ on v for CVs at Lapachol@aplygorskite-Modified Electrodes Immersed in 0.25 M HAc/NaAc Aqueous Solution at pH 4.75. Squares: Cathodic Peak Corresponding to the Reduction of Lapachol to Dihydrolapachol; Solid Squares: Anodic Peak Corresponding to the Oxidation of Lapachol to Dehydrolapachone. Reproduced from Ref. [31] (Doménech-Carbó et al. *PhysChemChemPhys*. 2014, 16: 19024–19034), with Permission.

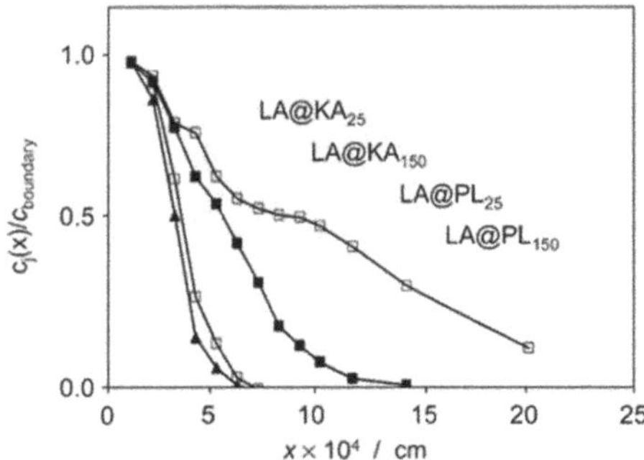

FIGURE 5.6 Concentration Profiles Calculated from CV Data, Using Eqs. (5.17) and (5.18) from Voltammetric Data in Conditions Such as in Figure. 5.7 for Unheated Lapachol@palygorskite (Solid Squares), Lapachol@palygorskite Heated at 150°C (squares), Unheated Lapachol@kaolinite (Solid Triangles) and Lapachol@kaolinite Heated at 150°C, Taking $D = 1.1 \times 10^{-11}$ cm^{2}s^{-1}. Reproduced from Ref. [31] (Doménech-Carbó et al. *PhysChemChemPhys*. 2014, 16: 19024–19034), with Permission.

In this equation, ΔE is the potential range where the signal appears in the voltammogram. Using voltammetric data at different scan rates, one can estimate the $c_k(x)/c(0)$ ratio taking a reasonable value for the diffusion coefficient. Figure 5.6 presents the concentration profiles calculated for different lapachol@clay specimens taking a value for the diffusion coefficient of 1.1×10^{-11} cm^{2}s^{-1}, estimated from averaged CA data in [30,31]. As expected, in the case of kaolinite, a laminar clay, voltammetric data indicate a fast concentration decay denoting that the organic molecules necessarily must be adsorbed onto the surface of the clay crystals. In the case of palygorskite, a channeled clay, the concentration profiles show a stepped decay suggesting a nonuniform concentration of guest species. The concentration decay is clearly faster for the unheated specimen than for the heated one above 100°C. This agrees with the recognized need to eliminate zeolitic water to prepare stable dye@palygorskite or dye@sepiolite hybrids (vide infra).

It should be emphasized, however, that the above calculations involve a simplified view of the structure and species distribution of aluminosilicates because an ideal, semi-infinite diffusion through homogeneous zeolite particles is assumed. A more accurate description should consider a more precise description of the zeolite topology where truncated cages, fractures, etc. provide an opportunity for the access of bulky electrolyte counterions to regions beyond the ideal external surface of crystals.

5.5 SPECIATION: THE MAYA BLUE PROBLEM

The study of Maya blue (MB), an ancient nanostructured organic-inorganic hybrid material, illustrates the capabilities of electrochemical methods to obtain information on microporous systems and, in particular, face speciation problems. MB is a pigment widely used in wall paintings, pottery, and sculptures (see Figure 5.7) by the ancient Mayas and other Mesoamerican people during centuries [32]. This material claimed attention by its peculiar palette, ranging from a bright

FIGURE 5.7 Image of Maya Wall Paintings in the Substructure I of Calakmul (Campeche, México), Early Classical Maya Period. Photograph from María Luisa Vázquez de Agredos Pascual. Reproduced from Ref. [32] (Doménech-Carbó et al. *Anal. Chem.* 2007, 79: 2812–2821), with Permission.

turquoise to a dark greenish-blue, its enormous stability with the attack of acids, alkalis, organic reagents, and biodegradation. The nature of MB, however, has challenged chemists for decades, and, despite intensive research, several controversial aspects remain unsolved.

Shepard (1962) first introduced the idea of MB being an unusual material consisting of a dye attached to certain Yucatán clays. Van Olphen (1967) prepared synthetic specimens with chromatic and chemical properties analogous to those of MB by crushing palygorskite—a local clay—and indigo, while Arnold (1967) identified palygorskite in Yucatán clays and localized sites for its extraction [33].

Currently, it is known that MB is a material resulting from the attachment of indigo, a blue dye extracted from leaves of *añil* or *xiuquitlitl* (*Indigofera suffruticosa* and other species) to the clay matrix of palygorskite, a fibrous phyllosilicate of ideal composition $(Mg,Al)_4Si_8(O,OH,H_2O)_{24} \cdot nH_2O$, whose structure can be described as a continuous set of layers formed by two-dimensional tetrahedral and octahedral sheets. The tetrahedral and octahedral mesh gives rise to a series of rectangular tunnels of dimensions 6.4×3.7 Å. Palygorskite crystals are crossed by zeolite-like channels and permeated by weakly bound, non-structural (zeolitic) water. Magnesium and aluminum cations complete their coordination with tightly bound water molecules (structural water).

The nature of the indigo-palygorskite association and the reasons for the hue and durability of Maya Blue have become controversial. Application of solid-state electrochemical methods, supported by visible and infrared spectroscopies and nuclear magnetic resonance data, led Doménech-Carbó et al. [34] to insert a new piece into the MB puzzle—dehydroindigo. Comparison of SWVs of indigo and Maya Blue samples (see Figure 5.8) permitted to unambiguously detect the presence of the dye associated to the palygorskite matrix in Substructure II-C in the archaeological site of Calakmul dated in the Early Classical period (440–450 BC), providing evidence on the use of this pigment 700–750 years prior the date accepted at the time.

The voltammograms in Figure 5.8, recorded for solid samples attached to paraffin-impregnated graphite electrodes in contact with aqueous acetate buffer, display peaks at 0.45 and –0.30 V vs.

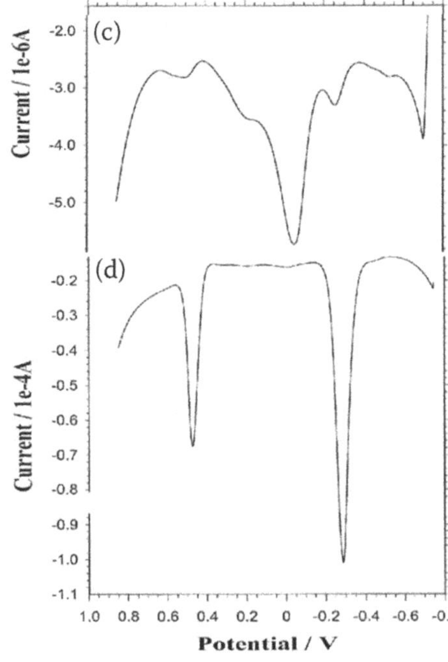

FIGURE 5.8 SWVs of (a) Indigo Microparticles, and (b) Maya Blue Sample from Calakmul, Substructure II-C, Immersed into 0.50 M Acetate Buffer, pH 4.85. Potential Scan Initiated at –750 mV in the Positive Direction. Potential Step Increment 4 mV; Square Wave Amplitude 25 mV; Frequency 5 Hz. The Sample, Dated in the Late Preclassical Period, May be the Most Ancient Sample of Maya Blue Currently Detected. Adapted from Ref. [32] (Doménech-Carbó et al. *Anal. Chem.* 2007, 79: 2812–2821) with Permission.

Ag/AgCl corresponding, respectively, to the oxidation of indigo (IND = H_2X) to dehydroindigo (DHI = X) and the reduction of indigo to leucoindigo (LND = H_4X). These processes can be represented as (see Figure 5.9):

$$\{H_2X\}_{solid} \rightarrow \{X\}_{solid} + 2H^+_{aq} + 2e^- \tag{5.18}$$

$$\{H_2X\}_{solid} + 2H^+_{aq} + 2e^- \rightarrow \{H_4X\}_{solid} \tag{5.19}$$

Notice that, in both processes, charge conservation is ensured by the ingress/issue of protons to/from the palygorskite framework.

The common presence of indigo and dehydroindigo attached to the palygorskite matrix could be responsible for the peculiar hue of MB, as clearly suggested by spectral data [32–34]. Voltammetric data permitted estimation of thermochemical data) for the association of indigo, dehydroindigo, and leucoindigo (the reduced form of indigo) to the palygorskite framework from temperature-dependent electrochemical parameters [34].

The aforementioned data are consistent with the idea that the ancient Mayas probably were able to change the hue of the pigment by crushing indigo and palygorskite and submitting the system to different thermal treatments (heating at 100–190°C). Increasing temperature should result in an increase in the dehydroindigo/indigo ratio, intensifying the greenish hue of the pigment. Further studies allowed the proposal that different procedures (evolving along the Maya times) for the preparation of MB were probably used by the ancient Mayas, eventually involving the use of additives such as ochre [32] and the appearance of minority indigoid products like isatin and indirubin [34].

Microscopical examination of palygorskite and MB samples provide support for the above ideas. Figure 5.10a,b shows Atomic Force Microscopy (AFM) images of pristine palygorskite and a genuine Maya Blue sample from the archaeological site of Mulchic from the late classical period. While palygorskite crystals show an elongated shape, 0.5–1 μm-sized having fine fiber structures 30–60 nm-spaced, Maya Blue samples contain irregularly shaped palygorskite crystals divided in

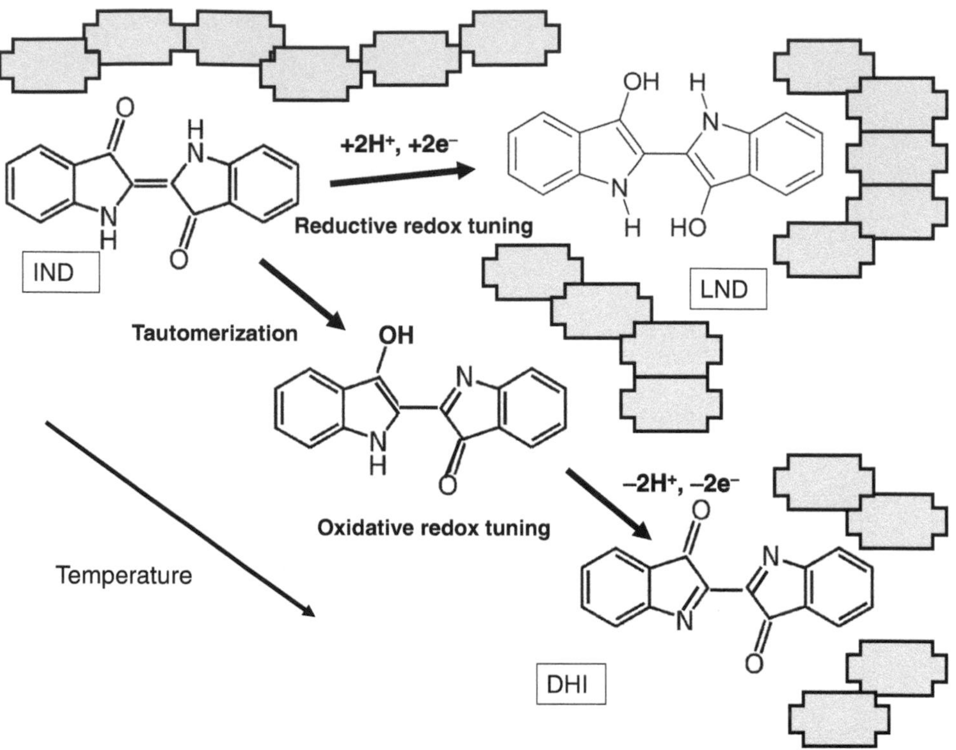

FIGURE 5.9 Structural Diagram for Possible Processes Involved in the Electrochemistry of MB. IND: Indigo, DHI: Dehydroindigo, LND: Leucoindigo.

almost square domains of rough texture. These features can be attributed to the effect of thermal treatments resulting in the loss of zeolitic water. In agreement with that idea, Transmission Electron Microscopy (TEM) offers different images of MB samples, from irregular-shaped palygorskite crystals to crystals crossed by a dense array of pores whose size ranges from 2 to even 20 nm (see Figure 5.10c,d). Since such pores can unambiguously be attributed to the segregation and evacuation of physisorbed and zeolitic water, the observed differences in their shape, size, and distribution on the surface of palygorskite crystals can be taken as indicative of the use of different preparation procedures for MB.

Interestingly, such textural differences can be correlated with the observed variation of electrochemical parameters so that a classification of Maya Blue voltammograms into different 'electrochemical types' can be made. Then, the application of multivariate chemometric techniques establishes definite relationships between samples from different archaeological sites. It was concluded that a ramified scheme applies to the evolution of MB preparation procedures, enabling the obtainment of a fine chronology of Maya history [32].

Figure 5.12 presents an example of how electrochemical techniques can aid to obtain structural information on hybrid systems. Here, topographic SECM images of synthetic MB specimens are prepared by heating indigo (1% w-w) with palygorskite and indigo (1% w-w) plus kaolinite mixtures (respectively labeled as IND@PAL_{140} and IND@KAO_{140}) at 140°C under application of different potentials to the clay substrate in contact with a $Fe(CN)_6^{4-}$ redox probe solution [35]. Kaolinite ($Al_2Si_2O_5(OH)_4$) is a layered silicate where, in contrast with palygorskite, there is no possibility of encapsulation of indigo molecules. In the SECM experiments, the tip potential was fixed at 0.35 V, positive enough to ensure the oxidation of $Fe(CN)_6^{4-}$ under diffusion-controlled conditions. In turn, potentials of 0 and 0.55 V were applied for short times to the substrate

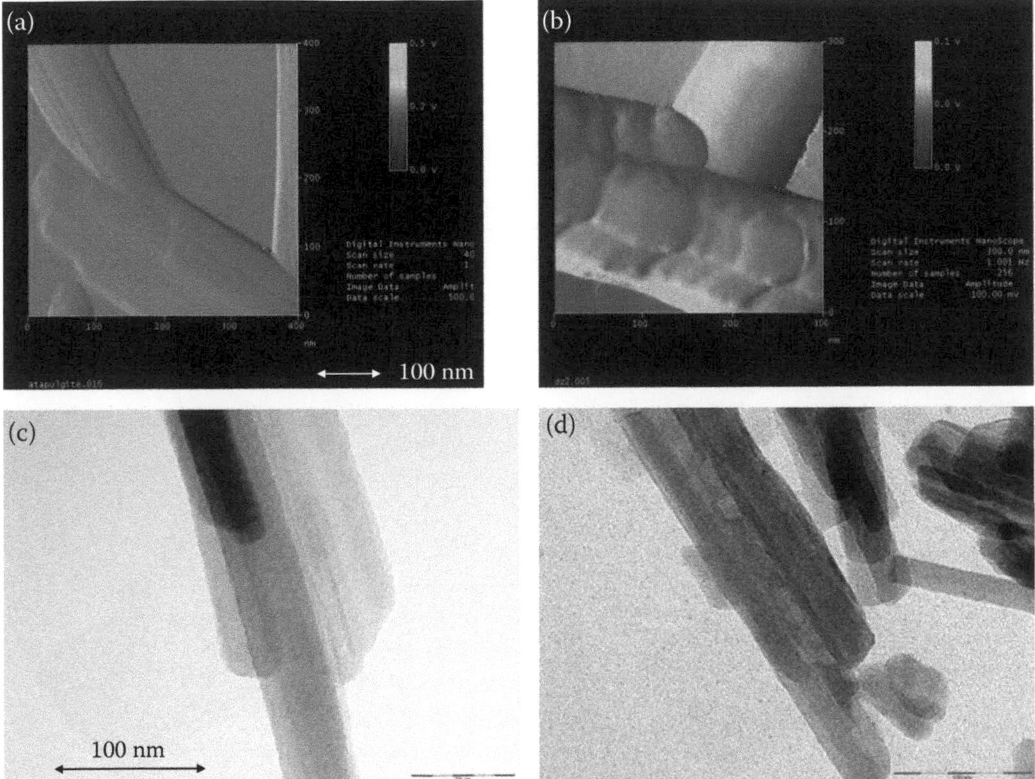

FIGURE 5.10 (a,b) AFM and (c,d) TEM Images of Pristine Palygorskite Crystals from Sacalum (a,c), a MB Sample from Dzibilnocac (Yucatán, Late Classical Maya Period) (b), and a MB sample from Mayapán, Postclassical Period (d). Reproduced from Ref. [34] (Doménech-Carbó et al. *J. Phys. Chem. B.* 2006, 110: 6027–6039), with Permission.

electrode containing microparticulate deposits of the $IND@PAL_{140}$ and $IND@KAO_{140}$ specimens. At this potential, indigo is electrochemically oxidized to dehydroindigo, as previously noted. In the case of palygorskite, no differences were observed in the topography of clay grains between the two applied substrate potentials, as expected in the case of a strong attachment of the dye to the clay (Figure 5.11a,b). In contrast, the kaolinite hybrid displayed significant topographic differences between the two substrate potentials (Figure 5.11c,d) in agreement with the 'external' location of indigo molecules in the inorganic host.

The nature of the indigo-palygorskite association, however, remains yet unsolved. Van Olphen (1967) suggested that indigo molecules are too large to enter the channels of palygorskite so they become—to any extent—sealed at their ends by dye molecules. Kleber et al. (1967) suggested that, in view of the dimensions of indigo molecules and palygorskite channels, partial or even deep penetration of indigo molecules into the palygorskite channels cannot be excluded [33]. Recent formulations include the formation of hydrogen bonds between the carbonyl and amino groups of indigo with edge silanol units of the clay, closing palygorskite channels, hydrogen bonding between indigo carbonyl and structural water, formation of hydrogen bonds between C=O and N-H groups of indigo molecules, and structural water molecules [36,37]. Apart from hydrogen bonding, however, van der Waals interactions and a direct interaction between indigo and clay octahedral cations—not mediated by structural water—could also play an important role in anchoring indigo molecules [36]. Electrochemical data [29,38] suggest that there are different topological redox isomers of indigo attached to the palygorskite matrix and that there is a possibility of a stabilization

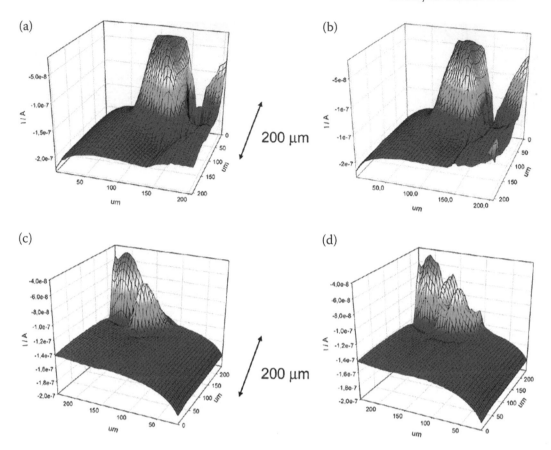

FIGURE 5.11 Topographic SECM Images of Synthetic (a,b) IND@PAL$_{140}$ and (c,d) IND@KAO$_{140}$MB Specimens in Contact with 5 mM $K_4Fe(CN)_6$ Redox Probe in 0.25 M Aqueous Acetate Buffer Solution at pH 4.75. The Tip Potential was Fixed at 0.35 V vs. Ag/AgCl and the Substrate Potential was of 0.0 V (a,c) and 0.55 V (b,d). Adapted from Ref. [35] (Doménech-Carbó et al. *Appl. Mater. Sci. Interfaces*. 5: 8134–8145), with Permission.

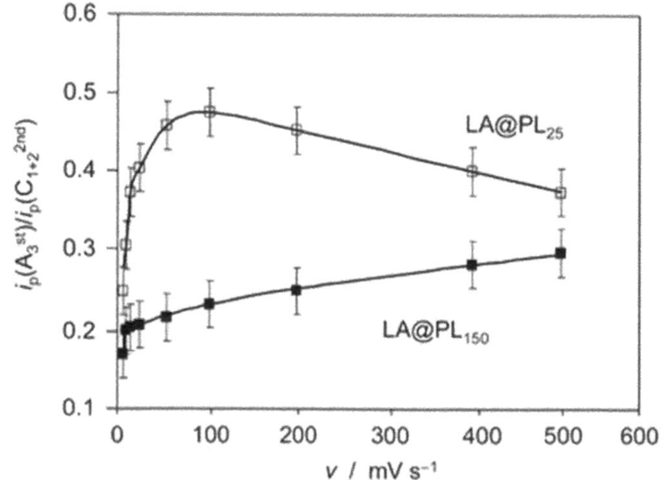

FIGURE 5.12 In-depth Variation of the Oxidized/Reduced Molar Ratio in Unheated Lapachol@palygorskite (Solid Squares), Lapachol@palygorskite Heated at 150°C (Squares) from CV Data. Reproduced from Ref. [31] (Doménech-Carbó et al. *PhysChemChemPhys*. 2014, 16: 19024–19034), with Permission.

of the tautomer of indigo (see Figure 5.9) in the palygorskite framework influencing the color of the pigment. Then, the preparation and electrochemistry of MB would involve tautomerization and redox tuning processes as schematized in Figure 5.9.

The key question is, in summary, to determine the coexistence of species in different oxidation states associated with the clay support as well as their in-depth distribution. The treatments described in Section 2.7 can be applied to microporous solids containing immobile electroactive centers distributed uniformly. For non-uniform distributions, the theoretical approach described in Section 5.5 can be extended to the case of immobile redox centers coexisting in two interconnected species, A and B, consisting of the oxidized and reduced forms of the guest. In the most favorable case, species A and B produce separate, diffusion-controlled voltammetric signals, Then, the corresponding $I/v^{1/2}$ ratio should satisfy the relationships [31]:

$$\frac{I_n^{\mathrm{A}}(x)}{v^{1/2}} = \frac{\sum_{j=1}^{n} g^{\mathrm{A}} c_j^{\mathrm{A}}(x)\delta_j}{\sum_{j=1}^{n}\delta_j} = \frac{\sum_{j=1}^{n} g^{\mathrm{A}} c_j^{\mathrm{A}}(x)\delta_j}{x_n} \tag{5.20}$$

$$\frac{I_n^{\mathrm{B}}(x)}{v^{1/2}} = \frac{\sum_{j=1}^{n} g^{\mathrm{B}} c_j^{\mathrm{B}}(x)\delta_j}{\sum_{j=1}^{n}\delta_j} = \frac{\sum_{j=1}^{n} g^{\mathrm{B}} c_j^{\mathrm{B}}(x)\delta_j}{x_n} \tag{5.21}$$

Using the precedent closure, the molar ratio between the forms A and B at the k-layer f_k^{A}, can be calculated as:

$$f_{\mathrm{k}}^{\mathrm{A}} = \frac{(c_{\mathrm{A}}/c_{\mathrm{B}})_{\mathrm{k}}}{(c_{\mathrm{A}}/c_{\mathrm{B}})_{\mathrm{boundary}}} = \frac{\left(\frac{x_{\mathrm{n+1}}}{\delta_{\mathrm{k}}}\right)\left[\frac{I_{\mathrm{n+1}}^{\mathrm{A}}(x)}{I_{\mathrm{lim}}^{\mathrm{A}}} - \frac{I_{\mathrm{n}}^{\mathrm{A}}(x)}{\mathrm{I}_{\mathrm{lim}}^{\mathrm{A}}}\right] - \left(\frac{I_{\mathrm{n}}^{\mathrm{A}}(x)}{I_{\mathrm{lim}}^{\mathrm{A}}}\right)}{\left(\frac{x_{\mathrm{n+1}}}{\delta_{\mathrm{k}}}\right)\left[\frac{I_{\mathrm{n+1}}^{\mathrm{B}}(x)}{I_{\mathrm{lim}}^{\mathrm{B}}} - \frac{I_{\mathrm{n}}^{\mathrm{B}}(x)}{I_{\mathrm{lim}}^{\mathrm{B}}}\right] - \left(\frac{I_{\mathrm{n}}^{\mathrm{B}}(x)}{I_{\mathrm{lim}}^{\mathrm{B}}}\right)} \tag{5.22}$$

Figure 5.13 shows the in-depth variation of the oxidized/reduced molar ratio in unheated lapachol@palygorskite (solid squares), lapachol@palygorskite heated at 150°C (squares) specimens from CV data using Eq. (5.22) [30].

All the expense of more accurate modeling, these results are illustrative of the inherent capability of electrochemical methods in providing information relevant to the knowledge of structure and reactivity of electroactive species incorporated into porous materials.

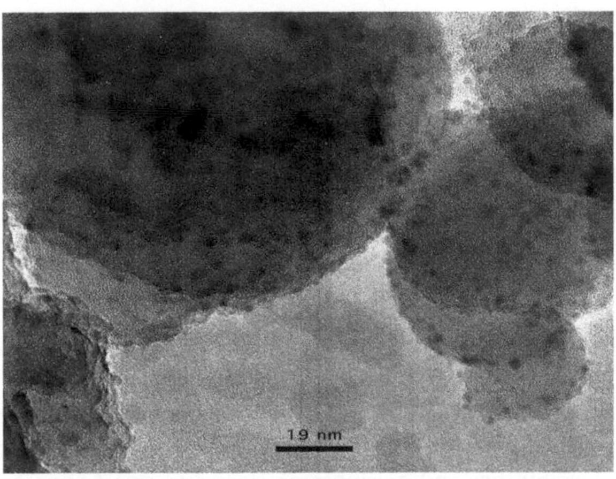

FIGURE 5.13 TEM Image for Iron Oxide Nanoparticles Generated in FeZSM-5 Zeolite. Reproduced from Ref. [39] (Doménech-Carbó et al. 2002. *J. Electroanal. Chem.* 2002, 519: 72–84), with Permission.

5.6 ELECTROACTIVE STRUCTURAL SPECIES

In previous sections, we have concentrated the attention on the electrochemistry of electroactive species incorporated as guests in the voids of porous aluminosilicate lattices. In most cases, however, electroactive species are constituents of the inorganic host. This is the case with FeZSM-5 zeolites, where framework iron centers are accompanied by isolated iron ions in the channels/cages, oligonuclear iron clusters, and nanoaggregated iron oxide [39]. The possibility of differentiated electroactive centers has been reported for vanadium silicalites [40] and titanium [41] silicalites and cobalt cordierites [42].

One of the possible ways to obtain nanoparticles is the application of hydrothermal treatments to aluminosilicate materials incorporating metal ions in their lattice. This is the case with Fe-, Co-, etc. –doped zeolites. Thus, FeZSM-5 materials are prepared from H-ZSM5 using solid and liquid ion-exchange methods, chemical vapor deposition of volatile iron compounds ($FeCl_3$), or hydrothermal synthesis of isomorphous substituted FeZSM-5 followed by calcinations and steam treatment. These treatments induce the migration of iron towards extra-framework positions by cleavage of Si-O-Fe bonds and dealumination of the zeolite framework. Figure 5.14 shows a TEM image for iron oxide nanoparticles generated in FeZSM-5 zeolite after calcinations and steam treatment [39]. Dark spots in this image correspond to homogeneously dispersed iron oxide nanoaggregates of 1–2 nm.

For our purposes, the relevant point to emphasize is that voltammetry of microparticles methodology permits discrimination between electron-transfer processes for different iron species—framework iron centers isolated, extra-framework iron ions in the more external zeolite sites, solid-state reduction of oligonuclear iron oxospecies, and reductive dissolution of extra-zeolitic iron oxide nanoparticles [39]. Notice that the different species display quite different voltammetric responses. In the case of framework Fe(III), centers substituting Si(IV) in SiO_4 units (resulting in $FeO_3(OH)$ units) can either be represented as [39,43]:

$$\{Fe^{III}O_3(OH)\}_{zb} + H^+_{aq} + e^- \rightarrow \{Fe^{II}O_2(OH)_2\}_{zb} \quad (5.23)$$

$$\{Fe^{III}O_3(OH)\}_{zb} + M^+_{aq} + e^- \rightarrow \{MFe^{II}O_3(OH)\}_{zb} \quad (5.24)$$

where zb represents the zeolite external boundary. In the first case, the reduction process involves the protonation of an oxo group while in the second, the reduction is accompanied by the ingress of a monovalent cation from the electrolyte into the zeolite channel/cage system. In turn, the reduction of iron oxide nanoparticles can be described in terms of a reductive dissolution process:

$$\{Fe^{III}O_x\} + 2xH^+_{aq} + (2x - 2)e^- \rightarrow Fe^{2+}_{aq} + xH_2O \quad (5.25)$$

Although this rich electrochemistry has not been extensively studied, there is a possibility of future developments concerning the exploitation of the electrocatalytic capabilities of the different centers. Recent research is concentrated on the anchorage of different nanostructured units (metal and metal oxide nanoparticles in particular) to aluminosilicate frameworks. This is the case with Pd-modified zeolite X acting as a catalyst for HER [44]. Here, the formation of nanostructured PdO on the outer surface of the zeolite (prepared by exchange of Na^+ ions by Pd^{2+} ions) appears to play a decisive role in the electrocatalytic process. Again, as illustrated by the Maya blue, the aluminosilicate framework not only acts as an inert host framework but also as a promoter of chemical transformations of the guest species. In this respect, the structure, pore (channel, cages) size of the support and distribution, and hardness of Brønsted and Lewis acid sites become significant [45]. Figure 5.15 shows an illustrative example reflected in the voltammetric response of different Pd^{2+}-supported catalysts. One can see in this figure that the cathodic LSVs of

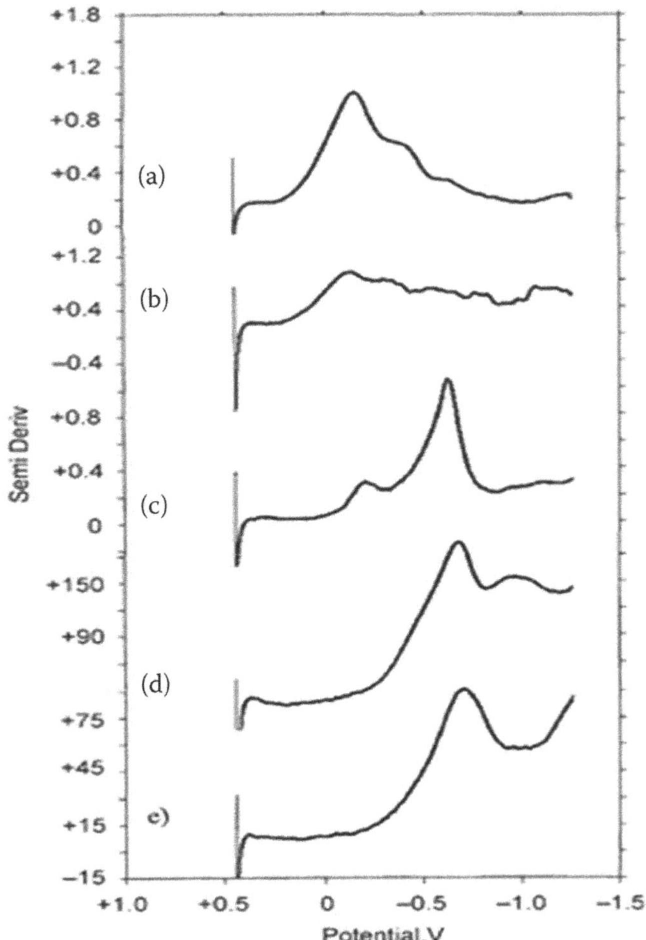

FIGURE 5.14 LSVs of Zeolite Modified Electrodes in Contact with 0.1 M LiClO4 Aqueous Solution. (a) $PdCl_2$–CsX; (b) $PdCl_2$–KX; (c) $PdCl_2$–natural Sepiolite; (d) $PdCl_2$–HBeta; (e) Dissolved $PdCl_2$. Reproduced from Ref. [45] (Corma et al. *New J. Chem.* 2004, 28: 361–365), with Permission.

$PdCl_2$–natural sepiolite and $PdCl_2$–HBeta zeolite, both lacking (or having low content of) in Al^{3+}, are close to that of a $PdCl_2$ solution in $LiClO_4$/water-electrolyte, suggesting a fast exchange of Pd^{2+} ions with the electrolyte. In contrast, the voltammograms of Al^{3+}-containing $PdCl_2$–CsX and $PdCl_2$–KX zeolites differ significantly, suggesting that not only pore/cage/channel site effects have to be accounted.

5.7 MESOPOROUS MATERIALS

The electrochemical response of mesoporous aluminosilicates has claimed attention because of their potential applications in catalysis, separation, adsorption, sensing, and microelectronics. Such materials (MCM-41 periodic mesoporous silica as a representative example) provide much less exigent size restrictions than zeolites with regard to ion diffusion, the electrochemical response being dependent on (i) the electrochemical properties of the guest, (ii) its concentration, (iii) the size of the separation between cavities able to attach electroactive guest species, (iv) the type of association of the electroactive species with the aluminosilicate framework, and (v) the—more or less—ordered nature of the mesoporous host and the hierarchical distribution of pores. As in the case of zeolites, however, charge conservation requires the ingress/issue of charge-balancing electrolyte ions coupled with electron transfer steps. As described in Chapter 3, the advance of the redox reaction across the porous materials involves electron hopping between immobile redox centers coupled with the transport of mobile ions through the pore/channel system of the aluminosilicate framework. Obviously, there is an opportunity for charge transport associated with the motion of redox-mediators. Eventually, charge transport could involve proton hopping between silanol groups and acid sites.

Given the adsorbent properties of mesoporous aluminosilicates, these materials can be used for selectively preconcentrating species in solution, following a scheme also used in zeolites. Accumulation is driven by the ease of transport of the analyte to binding sites so that efficiency is high when using materials displaying high specific surface area and regular channels of tailor-made monodispersed dimensions. The regular tridimensional structure of systems like MCM-41 leads to an easier way of guest species to bind with host binding sites relative to non-ordered systems such as amorphous silica gel [46]. However, the mesopore/macropore shape and size distribution, the presence of Brønsted and Lewis acid sites, etc. can modulate the electrochemical response of the guest species and their electrocatalytic properties, as exemplified in Section 4.5 of the precedent chapter.

The electrocatalytic abilities of mesoporous aluminosilicates with and without incorporation of guests and/or functionalization with active groups is receiving considerable attention [46]. The mesoporous framework can act as a reservoir to store (i.e. to preconcentrate) analytes, enhancing the sensitivity of the determinations. This effect is added to the selectivity resulting from the specificity of the analyte/guest interactions and charge/size selectivity effects. Here, apart from the aforementioned factors, the modified electrode configuration (monolayer, multilayer, etc.) can significantly influence the final electrochemical response. It should be noted that chemical modification of the mesoporous host, for instance by thiol groups, can provide not only functional attachment to electrode surfaces, but also specific binding sites for preconcentration of selected analytes.

5.8 FINAL REMARKS

Aluminosilicates display a rich electrochemistry with interesting applications in electrocatalysis and sensing as well as in energy storage and production due to their capability to store guest species. Much of this electrochemistry is focused on the electrocatalytic properties of these materials. Here, two main effects can be accounted for: the site isolation and the docking effects. By the first token, the physical separation of active guest species into the pores/cages/channels of the porous/channeled aluminosilicates minimizes lateral interactions, facilitating the substrate/catalyst direct reaction. Additionally, the confinement of the guest into the cavities of the host favors the adoption of convenient conformations to react with the substrate.

The role of the host, however, is not merely that of a rigid superstructure, but also to interact with the guest yielding chemical changes. These may occur in the guest such as isomerization and redox tuning, but also in the host, incorporating/releasing mobile ions in/from channels, cages, etc., and/or substituting structural units. Future research should emphasize the detailed knowledge of the relations between the composition/structure of the aluminosilicate framework and the reactivity and electrochemistry of attached guest species.

REFERENCES

[1]. Beck, J.S.; Vartuli, J.C.; Roth, W.J.; Leonowicz, M.E.; Kresge, C.T.; Smith, K.D.; Chu, C.T.-W.; Olson, D.H.; Sheppard, E.W.; McCullen, S.B.; Higgins, J.B.; Schlenker, J.L. 1992. A new family of mesoporous molecular sieves prepared with liquid crystal templates. *Journal of the American Chemical Society*. 114: 10834–10843.

[2]. Wu, S.-H.; Mou, C.-Y.; Lin, H.-P. 2013. Synthesis of mesoporous silica nanoparticles. *Chemical Society Reviews*. 42: 3862–3875.
[3]. Breck, D.W. 1974. *Zeolite Molecular Sieves: Structure, Chemistry, and Use*. Wiley-Interscience, New York.
[4]. Walcarius, A. 2003. *Implication of Zeolite Chemistry in Electrochemical Science and Applications of Zeolite-Modified Electrodes in Handbook of Zeolite Science and Technology*. Marcel Dekker, New York, chapter 14.
[5]. Ghosh, P.K.; Bard, A.J. 1983. Clay-modified electrodes. *Journal of the American Chemical Society*. 105: 5691–5693.
[6]. Rolison, D.R.; Bessel, C.A. 2000. Electrocatalysis and charge-transfer reactions at redox-modified zeolites. *Accounts of Chemical Research*. 33: 737–744.
[7]. Walcarius, A. 2013. Mesoporous materials and electrochemistry. *Chemical Society Reviews*. 42: 4098–4140.
[8]. Corma, A.; Fornes, V.; Martín-Aranda, R.M.; García, H.; Primo, J. 1990. Zeolites as base catalysts: Condensation of aldehydes with derivatives of malonic esters. *Applied Catalysis*. 59: 237–248.
[9]. Shaw, B.R.; Creasy, K.E.; Lanczycki , C.J.; Sargeant, J.A.; Tirhado, M. 1988. Voltammetric response of zeolite-modified electrodes. *Journal of the Electrochemical Society*. 135: 869–876.
[10]. Dutta, P.K.; Ledney, M. 1997. Charge-transfer processes in zeolites: Toward better artificial photosynthetic models. *Progress in Inorganic Chemistry*. 44: 209–271.
[11]. Doménech-Carbó, A. 2015. Theoretical scenarios for the electrochemistry of porous silicate-based materials: An overview. *Journal of Solid State Electrochemistry*. 19: 1887–1903.
[12]. Bessel, C.A.; Rolison, D.R. 1997. Topological redox isomers: Surface chemistry of zeolite-encapsulated Co(salen) and [Fe(bpy)3] $^{2+}$ complexes. *Journal of Physical Chemistry B*. 101: 1148–1157.
[13]. Bessel, C.A.; Rolison, D.R. 1997. Electrocatalytic reactivity of zeolite-encapsulated Co(salen) with benzyl chloride. *Journal of the American Chemical Society*. 119: 12673–12674.
[14]. Doménech-Carbó, A.; Formentín, P.; García, H.; Sabater, M.J. 2002. On the existence of different zeolite-associated topological redox isomers. Electrochemistry of Y zeolite-associated Mn(salen)N_3 complex. *Journal of Physical Chemistry B*. 106: 574–582.
[15]. Doménech-Carbó, A.; Doménech-Carbó, M.T.; García, H.; Galletero, M.S. 1999. Electrocatalysis of neurotransmitter catecholamines by 2,4,6-triphenylpyrylium ion immobilized inside zeolite Y supercages. *Journal of the Chemical Society Chemical Communications*. 2173–2174.
[16]. DeWilde, W.; Peeters, G.; Lunsford, J.H. 1980. Synthesis and spectroscopic properties of tris(2,2′-bipyridine)ruthenium(II) in zeolite Y. *Journal of Physical Chemistry*. 84: 2306–2310.
[17]. Hanson, R.M. 1991. The synthetic methodology of nonracemic glycidol and related 2,3-epoxy alcohols. *Chemical Reviews*. 91: 437–475.
[18]. Ledney, M.; Dutta, P.K. 1995. Oxidation of water to dioxygen by intrazeolite Ru(bpy)33+. *Journal of the American Chemical Society*. 117: 7687–7695.
[19]. Doménech-Carbó, A.; Ferrer, B.; Fornés, V.; García, H.; Leyva, A. 2005. Ship-in-a-bottle synthesis of triphenylamine inside faujasite supercages and generation of the triphenylamminium radical ion. *Tetrahedron*. 61: 791–796.
[20]. Barrer, R.M. 1978. *Zeolites and Clay Minerals as Sorbents and Molecular Sieves*. Academic Press, London.
[21]. Chretien, M.N.; Cosa, G.; García, H.; Scaiano, J.C. 2002. Increasing the life expectancy of carbanions by zeolite inclusion. *Chemical Communications*. 2154–2155.
[22]. Doménech-Carbó, A.; García, H.; Marquet, J.; Herance, J.R. 2005. Electrochemical monitoring of compartmentalization effects in the stability of Mesenheimer anions supported in hydrotalcite and X and Y zeolites. *Journal of the Electrochemical Society*. 152: J74–J81.
[24]. Li, J-W.; Pfanner, K.; Calzaferri, G. 1995. Silver-zeolite-modified electrodes: An intrazeolite electron transport mechanism. *Journal of Physical Chemistry*. 99: 2119–2126.
[23]. Turro, N.; Garcia-Garibay, M. 1991. Thinking Topologically About Photochemistry in Organized Media, in Photochemistry in Organized and Constrained Media in Ramamurthy, V. Ed. VCH, New York, pp. 1–38.
[25]. Scaiano, J.C.; García, H. 1999. Intrazeolite photochemistry: Toward supramolecular control of molecular photochemistry. *Accounts of Chemical Research*. 21: 783–793.
[26]. Doménech, A.; García, H.; Alvaro, M.; Carbonell, E. 2003. Study of redox processes in zeolite Y-associated 2,4,6-triphenylthiopyrylium ion by square wave voltammetry. *Journal of Physical Chemistry B*. 107: 3040–3050.
[27]. Vitale, M.; Castagnola, N.B.; Ortins, N.J.; Brooke, J.A.; Vaidyalingam, A.; Dutta, P.K. 1999. Intrazeolitic photochemical charge separation for $Ru(bpy)_3^{2+}$–bipyridinium system: Role of the zeolite structure. *Journal of Physical Chemistry B*. 103: 2408–2416.

[28]. Bruhwiler, D.; Calzaferri, G. 2004. Molecular sieves as host materials for supramolecular organization. *Microporous and Mesoporous Materials*. 72: 1–23.

[29]. Hashimoto, S.; Uehara, K.; M. Sogawa, K.; Takada, H.; Fukumura, H. 2006. Application of time- and space-resolved fluorescent spectroscopy to the distribution of guest species into micrometer-sized zeolite crystals. *Physical Chemistry Chemical Physics*. 8: 1451–1458.

[30]. Doménech-Carbó, A.; Doménech-Carbó, M.T.; Sánchez del Río, M.; Vázquez de Agredos Pascual, M.L. 2009. Comparative study of different indigo-clay Maya blue-like systems using the voltammetry of microparticles approach. *Journal of Solid State Electrochemistry*. 13: 869–878.

[31]. Doménech-Carbó, A.; Martini, M.; Valle-Algarra, F.M. 2014. Determination of the depth profile distribution of guest species in microporous materials using the voltammetry of immobilized particles: Application to lapachol attachment to palygorskite and kaolinite. *Physical Chemistry Chemical Physics*. 16: 19024–19034.

[32]. Doménech-Carbó, A.; Doménech-Carbó, M.T.; Vázquez de Agredos Pascual, M.L. 2007. Chemometric study of Maya blue from the voltammetry of microparticles approach. *Analytical Chemistry*. 79: 2812–2821.

[33]. Doménech-Carbó, A.; Holmwood, S.; Di Turo, F.; Montoya, N.; Valle-Algarra, F.M.; Edwards, H.G.M.; Doménech-Carbó, M.T. 2019. On the composition and color of Maya blue: A reexamination of literature data based on the dehydroindigo model. *Journal of Physical Chemistry C*. 123: 770–782. See references herein on the historical studies and controversies on Maya blue.

[34]. Doménech-Carbó, A.; Doménech-Carbó, M.T.; Vázquez de Agredos Pascual, M.L. 2006. Dehydroindigo: A new piece into the Maya blue puzzle from the voltammetry of microparticles approach. *Journal of Physical Chemistry B*. 110: 6027–6039.

[35]. Doménech-Carbó, A.; Valle-Algarra, F.M.; Doménech-Carbó, M.T.; Domine, M.E.; Osete-Cortina, L.; Gimeno-Adelantado, J.V. 2013. Redox tuning and species distribution in Maya blue-type materials: A reassessment. *Applied Materials Science & Interfaces*. 5: 8134–8145.

[36]. Fois, E.; Gamba, A.; Tilocca, A. 2003. On the unusual stability of Maya blue paint: Molecular dynamics simulations. *Microporous and Mesoporous Materials*. 57: 263–272.

[37]. Giustetto, R.; Llabrés i Xamena, F.X.; Ricchiardi, G.; Bordiga, S.; Damin, A.; Gobetto, R.; Chierotti, M.R. 2005. Maya blue: A computational and spectroscopic study. *Journal of Physical Chemistry B*. 109: 19360–19368.

[38]. Doménech-Carbó, A.; Doménech-Carbó, M.T.; Sánchez del Río, M.; Goberna, S.; Lima, E. 2009. Evidence of topological indigo/dehydroindigo isomers in Maya blue-like complexes prepared from palygorskite and sepiolite. *Journal of Physical Chemistry C*. 113: 12118–12131.

[39]. Doménech-Carbó, A.; Pérez, J.; Ribera, A.; Mul, G.; Kapteijn, F.; Arends, I.W.C.E. 2002. Electrochemical characterization of iron sites in ex-framework FeZSM-5. *Journal of Electroanalytical Chemistry*. 519: 72–84.

[40]. Grygar, T.; Čapek, L.; Adam, J.; Machovič, V. 2009. V(V) species in supported catalysts: Analysis and performance in oxidative dehydrogenation of ethane. *Journal of Electroanalytical Chemistry*. 633: 127–133.

[41]. de Castro-Martins, S.; Khouzami, S.; Fuel, A.; Bentaarit, Y.; El Murr, N.; Sellami, A. 1993. Characterization of titanium silicalite using TS-1-modified carbon paste electrodes. *Journal of Electroanalytical Chemistry*. 350: 15–28.

[42]. Doménech-Carbó, A.; Torres, F.J.; Alarcón, J. 2004. Electrochemical characterization of cobalt cordierites attached to paraffin-impregnated graphite electrodes. *Journal of Solid State Electrochemistry*. 8: 127–137.

[43]. Pérez-Ramírez, J.; Mul, G.; Kapteijn, F.; Moulijn, J.A.; Overweg, A.R.; Doménech-Carbó, A.; Ribera, A.; Arends, I.W.C.E. 2002. Physicochemical characterization of isomorphously substituted Fe-ZSM5 during activation. *Journal of Catalysis*. 207: 113–126.

[44]. Vasić, M.; Čebela, M.; Pašti, I.; Amaral, L.; Hercigonja, R.; Santos, D.M.F.; Šljukić, L. 2017. Efficient hydrogen evolution electrocatalysis in alkaline medium using Pd-modified zeolite X. *Electrochimica Acta*. 259: 882–892.

[45]. Corma, A.; García, H.; Primo, A.; Doménech-Carbó, A. 2004. A test reaction to assess the presence of Brönsted and the softness/hardness of Lewis acid sites in palladium supported catalysts. *New Journal of Chemistry*. 28: 361–365.

[46]. Walcarius, A.; Bessière, J. 1999. Electrochemistry with mesoporous silica: Selective mercury(II) binding. *Chemistry of Materials*. 11: 3010–3011.

6 Electrochemistry of Metal-Organic Frameworks

6.1 INTRODUCTION

Metal-Organic Frameworks (MOFs) and covalent organic frameworks are crystalline systems that can be described as infinite networks resulting from the bonding of metal ions—which act as coordination centers—with polyfunctional organic molecules. First prepared by Tomic [1] and Yaghi et al. [2], MOFs are assembled by joining inorganic metal nodes with organic linkers defining inorganic secondary building units (SBUs), as illustrated in Figure 6.1 [3]. The organic linkers are constituted by a variety of conjugate bases of polycarboxylic acids, typically terephthalic acid, and other units such as 4,4′-bipyridine; some representative examples are shown in Figure 6.2 [3,4]. These materials inspired the preparation of covalent organic frameworks (COFs) that are built from organic units with covalent bonds [5].

MOFs and COFs are robust frameworks that can be regarded as coordination polymers that combine crystallinity and high porosity [6]. The singular porosity of MOFs allows a significant redox conductivity that, in contrast with zeolites and other microporous aluminosilicates, can involve all units of the material. This is the case with Cu^{2+}- and Zn^{2+}-based MOFs with terephthalic acid; in these materials, both metal centers and organic units are potentially electroactive when in contact with suitable electrolytes.

The electrochemistry of MOFs should, in principle, be based on redox processes for both metal ions and polyfunctional organic molecules, coupled with the ingress/issue of electrolyte ions in/from the MOF lattice. In view of the large ionic permeability of MOFs, one can expect that an internal diffusion occurs. This means that the redox reactivity can be significantly (and rapidly) extended along the crystal. Accordingly, redox conductivity should be heavily influenced by the structure of the material and possible structural conditioning (anisotropy, reticular defects). Conversely, cooperative, large-scale structural effects (crystalline domains, secondary phases) might be produced as a result of the application of electrochemical inputs.

In this chapter, the electrochemistry of MOFs will be studied with particular attention to their peculiarities with other microporous materials. Two general types of electrochemical processes will be considered: those involving ion insertion-driven electrochemical process which, in principle, does not demand sharp structural changes, and those involving formation of metal particles, requiring the formation of a new phase.

Most MOFs and COFs can be considered as electrical insulators, with conductivities below 10^{-10} S cm^{-1}. The electrochemical process involves electron and ion transport. The former can occur in three modes [5]: through-bond, through-space, and through guest. Through bond charge transport can be enhanced by varying the symmetry, frontier orbital overlap, and energy match between the coordinately/covalently bonded components [7]. The through-space conduction is associated with π–π stacking of aromatic moieties, typically present in laminar structures [8]. The thorough guest conduction is achieved by the intercalation of conductive polymers or conjugated molecules able to bind with metal centers [9].

6.2 MOFS ELECTROCHEMISTRY: AN OVERVIEW

The first remarkable feature is the significant dependence of the electrochemical response of MOFs on the size of electrolyte ions [10]. This is illustrated in Figure 6.3 for the reduction of a deposit of

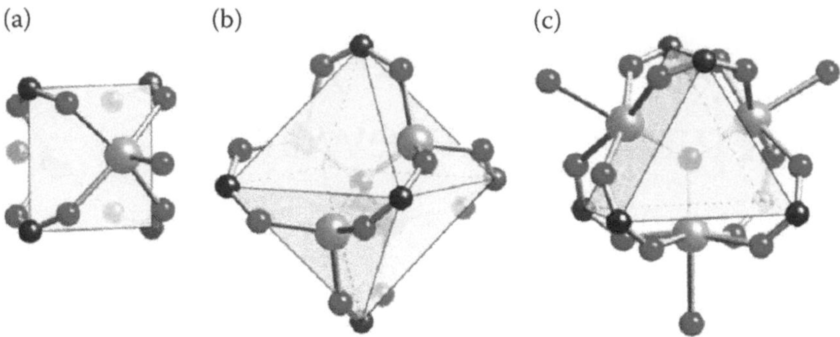

FIGURE 6.1 Typical Inorganic Secondary Building Units (SBUs) Occurring in Metal Carboxylates. (a) Square "Paddlewheel", with Two Terminal Ligand Sites; (b) Octahedral "Basic Zinc Acetate" Cluster; (c) Trigonal Prismatic Oxo-centered Trimer with Three Terminal Ligand Sites. Metals: Blue Spheres, Carbon: Black Spheres, Oxygen: Red Spheres, Nitrogen: Green Spheres. Reproduced from Ref. [3] (Rowsell and Yaghi, *Micropor. Mesopor. Mater.* 2004, 73: 3–14), with Permission.

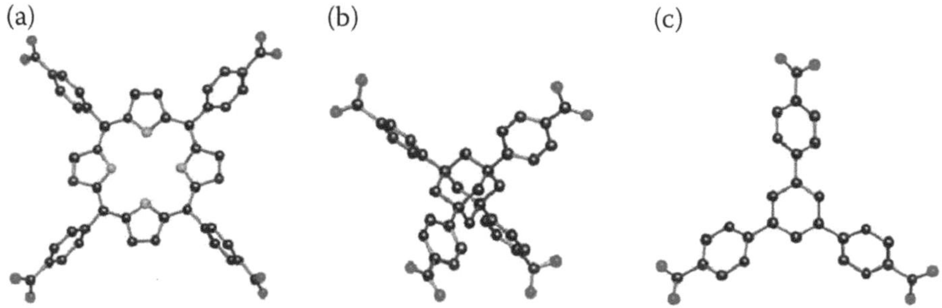

FIGURE 6.2 Examples of Organic SBUs in MOFs. Conjugate Bases of: (a) Square Tetrakis(4-carboxyphenyl)porphyrin; (b) Tetrahedral Adamantane-1,3,5,7-tetracarboxylic acid; (c) Trigonal 1,3,5-tris(4-Carboxyphenyl)benzene. Metals: Blue Spheres, Carbon: Black Spheres, Oxygen: Red Spheres, Nitrogen: Green Spheres. Reproduced from Ref. [3] (Rowsell and Yaghi, *Micropor. Mesopor. Mater.* 2004, 73: 3–14), with Permission.

Cu-MOF immersed into different MeCN electrolytes. In contact with Bu_4NPF_6/MeCN, weak reduction peaks were obtained in the 0.60 to –1.85 V vs. AgCl/Ag potential region. In the presence of Et_4NClO_4/MeCN, reduction peaks at 0.20 and –0.65 V were recorded. Upon immersion into $LiClO_4$/MeCN electrolytes, the above peaks are enhanced and exhibit peak splitting, while a prominent cathodic peak appears at –1.75 V. These features can be rationalized assuming that the reduction of Cu^{2+} centers existing in the MOF involves the coupled ingress of electrolyte cations into the solid microporous lattice. This ingress of such cations should be significantly size-hindered so that bulky Bu_4N^+ ions cannot enter the MOF framework. In contact with Bu_4N^+/MeCN electrolytes, only ill-defined cathodic waves, which can be attributed to the reduction of external Cu^{2+} centers, appear. On the contrary, for Et_4N^+/MeCN and Li^+/MeCN electrolytes, the ingress of Li^+ ions and, partially, Et_4N^+ ions, should be allowed, permitting the record of well-defined voltammetric peaks corresponding to the propagation of the redox reaction to the Cu-MOF bulk. Consistently, the net amount of charge transferred was found to be dependent on the electrolyte concentration, as expected for ion insertion solids (see Chapter 3).

This electrochemistry is focused on the reduction of Cu(II) centers and can be described in terms of different electrochemical pathways. Among other possibilities, we can hypothesize the:

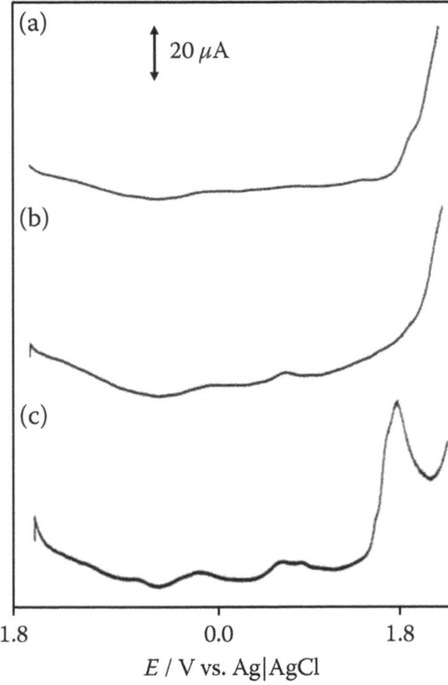

FIGURE 6.3 SWVs of Cu-MOF Immersed into: (a) 0.10 M Bu_4NPF_6/MeCN, (b) 0.10 M Et_4NClO_4/MeCN, and (c) 0.10 M $LiClO_4$/MeCN. Potential Scan Initiated at 1.65 V vs. AgCl/Ag in the Negative Direction. Potential Step Increment 4 mV; Square Wave Amplitude 25 mV; Frequency 5 Hz. Reproduced from Ref. [11] (Doménech-Carbó et al. *J. Phys. Chem. C.* 2007, 111: 13701–13711), with Permission.

a. Topotactic solid-state transformation coupled with the ingress of electrolyte cations, M^+, into the solid:

$$\{Cu^{II}R\}_{n\ solid} + nM^+_{sol} + ne^- \rightarrow \{Cu^IR^- \cdots M^+\}_{n\ solid} \tag{6.1}$$

b. Proton-assisted topotactic solid-state transformation involving the protonation of organic groups in the solid lattice:

$$\{Cu^{II}R\}_{n\ solid} + qnH^+_{sol} + qne^- \rightarrow \{Cu^{II}(RH_q)\}_{n\ solid} \tag{6.2}$$

c. Reductive reorganization involving both metal and organic units:

$$\{Cu^{II}R\}_{n\ solid} + (n - m)M^+_{sol} + qmH^+_{sol} + (n + qm - m)e^- \rightarrow \{Cu^IR^- \cdots M^+\}_{n-m}\{Cu^{II}(RH_q)\}_{m\ solid} \tag{6.3}$$

d. Reductive dissolution process:

$$\{Cu^{II}R\}_{n\ solid} + ne^- \rightarrow nCu^+_{sol} + nR^{2-}_{sol} \tag{6.4}$$

In all these expressions, { } denote—as customary along the text—solid phases while sol represents solvated species existing in solution phase. Of course, anion-insertion processes follow similar schemes. Figure 6.4 shows an example [11] of the sensitivity of the voltammetric response of organic polyfunctional units to the composition of the supporting electrolyte. In this figure, SWVs initiated at –2.05 V in the positive direction of Cu-MOF deposits in contact with 0.10 M Bu_4NPF_6/MeCN, Et_4NClO_4/MeCN, and $LiClO_4$/MeCN electrolytes are shown. In these voltammograms, the

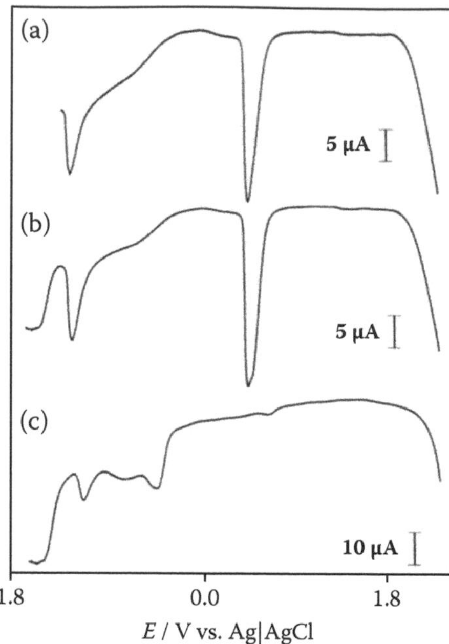

FIGURE 6.4 Square Wave Voltammograms of Cu-MOF Immersed into: (a) 0.10 M Bu_4NPF_6/MeCN, (b) 0.10 M Et_4NClO_4/MeCN, and (c) 0.10 M $LiClO_4$/MeCN. Potential Scan Initiated at –2.05 V vs. AgCl/Ag in the Positive Direction. Potential Step Increment 4 mV; Square Wave Amplitude 25 mV; Frequency 5 Hz. Reproduced from Ref. [11] (Doménech-Carbó et al. *J. Phys. Chem. C.* 2007, 111: 13701–13711), with Permission.

prominent anodic peak at +0.05 V corresponds to the stripping oxidation of Cu metal generated by electrochemical reduction of Cu^{2+} centers at the negative potentials applied at the beginning of the voltammetric scan. This peak results in the release of Cu^+ ions to the solution (Cu(I) is stabilized in MeCN solution), further oxidized to Cu^{2+} at +1.6 V. Remarkably, a prominent Cu stripping was obtained in contact with Bu_4NPF_6/MeCN and M Et_4NClO_4/MeCN electrolytes at ca. –0.4 V, but not with $LiClO_4$/MeCN. This difference can be attributed to a different metal deposition mode associated with the different permeability of the material to cation transport (*vide infra*).

Additional anodic peaks located in the 0.60 to 1.80 V potential region are attributable to the oxidation of polyfunctional organic groups. Such oxidation processes should involve the ingress of electrolyte counteranions coupled with the issue of electrons into/from the MOF lattice:

$$\{Cu^{II}R\}_{n\ solid} + qnX^-_{sol} \rightarrow \{Cu^{II}R^{ox\ q+} \cdots qX^-\}_{n\ solid} + qne^- \tag{6.5}$$

Now, in the presence of bulky PF_6^- anions, only Cu-associated oxidation peaks appear (Figure 6.4a); whereas in the presence of ClO_4^- ions, additional anodic peaks are recorded (Figure 6.4b,c). Differences in these last processes in $LiClO_4$/MeCN and Et_4NClO_4/MeCN electrolytes are attributable to the coexistence of a competing mechanism, involving the issue of cations that previously entered the MOF material as a result of reductive processes described by Eq. (6.1). This second oxidative pathway can be represented as:

$$\{Cu^IR^- \cdots M^+\}_{n\ solid} \rightarrow \{Cu^IR^{ox}\}_n + nM^+_{sol} + ne^- \tag{6.6}$$

Simultaneously, the oxidation of Cu(I) to Cu(II) can occur via competing pathways:

$$\{Cu^IR^- \cdots M^+\}_{n\ solid} + nX^-_{sol} \rightarrow \{Cu^{II}R \cdots M^+ \cdots X^-\}_{n\ solid} + ne^- \tag{6.7}$$

$$\{Cu^IR^- \cdots M^+\}_{n\ solid} \rightarrow \{Cu^{II}R\}_{n\ solid} + nM^+_{sol} + ne^- \tag{6.8}$$

The coexistence of different oxidative pathways would be responsible for the peak splitting in Figure 6.4c, illustrating the complexity attained from MOF electrochemistry.

6.3 ELECTROCHEMISTRY OF MOFS INVOLVING TOPOTACTIC TRANSFORMATIONS

The occurrence of reversible ion insertion processes without structural transformation has been characterized for different MOFs. The compound MIL-53(Fe), of formula $Fe(OH)_{0.8}F_{0.2}(BDC)$, is constituted by infinite chains of Fe(III) centers that are partially reduced to Fe(II) ones with Li^+ insertion [12]:

$$\begin{aligned}&\{Fe^{III}(OH)_{0.8}F_{0.2}(BDC)\}_{solid} + xLi^+_{sol} + xe^- \rightarrow\\&\{Li_xFe^{III}_{1-x}Fe^{II}_x(OH)_{0.8}F_{0.2}(BDC)\cdots xLi^+\}_{solid}\end{aligned} \tag{6.9}$$

where BDC = 1,4-Benzenedicarboxylate. Insertion of multiple-valent cations has also been achieved [13]. Topotactic oxidative insertion of anions has been reported for Fe_2(DOBPDC) [14]. The electrochemical process can be represented as:

$$\{Fe^{II}_2(DOBPDC)\}_{solid} + xA^-_{sol} \rightarrow \{Fe^{III}_xFe^{II}_{2-x}(DOBPC)\cdots xA^-\}_{solid} + xe^- \tag{6.10}$$

where DOBPDC = 4,4′-Dioxidobiphenyl-3,3′-dicarboxylate and $A^- = BF_4^-$, PF_6^-, $BArF^-$, (Bis(trifluoromethylsulfonyl)imide)$^-$.

Cu(2,7- 2,7-Anthraquinonedicarboxylate) is an example of independent redox activity on both metal nodes and ligands where the anthraquinone unit exhibits two reversible reduction steps. Added to the Cu(II)/Cu(I) couple, the overall process involves the uptake of three Li^+ ions per unit [15].

6.4 ELECTROCHEMISTRY OF MOFS INVOLVING METAL DEPOSITION

Figure 6.5 shows a typical CV for Cu-MOF in contact with aqueous acetate buffer. Here, two overlapping cathodic peaks at –0.21 and –0.46 V, preceding an ill-defined wave near –1.0 V, appear. In the reverse scan, a typical stripping peak at –0.37 V is recorded. This peak, associated with the oxidation of copper metal electrochemically formed, disappears when the potential is switched at potentials ca. –0.25 V, but is clearly recorded for switching potentials more than –0.35 V. As a result, it is concluded that Cu(II) centers are reduced to Cu(I) ones at potentials less negative than –0.50 V. At more negative potentials, metallic copper is formed. In the case of Zn(MOF), the direct two-electron reduction to metallic Zn is evidenced by the prominent stripping peak corresponding to the oxidative dissolution of Zn that can be seen in Figure 6.6.

Description of the overall electrochemical processes requires consideration of at least two steps: (i) reduction of solid materials coupled with ion insertion, and (ii) formation of metal deposits via nucleation and nuclei growth/diffusion processes. With this regard, the modeling of the propagation of electrochemical reactions through ion-permeable solids described in Section 3.6 can be combined with different models for metal electrodeposition, namely, two-, three-dimensional, and layer-by-layer electrodeposition. Among others, the model of Sharifker and Hills [16] considers two extreme cases involving three-dimensional instantaneous nucleation and progressive nucleation. In the first case, predicted CA curves are given by:

$$I = \frac{nAFcD^{1/2}}{\pi^{1/2}t^{1/2}}[1 - \exp(-N\pi kDt)] \tag{6.11}$$

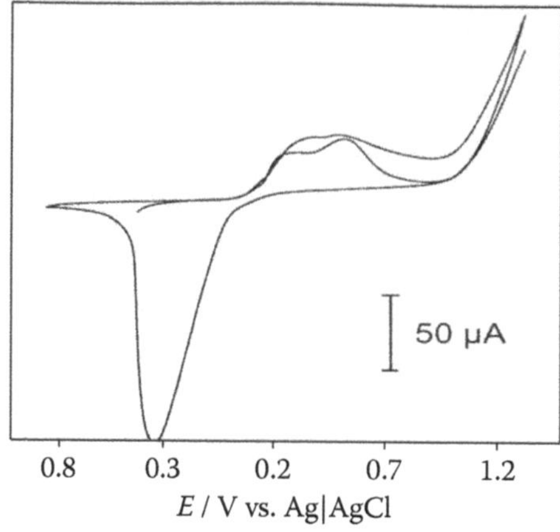

FIGURE 6.5 CV of a Polymer Film Electrode Modified with Cu-MOF Immersed into 0.25 M HAc + 0.25 M NaAc, pH 4.85. Potential Scan Rate 50 mV s^{-1}. Reproduced from Ref. [11] (Doménech-Carbó et al. *J. Phys. Chem. C.* 2007, 111: 13701–13711), with Permission.

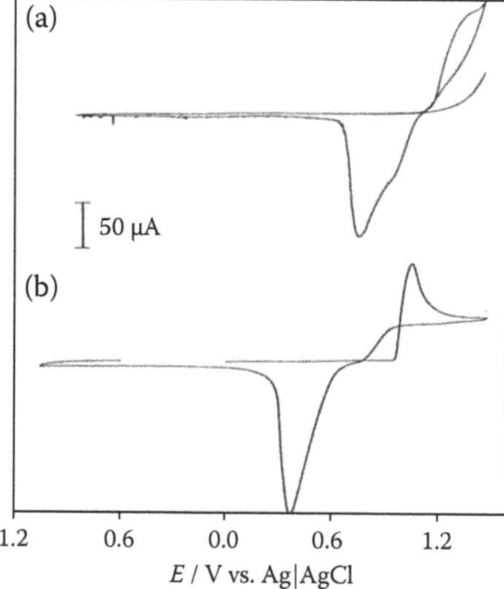

FIGURE 6.6 CVs of (a) PFE Modified with Zn-MOF Immersed into 0.25 M HAc + 0.25 M NaAc, pH 4.85, (b) an Unmodified GCE in Contact with a 1.0 mM Solution of Zn $(NO_3)_2$ in the Above Electrolyte. Potential Scan Rate 50 mV s^{-1}. Reproduced from Ref. [11] (Doménech-Carbó et al. *J. Phys. Chem. C.* 2007, 111: 13701–13711), with Permission.

while for progressive nucleation, it is predicted as

$$I = \frac{nAFcD^{1/2}}{\pi^{1/2}t^{1/2}}[1 - \exp(-AN^{*}\pi kDt^{2}/2)] \tag{6.12}$$

In these equations, A represents the electrode area, D the diffusion coefficient of the electroactive ions in the bulk of the solution and N the number of nuclei. For layer-by-layer deposition with instantaneous nucleation, CA curves can be described by [17]:

$$\ln \ln\left[1 - \frac{\pi^{1/2}}{nFcAD^{1/2}}It^{1/2}\right] - 1 = \ln(N^{*}\pi kD) + \ln t \tag{6.13}$$

while for progressive nucleation the following current-time variation is predicted as:

$$\ln \ln\left[1 - \frac{\pi^{1/2}}{nFcAD^{1/2}} It^{1/2}\right] - 1 = \ln(N^* \pi k DA/2) + 2 \ln t \qquad (6.14)$$

In these equations, N^* represents the density of nuclei (cm^{-2}) and k is a material-dependent rate constant.

Such models tested by CA data for Cu-MOF in contact with aqueous acetate buffer revealed the existence of different current/time responses depending on the applied potential [11]. At increasingly negative applied potentials, such responses can be associated to diffusion-controlled ion-insertion process, a three-dimensional nucleation/growth deposition, and a layer-by-layer deposition process, respectively, in agreement with AFM and TEM observations. From such data, the diffusion coefficient for electrons was calculated as $D_e = 2.0 \times 10^{-11}$ cm^2s^{-1}, whereas the diffusion coefficients for Li^+, Na^+, and K^+ ions were: $D_M(Li^+) = 1.4 \times 10^{-7}$, $D_M(Na^+) = 4.4 \times 10^{-8}$, and $D_M(K^+) = 4.1 \times 10^{-8}$ cm^2s^{-1}.

6.5 MOFS AT THE MESOSCOPIC SCALE

AFM examination of the metallic deposits produced by electrochemical reduction of Cu(MOF) and Zn(MOF) revealed significant differences seen in Figure 6.7. Pristine Cu-MOF particles appeared as elongated crystals 100–200 nm-sized with a corrugated surface texture (Figure 6.7a). Upon application of a constant potential step at –250 mV for five minutes, a contraction of the crystals was observed, accompanied by a small variation in their profile. As seen in Figure 6.7b, the application of potential stages at more potentials negative enough to promote the reduction to Cu metal, a different series of marked transversal structures 10–15 nm-sized and 45–50 nm regularly

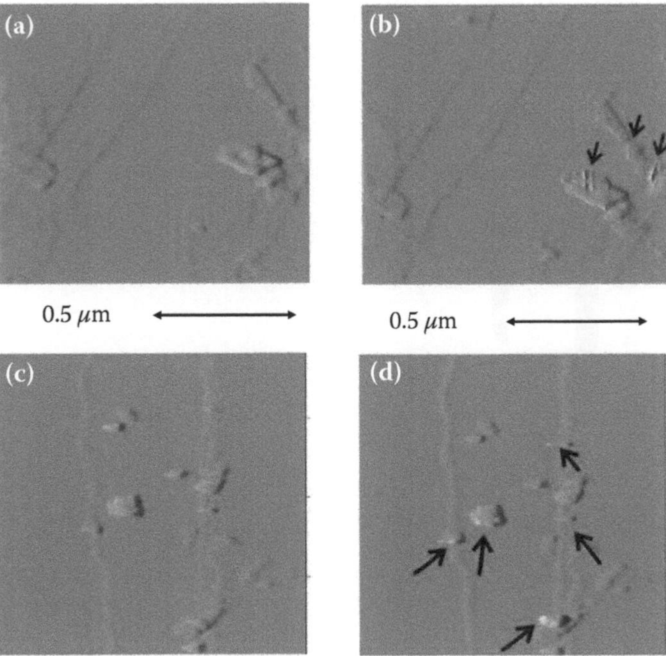

FIGURE 6.7 AFM Images Recorded on a Cu-MOF Modified Electrode in Contact with 0.50 M Acetate Buffer (pH 4.85): (a) Pristine Sample; (b) After Application of a Potential Step at –0.85 V for 5 min, and Zn-MOF Modified Electrode Immersed in the Same Electrolyte: (c) Pristine Sample; (d) After Application of a Potential Step at –0.94 V for five minutes. The Arrows Mark the Metallic Features Formed After Application of Cathodic Potential Inputs. Reproduced from Ref. [11] (Doménech-Carbó et al. *J. Phys. Chem. C.* 2007, 111: 13701–13711), with Permission.

spaced appear. Such transversal sheets can unambiguously be attributed to Cu metal deposits because: (i) such sheets increased in thickness by prolonging the time duration of the potential step and/or applying more negative potentials, (ii) in returning the potential at values close to –0.35 V, the above features entirely disappear. If more negative potentials are reached, the formation of external deposits of metal particles, similar to those obtained in conventional electrodeposition experiments from Cu^{2+} ions in solution, is observed.

In the case of Zn(MOF), there is a growth of the metallic phase in the border of the irregular crystals. As seen in Figure 6.7c, the deposit of Zn-MOF consisted of irregular 20–30 nm sized particles. Upon application of potentials of –0.94 V, the crystals show a set of small circular-like surface features, size ca. 10 nm, growing in the MOF crystal body, as seen in Figure 6.7d. Application of potentials of –1.25 V (or more negative ones) produced similar features with the parallel appearance of external Zn deposits. This behavior is comparable to that reported by Hasse and Scholz for the electrochemical reduction of litharge (PbO) to Pb metal described in terms of a proton-assisted topotactic transformation [18]. Notice that this behavior agrees with that expected from the solid-state electrochemistry modeling summarized in Section 3.6—the reduction of the M-MOF crystals starts at the particle/base electrode/electrolyte three-phase junction further expanding via the formation of a metallic phase in the vicinity of this junction.

Figure 6.8 compares a pictorial representation of the metal deposition mode with an AFM image of a crystal of Cu-MOF after being submitted to a cathodic input resulting in the formation of bands of metallic copper coring the crystal. These features suggest that the propagation of the redox reaction follows structural guidelines (probably dislocations, cracks, and crevices) determining the preferential alignment of the metallic phase conjointly with the kinetics of the nucleation and growth phenomena.

Again, the phenomena occurring at the so-called mesoscopic scale (see Section 3.9) becomes influential to the electrochemical properties of the porous materials. Another example can be seen

(a)

(b)

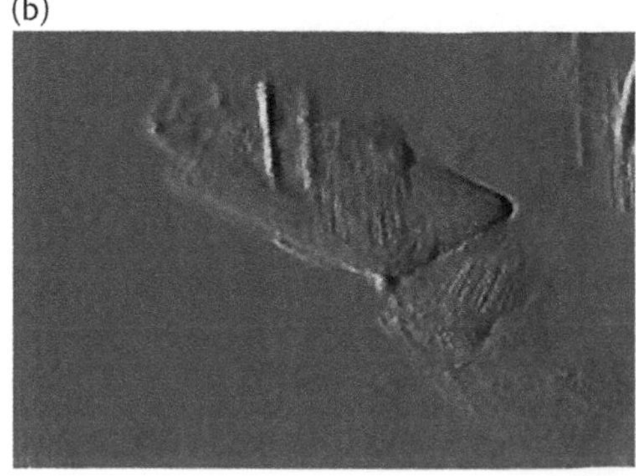

FIGURE 6.8 Pictorial Representation of Metal Deposition Mode on Cu-MOF (a) and AFM Image of a Crystal of This Compound on a Graphite Plate After Being Submitted to a Potential Step at –0.85 V for 5 in Contact with Aqueous Acetate Buffer at pH 4.85. Reproduced from Ref. [11] (Doménech-Carbó et al. *J. Phys. Chem. C.* 2007, 111: 13701–13711), with Permission.

in Figure 6.9, where TEM images recorded for Cu-MOF crystals (a) before and (b) after application of a constant potential of –1.0 V for five minutes while in contact with 0.50 M acetate buffer at pH 4.85 are shown. After application of the reductive potential input, the parent cuboid crystals 1–1.5 μm sized present their surface divided into rectangular domains separated by holes. These images can be rationalized assuming that a fast reductive dissolution with the issue of Cu^{2+} ions from the Cu-MOF lattice with concomitant formation of external deposits of Cu metal occurs. As before, privileged crystallographic directions are involved, so a regular distribution of holes results.

6.6 APPLICATIONS

The high specific surface area and large ion permeability, combined with the presence of redox-active centers, make MOFs interesting materials for electrochemical sensing and electrocatalysis. By the first token, size selectivity can be combined with the coordinating ability of redox-active centers. The low electrical conductivity of most MOFs, however, limits their sensing performance [19]. For this reason, MOFs are combined with a variety of other functional materials with high electrical conductivity. These include graphene and its derivatives (graphene oxide, GO, reduced graphene oxide, rGO) and carbon nanotubes (CNTs) [20], but other strategies involve the incorporation of guest molecules [21] or the formation of conducting polymers within the MOF channels [9].

Another line of research is devoted to synthesizing MOFs with high electrical conductivity [22]. This is of interest also for gas sensing applications, prompted by the large capacity and selective gas adsorption properties of MOFs. Field-effect transistors (FETs) are increasingly studied for this purpose. Here, the conductance change in the material forming the channel as a result of gas adsorption is the key to the sensing process [23].

Electrocatalytic applications of MOFs have been directed to increase the rate of HER, ORR, and OER processes. Again, hybridization with other materials is exploited. One interesting possibility is the insertion of GO into the Cu-MOF as an integral component [24]. This strategy improves the limitation in local charge transport in MOFs by the pore aperture-defined size-exclusion effects, GO linking MOF nodes. Figure 6.10 shows an example of a bifunctional catalyst formed by

FIGURE 6.9 TEM Images Recorded for Cu-MOF crystals (a) Before, and (b) After Application of a Constant Potential of –1.0 V for Five Minutes while in Contact with 0.50 M Acetate Buffer (pH 4.85). Reproduced from Ref. [11] (Doménech-Carbó et al. *J. Phys. Chem. C.* 2007, 111: 13701–13711), with Permission.

Co-MOF with CNTs [25]. This catalyst exhibits high catalytic effects on OER and ORR processes, as illustrated by the LSVs in Figure 6.10, corresponding to the OER curves recorded at Co-MOF@CNTs with different loadings. The best catalyst, Co-MOF@CNTs (5 wt% CNTs), lowers the onset potential (see Section 4.4) by approximately 200 mV relative to the CNTs. In turn, the composite material improves the performance of the Co-MOF alone by ca. 50 mV.

The relevant properties of MOFs have prompted their use for catalysis, gas storage, and separation [26], as well as fuel cells [27], lithium batteries [28] supercapacitors [29], electrochromic devices [30], drug delivery [31], and bioimaging [32]. Conversely, MOFs can be synthesized electrochemically, as described by Mueller et al. [27]. MOFs are indirectly related to other electrochemical applications acting as a template for the synthesis of porous carbon to be applied as double-layer electrochemical capacitors [33]. In summary, the capability of MOFs to promote electrode reactions is based on [3]: (i) the strong adsorbability to the reactant; (ii) the capability of deposition of insoluble, inert products to be stored in the nanopores, inducing the growth of the products into uniform nanostructures; (iii) the dispersion of catalytic centers in the porous framework for efficient or even synergistic catalysis; and (iv) the possibility to establish systematic structure–reactivity relationships, prompting the synthesis of MOFs with definite properties.

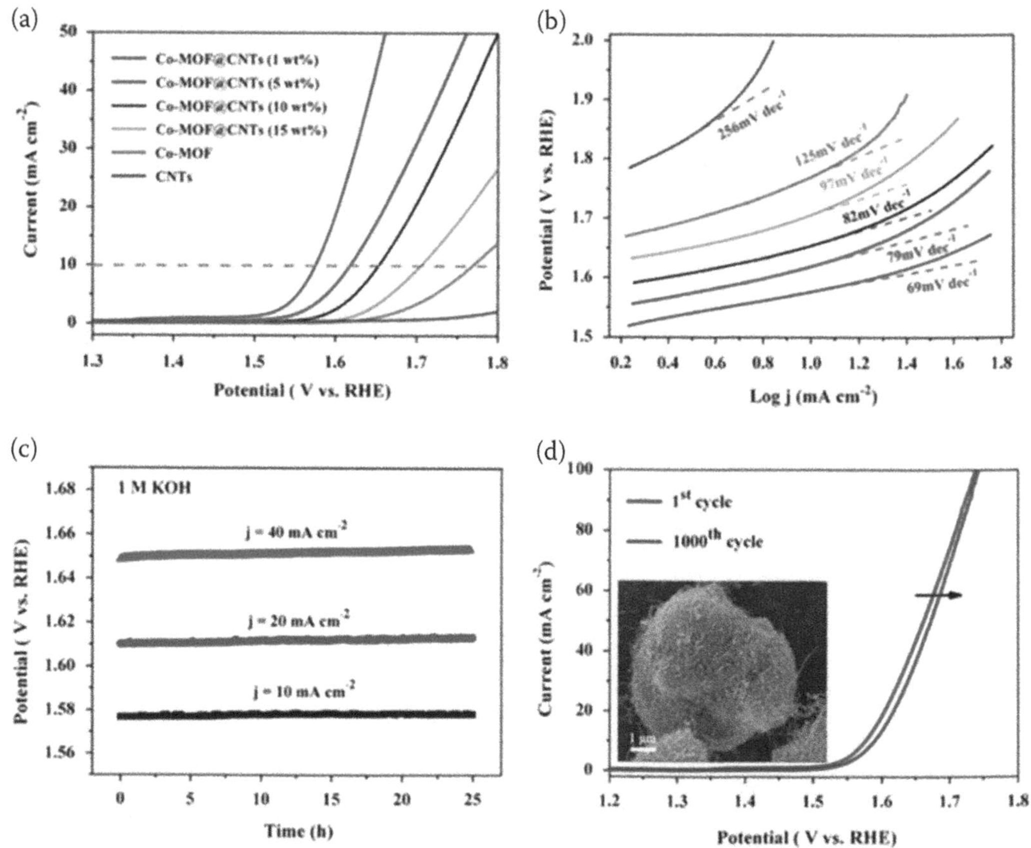

FIGURE 6.10 (a) LSV Curves of OER on CNTs, Co-MOF, Co-MOF@CNTs (15 wt %), Co-MOF@CNTs (10 wt %), Co-MOF@CNTs (5 wt %), and Co-MOF@CNTs (1 wt %) in 1.0 M KOH. Potential Scan Rate 5 mV s^{-1}. (b) Tafel Plots; (c) Chronopotentiometric Curves of Co-MOF@C at Different Current Densities; (d) Durability Test for the Co-MOF@CNTs (5 wt%) After 1,000 Cycles (Inset: TEM of the Co-MOF@CNTs (5 wt%) After the Stability Test. From Ref. [25] (Fang et al. *J. Power Sources*. 2016, 326: 50–59), with Permission.

REFERENCES

[1]. Tomic, E.A.J. 1965. Glass transition temperatures of poly(methyl methacrylate) plasticized with low concentrations of monomer and diethyl phthalate. *Applied Polymer Science*. 9: 3745–3818.

[2]. Yaghi, O.M.; Li, H.; Groy, T.L. 1996. Construction of porous solids from hydrogen-bonded metal complexes of 1,3,5-benzenetricarboxylic acid. *Journal of the American Chemical Society*. 118: 9096–9101.

[3]. Rowsell, J.L.C.; Yaghi, O.M. 2004. Metal–organic frameworks: A new class of porous materials. *Microporous and Mesoporous Materials*. 73: 3–14.

[4]. Maurin, G.; Serre, C.; Cooper, A.; Fereyd, G. 2017. The new age of MOFs and their porous-related solids. *Chemical Society Reviews*. 46: 3104–3107.

[5]. Zhou, J.; Wang, B. 2017. Emerging crystalline porous materials as a multifunctional platform for electrochemical energy storage. *Chemical Society Reviews*. 46: 6927–6945.

[6]. Kitagawa, S.; Kitaura, R.; Noro, S. 2004. Functional porous coordination polymers. *Angewandte Chemie International Edition*. 43: 2334–2375.

[7]. Narayan, T.C.; Miyakai, T.; Seki, S.; Dinca, M. 2012. High charge mobility in a tetrathiafulvalene-based microporous metal–organic framework. *Journal of the American Chemical Society*. 134: 12932–12935.

[8]. Wan, S.; Gandara, F.; Asano, A.; Furukawa, H.; Saeki, A.; Dey, S.K.; Liao, L.; Ambrogio, M.W.; Botros, Y.Y.; Duan, X.; Seki, S.; Stoddart, J.F.; Yaghi, O.M. 2011. Covalent organic frameworks with high charge carrier mobility. *Chemistry of Materials*. 23: 4094–4097.

[9]. Le Ouay, B.; Boudot, M.; Kitao, T.; Yanagida, T.; Kitagawa, S.; Uemura, T. 2016. Nanostructuration of PEDOT in porous coordination polymers for tunable porosity and conductivity. *Journal of the American Chemical Society*. 138: 10088–10091.

[10]. Doménech, A.; García, H.; Doménech, M.T.; Llabrés-i-Xamena, F. 2006. Electrochemical nanometric patterning of MOF particles: Anisotropic metal electrodeposition in Cu/MOF. *Electrochemistry Communications*. 8: 1830–1834.

[11]. Doménech-Carbó, A.; García, H.; Doménech-Carbó, M.T.; Llabrés-i-Xamena, F. 2007. Electrochemistry of metal-organic frameworks: A description from the voltammetry of microparticles approach. *Journal of Physical Chemistry C*. 111: 13701–13711.

[12]. Ferey, G.; Millange, F.; Morcrette, M.; Serre, C.; Doublet, M.-L.; Greneche, J.-M.; Tarascon, J.-M. 2007. Mixed-valence Li/Fe-based metal–organic frameworks with both reversible redox and sorption properties. *Angewandte Chemie International Edition*. 46: 3259–3263.

[13]. Aubrey, M.L.; Long, J.R. 2015. A dual–ion battery cathode via oxidative insertion of anions in a metal–organic framework. *Journal of the American Chemical Society*. 137: 13594–13602.

[14]. Wang, R.Y.; Shyam, B.; Stone, K.H.; Weker, J.N.; Pasta, M.; Lee, H.-W.; Toney, M.F.; Cui, Y. 2015. Reversible multivalent (monovalent, divalent, trivalent) ion insertion in open framework materials. *Advanced Energy Materials*. 5: artic. 1401869.

[15]. Zhang, Z.; Yoshikawa, H.; Awaga, K. 2014. Monitoring the solid-state electrochemistry of Cu(2,7-AQDC) (AQDC = anthraquinone dicarboxylate) in a lithium battery: Coexistence of metal and ligand redox activities in a metal–organic framework. *Journal of the American Chemical Society*. 136: 16112–16115.

[16]. Sharifker, B.; Hills, G. 1983. Theoretical and experimental studies of multiple nucleation. *Electrochimica Acta*. 28: 879–889.

[17]. Milchev, A.; Heerman, L. 2003. Electrochemical nucleation and growth of nano- and microparticles: Some theoretical and experimental aspects. *Electrochimica Acta*. 48: 2903–2913.

[18]. Hasse, U.; Scholz , F. 2001. In situ atomic force microscopy of the reduction of lead oxide nanocrystals immobilized on an electrode surface. *Electrochemistry Communications*. 3: 429–434.

[19]. Fang, X.; Zong, B.; Mao, S. 2018. Metal–organic framework-based sensors for environmental contaminant sensing. *Nano-Micro Letters*. 10: artic. 64.

[20]. Wang, X.; Wang, Q.X.; Wang, Q.H.; Gao, F.; Gao, F.; Yang, Y.Z.; Guo, H.X. 2014. Highly dispersible and stable copper terephthalate metal–organic framework-graphene oxide nanocomposite for an electrochemical sensing application. *ACS Applied Materials & Interfaces*. 6: 11573–11580.

[21]. Kung, C.-W.; Otake, K.; Buru, C.T.; Goswami, S.; Cui, Y.; Hupp, J.T.; Spokoyny, A.M.; Farha, O.K. 2018. Increased electrical conductivity in a mesoporous metal–organic framework featuring metalla-carboranes guests. *Journal of the American Chemical Society*. 140: 3871–3875.

[22]. Campbell, M.G.; Sheberla, D.; Liu, S.F.; Swager, T.M.; Dinca, N. 2015. Cu(3)(hexaiminotriphenylene)(2): An electrically conductive 2D metal–organic framework for chemiresistive sensing. *Angewandte Chemie International Edition*. 54: 4349–4352.

[23]. Mao, S.; Chang, J.; Pu, H.; Lu, G.; He, Q.; Zhang, H.; Chen, J. 2017. Two dimensional nanomaterial-based field-effect transistors for chemical and biological sensing. *Chemical Society Reviews*. 46: 6872–6904.
[24]. Jahan, M.; Liu, Z.; Loh, K.P.A. 2013. Graphene oxide and copper-centered metal organic framework composite as a tri-functional catalyst for HER, OER, and ORR. *Advanced Functional Materials*. 23: 5363–5372.
[25]. Fang, Y.; Li, X.; Li, F.; Lin, X.; Tian, M.; Long, X.; An, X.; Fu, Y.; Jin, J.; Ma, J. 2016. Self-assembly of cobalt-centered metal organic framework and multiwalled carbon nanotubes hybrids as a highly active and corrosion-resistant bifunctional oxygen catalyst. *Journal of Power Sources*. 326: 50–59.
[26]. Janiak, C. 2003. Engineering coordination polymers towards applications. *Journal of the Chemical Society Dalton Transactions*. 2781–2804.
[27]. Mueller, U.; Schubert, M.; Teich, F.; Puetter, H.; Schierle-Arndt, K.; Pastré, J. 2006. Metal-organic frameworks-prospective industrial applications. *Journal of Materials Chemistry*. 16: 626–636.
[28]. Li, X.; Cheng, F.; Zhang, S.; Chen, J. 2006. Shape-controlled synthesis and lithium-storage study of metal-organic frameworks Zn4O(1,3,5-benzenetribenzoate)2. *Journal of Power Sources*. 160: 542–547.
[29]. Simon, P.; Gogotsi, Y. 2008. Materials for electrochemical capacitors. *Nature Materials*. 7: 845–854.
[30]. Mortimer, R.J. 2011. Electrochromic materials. *Annual Reviews of Material Research.* 41: 241–268.
[31]. Horcajada, P.; Serre, C.; Vallet-Regi, M.; Sebban, M.; Taulelle, F.; Ferey, G. 2006. Metal–organic frameworks as efficient materials for drug delivery. *Angewandte Chemie International Edition*. 45: 5974–5978.
[32]. Park, K.M.; Kim, H.; Murray, J.; Koo, J.; Kim, K. 2017. A facile preparation method for nanosized MOFs as a multifunctional material for cellular imaging and drug delivery. *Supramolecular Chemistry*. 29: 441–445.
[33]. Liu, B.; Shioyama, H.; Akita, T.; Xu, Q. 2008. Metal-organic frameworks as a template for porous carbon synthesis. *Journal of the American Chemical Society*. 130: 5390–5391.

7 Electrochemistry of Porous Metals and Anodic Metal Oxide Films

7.1 INTRODUCTION

Porous metals are widely used in electrochemistry as high surface area working electrodes, current collectors, and as substrates for electrode modification. The preparation of a variety of porous metal architectures with different porosity length scales and various degrees of ordering or periodicity, is receiving considerable attention due to the electrocatalytic applications of these materials [1]. The effective utilization of the catalysts is conditioned by several factors, including (a) mass-specific surface area; (b) mechanical, chemical, and electrochemical stability (corrosion resistance); (c) interfacial interactions with the catalyst support; (d) regime of mass transport of reactants and product species between the electrode surface and the electrolyte bulk; (e) poisoning of the catalyst due to the interaction with reactants, products or matrix species; and (f) catalysts losses due to secondary electrochemical processes [2].

Obviously, adopting porous electrodes enhances the effective surface area, but this can be complemented with several other strategies to improve catalyst efficiency—the use of metal nanoparticles, formation of nanocomposites, and generation of metal oxides. From a synthetic point of view, porous metals can be prepared by directed chemical routes, using sacrificial templates [3], dealloying approaches, or macropore formation methods, among others [4]. However, the minimization of structural defects is a common problem currently faced by additive manufacturing methods, including 3D printing [5,6].

From the structural point of view, metallic porous materials can be divided into stochastic and ordered depending on the randomness of pore size and arrangement [1]. Figure 7.1 shows a representation of different types of metallic porous structures accompanied by a qualitative evaluation of the application of traditional and additive manufacturing techniques to their fabrication. Stochastic materials include foams and sponges while ordered materials forming micro- or nanolattices can be divided into periodic and aperiodic. In all cases, to achieve high specific surface area and facilitate diffusion, the porous structures should ensure high connectivity, minimizing the number of isolated cavities [7]. Figure 7.2 illustrates the porous structure of different Ni-Fe foams acting as OER catalysts [8]. This is illustrative of one of the widely used strategies to build electrocatalysts: the formation of oxide layers. Here, Ni-Fe foams are submitted to thermal treatments forming a fine oxidized coating that displays catalytic effects on OER. As an alternative, the electrochemical oxidation (anodization) of metal surfaces aimed to obtain efficient catalysts.

Most of the electrochemical applications of porous metals involve some modification via incorporation of metal or metal oxide nanoparticles, conducting polymers, carbonaceous materials, etc. In this chapter, we will treat the systems formed by anodization of metal substrates to form oxidic porous films that have an important role in electrochemical sensing and catalysis.

7.2 ELECTROCHEMISTRY OF NOBLE METALS

The aqueous electrochemistry of the majority of metals is dominated by the oxidative dissolution processes where the metal surfaces are stripped. These processes, favored by the presence of complexing species in the electrolyte solution, obviously condition their applications for sensing, electrosynthesis, energy production, and storage, etc. because the metal stability is limited to a

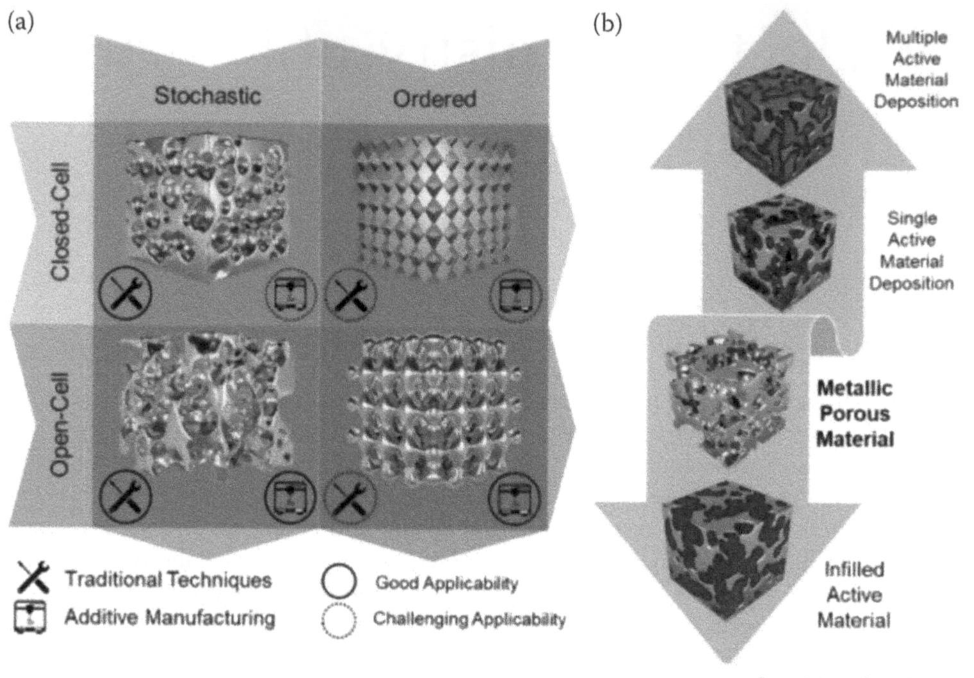

FIGURE 7.1 Types of Metallic Porous Structures and Qualitative Evaluation of the Application of Traditional and Additive Manufacturing Techniques to the Fabrication of Each Material Type. Reproduced from Ref. [1] (Egorov and O'Dwyer, *Curr. Opin. Electrochem.* 2020, 21: 201–208), with Permission.

relatively narrow range of potentials. The behavior of several so-called noble metals, however, is different and they remain stable in a wide range of potentials.

A paradigmatic example is the electrochemistry of gold in acidic media. Figure 7.3 shows a typical initial anodic scan CV of a microcrystalline gold electrode in contact with air-saturated 0.1 M H_2SO_4 aqueous solution. The voltammogram shows an anodic wave at ca. 1.2 (A_{Au}) that precedes the rising current for the OER process accompanied—in the subsequent cathodic scan—by a tall peak at ca. 0.85 V (C_{Au}), which precedes the ORR wave at –0.15 V and the HER rising current beyond –0.45 V. This voltammetry is interpreted in terms of the formation of a gold oxide monolayer (A_{Au} process) that is subsequently reduced to gold metal (C_{Au} process). For simplicity, the A_{Au} process can be represented as [9]:

$$\{Au\}_{solid} + 3H_2O \rightarrow \{Au(OH)_3\}_{solid} + 3H^+_{aq} + 3e^- \quad (7.1)$$

This hydrated oxide monolayer can experience different processes [10]: dehydration to Au_2O_3:

$$2\{Au(OH)_3\}_{solid} \rightarrow \{Au_2O_3\}_{solid} + 3H_2O \quad (7.2)$$

and/or successive electrochemical reactions at more positive potentials:

$$\{Au(OH)_3\}_{solid} + \{Au\}_{solid} + H_2O \rightarrow 2\{AuO(OH)\}_{solid} + 3H^+_{aq} + 3e^- \quad (7.3)$$

$$5\{AuO(OH)\}_{solid} \rightarrow 2\{Au_2O_3\}_{solid} + \{Au\}_{solid} + 2O_2 + 5H^+_{aq} + 5e^- \quad (7.4)$$

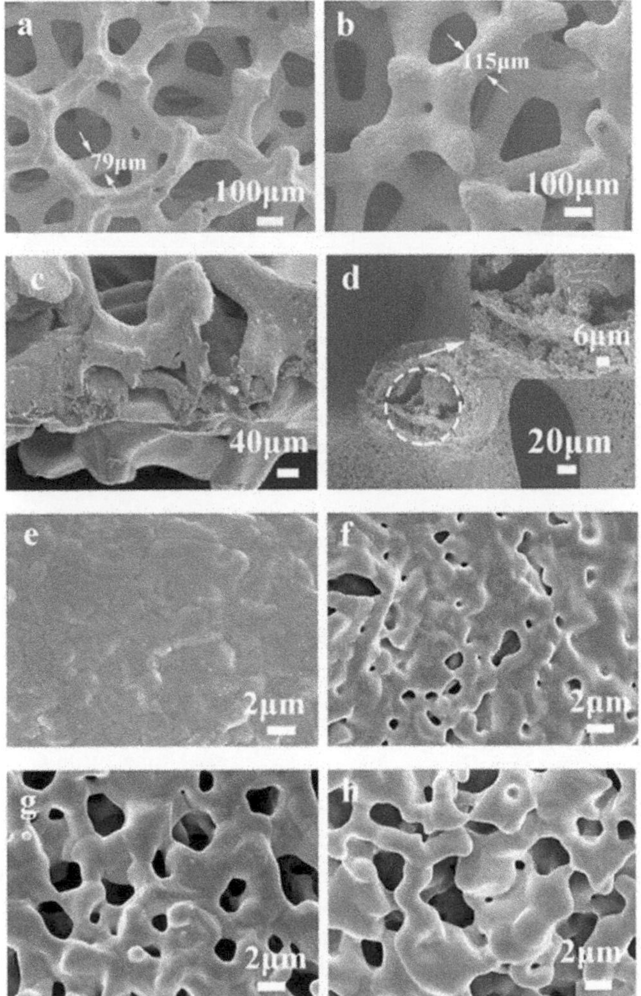

FIGURE 7.2 SEM Images of NiFe Foams Oxidized at Different Temperatures (a) Initial and (b) Redox-NiFe-800; Cross-Section SEM Images of (c) Initial Foam and (d) Redox-NiFe-800. Magnified SEM Images of (e) Pristine NiFe, (f) Redox-NiFe-700, (g) Redox-NiFe-800 and (h) Redox-NiFe-900, Respectively. Reproduced from Ref. [8] (Wu et al. *Electrochim. Acta.* 2019, 309: 415–423), with Permission.

As a result, Au_2O_3 is formed. The gold(III) oxide is then reduced to Au via the process C_{Au}:

$$\{Au_2O_3\}_{solid} + 6H^+_{aq} + 6e^- \rightarrow 2\{Au\}_{solid} + 3H_2O \tag{7.5}$$

at potentials clearly different from where the process A_{Au} takes place. However, the gold electrochemistry is complicated by the process described by Eq. (7.4); it can be viewed as producing a catalytic effect on OER, already seen as involving intermediate hydroperoxy species [9,10].

The voltammetric response, however, is quite sensitive to the surface treatment of the gold electrode so that the gold oxidation signal often appears like two or three overlapping peaks whose relative height varies with the number of cycles—this feature can be related to the oxide formation on the exposed crystal planes (Au(100), Au(110) and Au(111) faces, mainly) that are oxidized at different potentials [11]. The electrochemical turnovers produce the surface restructuring by

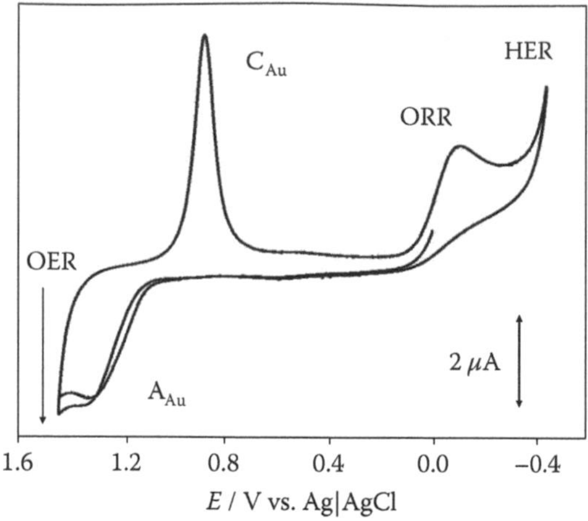

FIGURE 7.3 CV of a Polycrystalline Gold Electrode Immersed into Air-Saturated 0.10 M H_2SO_4 Aqueous Solution. Potential Scan Rate 50 mV s^{-1}.

potential cycling due to gold oxide formation/reduction, gold dissolution/re-deposition, and/or diffusion of dissolved species into bulk solution.

In this electrochemistry, adsorbed oxygen species play an important role because it has been proposed that the early stage of oxidation of noble metal surfaces involves the formation of adsorbed hydroxyl radicals [12,13]:

$$H_2O \rightarrow HO\cdot_{ads} + H^+_{aq} + e^- \quad (7.6)$$

Electrochemical gold dissolution is a process to be viewed within this context. It has been proposed that it proceeds through two pathways: at low anodic potentials, the dissolution occurs by place-exchange between metal and adsorbed hydroxyl/oxygen ions; at higher potentials, gold dissolution is superimposed to the OER process [14]. Figure 7.4 shows a schematic representation of such processes. First, there is the formation of a layer of adsorbed O/OH species that protects gold from dissolution. This passivation is accompanied by place-exchange between O/OH and Au ions, resulting in de-passivation and re-passivation possibilities during the anodic process. The subsequent reduction involves desorption of reversibly adsorbed O/OH and partial gold dissolution. As a result, the topology of the remaining gold surface is significantly altered with respect to the parent surface.

In the presence of a complexing agent—in particular, chloride ions—the oxidative dissolution of gold is more pronounced. For simplicity, this can be represented as [15]:

$$\{Au\}_{solid} + 4Cl^-_{aq} \rightarrow AuCl^-_{4aq} + 3e^- \quad (7.7)$$

Another significant aspect in the electrochemistry of noble metals is the generation of active sites via chemical or electrochemical pathways [16,17]. Roughly, one can distinguish three types of gold atoms near the metal surface: (a) bulk lattice atoms occupying ordered lattice sites; (b) regular surface atoms having a lower coordination number; and (c) displaced surface atoms having an unusually low coordination number as a result of their location on surface defect sites and/or being place-exchanged with surface oxide. The third group corresponds to active gold sites whose oxidation, which can be represented as

$$Au^* \xrightarrow{-3e^-} Au^{3+} \xrightarrow{+(n+3)H_2O} Au^{3+}\cdot(n+3)H_2O \xrightarrow{-3H^+} Au(OH)_3\cdot nH_2O \quad (7.8)$$

occurs at lower potentials than bulk gold.

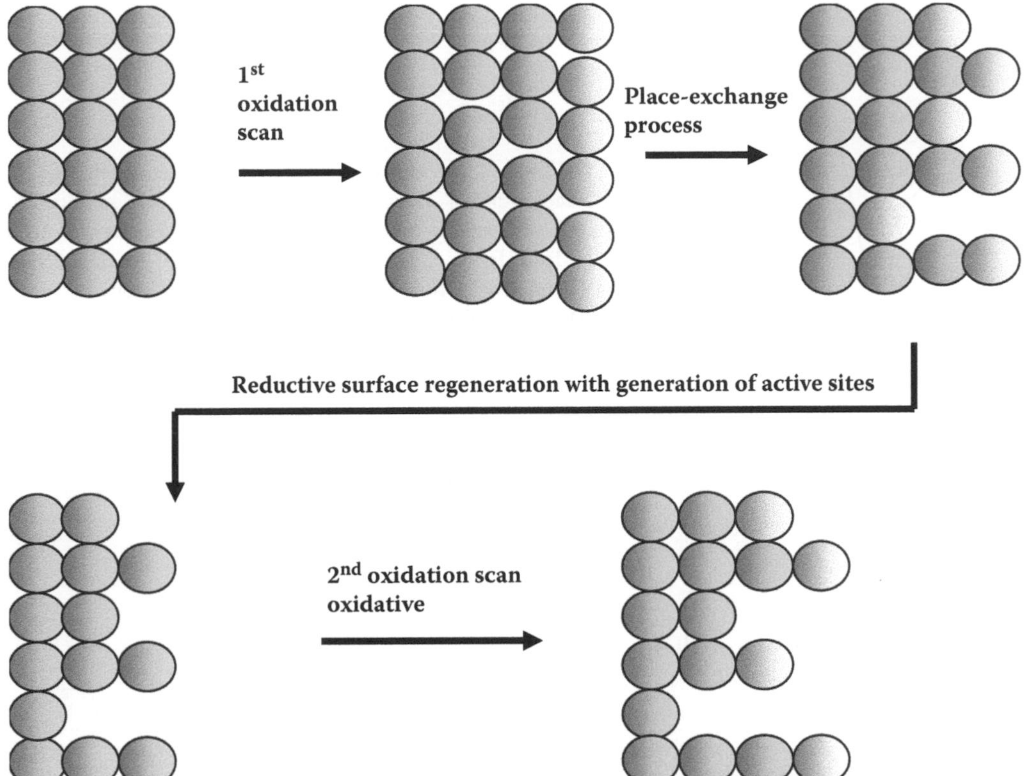

FIGURE 7.4 Model of the Oxidation/Reduction Cycles of Gold Electrodes in Contact with Aqueous Electrolytes. Gold in the Double-Layer Potential Range Acquires Reversibly Adsorbed O/OH Ions on the Surface at Potentials Around 1.3 V. This Process is Followed by Place-Exchange Between O/OH and Au Ions. In the Subsequent Cathodic Step (see Figure 7.2) There is Desorption of Adsorbed O/OH with Concomitant De-passivation and Partial Gold Dissolution. As a Result, the Roughened Surface Gold Surface Becomes Clearly Different from the Original One.

Figure 7.5 shows first and second LSVs recorded for a nanosample of *astater* coin minted in Cartago in 310–290 BCE attached to a graphite electrode in contact with 0.10 M HCl [18]. In the initial anodic scan, a unique oxidation peak ca. 1.0 V vs. Ag/AgCl is recorded. After this, a cathode scan was carried out. The intensity of the gold oxidation peak decreases in the second scan due to the partial oxidative dissolution of the sample, whereas a second anodic signal ca. 0.8 V appears. This second signal is interpreted as due to the oxidation of gold active sites generated as a result of the dissolution/reconstruction of the gold lattice illustrated in Figure 7.4.

The formation of active gold sites via a place-exchange process with oxygen surface atoms determines the penetration of such atoms in deeper layers and is involved in the spillover process [19]. The role of the surface properties (fractal surfaces, for instance [20]), including such active sites in electrocatalysis, is a matter for current research [17,21].

7.3 ELECTROCATALYSIS OF NOBLE METALS

Among the processes widely studied with Pt, Au, etc. electrodes, the ORR and OER in aqueous electrolytes have particular interest. In both cases, the formation of adsorbed oxygenated species is considered as essential in determining the catalyst performance [22]. The proposed ORR mechanisms can be resumed in three pathways:

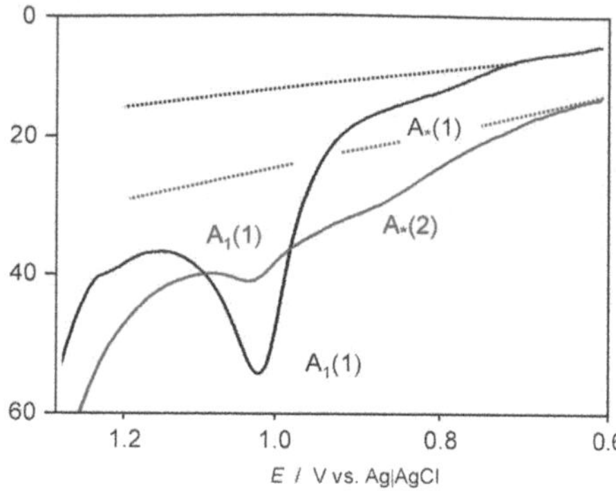

FIGURE 7.5 Anodic Scan LSVs Recorded for a Nanosample of *Astater* Coin Minted in Cartago in 310–290 BCE Attached to a Graphite Electrode, in Contact with 0.10 M HCl. Black Line: First Scan; Red Line: Second LSVs. Potential Scan Rate 50 mV s^{-1}. Reproduced from Ref. [18] (Doménech-Carbó and Scholz, *Acc. Chem. Res.* 2019, 52: 400–406), with Permission.

a. Association mechanism involving the formation of adsorbed OOH, which breaks into O and OH. The associative mechanism in alkaline media can be represented as:

$$O_2 + H_2O + e^- \rightarrow HO_2{\cdot}_{ads} + OH^-_{aq} \quad (7.9)$$

$$HO_2{\cdot}_{ads} + e^- \rightarrow O{\cdot}_{ads} + OH^-_{aq} \quad (7.10)$$

$$O{\cdot}_{ads} + H_2O + e^- \rightarrow HO{\cdot}_{ads} + OH^-_{aq} \quad (7.11)$$

$$HO{\cdot}_{ads} + e^- \rightarrow OH^-_{aq} \quad (7.12)$$

b. Dissociation/recombination mechanism that proceeds through the breaking of the O=O bond yielding $O{\cdot}_{ads}$,

$$\tfrac{1}{2}O_2 \rightarrow O{\cdot}_{ads} \quad (7.13)$$

subsequently yielding $HO{\cdot}_{ads}$ via one proton, one electron reduction, subsequently given OH^-_{aq}, as described in Eq. (7.12).

c. Peroxo mechanism involving two successive electron-transfer steps, leading to $HO_2{\cdot}_{ads}$ and to H_2O_{2ads}, the last breaking into $HO{\cdot}_{ads}$.

From the phenomenological point of view, electrochemical data differentiate between direct four-electron reduction of O_2 to H_2O and stepped reduction through two two-electron steps involving intermediate formation of H_2O_2. Here, there are competing pathways of H_2O_2 desorption and H_2O_2 dissociation influenced, among other factors, by the presence of spectator species that can adsorb onto the electrode surface [23].

The catalytic performance is sensitive to the adsorption of electrolyte species on Pt(*hkl*) planes and a variety of factors. Much research has been focused on deposits of metal nanoparticles, including different compositions (bimetallic, etc.) and structures (core-shell, etc.). The above ORR mechanisms are correlated with those of the OER. Two main mechanisms have been proposed for this reaction on metal surfaces, both involving the formation of metal-hydroxyl surface species in active (M^{δ}) metal centers [24]:

$$\{M\}_{surf} + H_2O \rightarrow \{M^{\delta} - OH\}_{surf} + H^+_{aq} + e^- \quad (7.14)$$

$$\{M^{\delta} - OH\}_{surf} \rightarrow \{M^{\delta} - O\}_{surf} + H^+_{aq} + e^- \quad (7.15)$$

The first pathway consists of a direct recombination of oxygen atoms to give O_2,

$$\{M^{\delta} - OH\}_{surf} + \{M^{\delta} - OH\}_{surf} \rightarrow \{M^{\delta} - O\}_{surf} + \{M\}_{surf} + H_2O \quad (7.16)$$

$$\{M^{\delta} - O\}_{surf} + \{M^{\delta} - O\}_{surf} \rightarrow \{M\}_{surf} + \{M\}_{surf} + O_2 \quad (7.17)$$

while the second involves two new electron transfer processes:

$$\{M^{\delta} - O\}_{surf} + H_2O \rightarrow \{M^{\delta} - OOH\}_{surf} + H^+_{aq} + e^- \quad (7.18)$$

$$\{M^{\delta} - OOH\}_{surf} \rightarrow \{M\}_{surf} + O_2 + H^+_{aq} + e^- \quad (7.19)$$

Figure 7.6 presents a scheme summarizing the ORR and OER pathways [22]. The catalytic performance appears to be clearly dependent on the adsorption of oxygen species, in contrast with ORR, where only adsorption appears to be influential; in OER the subsurface processes are important [22]. There is a compromise between the oxygen adsorption and the formation of HO_{2ads} and the OH_{ads} desorption rates, so Pt and Pd are the best catalysts for ORR.

There is a variety of factors influencing the catalytic OER at metal electrodes. Among others, studies on single-crystal data reveal that the formation of surface oxide and O_2 evolution are plane-sensitive, increasing in the order Au(110) > Au(111) > Au(100) in the case of gold [10]. Similarly, the porosity and pore structure of the catalyst is of considerable importance—in particular, the macro- and mesoporosity because the microporous fraction (pore size <2 nm) of the total surface is relatively inaccessible by reactants during electrochemical reactions [25]. In turn, the pore distribution influences the detachment of O_2 because the formation of gas bubbles can block parts of the electroactive surface area.

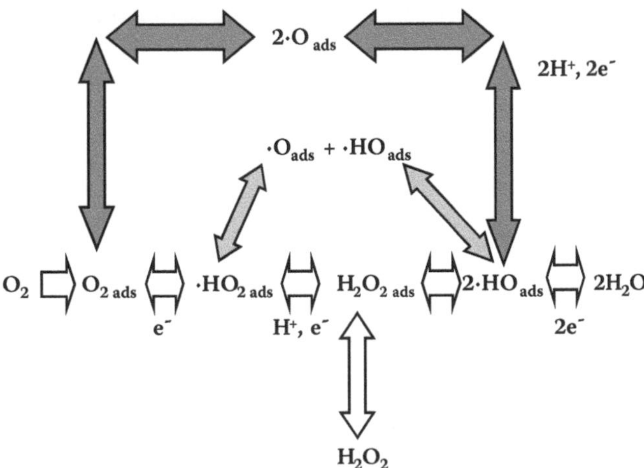

FIGURE 7.6 Scheme Representing the Different Mechanisms Proposed for the ORR and OER Process. White Arrows: Peroxo Pathway; Pale Grey Arrows: Associative Pathway; Dark Grey Arrows: Dissociative Pathway.

Recent studies have revealed the possibility of Pt(II) and Pt(IV) species in the metal surface [26] and underlined the importance of the surface and subsurface composition and structure [25]. Figure 7.7 depicts a scheme for the catalytic mechanism and phase structure for OER on platinum under alkaline conditions [26]. At the open circuit potential (OCP), the Pt electrode has a subsurface layer of Pt^{δ}–OH_{ads} and a surface layer of $Pt(OH)_2$ and PtO. At potentials positive enough to promote OER, an additional layer of PtO_2 is developed in the outermost region, while the proportion of Pt^{δ}–OH_{ads} increases significantly. According to this modeling, the formation of the Pt^{δ}–OH_{ads} layer appears as the rate-determining step of the overall process. This is accompanied by the formation of an external oxidic layer of high surface area where OH^- ions need to diffuse to activate the OER.

Figure 7.8 shows two examples of Volcano plots for the electrocatalytic reduction of CO_2 to CO and CH_4 at metal oxide surfaces. The representation in Figure 7.8a corresponds to the dependence of the partial current density for CO_2 reduction at −0.8 V vs. the binding strength of CO and metal surfaces, whereas the plot in Figure 7.8b corresponds to plots of onset potentials for CO_2 reduction and methane/methanol [27].

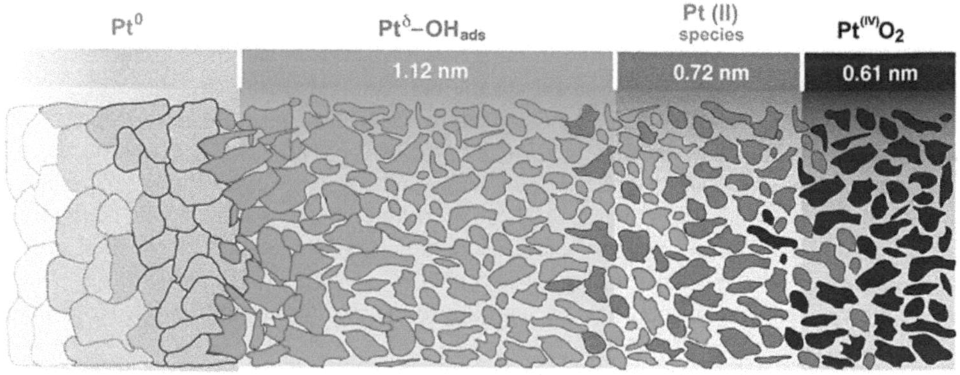

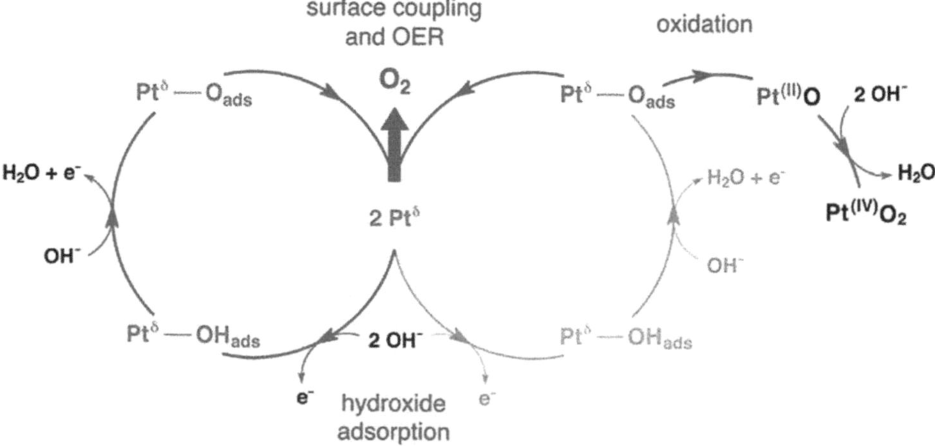

FIGURE 7.7 Catalytic Mechanism and Phase Structure for OER on Platinum Under Alkaline Conditions. Reproduced from Ref. [26] (Favaro et al. *J. Mater. Chem. A*. 2017, 5: 11634–11643), with Permission.

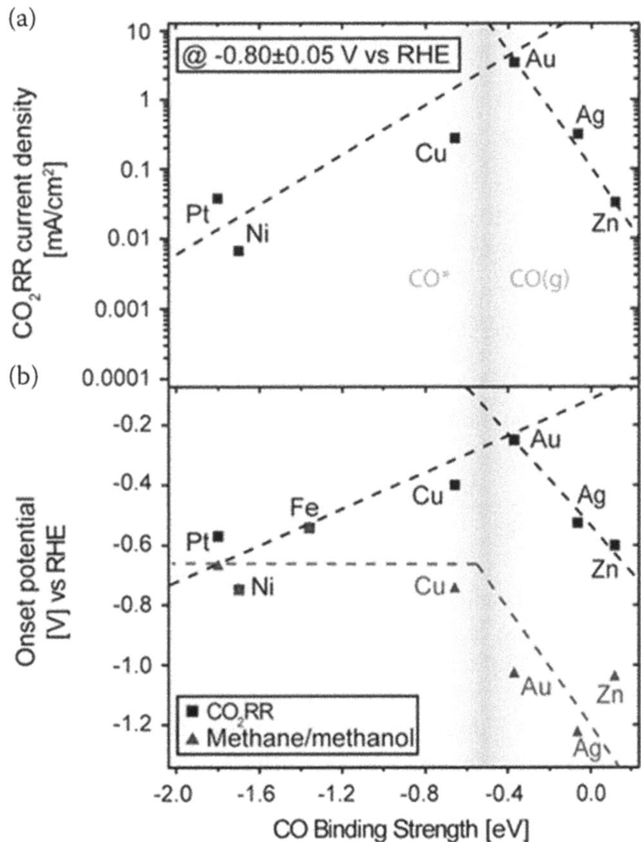

FIGURE 7.8 Volcano Plots of (a) Partial Current Density for CO_2 Reduction at –0.8 V vs. CO Binding Strength, and (b) Distinct Onset Potentials vs. CO Binding Energy: The Overall CO_2 Reduction and Reduction to Methane or Methanol. Reproduced from Ref. [27] (Kuhl et al. *J. Am. Chem. Soc.* 136: 14107–14113), with Permission.

7.4 POROUS ANODIC METAL OXIDE FILMS

Porous films of more or less hydrated metal oxides are electrochemically formed by surface oxidation of metallic electrodes immersed in suitable electrolytes. Such porous anodic metal oxide films can be regarded as technological materials with a variety of applications (*vide infra*).

There are several models to describe the growth of oxide layers on metal surfaces [28]. The pore resistance formulation assumes that the process is controlled by the ohmic resistance of the passive layer. Under conditions of two-dimensional growth, the thickness of the film remains constant while the covered surface area increases. Then, the peak current measured in an LSV experiment will satisfy the relationship [29]:

$$I_p = A(1 - \vartheta_p)\sqrt{\frac{nF\rho\sigma v}{M}} \quad (7.20)$$

where A is a geometrical surface area of the electrode covered by a passivating compound of molecular mass M and density ρ, σ represents the specific conductivity of the electrolyte, and θ_p is the fraction of covered surface at I_p.

To consider three-dimensional growth of the anodic film, one can assume that the rate of the film thickening is proportional to the overpotential relative to the OCP [28]

$$\frac{\mathrm{d}\delta(t)}{\mathrm{d}t} = \lambda(E - E_{\mathrm{OCP}}) \tag{7.21}$$

where $\delta(t)$ is the film thickness at time t and λ a proportionality constant. Accordingly, the current will depend on the rate of growth of both the coverage ($\theta(t)$) and thickness of the film:

$$I = \frac{nF\rho A}{M}\left[\delta(t)\frac{\mathrm{d}\vartheta(t)}{\mathrm{d}t} + \vartheta(t)\frac{\mathrm{d}\delta(t)}{\mathrm{d}t}\right] \tag{7.22}$$

Through some manipulations, the model leads to [28]:

$$\frac{v}{I_{\mathrm{p}}^{2}} = \frac{M}{nF\rho oA^{2}(1 - \vartheta_{\mathrm{p}})^{2}} + \frac{2\lambda(1 - 2\vartheta_{\mathrm{p}})}{oA(1 - \vartheta_{\mathrm{p}})^{2}}\frac{(E_{\mathrm{p}} - E_{\mathrm{OCP}})}{I_{\mathrm{p}}} \tag{7.23}$$

Accordingly, if experiments carry out different potential scan rates, a plot of v/I_p^2 vs. $(E_p–E_{OCP})/I_p$ should give a sloping straight line in the case of three-dimensional growth. For two-dimensional growth, Eq. (7.23) reduces to (7.20) and the plot of v/I_p^2 vs. $(E_p–E_{OCP})/I_p$ should give a horizontal straight line.

Among others, porous anodic alumina films have found application in nanotechnology due to their nanometer-scale porous structure and sizes of particles forming the pore wall material. Such films are formed by galvanostatic anodization of Al electrodes immersed into solutions of, for instance, H_2SO_4 plus $Al_2(SO_4)_3 \cdot 18H_2O$, under vigorous stirring. The phenomena inside the pores during Al anodizing are complex and greatly affect the mechanism of film growth. The pore generally broadens towards the film surface as a result of the pore wall chemical dissolution reaction by the electrolyte inside the pores during film growth [30,31], but pores can also open towards the surface owing to the similar open-circuit oxide dissolution, which is a first-order reaction with respect to proton activity and is hindered by incorporated electrolyte anions.

Figure 7.9 shows a schematic representation of migration processes accompanying the formation of porous oxide films by anodization of metals, illustrated here for the case of aluminum. Two layers can be distinguished—that adjacent to the metal where O^{2-} and Al^{3+} are the only migrating species, and the outer layer (mixed or barrier layer) where H^+ and OH^- also migrate and electrolyte ions are embodied [32,33]. Al is consumed according to Faraday's law, oxide forms in the metal/oxide interface, and O^{2-} and Al^{3+} migrate through vacancies in the pure oxide and barrier layers—ac-accompanied by migration of H^+ and OH^- ions (*vide infra*). It should be noted, however, that different mechanisms of anodic oxide film formation are operative in acidic and alkaline media [32].

According to Moon and Pyun [34], the capacitance, C_{ox}, and resistance, R_{ox}, of the oxide film depend on its thickness, δ_{ox}, as:

$$C_{\mathrm{ox}} = \frac{\varepsilon_{\mathrm{o}}\varepsilon_{\mathrm{ox}}{}'A_{\mathrm{ox}}}{\delta_{\mathrm{ox}}} \tag{7.24}$$

$$R_{\mathrm{ox}} = \frac{\delta_{\mathrm{ox}}}{\sigma_{\mathrm{ox}}A_{\mathrm{ox}}} \tag{7.25}$$

In the above equations, A_{ox} represents the exposed area of the specimen, σ_{ox} the film conductivity, ε_o is the vacuum permittivity, and ε_{ox}' the relative permittivity of the metal oxide. Experimental

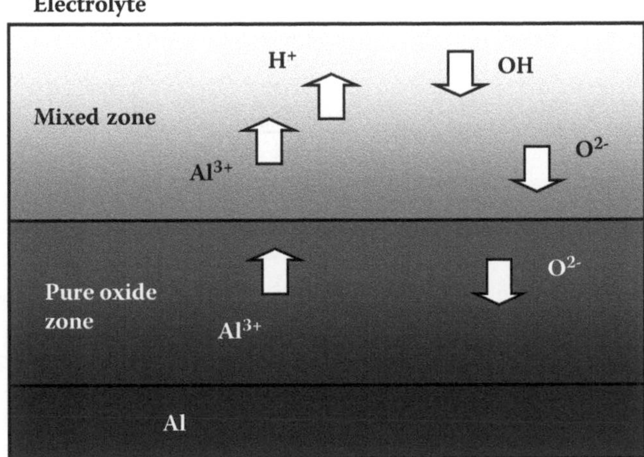

FIGURE 7.9 Schematic Diagram Shown Migration Processes Accompanying the Formation of Porous Oxide Films by Anodization of Al Metal.

data for Al anodization [34] indicate that the film resistance increases linearly with film formation potential in alkaline solution. However, in acidic media, the resistance of the oxide film decreases by increasing the film formation potential. This can be attributed to an increase in the vacancy concentration within the oxide film, which could be explained by a field-assisted dissolution of anodic oxide film at the oxide/solution interface generating aluminum vacancies:

$$Al_{Al} \rightarrow Al^{3+}(aq) + V'''_{Al} \tag{7.26}$$

In the above equation, Al_{Al} represents the 'normal' aluminum ion in the regular site of the oxide film and V'''_{Al} represents the negatively charged aluminum vacancy in the oxide film. Here, the Kröger-Vink notation for representing point defects in solids is used (for a more detailed view, see Chapter 13). In alkaline media, the generation of aluminum vacancies can be explained by the occurrence of a process involving water molecules adsorbed on the oxide film:

$$H_2O\ (ads) \rightarrow O_O + 2/3V'''_{Al} + 2H^+(aq) \tag{7.27}$$

Here, O_O represents normal oxygen ion in the regular site of the oxide film. Since water is an uncharged species, its adsorption on the oxide film is scarcely influenced by the applied electric field, thus resulting in a small change in the concentration of aluminum vacancies with the applied potential.

The current (I) measured during the growth of the porous film can be taken as the sum of the ionic current, I_{ion}, due to the oxidation of the metal on the metal/oxide interface and the electronic current, I_{el}, due to Faradaic processes occurring at the oxide/electrolyte interface. The former can be described as the sum of the formation current density, I_{form}, associated with oxide formation, and the dissolution current, I_{dis}, related to the dissolution of metal ions into the electrolyte at the pore bottom. Then,

$$I = I_{ion} + I_{el} = I_{form} + I_{dis} + I_{el} \tag{7.28}$$

The average currents can be evaluated from weight measurements by applying Faraday's law using the relationships [35]:

$$(I_{\text{ion}})_{\text{av}} = \frac{z(\Delta m)_{\text{M}}F}{S_{\text{m}}M_{\text{M}}\Delta t} \tag{7.29}$$

$$(I_{\text{form}})_{\text{av}} = \frac{z(\Delta m)_{\text{MO}}F}{S_{\text{m}}M_{\text{MO}}\Delta t} \tag{7.30}$$

$$(I_{\text{dis}})_{\text{av}} = (I_{\text{ion}})_{\text{av}} - (I_{\text{form}})_{\text{av}} \tag{7.31}$$

where S_m is the apparent surface area of the sample, $(\Delta m)_M$ and $(\Delta m)_{MO}$ are the masses of consumed metal and metal oxide formed, respectively, at a time Δt, and z is the ionic charge. Porosity, P, defined as the pore volume (S_p)/total volume (S_m) ratio, can be approached as [36]:

$$P = \frac{S_{\text{p}}}{S_{\text{m}}} = \frac{I_{\text{dis}}}{I_{\text{ion}}} \tag{7.32}$$

Regular films of low porosity are desired for mechanical applications, while high pore surface density, high pore base diameter (probably), thickness, and real surface are desired for catalysis. The growth of low porosity films is favored by low temperatures, low electrolyte concentrations, and high current densities. Sulfate additives like $Al_2(SO_4)_3$, $MgSO_4$, Na_2SO_4, however, favor the appearance of pitting, which is due to the non-uniform growth of film, yielding to undesired porosity, roughness, cracks, etc. Current developments attempt to obtain regular grown, low enough porosity, and high surface density and base diameter of the pores and build of multiple oxide nanoarchitectures.

7.5 IMPEDANCE ANALYSIS OF METAL OXIDE LAYERS

Given the capability of EIS to differentiate multiple interfaces based on the frequency response to a modulated potential, this technique is extensively used to characterize metal/metal oxide systems—particularly in devices devoted to energy production and storage [37]. Figure 7.10 shows a set of equivalent circuits commonly used to describe EIS data of metal oxide electrodes and the corresponding Nyquist plots for different values of the impedance parameters. The most single case (A) is the Randles circuit, for which the Nyquist plot consists of a semicircle arc. If the capacitor ($\beta = 1$ in Eq. (1.22)) is substituted by a CPE (i.e. a nonideal capacitor, $0 < \beta < 1$), a depressed semicircle is recorded. The (minus) phase angle presents a maximum at an angular frequency ω equal to $1/R_1C_1$.

The second model (B) includes one (or more) additional parallel RC parallel unit. This is often attributed to the presence of surface passivation layers on metal oxide electrodes to account for charge transport through the outside layer before reaching the base electrode [38]. In model C, the second parallel RC circuit is embedded within the former. The physical interpretation of this widely used equivalent circuit is to some extent controversial. It has been attributed to the presence of defect states in semiconducting metal oxides [39] and contact resistance associated with the difference between the work functions of the metal and the metal oxide [40,41].

The model D is obtained by adding a Warburg element (see Section 1.11) in series with the charge transfer resistance to the Randles circuit. This term accounts for the diffusion of electroactive species to the electrode surface and/or the diffusion charge-balancing counterions through the porous metal oxide. This element can also be associated with the motion of mobile carriers (electrons and holes) through conduction and valence bands in semiconductor metal oxides. The Warburg diffusion in a layer of finite thickness can be differently expressed depending on the "open" and "shorted" contacts at the solid/electrolyte interface. The respective expressions are:

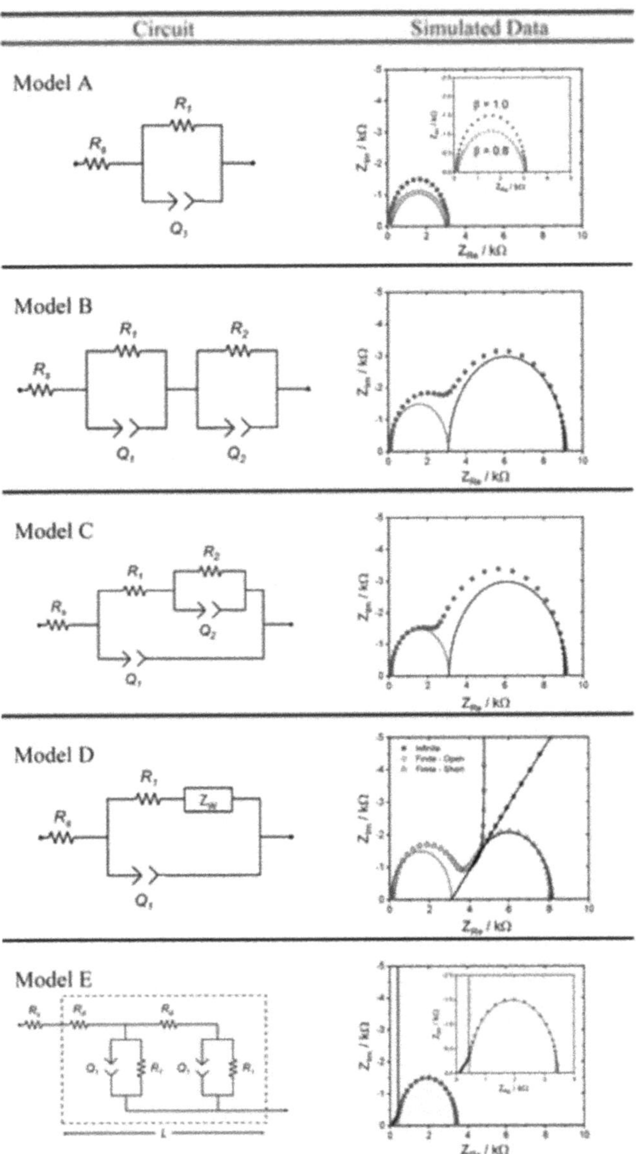

FIGURE 7.10 Summary of Equivalent Circuits Frequently Used to Describe EIS Data of Metal Oxide Electrodes. Numerical Values of Impedance Parameters: $R_1 = 3$ kΩ, $R_2 = 6$ kΩ, $Q_1 = 1$, μF $s^{\beta-1}$, $Q_2 = 10$ μF $s^{\beta-1}$, $R_d = 1$ kΩ, $\sigma = 5$ kΩ $s^{-1/2}$, $L/D^{1/2} = 1$ $s^{1/2}$, $\beta = 1.0$ Unless Otherwise Indicated. Reproduced from Ref. [37] (Bredar et al. *ACS Appl. Energ.Mater.* 2020, 3: 66–68), with Permission.

$$Z_W = \frac{\sigma_W \sqrt{2}}{(j\omega)^{1/2}} \coth\left(\frac{L}{D^{1/2}}(j\omega)^{1/2}\right) \tag{7.33}$$

$$Z_W = \frac{\sigma_W \sqrt{2}}{(j\omega)^{1/2}} \tanh\left(\frac{L}{D^{1/2}}(j\omega)^{1/2}\right) \tag{7.34}$$

where L represents the thickness of the diffusion layer and σ_W is given by Eq. (1.24).

The model E is applied to describe porous metal oxide semiconductors where thin layer diffusion of mobile carriers occurs [42]. This is represented by a resistance R_d in series with a parallel R_1/Q_1 circuit accompanied by distributed r_d and r_1/q_1 elements. These last elements describe the porosity of the system where there are electrolyte/oxide and oxide/base metal interfaces. These elements are defined as $r_d = R_d/L$, $r_1 = R_1L$ and $q_1 = Q_1/L$ [43], and it is assumed that the charge transfer processes at the solid/electrolyte interface are coupled with the diffusion of carriers by assuming that these have a steady-state concentration [42]. Figure 7.11 depicts the equivalent circuit models used to describe impedance spectra of mesoporous TiO_2 electrodes upon application of different potentials giving rise to insulator, semiconductor, and metallic behaviors [43]. This approach will be complemented in Chapter 15 upon describing the photoelectrochemistry of semiconducting materials.

7.6 PSEUDOCAPACITIVE BEHAVIOR

The later generation of capacitors (supercapacitors) include electrochemical double-layer capacitors (EDLCs) and pseudocapacitors that will be studied in Chapter 14. In the first group, the charge is stored by adsorption of electrolyte ions onto the electrode surface; in the second, charge storage is associated with redox reactions occurring at the electrode surface. Figure 7.12 shows an example of pseudocapacitive behavior in metal oxides, corresponding to CV of RuO_2 electrode in aqueous H_2SO_4 and acidified KCl electrolytes [44]. Apart from the rising current at the extreme of positive potentials—due to the OER process—the voltammograms exhibit a rectangular-type profile where, only

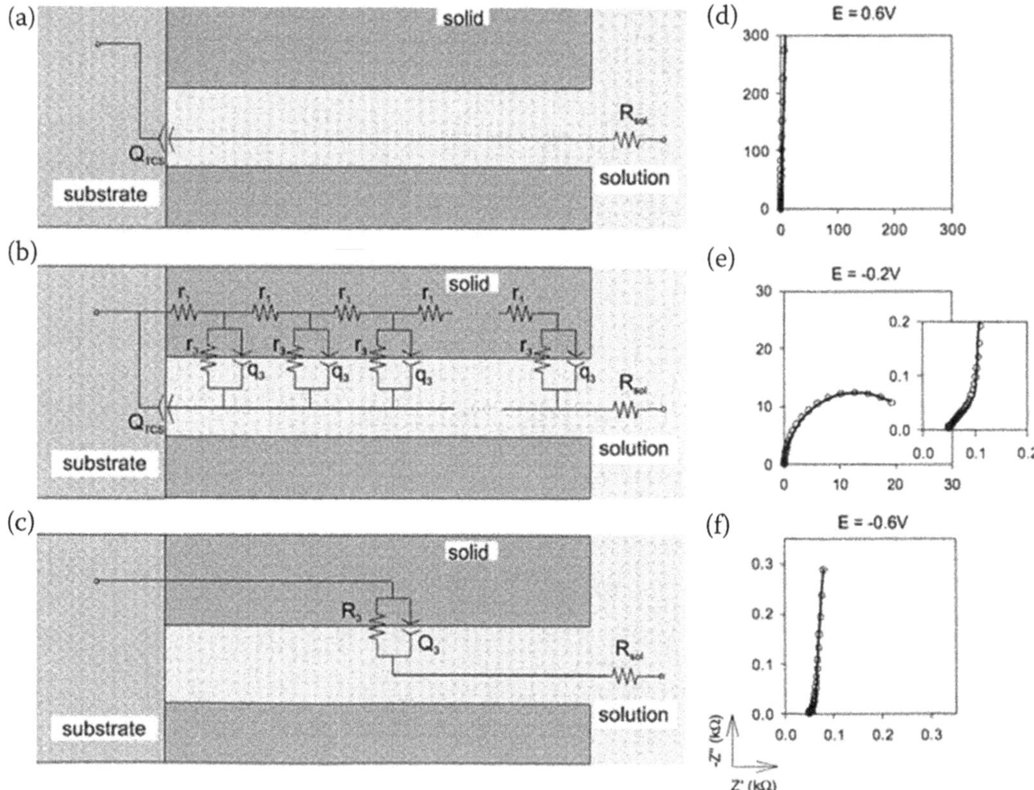

FIGURE 7.11 Equivalent Circuit Models Used to Describe Impedance Spectra and Experimental Data Obtained for Mesoporous TiO_2 Electrodes. (a) Insulator (Anodic Potentials), (b) Semiconductor, and (c) Metal (Cathodic Potentials). Nyquist Plots (d–f) Reproduced from Ref. [43] (Fabregat-Santiago et al. *J. Phys.Chem. B.* 2002, 106: 334 – 339), with Permission.

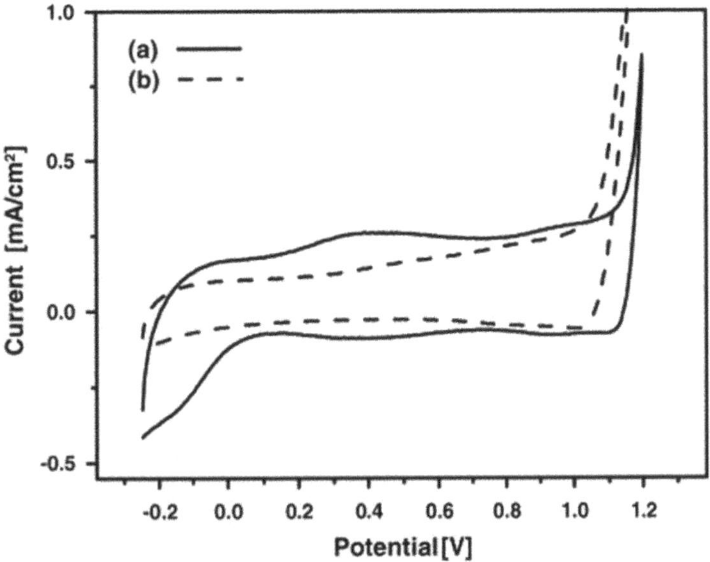

FIGURE 7.12 CV of RuO_2 Electrode in Aqueous (a) H_2SO_4 (pH = 2.04) and (b) KCl (pK = 2.00) Electrolytes. Potential Scan Rate 10 mV s^{-1}. Reproduced from Ref. [44] (Wen et al. *Electrochim. Acta.* 2004, 50: 849–855), with Permission.

in the case of H_2SO_4, weak cathodic/anodic waves appear. The capacitances varied linearly with the pH and pK^+, respectively, thus implying that the charging processes follow a first-order kinetics relative to proton and potassium ion concentrations [44]. The charging/discharging rate depends on the size of the hydrated cation and its mobility as well as on the rate of adsorption/desorption at the surface sites.

These processes have been attributed to the injection of hydrated protons at the oxide/solution interface into the active metal sites of the oxide surface. In the case of MnO_2, the pseudocapacitive processes can be represented as [44,45]:

$$\{MnO_2\}_{surf} + \delta H^+_{aq} + \delta e^- \rightarrow \{MnO_{2-\delta}(OH)_\delta\}_{surf} \quad (7.35)$$

$$\{MnO_2\}_{surf} + \delta K^+_{aq} \rightarrow \delta e^- \rightarrow MnO_{2-\delta}(OK)_\delta\}_{surf} \quad (7.36)$$

Experimental data reveals that, in contrast with electroactive species in the solution phase, the charge density (q) determined from the area under the voltammetric peaks decreases with increasing potential scan rate. This can be attributed to the progressive exclusion of less accessible regions of the oxide film to the penetration of the electrolyte protons.

A simple model can be made assuming that q can be divided into two "outer" (q_o) and "inner" (q_i) contributions. At high scan rates ($\nu \rightarrow \infty$), the measured charge density tends to vary linearly with $\nu^{-1/2}$. These conditions correspond to the confinement of the electrochemical process to the external region of the coating so that [46,47]:

$$q = q_o + \frac{A}{\sqrt{\nu}} \quad (7.37)$$

where A is a constant characterizing each oxide coating. In turn, at low scan rates ($\nu \rightarrow 0$), the measured voltammetric charge densities tends to a variation with potential scan rate that can be expressed as:

$$\frac{1}{q} = \frac{1}{q_o + q_i} + B\sqrt{v} \tag{7.38}$$

B being a second oxide-characteristic constant. Accordingly, voltammetric data in a sufficiently wide interval of scan rates allow us to determine the values of q_o and q_i, representative of the outer and inner components of the pseudocapacitive behavior of the metal oxide.

In the pseudocapacitive region of potentials, the EIS responses can generally be described using the models in Figure 7.10. Remarkably, the impedance behavior is quite sensitive to the bias potential applied during the record of the impedance spectra. In particular, when the applied potentials fall in the OER region (i.e. in the region of high positive potentials, see Figure 7.12), a new capacitive loop appears. This is interpreted in terms of the formation of a resistive interlayer between the active oxide layer and the metallic substrate [47].

7.7 APPLICATIONS OF POROUS METALS AND OXIDIC POROUS LAYERS

Sensing at gold and platinum electrodes has a long tradition in electroanalytical chemistry. Recent advances include porous metals and alloys; for instance, Cu, Cu-Ni, and Rh-modified Cu porous layers for nitrate and nitrite sensing [48]. The sensing strategies exploit both direct and indirect methods of metal electrodeposition [49].

As previously noted, porous anodic metal oxide films have a wide variety of applications, among others, in anti-corrosion [50], rechargeable batteries [51] and templates for metal nanowires, and synthesis of porous carbons and nanostructured catalysts. Lastly, there is interest in porous anodic membranes, which can be obtained after dissolving the remaining metal and oxide barrier film separating metal from the pore bottom. This kind of membrane is of interest by their possible use in aggressive environments, where the most common polymeric membranes do not resist [52].

REFERENCES

[1]. Egorov, V.; O'Dwyer, C. 2020. Architected porous metals in electrochemical energy Storage. *Current Opinion in Electrochemistry*. 21: 201–208.

[2]. Pilapil, B.K.; van Drunen, J.; Makonnen, Y.; Beauchemin, D.; Jerkiewicz, G.; Gates, B.D. 2017. Ordered porous electrodes by design: Toward enhancing the effective utilization of platinum in electrocatalysis. *Advanced Functional Materials*. 27: artic. 1703171.

[3]. Osiak, M.; Geaney, H.; Armstrong, E.; O'Dwyer, C. 2014. Structuring materials for lithium-ion batteries: Advancements in nanomaterial structure, composition, and defined assembly on cell performance. *Journal of Materials Chemistry A*. 2: 9433–9460.

[4]. Huang, A.; He, Y.; Zhou, Y.; Yang, Y.; Zhang, J.; Luo, L.; Mao, Q.; Hou, D.; Yang, J. 2019. A review of recent applications of porous metals and metal oxide in energy storage, sensing and catalysis. *Journal of Materials Science*. 54: 949–973.

[5]. Singh, S.; Bhatnagar, N. 2018. A survey of fabrication and application of metallic foams (1925–2017). *Journal of Porous Materials*. 25: 537–554.

[6]. Yeo, S.J.; Oh, M.J.; Yoo, P.J. 2019. Structurally controlled cellular architectures for high-performance ultra-lightweight materials. *Advanced Materials*. 31:artic. 1803670.

[7]. Schaedler, T.A.; Carter, W.B. 2016. Architected cellular materials. *Annual Reviews on Materials Research*. 46: 187–210.

[8]. Wu,Y.; Su, R.; Li, Y.; Wang, Z.; Lü, Z.; Xu, L.; Wei, B. 2019. Redox-sculptured dual-scale nickel-iron foams for efficient water oxidation. *Electrochimica Acta*. 309: 415–423.

[9]. Doyle, R.L.; Lyons, M.E.G. 2014. The mechanism of oxygen evolution at superactivated gold electrodes in aqueous alkaline solution. *Journal of Solid State Electrochemistry*. 18: 3271–3286.

[10]. Diaz-Morales, O.; Calle-Vallejo, F.; de Munck, C.; Koper, M.T.M. 2013. Electrochemical water splitting by gold: Evidence for an oxide decomposition mechanism. *Chemical Science*. 4: 2334–2343.

[11]. Jeyabharathi, C.; Ahrens, O.; Hasse, U.; Scholz, F. 2016. Identification of low-index crystal planes of polycrystalline gold on the basis of electrochemical oxide layer formation. *Journal of Solid State Electrochemistry*. 20: 3025–3031.

[12]. Conway, B.E. 1995. in Davidson, S.G., Ed. *Progress in Surface Science*, vol. 49. Pergamon, New York, p. 331.

[13]. Burke, L.D.; Nugent, P.F. 1997. The electrochemistry of gold: I. The redox behaviour of the metal in aqueous media. *Gold Bulletin*. 30: 43–53.

[14]. Cherevko, S.; Topalov, A.A.; Zeradjanin, A.R.; Katsounaros, I.; Mayrhofer, K.J.J. 2013. Gold dissolution: Towards understanding of noble metal corrosion. *RSC Advances*. 3: 16516–16527.

[15]. Herrera-Gallego, J.; Castellano, C.E.; Calandra, A.; Arvia, A.J. 1975. The electrochemistry of gold in acid aqueous solutions containing chloride ions. *Journal of Electroanalytical Chemistry*. 66: 207–230.

[16]. Burke, L.D.; O'Mullane, A.P. 2000. Generation of active surface states of gold and the role of such states in electrocatalysis. *Journal of Solid State Electrochemistry*. 4: 285–297.

[17]. Nowicka, A.; Hasse, U.; Sievers, G.; Donten, M.; Stojek, Z.; Fletcher, S.; Scholz, F. 2010. Selective knockout of gold active sites. *Angewandte Chemie International Edition*. 49: 3006–3009.

[18]. Doménech-Carbó, A.; Scholz, F. 2019. Electrochemical age determinations of metallic specimens – Utilization of the corrosion clock. *Accounts of Chemical Research*. 52: 400–406.

[19]. Chaparro, A.M. 2011. Study of spillover effects with the rotating disk electrode. *Electrochimica Acta*. 58: 691–698.

[20]. Zhao, W.; Xu, J.-J.; Shi, C.-G.; Chen, H.-Y. 2006. Fabrication, characterization and application of gold nano-structured film. *Electrochemistry Communications*. 8: 773–778.

[21]. Jeyabharathi, C.; Hasse, U.; Ahrens, P.; Scholz, F. 2014. Oxygen electroreduction on polycrystalline gold electrodes and on gold nanoparticle-modified glassy carbon electrodes. *Journal of Solid State Electrochemistry*. 18: 3299–3306.

[22]. Katsounaros, I.; Cherevko, S.; Zeradjanin, A.R.; Mayrhofer, K.J.J. 2014. Oxygen electrochemistry as a cornerstone for sustainable energy conversion. *Angewandte Chemie International Edition*. 53: 102–121.

[23]. Katsounaros, I.; Schneider, W.B.; Meier, J.C.; Benedikt, U.; Biedermann, P.U.; Cuesta, A.; Auer, A.A.; Mayrhofer, K.J.J. 2013. The impact of spectator species on the interaction of H_2O_2 with platinum – implications for the oxygen reduction reaction pathways. *Physical Chemistry Chemical Physics*. 15: 8058–8068.

[24]. Yeo, B.S.; Klaus, S.L.; Ross, P.N.; Mathies, R.A.; Bell, A.T. 2010. Identification of hydroperoxy species as reaction intermediates in the electrochemical evolution of oxygen on gold. *Chem Phys Chem*. 11: 1854–1857.

[25]. Conway, B.E. 1995. Electrochemical oxide film formation at noble metals as a surface-chemical process. *Progress in Surface Science*. 49: 331–452.

[26]. Favaro, M.; Valero-Vidal, C.; Eichhom, J.; Toma, F.M.; Ross, P.N.; Yano, J.; Liu, Z.; Crumlin, E.J. 2017. Elucidating the alkaline oxygen evolution reaction mechanism on platinum. *Journal of Materials Chemistry A*. 5: 11634–11643.

[27]. Kuhl, K.P.; Hatsukade, T.; Cave, E.R.; Abram, D.N.; Kibsgaard, J.; Jaramillo, T.F. 2014. Electrocatalytic conversion of carbon dioxide to methane and methanol on transition metal surfaces. *Journal of the American Chemical Society*. 136: 14107–14113.

[28]. Conway, B.E.; Barnett, B.; Angerstein-Kozlowska, H.; Tilak, B.V.A. 1990. Surface-electrochemical basis for the direct logarithmic growth law for initial stages of extension of anodic oxide films formed at noble metals. *Journal of Chemical Physics*. 93: 8361–8373.

[29]. Noskov, A.V.; Grishina, E.P. 2008. Note on the theory of three-dimensional growth of the porous passivating layers on metals. *Journal of Solid State Electrochemistry*. 12: 203–206.

[30]. Birss, V.I.; Wright, G.A. 1982. The kinetics of silver iodide film formation on the silver anode. *Electrochimica Acta*. 27: 1429–1433.

[31]. Patermarakis, G. 2006. Aluminium anodizing in low acidity sulphate baths: Growth mechanism and nanostructure of porous anodic films. *Journal of Solid State Electrochemistry*. 10: 211–222.

[32]. Patermarakis, G.; Chandrinos, J.; Masavetas, K. 2007. Formulation of a holistic model for the kinetics of steady state growth of porous anodic alumina films. *Journal of Solid State Electrochemistry*. 11: 1191–1204.

[33]. Li, Y.; Qin, Y.; Jin, S.; Hu, X.; Ling, Z.; Liu, Q.; Liao, J.; Chen, C.; Shen, Y.; Jin, L. 2015. A new self-ordering regime for fast production of long-range ordered porous ordered aluminum oxide films. *Electrochimica Acta*. 178: 11–17.

[34]. Moon, S.-M.; Pyun, S.-I. 1998. Growth mechanism of anodic oxide films on pure aluminium in aqueous acidic and alkaline solutions. *Journal of Solid State Electrochemistry*. 2: 156–161.

[35]. Bocchetta, P.; Conciauro, F.; Di Quarto, F. 2007. Nanoscale membrane electrode assemblies based on porous anodic alumina for hydrogen-oxygen fuel cell. *Journal of Solid State Electrochemistry*. 11: 1253–1261.

[36]. Sunseri, C.; Spadaro, C.; Piazza, S.; Volpe, M. 2006. Porosity of anodic alumina membranes from electrochemical measurements. *Journal of Solid State Electrochemistry*. 10: 416–421.

[37]. Bredar, A.R.C.; Chown, A.L.; Burton, A.R.; Farnum, B.H. 2020. Electrochemical impedance spectroscopy of metal oxide electrodes for energy applications. *ACS Applied Energy Materials*. 3: 66–98.

[38]. Zhang, W.; Richter, F.H.; Culver, S.P.; Leichtweiss, T.; Lozano, J.G.; Dietrich, C.; Bruce, P.G.; Zeier, W.G.; Janek, J. 2018. Degradation mechanisms at the $Li_{10}GeP_2S_{12}/LiCoO_2$ cathode interface in an all-solid-state lithium-ion battery. *ACS Applied Materials & Interfaces*. 10: 22226–22236.

[39]. Klahr, B.; Gimenez, S.; Fabregat-Santiago, F.; Hamann, T.; Bisquert, J. 2012. Water oxidation at hematite photoelectrodes: The role of surface states. *Journal of the American Chemical Society*. 134: 4294–4302.

[40]. Herraiz-Cardona, I.; Fabregat-Santiago, F.; Renaud, A.; Julián-López, B.; Odobel, F.; Cario, L.; Jobic, S.; Giménez, S. 2013. Hole conductivity and acceptor density of P-type $CuGaO_2$ nanoparticles determined by impedance spectroscopy: The effect of Mg doping. *Electrochimica Acta*. 113: 570–574.

[41]. Venkatraman, M.S.; Cole, I.S.; Emmanuel, B. 2011. Corrosion under a porous layer: A porous electrode model and its implications for self-repair. *Corrosion Science*. 56: 8192–8203.

[42]. Bisquert, J. 2002. Theory of the impedance of electron diffusion and recombination in a thin layer. *Journal of Physical Chemistry B*. 106: 325–333.

[43]. Fabregat-Santiago, F.; Garcia-Belmonte, G.; Bisquert, J.; Zaban, A.; Salvador, P. 2002. Decoupling of transport, charge storage, and interfacial charge transfer in the nanocrystalline TiO_2/electrolyte system by impedance methods. *Journal of Physical Chemistry B*. 106: 334–339.

[44]. Wen, S.; Lee, J.-W.; Yeo, I.-H.; Park, J.; Mho, S.-I. 2004. The role of cations of the electrolyte for the pseudocapacitive behavior of metal oxide electrodes, MnO_2 and RuO_2. *Electrochimica Acta*. 50, 849–855.

[45]. Ardizzone, S.; Fregonara, G.; Trasatti, S. 1990. "Inner" and "outer" active surface of RuO_2 electrodes. *Electrochimica Acta*. 35: 236–237.

[46]. De Pauli, C.P.; Trasatti, S. 1995. Electrochemical surface characterization of IrO_2 + SnO_2 mixed oxide electrocatalysts. *Journal of Electroanalytical Chemistry*. 396: 161–168.

[47]. Faria, E.R.; Ribeiro, F.M.; Franco, D.V.; Da Silva, L.M. 2018. Fabrication and characterisation of a mixed oxide-covered mesh electrode composed of $NiCo_2O_4$ and its capability of generating hydroxyl radicals during the oxygen evolution reaction in electrolyte-free water. *Journal of Solid State Electrochemistry*. 22: 1289–1302.

[48]. Comisso, N.; Cattarin, S.; Guerriero, P.; Mattarozzi, L.; Musíani, M.; Vázquez-Gómez, L.; Verlato, E. 2016. Study of Cu, Cu-Ni and Rh-modified Cu porous layers as electrode materials for the electroanalysis of nitrate and nitrite ions. *Journal of Solid State Electrochemistry*. 20: 1139–1148.

[49]. Kanyanee, T.; Fletcher, P.J.; Madrid, E.; Marken, F. 2020. Indirect (hydrogen-driven) electrodeposition of porous silver onto a palladium membrane. *Journal of Solid State Electrochemistry*. 24: 2789–2796.

[50]. Diggle, J.W.; Downie, T.C.; Goulding, C.W. 1969. Anodic oxide films on aluminium. *Chemical Reviews*. 69: 365–405.

[51]. Mozalev, A.; Magaino, S.; Imai, H. 2001. The formation of nanoporous membranes from anodically oxidized aluminium and their application to Li rechargeable batteries. *Electrochimica Acta*. 46: 2825–2834.

[52]. Asoh, H.; Nishio, K.; Nakao, M.; Tamamura, T.; Masuda, H. 2001. Conditions for fabrication of ideally ordered anodic porous alumina using pretextured Al. *Journal of the Electrochemical Society*. 148: B152–B156.

8 Electrochemistry of Porous Oxides and Related Materials

8.1 OVERVIEW

Metal oxides are widely employed as catalysts and electrocatalysts [1]. The electrochemistry of metal oxides in contact with aqueous electrolytes is dominated by reductive/oxidative dissolution processes and proton-assisted reductions to the corresponding metals, processes that can be characterized and studied using the VIMP methodology [2]. As such processes involve disintegration of the porous structure of the material, electrochemically assisted dissolution processes will be taken only tangentially here.

Most oxide materials adopt microporous structures formed by oxometal units tailored to form cage and tunnel cavities. A significant part of the electrochemistry of metal oxides in contact with aqueous electrolytes consists of surface redox reactions involving tightly bound hydrated oxometal groups. Porous oxides, however, can also undergo electron transfer processes coupled with ion insertion/issue to/from the electrolyte, with the possibility of adopting interstitial positions or binding to metal centers via the formation of hydroxo and aqua groups. Apart from the above Faradaic processes, porous metal oxides and related materials display a rich capacitive and pseudocapacitive electrochemistry.

The electrochemistry of porous solids in contact with non-aqueous electrolytes is receiving growing attention due to the possibility of promotion of ion insertion processes relevant for electrical energy production and storage. These processes will be studied in Chapter 14. Related materials comprise, among others, layered double hydroxides and compounds based on polyoxometalate groups. Layered double hydroxides form lamellar structures entrapping charge-balancing anions. Polyoxovanadates, molybdates, and tungstates are representative of polyoxometalate compounds where complicated geometries are obtained from MO_6 octahedral units containing charge-balancing ions occupying cavities. Finally, a fourth group of materials of electrochemical interest is that obtained by doping the above and related materials. Typically, electroactive ions substitute parent metal centers in the lattice of the original material.

8.2 ELECTROCHEMISTRY OF METAL OXIDES AND METAL OXOHYDROXIDES

A number of metal oxides can be described as porous materials. For instance, porous manganese oxides define octahedral molecular sieves that have been introduced in the last years as possible materials for batteries, separations, and chemical sensing [3]. Interestingly, metal oxides and related materials can eventually be obtained electrochemically. This is the case with MnO_2, which is deposited on solid electrodes upon oxidation of Mn^{2+} salts in aqueous media. The two basic MnO_2 forms, pyrolusite and ramsdellite, are constituted with MnO_6 octahedral units with edge or corner-sharing resulting in 1 × 1 (pyrolusite) or 1 × 2 (ramsdellite) tunnels. The oxidation process can be represented as:

$$Mn^{2+}_{aq} + 2H_2O \rightarrow \{MnO_2\}_{solid} + 4H^{+}_{aq} + 2e^{-} \tag{8.1}$$

The reduction of MnO_2, however, is a complex process, involving proton insertion into the crystal lattice with the formation of a scarcely conductive thin surface layer of MnOOH, further reduced to Mn^{2+} in the solution, the film limiting the reduction rate of MnO_2 [4,5].

These processes can be represented as:

$$\{MnO_2\}_{solid} + H^+_{aq} + e^- \rightarrow \{MnOOH\}_{solid} \tag{8.2}$$

$$\{MnOOH\}_{solid} + 3H^+_{aq} + e^- \rightarrow Mn^{2+}_{aq} + 2H_2O \tag{8.3}$$

Among the first studied synthetic porous oxides were the so-called manganese octahedral molecular sieves (OMS) with the structures of todorokite (OMS-1) and hollandite (OMS-2) [6]. The potassium form of the mineral hollandite (cryptomelane), KMn_8O_{16}, includes one-dimensional tunnels among rigid MnO_2 framework composed of edge-shared and corner-shared MnO_6 octahedra with a tunnel size of 4.6 × 4.6 Å, while the composition of OMS-1 is $Mg_{0.98-1.35}Mn^{II}{}_{1.89-1.94}Mn^{IV}{}_{4.38-4.54}O_{12}\cdot(4.47-4.55)$ H_2O and its structure defines cavities of size 6.9 Å. The electrochemistry of such materials is complicated by the fact that both Mn(IV) and Mn(II) centers coexist. The mixed valence of manganese makes this material a good semiconductor and oxidation catalyst, and the possibility of ion intercalation modulates their structural and catalytic properties [7].

Another example of the complex electrochemistry of porous oxides is V_2O_5. Figure 8.1a shows an SWV of a deposit of solid V_2O_5 attached to graphite electrode in contact with aqueous phosphate buffer at pH 7.0. The voltammograms show two main cathodic peaks at 0.65, 0.34 V, accompanied by weak, overlapping signals at 0.15 and –0.08 V. In the presence of Li^+ ions (Figure 8.1b), the peak at –0.08 V becomes significantly enhanced. This electrochemistry can be described [8] in terms of the superposition of reductive dissolution processes:

$$\{V^V_2O_5\}_{solid} + 6H^+_{aq} + 2e^- \rightarrow 2V^{IV}O^{2+}_{aq} + 3H_2O \tag{8.4}$$

and proton-assisted and metal cation-assisted solid-state transformation processes:

$$\{V^V_2O_5\}_{solid} + 2H^+_{aq} + 2e^- \rightarrow \{V^{IV}_2O_4\}_{solid} + H_2O \tag{8.5}$$

$$\{V^V_2O_5\}_{solid} + xM^+_{aq} + xe^- \rightarrow \{V^V_{2-x}V^{IV}_xO_5(M^+)_x\}_{solid} \tag{8.6}$$

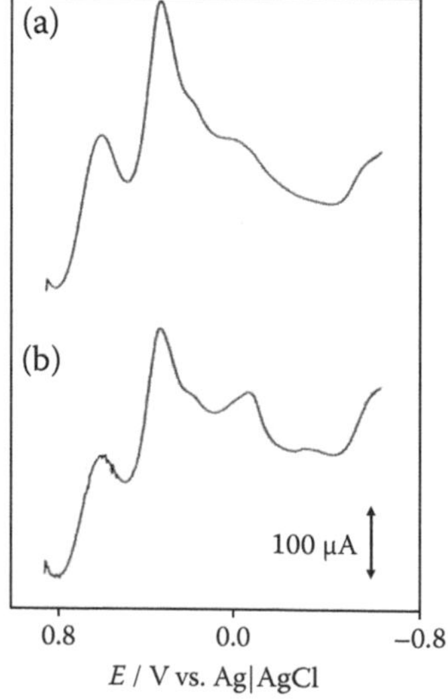

FIGURE 8.1 SWVs for V_2O_5 Attached to Paraffin-Impregnated Graphite Electrode Immersed into: (a) 0.50 M Phosphate Buffer, and (b) 0.50 M Phosphate Buffer Plus 0.05 M $LiNO_3$. Potential Step Increment 4 mV; Square Wave Amplitude 25 mV; Frequency 5 Hz.

To describe the electrochemistry of transition metal oxides in contact with alkaline aqueous electrolytes, it is generally assumed that surface aqua- and oxo-metal species have low degrees of bridging oxygen coordination to the bulk oxide lattice [9,10]. A widely studied case is that of RuO_2. Data for RuO_2-poly(vinyl chloride) (PVC) film electrode in strongly alkaline media also show different couples displaying symmetrical peak shapes, nearly equal anodic and cathodic peak heights, linear increase of anodic peak current with potential scan rate, anodic peak potential, and electrode-rotation independent profiles using RDE. These can be attributed to the Ru (VII)Ru(VI), Ru(VI)/Ru(IV), and Ru(IV)/Ru(III) couples. The variation of peak potentials with the pH can lead to an estimate of the number of OH^- ions per electron, s, involved in each electron transfer step. As in the case of processes in the solution phase, reversible systems should yield linear E_p vs. pH plots of slope $[\partial E_p/\partial pH] = 59\ s/n$. Based on such plots, Dharuman and Chandrasekara Pillai [11] proposed the following for RuO_2-PVC film electrodes in contact with NaOH solutions:

$$\{(-O-)_2Ru^{IV}(OH)_2(H_2O)_2\}_{solid} + 2OH^-_{aq} \rightarrow \{(-O-)_2Ru^{VI}(OH)_4\}_{solid} + 2H_2O + 2e^- \quad (8.7)$$

$$\{(-O-)_2Ru^{VI}(OH)_4\}_{solid} + OH^-_{aq} \rightarrow \{(-O-)_2Ru^{VII}O(OH)_3\}_{solid} + H_2O + e^- \quad (8.8)$$

$$\{(-O-)_2Ru^{III}O(H_2O)_3\}_{solid} + 1.25OH^-_{aq} \rightarrow \{(-O-)_2Ru^{IV}O(OH)_{1.25}(H_2O)_{1.75}\}_{solid} + 1.25H_2O + e^- \quad (8.9)$$

$$\{(-O-)_2Ru^{IV}O(OH)_{1.25}(H_2O)_{1.75}\}_{solid} + 2.5OH^-_{aq} \rightarrow \{(-O-)_2Ru^{VI}O_{1.75}(OH)_{2.25}\}_{solid} + 2.5H_2O + 2e^- \quad (8.10)$$

$$\{(-O-)_2Ru^{VI}O_{1.75}(OH)_{2.25}\}_{solid} + 1.25OH^-_{aq} \rightarrow \{(-O-)_2Ru^{VII}O_3(OH)\}_{solid} + 1.25H_2O + e^- \quad (8.11)$$

It should be noted that, in general, two regions can be distinguished in metal oxides: the external oxide/electrolyte interface, and the internal oxide/electrolyte interface located inside the fissures and cracks. Obviously, this region becomes less accessible for OH^- ions than the first, so the charge passed decreases on increasing potential scan rate. Typically, preparation of this kind of materials uses thermal treatments at different temperatures, so that charges passed through voltammetric experiments increase by decreasing the temperature of preparation of the material. This can be rationalized on considering that by decreasing the oxide preparation temperature, high surface area material—having countless fissures, cracks, crevices, etc.—results [1]. Nuclear reaction analysis coupled with radiometric methods indicated that a thickness of about 50 nm below the surface is involved in the proton exchange process [12]. The sweep rate variation of electrical charge passed, q, measured from the area under voltammetric peaks, can be approached, at low potential scan rates to [13]:

$$1/q = 1/q^* + gv^{1/2} \quad (8.12)$$

q^* being the maximum surface charge arising due to infinitely slow OH^- ion exchange, and g a constant of proportionality. q^*, thus, denotes the charge related to the whole active surface. Complementarily, plots of q vs. $v^{-1/2}$ and extrapolation to $v \rightarrow \infty$ gives an ordinate at the origin of q^*_{outer}, representing the net amount of charge associated with the outer surface of the oxide being forwardly accessible to OH^- ion exchange.

$$q = q^*_{\text{outer}} + s\nu^{-1/2} \tag{8.13}$$

s being a second constant of proportionality. Then, the charge associated with the internal oxide/solution interface, q^*_{in} can easily be calculated as $q^*_{\text{in}} = q^* - q^*_{\text{outer}}$. Notice that the above treatment is equivalent to that represented by Eqs. (7.37) and (7.38).

Real systems deviate frequently from ideal behavior due to large uncompensated electrolyte resistance, slow kinetics of electron transfer, and site-to-site interactions [14]. Such deviations from Nernstian behavior can be expressed by an interaction term, r, which can be estimated from the variation of peak potential with potential scan rate:

$$I_p = \frac{n^2F^2A\Gamma\nu}{RT(4 - 2r\Gamma)} \tag{8.14}$$

In this equation, A denotes the geometrical area of the electrode and Γ the surface concentration of electroactive adsorption sites. This last parameter can be calculated from the determination of the charge pass associated to solid-state redox reaction, q_s, because $q_s = nFA\Gamma$, or from capacitance determination using the relationship [15]:

$$C = \frac{n^2F^2\Gamma}{RT(4 - 2r\Gamma)} \tag{8.15}$$

In this scheme, each oxide particle has an outer, hydrated catalytically active layer that consists of dangling oxometal surface groups. Remarkably, the reaction of surface metal centers can significantly influence the observed electrochemistry. For instance, nanoparticle mesoporous films of ceria, CeO_2, presumably form a new $CePO_4$ phase during electrochemical reduction of aqueous phosphate buffer solution [16].

8.3 ELECTROCHEMISTRY OF LAYERED HYDROXIDES AND RELATED MATERIALS

Layered double hydroxides (LDHs) of Al, Ni, Zn, and other metals are composed by rigid layers where anions and water molecules can move. Typical layered double hydroxides can be represented by the formula $[M_{1-x}M'_x(OH)_2][A^{n-}]_{x/n}{\cdot}mH_2O$, where M is a divalent metal cation, M′ could be a trivalent metal cation or a mix of trivalent and divalent metal cations, and A^{n-} represents an interlayer anion or mix of anions. Both divalent and trivalent cations are located at the center of octahedral composed of OH^- ions; $M(OH)_6$ and $M'(OH)_6$ units share edges forming two-dimensional layers while the interlayer space incorporates anions (and water molecules) to maintain electroneutrality. These materials exhibit a stable lamellar structure (see scheme in Figure 8.5) while $0.2 \le x \le 0.5$, and the interlamellar A^{n-} anions can be freely exchanged by foreign anions [17,18]. Figure 8.2 depicts a schematic representation of LDHs.

Due to its ability to accommodate anionic guests in the interlayer region, such materials have been tested for different applications—from catalysis to sensing and as electrode materials for Ni batteries—as described in Chapter 14. Most of this research is focused on hybrid materials and composites. This is the case depicted in Figure 8.3, where the CV response of $(Co_xNi_{(1-x)})_2Al\text{–}NO_3$ [19] is depicted. This response can be described in terms of a fast entrance of OH^- ions into the interlayer region [20], so the electrochemical process can be represented as:

$$\{\text{LDH} \cdots \text{M}^{\text{II}}\}_{\text{solid}} + \text{OH}^-_{\text{aq}} \rightarrow \{\text{LDH} \cdots \text{M}^{\text{III}} \cdots \text{OH}^-\}_{\text{solid}} + e^- \tag{8.16}$$

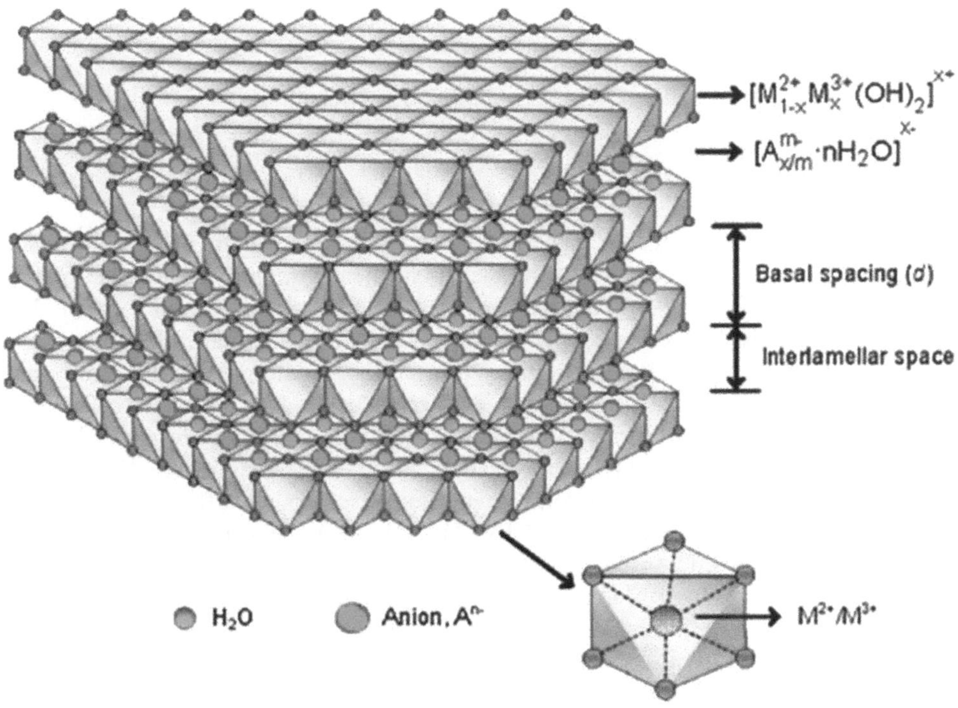

FIGURE 8.2 Schematic Representation of LDHs. Reproduced from Ref. [18] (Sarfraz and Shakir, *J. Energ. Storage*. 2017, 13, 103–122), with Permission.

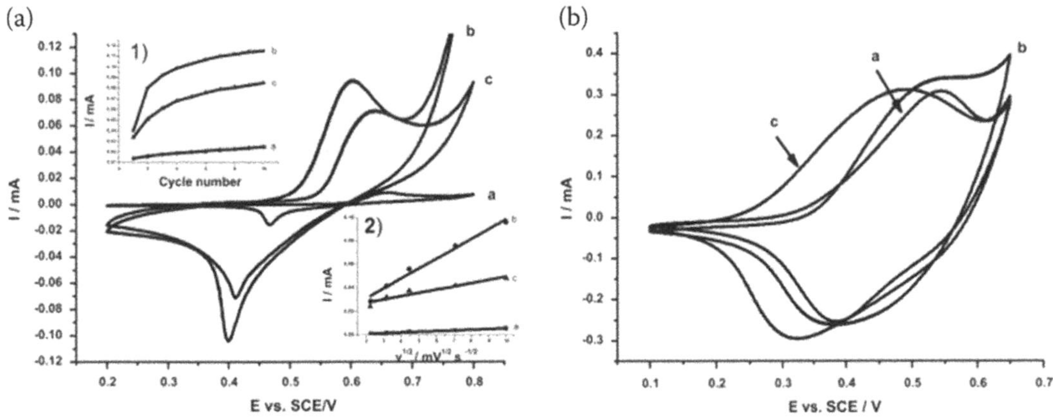

FIGURE 8.3 CVs of Thin films of the Extreme Components of the $(Co_xNi_{(1-x)})_2Al–NO_3$ Series Coated on Pt Electrodes in Contact with MOH Electrolytes. (A) $Ni_2Al–NO_3$; (B) $Co_2Al–NO_3$ ($M^+ = K^+$ (a), Na^+ (b), Li^+ (c)). Insets the Variation of Anodic Peak Current (I_{pa}) as a Function of Cycling Number (1) and $\nu^{1/2}$ (2). Reproduced from Ref. [19] (Vialat et al. *Electrochim. Acta*. 2013, 107: 599–610), with Permission.

Zn-Al LDHs can be prepared by electrodeposition from aqueous solutions of the nitrates of such metals [21]. Cathodic depositions, conducted at room temperature without stirring over noble metal-coated electrodes, were achieved by reducing nitrate ions to generate hydroxide ions on the working electrode, the optimal potential being −1.65 V vs. AgCl/Ag in 4 M KCl. The overall reactions involved are:

$$Zn^{2+}aq + 2OH^{-}aq \rightarrow Zn(OH)_{2\,solid} \tag{8.17}$$

$$Zn(OH)_{2\,solid} \rightarrow ZnO_{solid} + H_2O \tag{8.18}$$

$$NO_{3\,aq}^{-} + H_2O + 2e^{-} \rightarrow NO_{2\,aq}^{-} + 2OH_{aq}^{-} \tag{8.19}$$

Composition, in terms of the oxidation state of metal cations and the nature of intra-layer anions, can determine significant modifications in the electrochemical response. In the case of $[Ni_4Al(OH)_{10}]OH$, upon oxidation, the average oxidation state of Ni increases from 2 to ca. 3.7, as calculated from charge passed, denoting that Ni(III) and Ni(II) centers are formed. It has been proposed that, due to the high polarization ability of such ions, some protons drop from the hydroxyl and react with interlayer OH^{-} anions to give water molecules [22]. When the oxidized form is reduced, protons are removed from the interlayer water molecules. As a result, OH^{-} ions are easily interchanged with the electrolyte. However, if the interlayer anions are NO_3^{-}, the OH^{-} transport is significantly hindered by electrostatic repulsion, resulting in a different electrochemical performance between $[Ni_4Al(OH)_{10}]OH$ and $[Ni_4Al(OH)_{10}]NO_3$. Water relay and proton and anion migration play an important role in the electrochemistry of such materials, but the degree of crystallinity, crystal morphology, defects, and additives can also significantly influence their electrochemical properties [23]. With this regard, it should be noted that network-doping metal cations can enhance the electronic conductivity of the material, while the introduction of interlayer anions can modulate its ionic conductivity.

The electrochemistry of LDH-attached species has received minor attention compared to that of zeolite-associated species. The electrochemistry of bis(2-mercapto-2,2-diphenyl-ethanoate) dioxomolybdate(VI) complex, $[Mo^{VI}O_2(TBA)_2]^{2-}$ (TBA = $O_2CC(S)Ph_2$) attached to $Zn_{2.20}Al_{0.79}(OH)_6(NO_3)_{0.79}\cdot H_2O$ consisted of a two-electron reduction to Mo(IV) species (peak C_1) that subsequently experiences a reversible one-electron oxidation (peak A_1) to a Mo(V) analog (peak C_2), accompanied by other secondary signals (see Figure 8.4) [24].

Comparison of this voltammetric pattern with that of the non-LDH-attached species suggested that the complex, which in solid-phase exhibits a *cis*-dioxo Mo(VI) arrangement, is accommodated as a pillar in the interlamellar space of the host layered hydroxide a *fac*-trioxo Mo(VI) arrangement due to the disassociation of coordinated carboxylate groups as shown in Figure 8.5.

This disassociation would induce a sufficiently positive charge on the oxomolybdenum core to allow an oxygen atom transfer from a contiguous molecule of water with no change in the oxidation state of the metal [24]:

$$[Mo^{VI}O_2(TBA)_2]_{aq}^{2-} + 2H_2O + \{LDH \cdots NO_3^{-}\} \rightarrow \{LDH \cdots Mo^{VI}O_3(TBA)_2(H_2O)\} + 2H_{aq}^{+} + 2NO_{3\,aq}^{-} \tag{8.20}$$

Interestingly, in the last decades, there emerged a research line on the indirect synthesis of metal oxides and hydroxides based on the electrochemical generation of hydroxide ions in solutions containing metal ions [17]. Here, the hydroxide ions act as catalysts for sol-gel synthetic routes so that these processes can be considered as examples of electrochemically generated catalysts.

8.4 ELECTROCHEMISTRY OF POLYOXOMETALATES

It is well known that several transition metals form polyoxometalate ions in solution, eventually incorporating other elements. Thus, acidification of aqueous solutions of ammonium or sodium molybdates yields different so-called isopolyanions. As an example, condensation reactions for

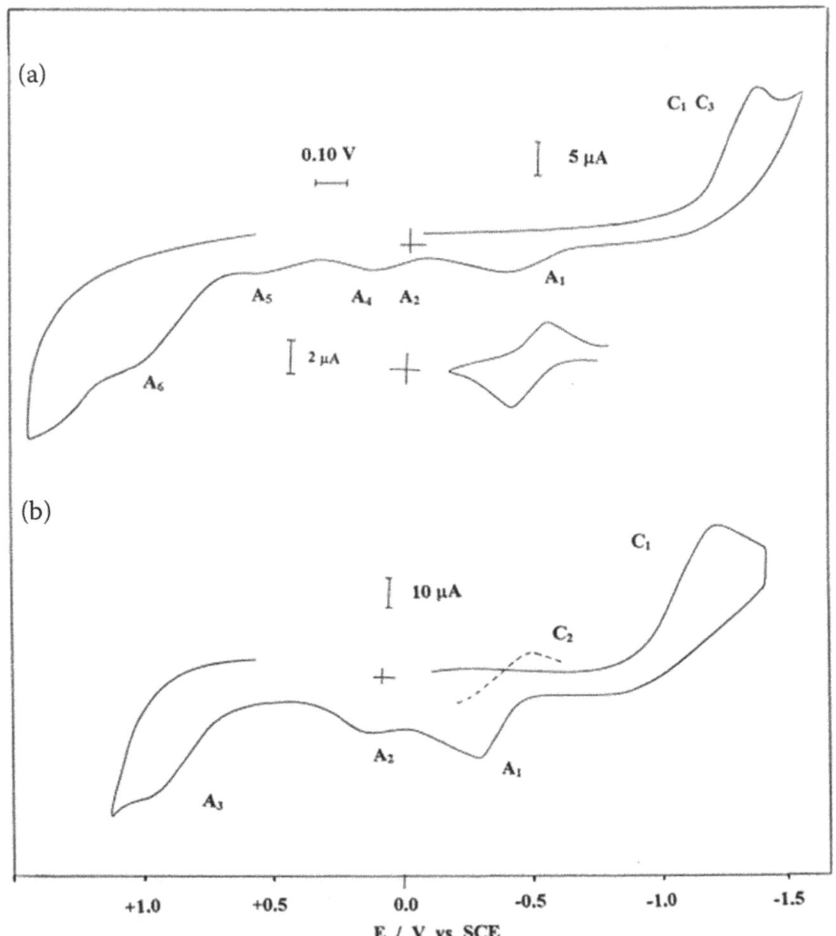

FIGURE 8.4 CVs HT–Mo(TBA) Coating Electrodes, Prepared from Lightly Ground Powder (a) Andabrasive Conditioning (b), Immersed into 0.50 M $NaClO_4$. Inset Presents the Region of Peaks C_2 and A_1 as Recorded in a Restricted Potential Scan. Potential Scan Rate 100 mV s^{-1}. From Ref. [24] (Doménech-Carbó et al. *J. Electroanal. Chem.* 1998, 458: 31–41), with Permission.

molybdate in an aqueous solution can be represented as a sequence of equilibria initiated by the formation of heptamolybdate ion:

$$7MoO_4{}^{2-}{}_{aq} + 8H^+_{aq} \rightarrow Mo_7O_{24}{}^{6-}{}_{aq} + 4H_2O \tag{8.21}$$

During condensation processes, oxo-anionic hetero-groups of p-block atoms ($PO_4{}^{3-}$, $SiO_4{}^{4-}$, etc.) can be incorporated into the structures, forming heteropolyanions ($PMo_{12}O_{40}{}^{3-}$, $SiW_{12}O_{40}{}^{3-}$). The architectures of most polyoxometalates (POMs) are based on specific structural types, such as the Lindquist ($M_6O_{19}{}^{2-}$), Keggin ($XM_{12}O_{40}{}^{3-}$), or Dawson (e.g., $X_2M_{18}O_{62}{}^{6-}$) metalates. When the condensation involves the acids (H_3PO_4, H_3BO_3) rather than their salts, hydrated forms can be obtained where the rigid, primary heteropolymetalate structure is surrounded by a secondary hydrate layer with a pseudo-liquid structure, and heteropolyacids are formed [25].

The electrochemistry of polyoxometalates in solution and attached to electrode surfaces has received considerable attention [26]. Typical multiple peak profiles were obtained, as depicted in

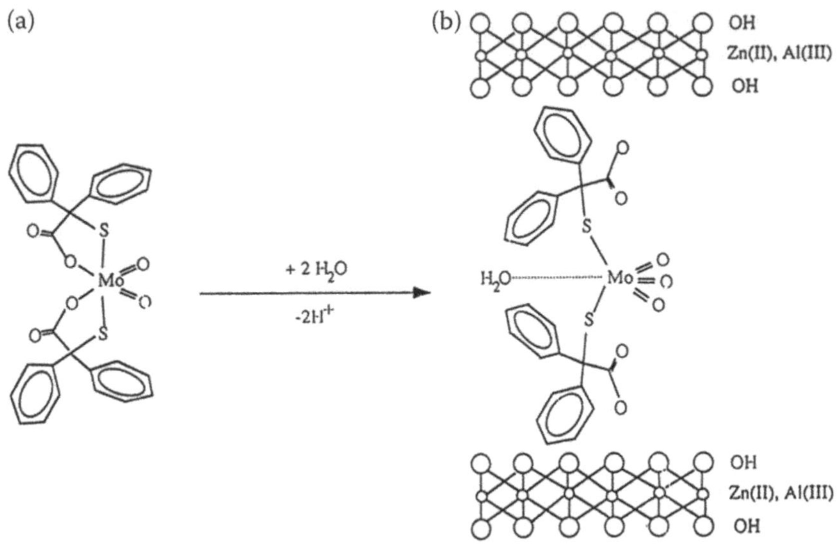

FIGURE 8.5 Scheme for the Possible Attachment of the bis(2-Mercapto-2,2-diphenyl-ethanoate) dioxomolybdate(VI) Complex, $[Mo^{VI}O_2(TBA)_2]^{2-}$ to Zn-Al Hydrotalcite Framework. From Ref. [24] (Doménech-Carbó et al. *J. Electroanal. Chem.* 1998, 458: 31–41), with Permission.

Figure 8.6, corresponding to CV of poly(3,4-ethylene dioxythiophene) (= PEDOT) films of $SiW_{11}Fe(H_2O)O_{39}^{5-}$ in contact with 0.2 M H_2SO_4 aqueous solution [27]. Here, three couples were obtained, corresponding to essentially reversible processes that can be represented as:

$$\{SiW^{VI}_{11}Fe^{III}(H_2O)O_{39}^{5-}\} + e^- \rightarrow \{SiW^{VI}_{11}Fe^{II}(H_2O)O_{39}^{6-}\} \quad (8.22)$$

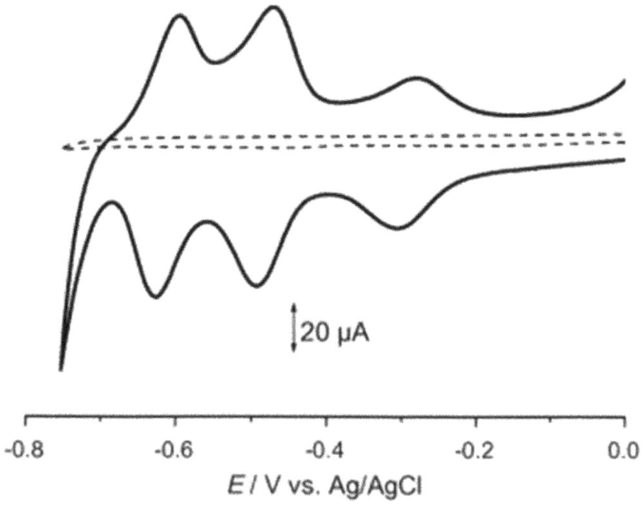

FIGURE 8.6 CV at Glassy Carbon Electrode Modified with $SiW_{11}Fe$–PEDOT and PEDOT (Dashed Line) of 1 mM $SiW_{12}O_{40}^{4-}$ in 0.2 M H_2SO_4 Aqueous Solution. Potential Scan Rate 100 mV s^{-1}. Reproduced from Ref. [27] (Fernandes et al. *J. Electroanal. Chem.* 2011, 660: 50–56), with Permission.

$$\{SiW^{VI}{}_{11}Fe^{II}(H_2O)O_{39}{}^{6-}\} + 2H^+_{aq} + 2e^- \rightarrow \{H_2SiW^{V}{}_2W^{VI}{}_9Fe^{II}(H_2O)O_{39}{}^{6-}\} \quad (8.23)$$

$$\{H_2SiW^{V}_2W^{VI}_9Fe^{II}(H_2O)O_{39}{}^{6-}\} + 2H^+_{aq} + 2e^- \rightarrow \{H_4SiW^{V}{}_4W^{VI}{}_7Fe^{II}(H_2O)O_{39}{}^{6-}\} \quad (8.24)$$

The peak currents are proportional to the potential scan rate, denoting the occurrence of surface-confined species. This response is characteristic of polyoxometalates, where quasi-reversible multi-step redox processes resulting in the formation of mixed-valence species and electron self-exchange between mixed-valence metal sites (e.g., $Mo^{VI,V}$) via electron hopping, occur. Extensive reduction of polyoxometalates leads to the formation of heteropolybrown species where such distinct ionic sites (e.g., W^{VI} and W^{IV}) can coexist. EIS spectra at different pH values and under application of bias potentials corresponding to the different couples are depicted in Figure 8.7. Nyquist plots are modeled with a Randles-type circuit incorporating a Warburg element. This last element relates directly to the flow of species at the electrode substrate and through the film. Analysis of impedance parameters suggested that there is great non-uniformity at the film/electrode interface [27].

The electrochemistry of Fe-, Ru-, and lanthanide-substituted heteropolyanions has also been studied. Now, metal-substituent- and POM-centered electrochemical signals can be recorded, as illustrated in the previous example. In particular, the iron-substituted heteropolyanions have large, negatively-charged densities favoring ion pairing with counteractions and protons. Then, the formal potentials for both the iron-localized and POM-localized electron transfer processes depend on the pH and counteraction concentration because of the competition between protonation and ion-pairing [26]. The electrochemistry of novel Fe_2 and Fe_4 clusters encapsulated in vacant polyoxotungstates via hydrothermal synthesis has been described in terms of Fe-localized and POM-localized signals [27,28].

Polyoxometalates have found application in homogeneous electrocatalysis but also after attachment to electrode surfaces via adsorption, entrapment into polymers, electrodeposition, or droplet evaporation. Multilayer assemblies can be obtained because of the strong interaction of polyoxometalates with cations [29]. For instance, polypyrrole coating doped by molybdophosphate anions can be used for corrosion prevention in carbon steels [30]. Structural protons in polyoxometalate single crystals can provide electroneutrality during redox processes, allowing electroanalytical investigation in the absence of an external electrolyte phase [31]. Within this group of materials, one can include the so-called metal oxide bronzes (typically tungsten bronzes). Starting from WO_3, electrochemical reduction processes yield intercalation materials with electrochromic properties [32].

8.5 ELECTROCHEMISTRY OF DOPED MATERIALS

The development of new synthetic routes has prompted the preparation of a variety of metal oxide type materials doped with different metal species. A significant part of research efforts in this field have been focused on doped materials for electrocatalysis, energy production and storage, and sensing. These applications will be treated in Chapters 12–14. In this chapter, the electrochemical characterization of doped metal-oxide-based materials will be briefly discussed.

This solid-state voltammetric response of doped oxides and related materials is characterized by (a) the response of doping species generally superimposed to that eventually displayed by the supporting material, and (b) thisresponse is limited to doping species confined to the surface sites in the crystal lattice. The interpretation of electrochemical data can be complicated, however, by the coexistence of solid solutions and multi-phase systems. For instance, in vanadium-doped zirconias ($V_xZr_{1-x}O_2$) at low vanadium loadings, V^{4+} ions substitute isomorphous Zr^{4+} ions in the lattice but, at relatively high vanadium loadings, some V^{5+} ions enter the solid network, promoting

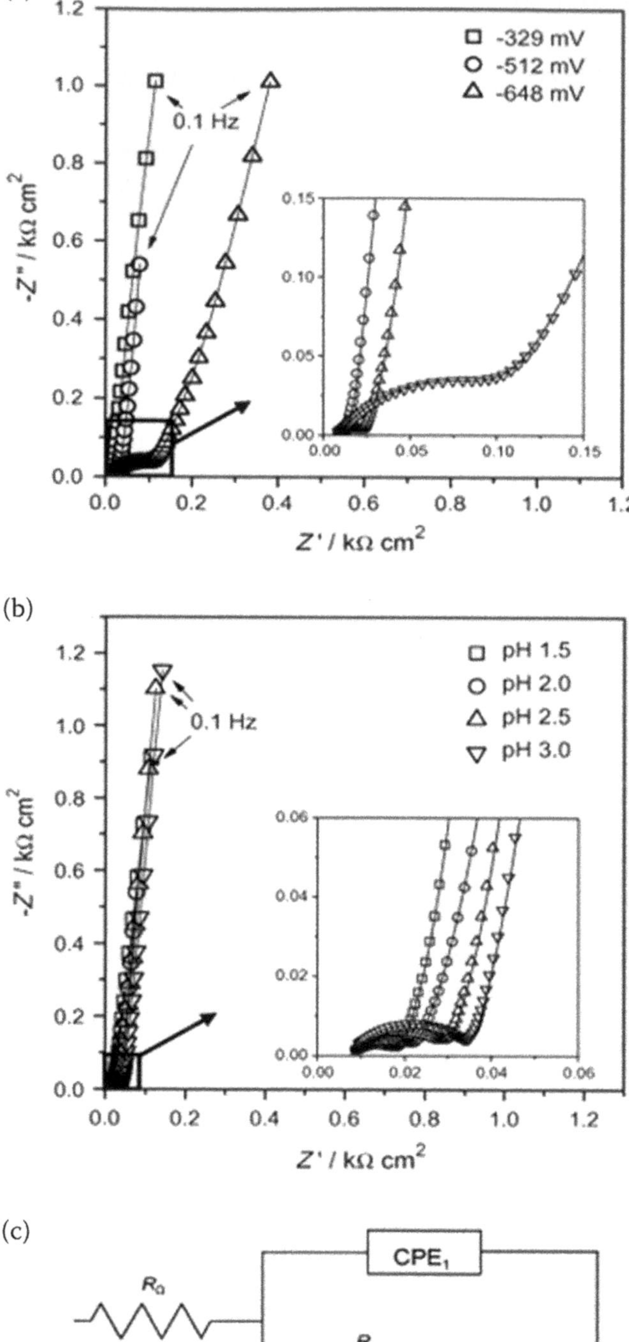

FIGURE 8.7 EIS Spectra (Nyquist Plots) of $SiW_{11}Fe$–PEDOT Modified Electrode in Na_2SO_4/H_2SO_4 Aqueous Buffer Solution, (a) at pH 2.0 and Different Applied Potentials; (b) at a Bias Potential of −512 mV vs. Ag/AgCl and Different pHs; (c) Equivalent Circuit Used to Model the Impedance Spectra. Reproduced from Ref. [27] (Fernandes et al. *J. Electroanal. Chem.* 2011, 660: 50–56), with Permission.

cation vacancies. At sufficiently larger vanadium loadings, finely dispersed vanadium oxide accompanies both monoclinic and tetragonal zirconias [33].

To rationalize this voltammetric response, it should be kept in mind that in monoclinic zirconias, Zr^{4+} ions (and V^{4+} ions in vanadium-doped zirconias) are seven-coordinated with three oxygen atoms at 2.07 Å and four oxygen atoms at 2.21 Å, whereas, in tetragonal zirconias, Zr^{4+} ions are eight-coordinated to four oxygens at 2.065 Å and four oxygens at 2.455 Å. As got electroactive V^{4+} centers in vanadium-doped, zirconias should be located in the external surface of crystals where truncated coordination exists; the coordination of surface metal centers should be completed with OH^- or H_2O units. Accordingly, voltammetric signals can be attributed to the oxidation of vanadium species located in different sites on the external surface of doped zirconia grains, and/or involving different electrochemical pathways. In alkaline media, electrochemical oxidation of vanadium centers can be represented by means the processes:

$$\{V_x^{IV}Zr_{1-x}O_2\}_{solid} + yOH^-_{aq} \rightarrow \{V^{IV}_{x-y}V^{V}_{y}Zr_{1-x}O_2(OH)_y\}_{solid} + ye^- \quad (8.25)$$

$$\{V_x^{IV}Zr_{1-x}O_2\}_{solid} + 2yH_2O \rightarrow \{V^{IV}_{x-y}Zr_{1-x}O_{2-4y}(OH)_{4y}\}_{solid} + yVO_{2\,aq}^{+} + ye^- \quad (8.26)$$

accompanied by solution-phase redox processes such as:

$$VO^+_{2\,aq} + 2H^+_{aq} + e^- \rightarrow VO^{2+}_{aq} + H_2O \quad (8.27)$$

Electrochemical data can provide information on the distribution of the doping species in different lattice sites and the oxidation state of doping centers [34]. Interestingly, electrochemical doping can be performed, for instance, incorporating Mn^{3+} ions into MnO_2 lattice [35].

An interesting aspect relatively unexplored is the existence of site-characteristic electrochemical responses in porous materials containing electroactive centers. As for electrochemical processes involving an interfacial electron transfer between the electroactive species and the electrode surface, it is reasonable to expect that the kinetics of that electron transfer should be conditioned by the structural arrangement of the electroactive center, resulting in discernible electrochemical responses. Additionally, the electrochemical response could also be influenced by the crystallinity, porosity, and distribution of electroactive centers in the porous material which refer to the so-termed mesoscopic scale.

An example is of this possibility is provided by magnesium-cobalt cordierites, $Co_xMg_{2-x}Al_4Si_5O_{18}$ ($0 < x < 2$), which can exist in different α- and β- forms characterized by six-membered rings of tetrahedrally coordinated cations linked through tetrahedral and octahedral polyhedra. In α-cordierite form, aluminum-containing tetrahedral units do not occupy definite positions, whereas, in β-cordierite, the aluminum-rich tetrahedral are located in fixed sites. Preparation from gel precursors yields a cordierite glass subsequently yielding a crystalline phase with μ-cordierite structure. This last is further converted, upon thermal treatment, into α- and β-cordierite, successively. Spectral data suggested that Co^{2+} ions, which occupied tetrahedral sites in the glasses, move into octahedral sites during crystallization [36].

Comparing voltammetric data for cobalt cordierites and CoO (with octahedral Co^{2+} coordination) and cobalt spinel ($CoAl_2O_4$, with tetrahedral Co^{2+} coordination) revealed a significant site-dependency of the electrochemical response of cobalt centers [37]. For 'pure' cobalt cordierite, the cathodic process in acidic media can be described as a more or less extensive reductive dissolution:

$$\{Co_2Al_4Si_5O_{18}\}_{solid} + 6H^+_{aq} + 4e^- \rightarrow 2\{Co\}_{solid} + 5SiO^{2-}_{3\,aq} + 4Al^{3+}_{aq} + 3H_2O \quad (8.28)$$

For magnesium-cobalt cordierites, however, the electrochemical process can be represented as a local reductive dissolution of cobalt centers with concomitant coordinative rearranging:

$$\{Co_xMg_{2-x}Al_4Si_5O_{18}\}_{solid} + 2xH^+ + 2xe^- \rightarrow x\{Co\}_{solid} + \{Mg_{2-x}Al_4Si_5O_{18-2x}(OH)_{2x}\}_{solid} \quad (8.29)$$

Obviously, the kinetic of the overall process will be influenced by both the partial reductive dissolution and the kinetics of the nucleation/growth of the deposit of Co metal. This process may involve intermediate cobalt species in solution occurring via solid-to-solid conversion to a metal phase, as described in Chapter 3. This electrochemistry probably includes oxidative dissolution and solid-state oxidation process involving chloride coordination of surface cobalt centers [37]:

$$\{Co^{II}_xMg_{2-x}Al_4Si_5O_{18}\}_{surf} + zCl^-_{aq} \rightarrow \{(Cl^-)_zCo^{III}zCo^{II}x - zMg_{2-x}Al_4Si_5O_{18}\}_{surf} + ze^- \quad (8.30)$$

An interesting implication of this site-characteristic electrochemistry is the possibility of promoting different catalytic effects. Electrocatalytic data for mannitol oxidation support the idea that octahedral Co(II) environments in 'ordered' crystalline structures yields more efficient and 'clean' electrocatalytic responses, but further research is needed to properly elucidate structural conditioning of electrocatalytic processes. Promising results have been obtained in the visualization of electroactive sites by means of scanning electrochemical microscopy (SECM). This technique can be used to monitor the highly localized electroactivity at titanium surfaces on which a native or anodically grown semiconducting oxide film has been deposited. Figure 8.8 shows a 300 × 300 μm SECM image of a TiO_2 electroactive site in a Ti/TiO_2 electrode in contact with 50 mM KBr plus 10 mM H_2SO_4 solution and the SECM tip electrode arrangement. A bias potential of 1.5 V was applied to the Ti/TiO_2 electrode (E_s) to oxidize Br^-, whereas the tip potential (E_t) was held at 0.0 V, sufficiently negative to reduce Br_2 at the mass-transport limited rate. The image was recorded after the current at the Ti/TiO_2 electrode had decayed to a steady-state value [38].

Such systems can provide selective effects due to size- and charge-exclusion effects in microporous solids, but also to symmetry effects associated with the coordinative arrangement of doping centers in metal oxides and related materials. Formation of surface-confined complexes between the substrate and the immobile redox-active center in the solid may involve exigent structural constraints, resulting in selectivity, including chiral selectivity that will be treated in Chapter 12.

8.6 ELECTROCATALYSIS

The use of porous metal oxides as electrocatalyst has received considerable attention [1]. As seen in previous chapters, most electrocatalytic processes of metal oxides in contact with alkaline aqueous electrolytes involve the formation of surface-confined metal hydroxo- and oxo-complexes. This is supported by the fact that surface oxy groups hydrate or hydroxylate in contact with aqueous media and that rapid redox reactions can occur owing to the simultaneous gain or loss of electrons and hydroxide ions. This situation can be reconciled with the general ideas in describing electrochemistry of nonconducting solids with possible ion insertion discussed in Chapter 3. This modeling states that there is surface redox activity even for: (a) highly restricted electron diffusion and (b) highly restricted cation diffusion. The second limiting case applies for insulating metal oxides; the reaction is confined to the particle/electrolyte interface. Then, the oxidation of a metal center in an oxide-type material can be represented as:

$$\{(-O-)_nM^{rd}(OH)_x(H_2O)_y\}_{solid} + zOH^-_{aq} \rightarrow \{(-O-)_nM^{ox}(OH)_{x+z}(H_2O)_{y-z}\}_{solid} + zH_2O + ze^- \quad (8.31)$$

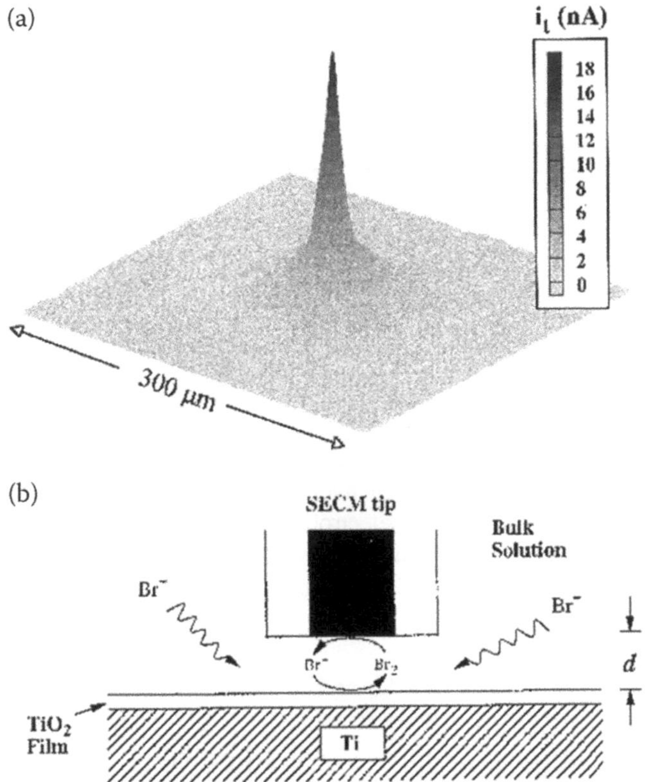

FIGURE 8.8 (a) A 300 × 300 μm SECM Image of a TiO_2 Electroactive Site in a Ti/TiO_2 Electrode in Contact with 50 mM KBr Plus 10 mM H_2SO_4 Solution. (b) Schematic Drawing Depicting the SECM Tip Positioned Above an Oxide-Covered Ti Electrode. The Tip Radius and Tip-to-Ti Electrode Separation are Drawn Approximately to Scale. The TiO_2 Film (~65 Å) is Much Thinner Than Drawn. The SECM Tip was Scanned at a Height, *d*, of ~5 μm Above the Ti/TiO_2 Electrode. Reproduced from Ref. [38] (Basame and White, *J. Phys. Chem. B.* 1998, 102: 9812–9819), with Permission.

In the above equation, it is assumed that the coordination number of the metal center remains unchanged upon oxidation. In the case of metal oxides, the eventual electron transport between contiguous metal centers should involve the coupled transport of hydroxy and/or oxo groups, namely:

$$\{M^{rd}(OH)_x\}_{surface,\ site\ 1} + \{M^{ox}(OH)_{x+z}\}_{surface,\ site\ 2} \rightarrow \{M^{ox}(OH)_{x+z}\}_{surface,\ site\ 1} + \{M^{rd}(OH)_x\}_{surface,\ site\ 2} \quad (8.32)$$

as well as within the metal centers of the subsurface region.

Since there is no possibility of substrate permeation within the metal oxide, the electrocatalytic processes will involve the formation of surface-confined adducts between the substrate and the metal centers in the surface of the particles of catalyst. A widely studied case is the oxidation of glucose in alkaline solution at different metal oxides. In the case of RuO_2-modified electrodes, oxometal surface groups act as mediators in a cyclic heterogeneous catalytic process [39] which can be represented as:

$$\{Ru^{VI}\} + \text{glucose} \rightarrow \{Ru^{IV} \cdots \text{glucanolactone}\} \rightarrow \{Ru^{IV}\} + \text{glucanolactone} \quad (8.33)$$

$$\{2\ Ru^{VII}\} + \text{glucose} \rightarrow \{2\ Ru^{VI} \cdots \text{glucanolactone}\} \rightarrow \{2\ Ru^{VI}\} + \text{glucanolactone} \quad (8.34)$$

the catalytic cycle is completed by the electrochemical oxidation of the Ru(IV) and Ru(VI) centers to Ru(VI) and Ru(VII).

It is assumed that glucose experiences a keto-enol transformation to form an anionic enol species:

$$R{-}CHOH{\cdot}CHO \leftrightarrow RCOH{=}CHOH \leftrightarrow RCOH{=}CHO^{-} + H^{+} \quad (8.35)$$

so that the electrostatic interaction between cationic metal centers and the anionic glucose form can be taken as an important factor for promoting glucose oxidation.

The electrochemical oxidation of sugars and alditols is a kinetically-controlled process where the rate-determining step is the abstraction of hydrogen from the carbon atom in the α-position with respect to the alcohol group, the overall electrochemical process being significantly influenced by molecular dimensions, preferred orientation, and steric hindrance [40]. NiO is typically used as catalyst. Following Casella et al. [41], one of the possible pathways can be represented by the following sequence of reactions:

$$\{NiO\} + OH^{-}_{aq} \rightarrow \{NiOOH\} + e^{-} \quad (8.36)$$

$$RCH_2OH_{aq} + \{NiO\} \rightarrow \{NiO(RCH_2OH)\}_{surf} \quad (8.37)$$

$$\{NiO(RCH_2OH)\}_{surf} + \{NiOOH\} \rightarrow \{NiO\} + \{NiO(R{\cdot}CHOH)\}_{surf} + H_2O \quad (8.38)$$

$$\{NiO(R{\cdot}CHOH)\}_{surf} + \{NiOOH\} \rightarrow \{NiO\} + \{NiO(RCOH)\}_{surf} + H_2O \quad (8.39)$$

This system can also catalyze the OER upon formation of active ·OH radicals in the oxide surface:

$$\{NiOOH\}_{solid} \rightarrow \{NiO\}_{solid} + HO{\cdot}_{ads} \quad (8.40)$$

$$2\ HO{\cdot}_{ads} \rightarrow H_2O + \tfrac{1}{2}\,O_2 \quad (8.41)$$

NiOOH being electrochemically regenerated via the process described in Eq. (8.36). Notice that, as previously noted, all these processes occur on the surface of the solid catalyst.

To study the kinetics of the catalytic process, Tafel representations from the current-potential data in the rising portion of the voltammetric curve can be used. It is assumed that, if the initial electron transfer is rate-determining, then a Tafel slope of 120 mV decade^{-1} would be predicted if the substrate were located at the outer Helmholtz plane (typically placed about 1 nm of the oxide surface). If the substrate is in a reaction plane closer to the oxide surface than the outer Helmholtz plane when it undergoes reaction, Tafel slopes of ca. 240–300 mV decade^{-1} are predicted [39].

8.7 ACTIVE SURFACE PHASES

As previously indicated, electrocatalytic processes at metal oxides and related materials are confined to the solid surface. Accordingly, it is essential to identify the active sites to elucidate the catalytic mechanisms. This information can be derived from different techniques,

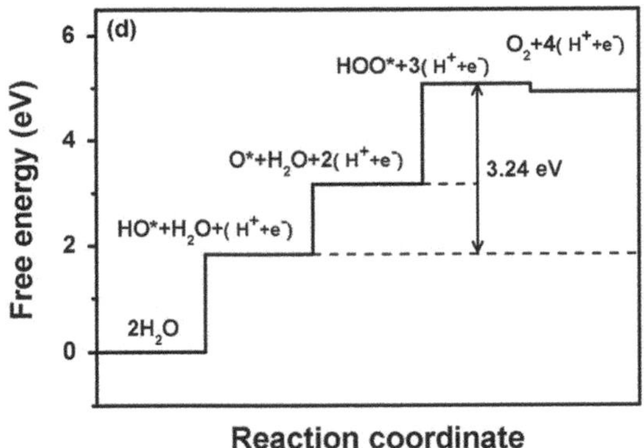

FIGURE 8.9 Free-Energy Diagram for the OER on the O Covered $MnO_2(110)$ at Zero Applied Potential at pH = 0 and T = 298 K. Reproduced from Ref. [44] (Su et al. *PhysChemChemPhysics*. 2012, 14: 1410–1422), with Permission.

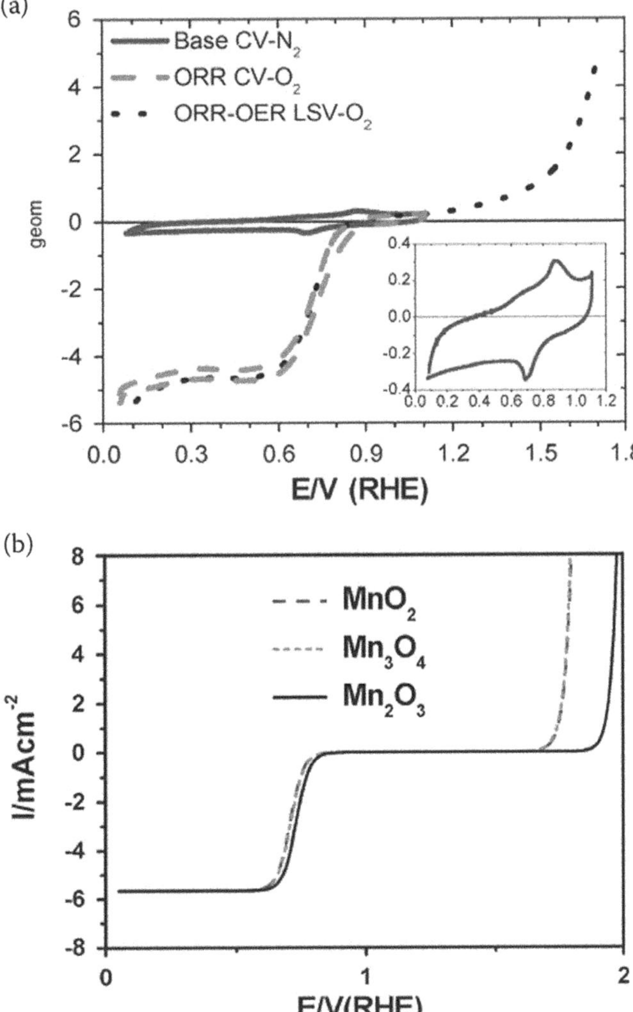

FIGURE 8.10 (a) CVs of an α-Mn_2O_3 Nanostructured Thin Film in 0.1 M KOH in N_2- (Also Inset) and O_2-Saturated Solutions. (b) Calculated Current Density for Mn_3O_4, Mn_2O_3, and MnO_2 from DFT Calculations. Reproduced from Ref. [44] (Su et al. *PhysChemChemPhysics*. 2012, 14: 1410–1422), with Permission.

combining electrochemical measurements with theoretical calculations based on density functional theory (DFT). In this context, much attention is focused on the evaluation of adsorption energies. The Gibbs free energy of adsorption of a given species (ΔG_{ads}) on a surface can be expressed as [42]:

$$\Delta G_{ads} = \Delta E_{DFT} + \Delta E_{ZP} - T\Delta S_{ads} + \Delta G_{ref} \tag{8.42}$$

where ΔS_{ads} is the change in entropy upon adsorption, ΔE_{ZP} is the change in zero-point energy, and ΔE_{DFT} is the adsorption energy *in vacuo* calculated by DFT methods. The corrections of zero-point energies are based on DFT calculations of the vibrational frequencies and standard thermochemical tables, while changes in entropy are calculated from the standard tables for gas-phase molecules [43,44]. These calculations permit construction of free energy diagrams as in Figure 8.9, corresponding to the OER at the surface of $MnO_2(110)$ [44].

The estimate of adsorption free energies permits an evaluation of the catalytic activities of metal oxides. In the case of different manganese oxides—Mn_3O_4, Mn_2O_3, and MnO_2—the possibility of acting as bifunctional catalysts for ORR and OER processes is illustrated in Figure 8.10.

Here, the experimental CVs recorded at an α-Mn_2O_3 nanostructured thin film in 0.1 M KOH, N_2- (also inset) and O_2-saturated solutions are compared with the calculated current density curves for such oxides [44]. In this case, the calculations suggested that the active surfaces for the ORR and the OER are, respectively, ·OH-covered Mn_2O_3 and ·O-covered MnO_2. Overall, these calculations discriminate more favorable mechanistic pathways (see Section 7.3) and understand surface phase changes during the reactions and a more rational design of catalyst structures.

REFERENCES

[1]. Trasatti, S. 1994. Transition metal oxides: versatile materials for electrocatalysis, in Lipkowski, J.; Ross, P.N. Eds. *The Electrochemistry of Novel Materials.*VCH, New York, pp. 207–295.

[2]. Scholz, F.; Schröder, U.; Gulabowski, R.; Doménech-Carbó, A. 2014. *Electrochemistry of Immobilized Particles and Droplets*, 2nd edit. Springer, Berlin.

[3]. Cai, J.; Liu, J.; Willis, W.S.; Suib, S.L. 2001. Framework doping of iron in tunnel structure cryptomelane. *Chemistry of Materials.* 13: 2413–2422.

[4]. Amarilla, J.M.; Tedjar, F.; Poinsignon, C. 1994. Influence of KOH concentration on the γ-MnO_2 redox mechanism. *Electrochimica Acta.* 39: 2321–2331.

[5]. Bodoardo, S.; Brenet, J.; Maja, M.; Spinelli, P. 1994. Electrochemical behaviour of MnO_2 electrodes in sulphuric acid solutions. *Electrochimica Acta.* 39: 1999–2004.

[6]. De Guzman, R.N.; Shen, Y.-F.; Shaw, B.R.; Suib, S.L.; O'Young, C.-L. 1993. Role of cyclic voltammetry in characterizing solids: Natural and synthetic manganese oxide octahedral molecular sieves. *Chemistry of Materials.* 10: 1395–1400.

[7]. Kitchaev, D.A.; Dacek, S.T.; Sun, W.; Ceder, G. 2017. Thermodynamics of phase selection in MnO_2 framework structures through alkali intercalation and hydration. *Journal of the American Chemical Society.* 139: 2672–2681.

[8]. Barrado, E.; Pardo, R.; Castrillejo, Y.; Vega, M. 1997. Electrochemical behaviour of vanadium compounds at a carbon paste electrode. *Journal of Electroanalytical Chemistry.* 427: 35–42.

[9]. Lyons, M.E.G.; Burke, L.D. 1987. Mechanism of oxygen reactions at porous oxide electrodes. Part 1.—Oxygen evolution at RuO_2 and $Ru_xSn_{1-x}O_2$ electrodes in alkaline solution under vigorous electrolysis conditions. *Journal of the Chemical Society Faraday Transactions I.* 83: 299–321.

[10]. Doubova, L.M.; Daolio, S.; De Battisti, A. 2002. Examination of single-crystal RuO_2 surfaces: Charge storage mechanism in H_2SO_4 aqueous solution. *Journal of Electroanalytical Chemistry.* 532: 25–33.

[11]. Dharuman, V.; Chandrasekara Pillai, K. 2006. RuO_2 electrode surface effects in electrocatalytic oxidation of glucose. *Journal of Solid State Electrochemistry.* 10: 967–979.

[12]. Lodi, G.; Zucchini, G.L.; De Battisti, A.; Giatti, A.; Battaglin, G.; Della Mea, G. 1991. Proton exchange in group VIII metal-oxide films. *Surface Science.* 251–252: 836–840.

[13]. Ardizzone, S.; Fregonara, G.; Trassati, S. 1990. "Inner" and "outer" active surface of RuO_2 electrodes. *Electrochimica Acta.* 35: 263–267.

[14]. Ilangovan, G.; Chandarasekara Pillai, K. 1997. Unusual activation of glassy carbon electrodes for enhanced adsorption of monomeric molybdate(VI). *Journal of Electroanalytical Chemistry*. 431: 11–14.
[15]. Smith, D.F.; Willman, K.; Kuo, K.; Murray R.W. 1979. Chemically modified electrodes: XV. Electrochemistry and waveshape analysis of aminophenylferrocene bonded to acid chloride functionalized ruthenium, platinum, and tin oxide electrodes. *Journal of Electroanalytical Chemistry*. 95: 217–227.
[16]. Cummings, C.Y.; Sott, S.J.; Bonné, M.J.; Edler, K.J.; King, P.M.; Mortimer, R.J.; Marken, F. 2008. Underpotential surface reduction of mesoporous CeO_2 nanoparticle films. Journal *of Solid State Electrochemistry*. 12: 1541–1548.
[17]. Walcarius, A. 2015. Electrochemically-assisted deposition by local pH tuning: A versatile tool to generate ordered mesoporous silica thin films and layered double hydroxide materials. *Journal of Solid State Electrochemistry*. 19: 1905–1931.
[18]. Sarfraz, M.; Shakir, I. 2017. Recent advances in layered double hydroxides as electrode materials for high-performance electrochemical energy storage devices. *Journal of Energy Storage*. 13: 103–122.
[19]. Vialat, P.; Leroux, F.; Taviot-Gueho, C.; Villemure, G.; Mousty, G. 2013. Insights into the electrochemistry of $(Co_xNi_{(1-x)})_2Al–NO_3$ layered double hydroxides. *Electrochimica Acta*. 107: 599–610.
[20]. Scavetta, E.; Berrettoni, M.; Nobili, F.; Tonelli, D. 2005. Electrochemical characterisation of electrodes modified with a Co/Al hydrotalcite-like compound. *Electrochimica Acta*. 50: 3305–3311.
[21]. Yarger, M.S.; Steinmiller, E.M.P.; Choi, K.-S. 2008. Electrochemical synthesis of Zn-Al layered double hydroxide (LDH) films. *Inorganic Chemistry*. 47: 5859–5865.
[22]. Lei, L.; Hu, M.; Gao, X.; Sun, Y. 2008. The effect of the interlayer anions on the electrochemical performance of layered double hydroxide electrode materials. *Electrochimica Acta*. 54: 671–676.
[23]. Ren, J.X.; Zhou, Z.; Gao, X.P.; Yan, J. 2006. Preparation of porous spherical α-$Ni(OH)_2$ and enhancement of high-temperature electrochemical performances through yttrium addition. *Electrochimica Acta*. 52: 1120–1126.
[24]. Doménech-Carbó, A.; Ribera, A.; Cervilla, A.; Llopis, E. 1998. Electrochemistry of hydrotalcite-suported bis(2-mercapto-2,2-diphenyl-ethanoate)dioxomolybdate complexes. *Journal of Electroanalytical Chemistry*. 458: 31–41.
[25]. Keita, B.; Nadjo, L. 2006. Electrochemistry of isopoly and heteropoly oxometalates, in Bard, A.J.; Stratmann, M.; Scholz, F.; Pickett, C.J. Eds. *Inorganic Chemistry. Encyclopedia of Electrochemistry*, vol. 7b. Wiley-VCH, Weinheim.
[26]. Sadakane, M.; Steckhan, E. 1998. Electrochemical properties of polyoxometalates as electrocatalysts. *Chemical Reviews*. 98: 219–237.
[27]. Fernandes, D.M.; Brett, C.M.A.; Cavaleiro, A.M.V. 2011. Preparation and electrochemical properties of modified electrodes with Keggin-type silicotungstates and PEDOT. *Journal of Electroanalytical Chemistry*. 660: 50–56.
[28]. Pichon, C.; Dolbecq, A.; Mialane, P.; Marrot, J.; Rivière, E.; Goral, M.; Zynek, M.; McCormac, T.; Borshch, S.A.; Zueva, E.; Sécheresse, F. 2008. Fe_2 and Fe_4 clusters encapsulated in vacant polyoxotungstates: Hydrothermal synthesis, magnetic and electrochemical properties, and DFT calculations. *Chemistry A European Journal*. 14: 3189–3199.
[29]. Ingersoll, D.; Kulesza, P.J.; Faulkner, L.R. 1994. Polyoxometallate-based layered composite films on electrodes. Preparation through alternate immersions in modification solutions. *Journal of the Electrochemical Society*. 141: 140–147.
[30]. Ohtsuka, T.; Iida, M.; Ueda, M. 2006. Polypyrrole coating doped by molybdo-phosphate anions for corrosion prevention of carbon steels. *Journal of Solid State Electrochemistry*. 10: 714–720.
[31]. Bard, A.J.; Inzelt, G.; Scholz, F. Eds. 2008. *Electrochemical Dictionary*. Springer, Berlin-Heidelberg.
[32]. Grandqvist, C.G. 1999. Progress in electrochromics: Tungsten oxide revisited. *Electrochimica Acta*. 44: 3005–3015.
[33]. Doménech-Carbó, A.; Alarcón, J. 2002. Electrochemistry of vanadium-doped tetragonal and monoclinic ZrO_2 attached to graphite/polyester composite electrodes. *Journal of Solid State Electrochemistry*. 6: 443–450.
[34]. Doménech-Carbó, A.; Torres, F.J.; Ruiz de Sola, E.; Alarcón, J. 2006. Electrochemical detection of high oxidation states of chromium (IV and V) in chromium-doped cassiterite and tin-sphene ceramic pigmenting systems. *European Journal of Inorganic Chemistry*. 638–648.
[35]. Chan, Z.M.; Kitchaev, D.A.; Nelson Weker, J.; Schnedermann, C.; Lim, K.; Ceder, G.; Tumas, W.; Toney, M.F.; Nocera, D.G. 2018. Electrochemical trapping of metastable Mn^{3+} ions for activation of MnO_2 oxygen evolution catalysts. *PNAS*. 115: E5261–E5268.

[36]. Villegas, M.A.; Alarcón, J. 2002. Mechanism of crystallization of Co-cordierites from stoichiometric powdered glasses. *Journal of the European Ceramic Society*. 22: 487–494.
[37]. Doménech-Carbó, A.; Torres, F.J.; Alarcón, J. 2004. Electrochemical characterization of cobalt cordierites attached to paraffin-impregnated graphite electrodes. *Journal of Solid State Electrochemistry*. 8: 127–137.
[38]. Basame, S.B.; White, H.S. 1998. Scanning electrochemical microscopy: Measurement of the current density at microscopic redox-active sites on titanium. *Journal of Physical Chemistry B*. 102: 9812–9819.
[39]. Lyons, M.E.G.; Fitzgerald, C.A.; Smyth, M.R. 1994. Glucose oxidation at ruthenium dioxide based electrodes. *Analyst*. 119: 855–861.
[40]. Konaka, R.; Terabe, S.; Kuruma, K. 1969. Mechanism of the oxidation reaction with nickel peroxide. *Journal of Organic Chemistry*. 34: 1334–1337.
[41]. Casella, I.G.; Cataldi, T.R.I.; Salvi, A.M.; Desimoni, E. 1993. Electrocatalytic oxidation and liquid chromatographic detection of aliphatic alcohols at a nickel-based glassy carbon modified electrode. *Analytical Chemistry*. 65: 3143–3150.
[42]. Doherty, A.P.; Vos, J.G. 1992. Electrocatalytic reduction of nitrite at an $[Os(bipy)_2(PVC)_{10}Cl]Cl$-modified electrode. *Journal of the Chemical Society Faraday Transactions*. 88: 2903–2907.
[43]. Hansen, H.A.; Man, I.C.; Studt, F.; Abild-Pedersen, F.; Bligaard, T.; Rossmeisl, J. 2010. Electrochemical chlorine evolution at rutile oxide (110) surfaces. *Physical Chemistry Chemical Physics*. 12: 283–290.
[44]. Su, H.-Y.; Gorlin, Y.; Man, I.C.; Calle-Vallejo, F.; Nørskov, J.K.; Jaramillo, T.F.; Rossmeisl, J. 2012. Identifying active surface phases for metal oxide electrocatalysts: A study of manganese oxide bi-functional catalysts for oxygen reduction and water oxidation catalysis. *Physical Chemistry Chemical Physics*. 14: 1410–1422.

9 Sulfides, Nitrides, Phosphides

9.1 INTRODUCTION

During the last decade, the search for novel technological materials has been stimulated by the increasing demand for sustainable and renewable energy sources. Electrochemical and solar-driven photoelectrochemical water splitting appears as promising methodologies, but they require the disposal of electrocatalysts to produce hydrogen (H_2) with high catalytic activity and low cost. Platinum and its alloys — treated in Chapter 4 — as well as different Pt-based composites, have been widely studied as electrocatalysts for HER, but their widespread use is limited by the scarcity of this element [1].

In this context, a series of novel electrocatalysts have emerged as suitable materials not only for hydrogen production but also for OER and ORR (see Chapter 7). These include metal chalcogenides, nitrides, phosphides, borides, and carbides [2] characterized by their capability to absorb solar radiation and quickly transfer interfacial charge facilitated by their nanoporous structure. Intensive research is currently performed in this field giving rise to an enormous variety of compounds and composite materials. In this chapter, a selection of the basic electrocatalysts will be presented.

9.2 MOLYBDENUM DISULFIDE AND RELATED MATERIALS

Molybdenum disulfide or molybdenum(IV) sulfide (MoS_2) was known for its use as an industrial hydrodesulfurization catalyst [3]. It is the most representative component of a series of transition metal dichalcogenides of general formula MX_2 (M = V, Mo, W, Ti, Zr, Hf, …; X = S, Se, Te) characterized by layered structures where sandwiched X-M-X layers are linked by Van der Waals interactions.

MoS_2, the best-known component of the family, has three common structure polymorphs labeled as 2 H, 3 R, and 1 T, whose physical properties can differ substantially. Figure 9.1 shows a scheme for the corresponding structures, all based on S-Mo-S layers built from edge-sharing MoS_6 trigonal prisms [1]. The 2 H form is the most stable thermodynamically and has two layers per unit cell stacked in hexagonal symmetry. The 3 R form has three layers per unit cell stacked in rhombohedral symmetry, whereas the metastable 1 T phase appears as monolayers obtained by exfoliation from bulk 2 H. In the 1 T form, the coordination of Mo is octahedral and contains one layer per unit cell.

The semiconducting character of 2 H MoS_2 with a thickness-modulated band gap between 1.3 and 1.8 eV prompts a variety of photophysical applications [4]. The electrocatalytic activity on HER was discovered to arise from the exposed planes on edges while the basal planes were catalytically inactive [5,6]. Figure 9.2 compares the Tafel region of LSV curves recorded for MoS_2 specimens with different density of S vacancies recorded in 0.5 M H_2SO_4 [7].

One can see that the electrocatalytic performance of MoS_2 is better than Au, but clearly inferior to that of Pt. Accordingly, intensive research is currently devoted to improving the HER performance of MoS_2 catalysts to maximize the exposed active sites. Strategies to improve the electrocatalytic activity of MoS_2 towards HER include:

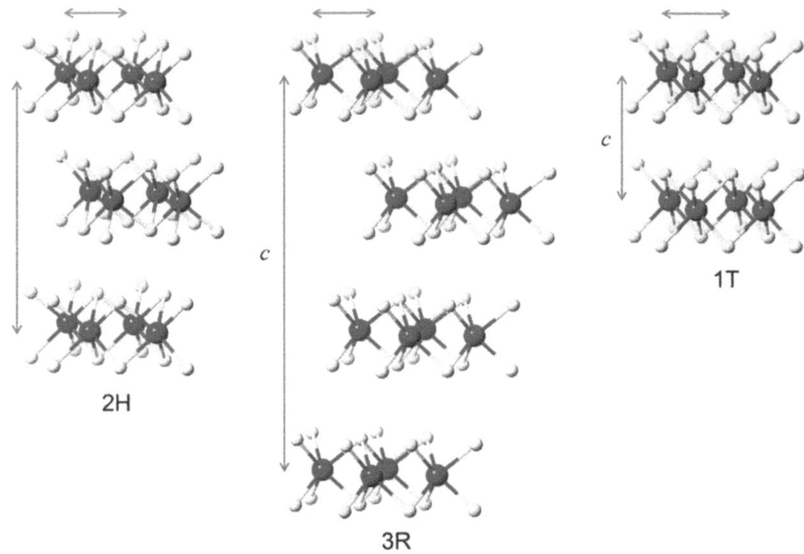

FIGURE 9.1 Scheme of the Structures of the 2 H, 3 R, and 1 T Forms of MoS_2. Reproduced from Ref. [1] (Ding et al. *Chem.* 2016, 1: 699–726), with Permission.

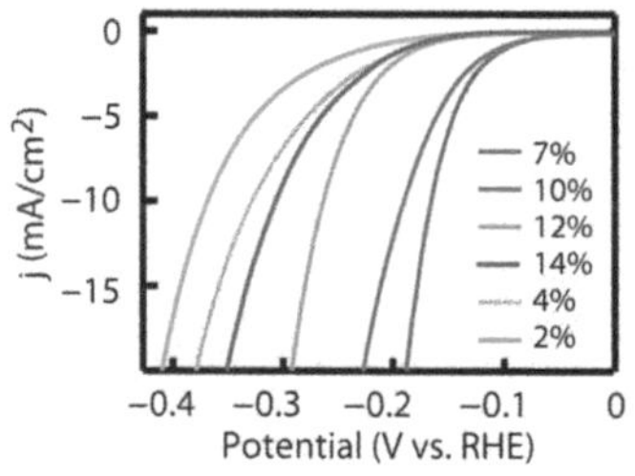

FIGURE 9.2 LSVs of MoS_2 Samples with Different Densities of S Vacancies in Contact with 0.5 M H_2SO_4. Potential Scan Rate 5 mV s^{-1}. As Customary, the Potentials Were Measured Relative to SCE Whose Potential Shift Was Calibrated to be −0.262 V vs. RHE. Reproduced from Ref. [7] (Li et al. *J. Am. Chem. Soc.* 2016, 138: 16632–16638), with Permission.

a. Increasing edge sites via formation of different nanoarchitecture (nanoparticles, vertical nanoflakes, nanowires) and mesoporous structures [8]. It has been shown that the catalytic activity depends on the length of the edge sites and grain boundaries [7].
b. Increasing defect density in the basal planes, particularly introducing sulfur vacancies [9]. This is supported by theoretical DFT calculations on the adsorption free energy of H atoms on different sites. Table 9.1 summarizes the results revealing the stability of the adsorbate on S vacancies [7]. The HER performance depends, however, not only on the density of sulfur vacancies but also on other factors such as the crystalline quality near the vacancies, so that partial repair of sulfur vacancies and other structural defects can modulate the electrocatalytic activity of the material [10]. Figure 9.2, corresponding to the polarization curves of MoS_2 specimens with different levels of sulfur vacancies in contact with 0.5 M H_2SO_4 [7], illustrates the significant effect of this factor on the electrocatalytic effect exerted on HER. Repair of sulfur vacancies does not restore the catalytic activity, as shown in Figure 9.3 [7]. Table 9.2 compares the values of the several parameters usually employed to evaluate the performance of HER catalysts. These include the onset potential, the Tafel slope, and the exchange current density (see Section 4.4). The overpotential at a current density of 10 mA cm^{-2} and the catalytic turnover frequency (TOF), which corresponds to the number of generated molecules

TABLE 9.1
Theoretical Adsorption Free Energies of H Atoms on Different MoS_2 Sites [7]

Site	ΔG°_{ads} (eV)
Mo edge	0.115
S vacancy (1 S)	–0.095
Grain boundary (8-4-4)	0.181
Grain boundary (12-4)	0.368
Grain boundary (5-7)	0.566
Basal plane	1.218

per active site per second, are also frequently used. These parameters are influenced by the catalyst loading and the electrochemically active surface area.

c. Tuning the electronic structure and phase engineering. The most obvious method is the preparation of 1 T MoS_2. This form has a metallic character and a large density of catalytic sites showing a significant enhancement of the catalytic performance [11]. In fact, it has been reported that the Volmer-Heyrovsky mechanism (see Chapter 4) is favored over the Volmer-Tafel mechanism, and the Heyrovsky step is the rate-limiting step for HER on 1T-MoS_2 [12,13]. Synthesis of amorphous MoS_x has been also reported to offer excellent electrocatalytic ability [14].
d. Doping and adatom incorporation. The introduction of oxygen atoms simultaneously with disorder engineering has been reported to produce an increased HER performance of MoS_2 nanosheet samples prepared in solution [15]. In turn, adsorption of atoms of several first-row transition metals and nonmetals, such as B, C, N, and O, can also tune HER activity at higher defect levels [16].

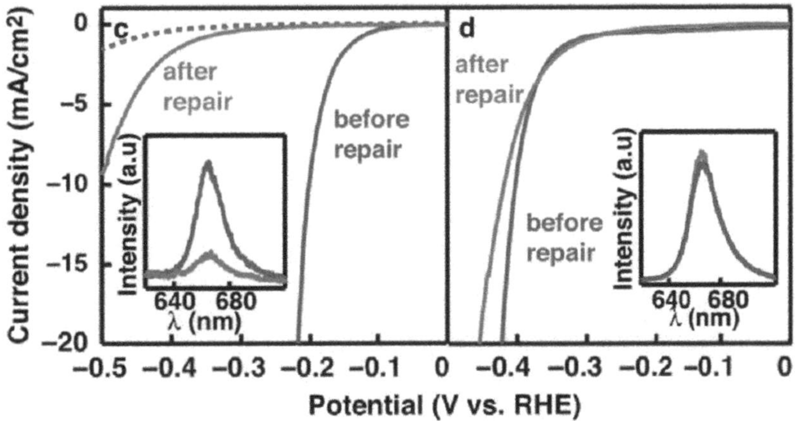

FIGURE 9.3 LSVs of MoS_2 Films (Left) and MoS_2 Flakes (Right) Before and After Repair of Sulfur Vacancies in Conditions Such as in Figure 9.2. The Dotted Curve Corresponds to the Polarization Curve of the Flake-Merged MoS_2 Film with Few Sulfur Vacancies. The Insets Depict the Corresponding Photoluminescence Spectral Curves. Reproduced from Ref. [7] (Li et al. *J. Am. Chem. Soc.* 2016, 138: 16632–16638), with Permission.

TABLE 9.2
Electrocatalytic Parameters on HER Reported for Different Catalysts [1]

Catalyst	E_{onset} (mV vs. RHE)	Tafel slope (mV decade^{-1})	j_o (mA cm^{-2})
Pt/C[a]	0	30	7.1×10^{-1}
MoS_2	−237	101	9.1×10^{-4}
$CoSe_2$	−50	48	8.4×10^{-3}
$MoS_2/CoSe_2$	−11	63	7.3×10^{-2}

[a] Commercial catalyst, wJohnson-Matthey, 20 wt% Pt/XC-72.

e. Ion intercalation. This appears during the conversion of 2H-MoS_2 to 1T-MoS_2 via chemical exfoliation using *n*-butyl lithium. Here, Li^+ ions are intercalated to form Li_xMoS_2 nanosheets [16].
f. Coupling with conductive scaffolds typically by association of MoS_2 nanostructures with carbonaceous nanostructures (graphene, carbon nanotubes, carbon nanofibers, carbon nanofoam, etc.) [17]. The enhanced catalytic performance can be attributed to the abundance of catalytic edge sites on the MoS_2 nanoarchitectures, and electrical coupling to the underlying conducting network that provides internal electron-transport channels from the less-conducting MoS_2 [18].
g. Formation of ternary compounds. Among others, combining CoS_x and MoS_x materials that are both electrocatalytically active (see Table 9.2) [19].

Currently, much effort is devoted to developing these preparative strategies in parallel to a better understanding of the electrocatalytic mechanisms produced by these materials. There is a plethora of new materials directly related to metal sulfides. Among others, the linking of Mo-S clusters with organic linkers. This follows to a great extent the MOF philosophy (see Chapter 6) seen in Figure 9.4 [20].

Although insufficiently studied, solid-state voltammetric techniques (VIMP) can also provide information on the structure and reactivity of these materials. Figure 9.5 shows, as an example, the SWVs of Co_9S_8, MoS_2, CoS_2, and a CoS_x-MoS_x in contact with 0.10 M H_2SO_4 [21]. Negative-going potential scans show a series of cathodic peaks that can be attributed to proton-assisted reduction processes of the type:

$$\{M^{IV}S_2\}_{solid} + xH^+_{aq} + xe^- \rightarrow \{M^{III}{}_xM^{IV}{}_{1-x}S_{2-x/2}\}_{solid} + (x/2)H_2S \tag{9.1}$$

$$\{M^{III}{}_xM^{IV}{}_{1-x}S_{2-x/2}\}_{solid} + xH^+_{aq} + xe^- \rightarrow \{M^{II}{}_xM^{IV}{}_{1-x}S_{2-x}\}_{solid} + (x/2)H_2S \tag{9.2}$$

Although a variety of mixed-valence and substoichiometric phases can be involved resulting in the complicated voltammetric profile recorded in the voltammograms, there is a possibility of discriminating between Co- and Mo-centered processes — these data are correlated with catalytic properties of such materials.

9.3 BORON NITRIDE AND RELATED MATERIALS

Boron nitride (BN) is an insulator with a wide bandgap (3.6–7.1 eV depending on preparation methods) [22] that can be reduced by forming thin layers by effect of the generation of B- and N-vacancies and impurity defects [23]. Due to its structural similarity with graphene, hexagonal BN

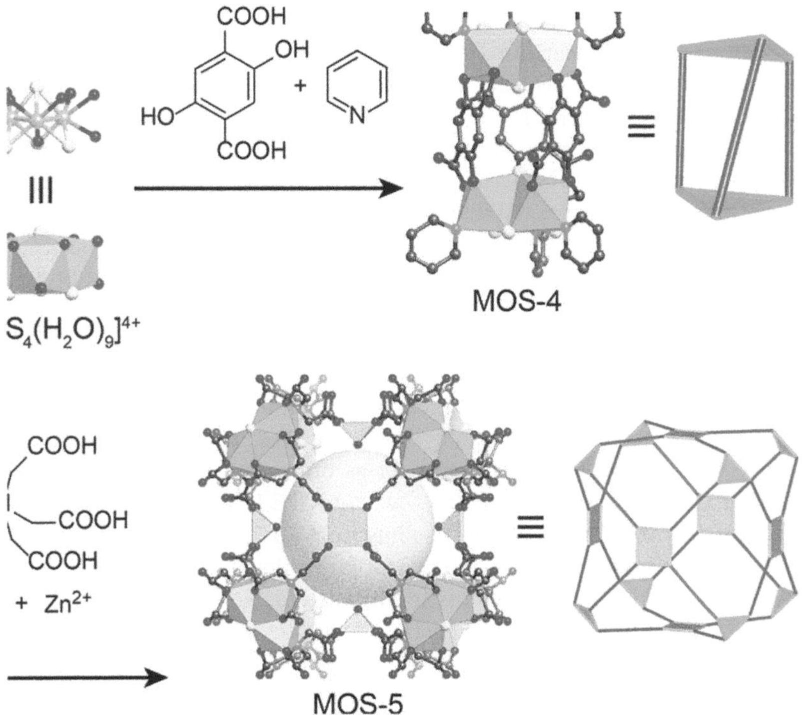

FIGURE 9.4 Scheme of the Preparation of Mo_3S_4 Clusters Connected by Organic Linkers into Cages. Reproduced from Ref. [20] (Ji et al. *J. Am. Chem. Soc.* 2018, 140: 13618–13622), with Permission.

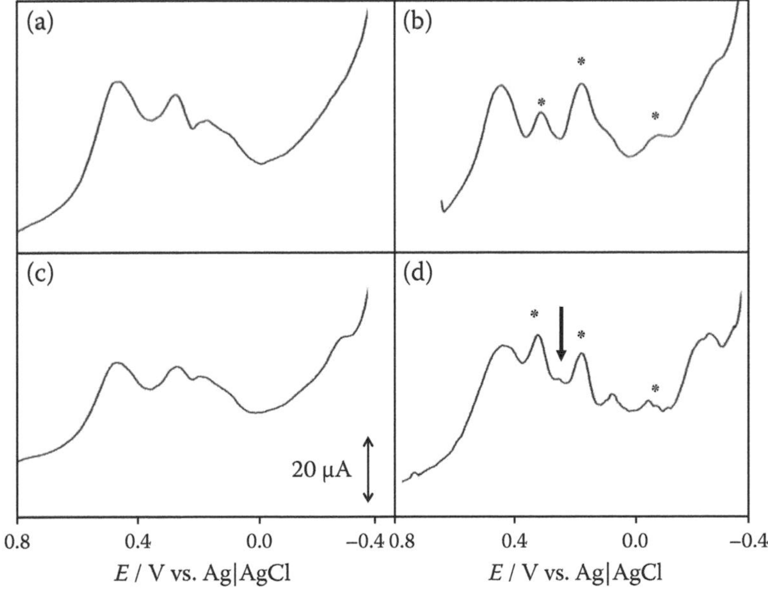

FIGURE 9.5 SWVs of: After Semi-derivative Convolution, of: (a) Co_9S_8, (b) MoS_2, (c) CoS_2, and (d) a CoS_x-MoS_x Catalyst Attached to Graphite Electrodes in Contact with 0.10 M H_2SO_4. Potential Scan Initiated at 0.80 V in the Negative Direction; Potential Step Increment 4 mV; Square Wave Amplitude 25 mV; Frequency 5 Hz. Mo-Centered Signals Are Marked by Asterisks. From Ref. [21] (Sorribes et al. *ACS Catalysis*. 2018, 8: 4545–4557), with Permission.

has received considerable attention in the last years. Hexagonal BN ultrathin monolayers supported on graphene or various transition metals exert a significant electrocatalytic effect on the ORR [24]. This application was inspired by the recognition of the catalytic properties of B- and N-doped carbon materials (see Chapter 10) [25].

In the case of BN monolayers on gold, there is a catalytic effect on ORR leading to the two-electron product of O_2 reduction, H_2O_2 [24]. Figure 9.6 illustrates this catalytic effect by comparing the LSVs at rotating electrodes (RDEs) of bare Au and glassy carbon (GCE) electrodes and the same covered by an ultrathin nanosheet BN layer. One can see that the electrocatalytic effect appears for ultrathin nanosheet BN on gold. The basis for this effect is the adsorption of O_2 by bridging two B atoms at the BN edge and bridging B and Au atoms at the perimeter interface between the BN island and the Au(111) surface, illustrated in Figure 9.7 [24].

It is pertinent to remark that, despite the lower porosity for conduction provided by aligned B and N atoms in the multilayer hexagonal BN [26], the formation of hybrid structures with carbon nanoarchitectures achieve increased electrocatalytic effects. This is the case of BN composites with multi-walled carbon nanotubes (MWCNTs) that act as bifunctional catalysts for both OER and ORR processes [27]. As shown in Figure 9.8a — where RDE LSVs for Pt/C catalyst, hexagonal BN, MWCNTs, and BN-MWCNTs composite films on GCE in O_2-saturated 0.1 M KOH are depicted — the composite material acts as an efficient bifunctional catalyst for ORR (now being a four-electron process) and OER. This capability is attributed to the merging of BN graphitic planes with few exfoliated graphene layers of MWCNTs yielding interfaces with favorable oxygen adsorption. A similar electrocatalytic effect has been promoted using $B_{12}N_{12}$ and $B_{60}N_{60}$ nanocages (see Figure 9.8) [28]. This effect can be associated to the sensitivity of both adsorption and dissociation of O_2 to the surface curvature of the nanostructured support [29] and the high stability of the BN clusters [30].

Transition metal nitrides have relatively high thermal stability and conductivity. This may be due to bonding and antibonding orbitals resulting from bonds between carbide or nitride 2p orbitals and metal d-orbital and associated vibrational properties [31]. These materials have attracted interest owing to their low density and large surface areas, with applications in electrochemical and

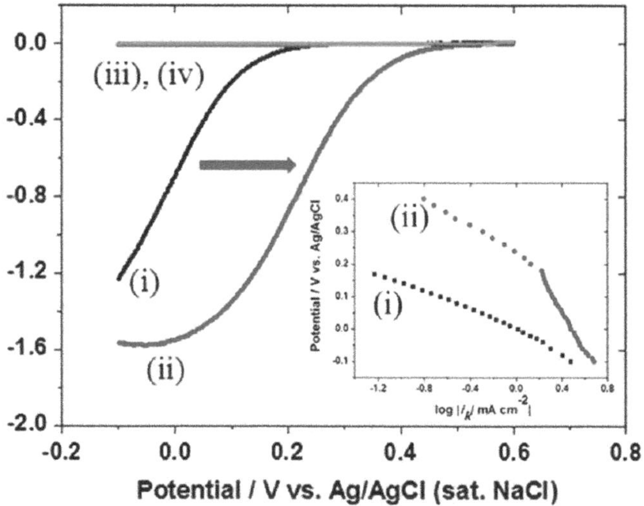

FIGURE 9.6 RDE LSVs of (i) Bare Au Electrode, (ii) Ultrathin Nanosheet BN on Au, (iii) Bare Glassy Carbon Electrode, (iv) Ultrathin Nanosheet BN on Glassy Carbon Electrode, All Immersed into O_2-Saturated 0.5 M H_2SO_4. Rotation Rate: 1500 rpm; Potential Scan Rate: 10 mV s^{-1}. Inset: Tafel Plot of Kinetic Currents at (i) Bare Au and (ii) Ultrathin Nanosheet BN on Au. Reproduced from Ref. [24] (Uosaki et al. *J. Am. Chem. Soc.* 2014, 136: 6542–6545), with Permission.

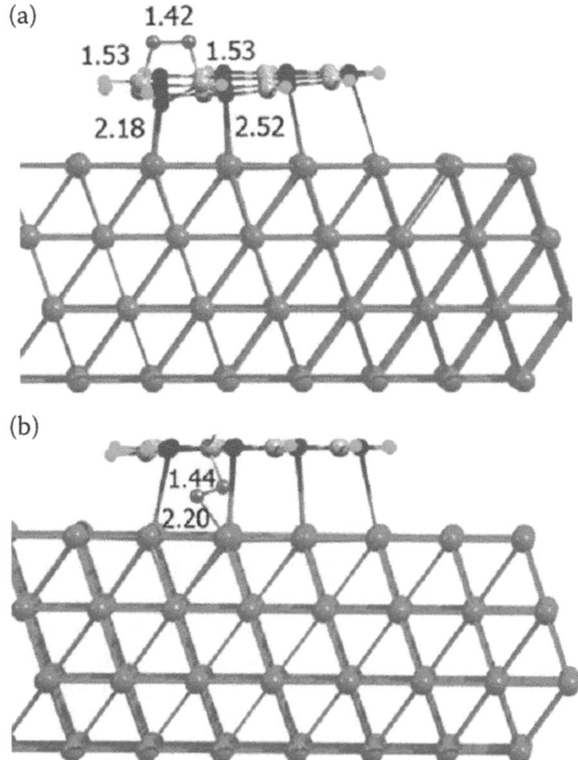

FIGURE 9.7 Optimized Structures (Distances in Å) Calculated for the Adsorption of O_2 on (a) B–B Bridge and (b) B–Au Bridge Configurations at the Edge of a Small Hexagonal BN Island on the Au(111) Surface. Reproduced from Ref. [24] (Uosaki et al. *J. Am. Chem. Soc.* 2014, 136: 6542–6545), with Permission.

photoelectrochemical sensing, catalysis and electrocatalysis, and energy storage [32,33]. It is pertinent to note that the properties of these compounds can be modulated by post-synthetic treatments. Sintering (converting powder compacts into strong, dense ceramic upon heating) notably influences properties like mechanical strength and dielectric response of the materials [32].

In particular, several transition metal nitrides have interesting electrocatalytic properties; for instance, mesoporous VN shows significant catalytic activity and reluctance to poisoning for formic acid oxidation, interesting in the construct anodes for direct formic acid fuel cells [34]. Molybdenum nitride (MoN_x) phases include disordered nitrogen vacancies, hexagonal phase δ_1-MoN (WC type), δ_2-MoN (NiAs type), δ_3-MoN (FeS type), γ-MoN_x (rock salt type), and α-MoN_2 (see Figure 9.8b). The possibility of hydrogen generation and storage within the material has been explored [33]. Many metal nitrides (VN, TiN, MoN, Fe2N, among others) and their compositedscomposites with carbon materials have been tested as materials for supercapacitors. Pseudocapacitance charge storage in alkaline aqueous solutions of TiN materials can be represented as [33]:

$$\{Ti^{III}N\}_{surf} + 2OH^-_{aq} + H_2O \rightarrow HTiO_3{}^-{}_{aq} + NH_{3aq} + e^- \quad (9.3)$$

Here, the pore sizes play an important role in the overall electrochemical/storage process. Accordingly, the preparation of 3D structures with stratified porosity facilitates channel interconnection, aiding electrolyte transportation and the accessibility of electrode materials [33].

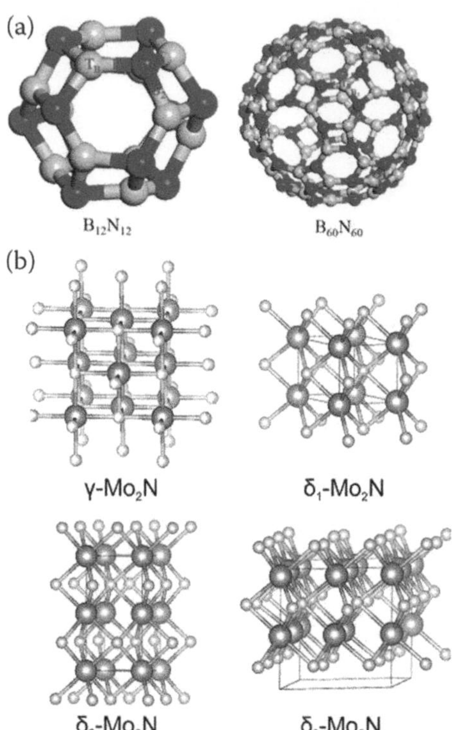

FIGURE 9.8 (a) Optimized Geometries of $B_{12}N_{12}$ and $B_{60}N_{60}$ Nanocages. Reproduced from Ref. [28] (Chen et al. *J. Phys. Chem. C.* 2016, 120: 28912–28916), with Permission, Fig. 1. (b) Several Mo_2N Structures. Reproduced from Ref. [33] (Shi et al. *Mater. Chem. Phys.* 2020, 245: artic 122533), with Permission.

9.4 METAL PHOSPHIDES

Several metal phosphides (CoP, MoP, WP, FeP, Ni_2P, Ni_5P_4, $Ni_{12}P_5$, among others) have been revealed as efficient catalysts for HER [36]. This application was inspired by the performance of transition metal phosphides as catalysts for dehydrosulfuration reactions. These processes involve the adsorption of organosulfur and H_2 molecule on the catalyst, followed by desorption of hydrocarbon fragments and the produced H_2S [37].

Regarding HER, the phosphorous centers act as proton acceptor sites while the metal centers act as hydride acceptor ones [38]. Theoretical studies suggest that the proton acceptor character of the P atom favors the process and that this effect is combined with the weakening of the metal-hydrogen bond. The electrocatalytic performance is reinforced by the high metallic/semi-conducting conductivity of these compounds. Then, it appears that the preferential adsorption of hydrogen atoms on P over the accompanying metal plays a key role in the catalytic effect exerted on HER. The complexity of the overall process — hydrogen adsorption, diffusion, recombination, and release from the catalyst surface — however, does not allow simple conclusions.

Several transition metal phosphides also show significant catalytic activity on OER. This is attributed to the *in-situ* formation of active oxo-hydroxides (MO(OH)) and phosphates in the surface of the catalyst [39]. However, surface oxidation, which easily occurs under ambient conditions, results in the formation of P(V) and metal oxide (high-valence state). These are non-catalytically active on HER so there is degradation of the catalyst with time [40].

To improve the performance of transition metal phosphide electrocatalysts, different strategies have been devised. These include:

a. Doping or alloying with metal or non-metal elements. Here, the P/M ratio appears to play an important role. There is a need to balance the increase of active sites promoted by high P content with the maintenance of high electrical conductivity favored by high M loadings.

b. A second strategy consists of non-metal (mainly with sulfide) alloying [41]. As with MoS_2, structural refinements such as those obtained in highly ordered mesoporous CoNiP [42], can tune the catalytic performance of the materials.
c. Preparation of hybrid materials with nanostructured carbons (graphene, rGO, CNTs) and metal phosphide nanoarchitectures. This is another widely investigated strategy with abundant literature [40].
d. Modification of surfaces and interfaces via formation of nanoporous structures. This strategy is aimed at the modulation of the wetting state of the catalyst surface. The nanoporous structure of the catalyst increases the number of exposed active sites and electrolyte transfer. Tuning the hydrophilicity and aerophobicity of the surface may facilitate water dissociation (and enhance mass transfer) and gas phase formation (avoiding gas pinning effects: adherence of gas bubbles to the catalyst surface), favoring both HER and OER processes. These effects appear to be preferred by hierarchical micro/nanoporous structuring the catalyst surface.
e. Formation of synergic interfaces with metal catalysts (Ni and Pt alloys), metal sulfides, etc. forming heterostructures.

Current research is focused on controlling the surface oxidation of transition metal phosphides, improving the preparation procedures and optimizing the formation of hybrids and nanostructured materials, coupling with light absorbers to develop photocatalysts, and developing OER catalysts based on non-precious metals [40].

9.5 METAL CARBIDES

Transition metal carbides have common properties with the corresponding nitrides [32,33,43,44]. The high melting point, considerable mechanical strength, and wear resistance of metal carbides are attributed not only to strong covalent carbon networks that exist within the structure but also to the mixed ionic-covalent character of the bonding. At least in several cases, such as TiC, metallic character exists between Ti-Ti; there is also ionic character associated to charge transfer occurring between Ti and C.

It is pertinent to remark that the lattice constant — representative of the crystal size — of metal carbides and nitrides is significantly larger than that of the respective metal. This implies a significant increment of the metal-metal bond distance. This would determine the contraction of the metal *d*-band with the concomitant increase in the density of states near the Fermi level [45]. As a result, the catalytic properties of transition metal carbides and nitrides become comparable to those of Pt and other noble metals.

The above properties are extensive to carbonitrides like WN_xC_y, which catalyzes hydrazine decomposition [46]. From the electrochemical point of view, TiC, Mo_2C, W_2C, and WC can be used as efficient HER electrocatalysts [47]. Here, the stability of the materials in contact with aqueous solutions plays a crucial role. Considering the redox stability of the corresponding metals, both Mo and W remain stable at potentials more negative than the reversible hydrogen potential. At more positive potentials, the metal surfaces are oxidized with the nature of the oxidation process depending on the pH of the solution. In strongly acidic media, the metals are oxidized to the corresponding MO_3 oxides at sufficiently positive potentials. The slightly oxidized metal surface is considered to be passivated. With further increase of the potential, the surface oxidation is extensive and sustained, and corrosion occurs. In both neutral and alkaline media, metal oxide species dissolve (surface oxidation/dissolution).

The stability zones of metals and metal carbides can be studied by chronopotentiometry titrations under quasi-static conditions [48] resulting in graphs like that in Figure 9.9 [49]. Assuming electrochemical reversibility, the slopes of the potential vs. pH curves estimate the proton (s)/electron (n) stoichiometry of the redox process occurring in the different pH regions as (potentials in mV at 298 K):

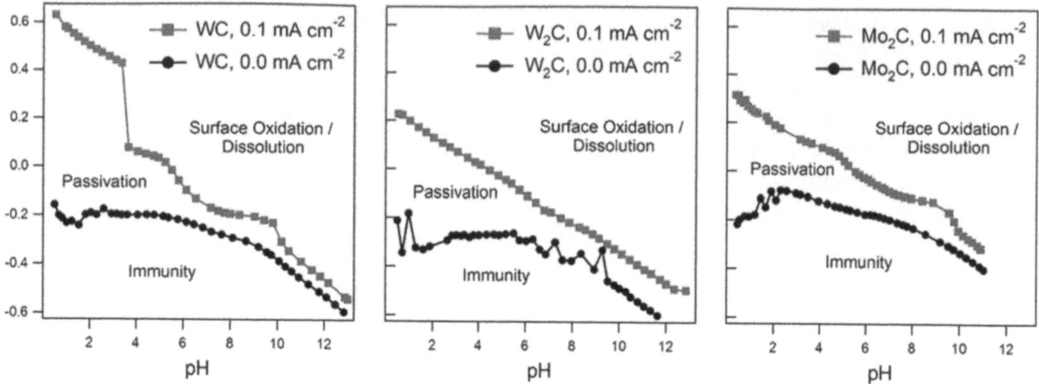

FIGURE 9.9 Chronopotentiometric Titration Curves Aimed to Map the Stability of WC, W_2C, and Mo_2C in Contact with 0.1 M Na_2SO_4 Solutions. Two Different Current Densities Were Used to Outline the Regions of Immunity, Passivation, and Surface Oxidation/Dissolution. Reproduced from Ref. [49] (Weidman et al. *J. Power Sources*. 2012, 202: 11–17), with Permission.

$$\frac{d(E)}{d(pH)} = -59.1\left(\frac{s}{n}\right) \tag{9.4}$$

The analysis of such plots distinguishes four types of processes [49]:

Oxidation reactions mediated by H^+ or OH^- resulting in the formation of solid oxides characterized by $s/n = 1$. Examples of these reactions are:

$$\{W\}_{solid} + 3H_2O \rightarrow \{WO_3\}_{solid} + 6H^+_{aq} + 6e^- \tag{9.5}$$

$$\{WC\}_{solid} + 5H_2O \rightarrow \{WO_3\}_{solid} + CO_{2aq} + 10H^+_{aq} + 10e^- \tag{9.6}$$

Oxidation processes that yield oxometal anions, characterized by $s/n > 1$, namely:

$$\{W\}_{solid} + 4H_2O \rightarrow {WO_4}^{2-}{}_{aq} + 8H^+_{aq} + 6e^- \tag{9.7}$$

$$\{WC\}_{solid} + 7OH^-_{aq} \rightarrow {WO_4}^{2-}{}_{aq} + {CO_3}^{2-}{}_{aq} + 7H^+_{aq} + 10e^- \tag{9.8}$$

Coupled chemical reactions (with no electron transfer). These are characterized by $s/n = \infty$. For instance:

$$\{WO_3\}_{solid} + 2OH^-_{aq} \rightarrow {WO_4}^{2-}{}_{aq} + H_2O \tag{9.9}$$

Oxidation reactions that are not mediated by H^+ or OH^-, characterized by $s/n = 0$.

$$\{Mo_2C\}_{solid} \rightarrow 2Mo^{3+}_{aq} + CO_{2aq} + 6e^- \tag{9.10}$$

Weidman et al. [49] concluded from such data that the corrosion of W and W_2C electrodes is kinetically limited by the formation of solid W-oxide surface species at all pHs, while the oxidation of WC and Mo_2C appears to be controlled by the oxidative dissolution of the surface oxides. The oxidative dissolution of metal carbides starts from a critical, pH-dependent potential. Table 9.3 summarizes reported data in acidic, neutral, and alkaline media [50].

Metal carbides can be electrochemically synthesized. Electrochemical syntheses can offer advantages relative to other synthetic procedures, by blocking the formation of secondary products and controlling the formation of the product at meso- and nano-sizes as well as its porosity. The methods are inspired by the synthesis of metals by electrolysis of oxides in molten salts. To prepare metal carbides, the metal oxide is sintered with carbon and submitted to electrolysis. It is interesting to mention that carbon acts not only as a precursor but also an accelerator of the electrochemical process by increasing the porosity and oxygen vacancies in the metal oxide during the sintering process [32].

9.6 OTHER COMPOUNDS

A variety of materials are currently investigated for different applications, with particular attention to water splitting. These include highly anisotropic layered inorganic compounds whose more representative example is graphite. Different layered systems are currently under study, namely, fluorographites and fluorographenes, graphanes, misfit systems, layered superconductors, and layered engineered van der Waals structures [51]. Among this extensive list, the layered allotropes of group 15 elements (P, As, Sb, and Bi, also called pnictogens) are receiving attention due to their semiconducting character and anisotropic properties. Pnictogens show a marked puckered structure with dative electron lone pairs located on the surface atoms, which results in semiconducting character and ability to adsorb and stabilize organic molecules through van der Waals interactions [52]. Black phosphorus, whose structure consists of two interconnected atomic layers, is particularly interesting [53]. As in the case of white phosphorus [54], the voltammetry of black phosphorus in contact with aqueous electrolytes is dominated by an apparently irreversible oxidation to P(V) species [51].

The phosphorene, arsenene, antimonene, and bismuthene derivatives have also been under study [55]. These materials have potential applications in catalysis, optoelectronics, electrical energy production, and storage. Figure 9.10 shows the CV of antimonene in contact with 1 M $NaClO_4$/ethylene carbonate plus diethyl carbonate electrolyte [56]. In the initial cathodic scan, a strong reduction peak at 0.35 V vs. Na^+/Na appears in the subsequent anodic scan, followed by overlapping oxidation peaks at 0.88 and 0.96 V. During the second cathodic scan, three peaks at 0.68, 0.55, and 0.39 V are recorded. This voltammetry can be interpreted in terms of multistep Na–Sb alloying reactions passing from Sb to NaSb and Na_3Sb phases, successively with their corresponding anodic desodiation reactions. The decrease in the cathodic peak current from the first to the subsequent cathodic scans can be attributed to electrolyte decomposition, forming solid-

TABLE 9.3
Critical Potential of the Electrochemical Oxidation of the Metal Carbides

Electrolyte	NbC	Nb_2C	VC	TiC
0.1 M HCl	0.9	0.6	0.6	0.6
0.1 M KCl	0.7	0.4	0.5	0.6
0.1 M NaOH	0.3	0.2	–0.4	0.1

From CVs at 100 mV s^{-1}; potentials relative to Ag/AgCl [50].

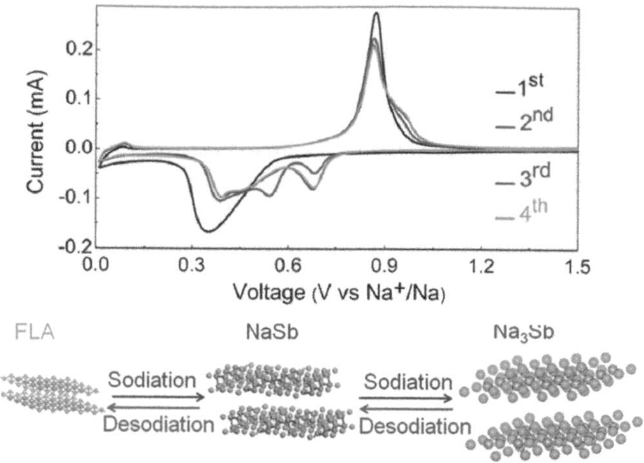

FIGURE 9.10 CV of 2D Few-Layer Antimonene in Contact with 1 M $NaClO_4$/Ethylene Carbonate Plus Diethyl Carbonate Electrolyte and Schematics for the Reductive/Oxidative Sodiation/Desodiation Processes. Potential Scan Rate 100 mV s^{-1}. Reproduced from Ref. [56] (Tian et al. *ACS Nano*. 2018, 12: 1887–1893), with Permission.

state interface layers on the antimonene. In summary, these 2D materials offer interesting possibilities in the aforementioned applications fields whose scope will be enhanced in the next years as a result of intensive research.

REFERENCES

[1]. Ding, Q.; Song, B.; Xu, P.; Jin, S. 2016. Efficient electrocatalytic and photoelectrochemical hydrogen generation using MoS_2 and related compounds. *Chem*. 1: 699–726.

[2]. Morales-Guio, C.G.; Hu, X.L. 2014. Amorphous molybdenum sulfides as hydrogen evolution catalysts. *Accounts of Chemical Research*. 47: 2671–2681.

[3]. Chianelli, R.R.; Siadati, M.H.; De la Rosa, M.P.; Berhault, G.; Wilcoxon, J.P.; Bearden, R.; Abrams, B.L. 2006. Catalytic properties of single layers of transition metal sflfide catalytic materials. *Catalysis Reviews*. 48: 1–41.

[4]. Lee, H.S.; Min, S.W.; Chang, Y.G.; Park, M.K.; Nam, T.; Kim, H.; Kim, J.H.; Ryu, S.; Im, S. 2012. MoS_2 nanosheet phototransistors with thickness-modulated optical energy gap. *Nano Letters*. 12: 3695–3700.

[5]. Jaramillo, T.F.; Jorgensen, K.P.; Bonde, J.; Nielsen, J.H.; Horch, S.; Chorkendorff, I. 2007. Identification of active edge sites for electrochemical H_2 evolution from MoS_2 nanocatalysts. *Science*. 317: 100–102.

[6]. Benck, J.D.; Hellstern, T.R.M.; Kibsgaard, J.M.; Chakthranont, P.; Jaramillo, T.F. 2014. Catalyzing the hydrogen evolution reaction (HER) with molybdenum sulfide nanomaterials. *ACS Catalysis*. 4: 3957–3971.

[7]. Li, G.; Zhang, D.; Qiao, Q.; Yu, Y.; Peterson, D.; Zafar, A.; Kumar, R.; Curtarolo, S.; Hunte, F.; Shannon, S.; Zhu, Y.; Yang, W.; Cao, L. 2016. All The Catalytic Active Sites of MoS2 for Hydrogen Evolution. *Journal of the American Chemical Society*. 138: 16632–16638.

[8]. Kong, D.S.; Wang, H.T.; Cha, J.J.; Pasta, M.; Koski, K.J.; Yao, J.; Cui, Y. 2013. Synthesis of MoS_2 and $MoSe_2$ films with vertically aligned layers. *Nano Letters*. 13: 1341–1347.

[9]. Li, H.; Tsai, C.; Koh, A.L.; Cai, L.L.; Contryman, A.W.; Fragapane, A.H.; Zhao, J.H.; Han, H.S.; Manoharan, H.C.; Abild-Pedersen, F.; Nørskov, J.K.; Zheng, X. 2016. Activating and optimizing MoS2 basal planes for hydrogen evolution through the formation of strained sulphur vacancies. *Nature Materials*. 15: 48–53.

[10]. Yin, Y.; Han, J.; Zhang, Y.; Zhang, X.; Xu, P.; Yuan, Q.; Samad, L.; Wang, X.; Wang, Y.; Zhang, Z.; Zhang, P.; Cao, X.; Song, B.; Jin, S. 2016. Contributions of phase, sulfur vacancies, and edges to the

hydrogen evolution reaction catalytic activity of porous molybdenum disulfide nanosheets. *Journal of the American Chemical Society*. 138: 7965–7972.

[11]. Lukowski, M.A.; Daniel, A.S.; Meng, F.; Forticaux, A.; Li, L.S.; Jin, S. 2013. Enhanced hydrogen evolution catalysis from chemically exfoliated metallic MoS_2 nanosheets. *Journal of the American Chemical Society*. 135: 10274–10277.

[12]. Tsai, C.; Chan, K.R.; Norskov, J.K.; Abild-Pedersen, F. 2015. Theoretical insights into the hydrogen evolution activity of layered transition metal dichalcogenides. *Surface Science*. 640: 133–140.

[13]. Tang, Q.; Jiang, D.-E. 2016. Mechanism of hydrogen evolution reaction on 1T-MoS_2 from first principles. *ACS Catalysis*. 6: 4953–4961.

[14]. Merki, D.; Fierro, S.; Vrubel, H.; Hu, X.L. 2011. Amorphous molybdenum sulfide films as catalysts for electrochemical hydrogen production in water. *Chemical Science*. 2: 1262–1267.

[15]. Xie, J.F.; Zhang, J.J.; Li, S.; Grote, F.; Zhang, X.D.; Zhang, H.; Wang, R.X.; Lei, Y.; Pan, B.C.; Xie, Y. 2013. Controllable disorder engineering in oxygen-incorporated MoS_2 ultrathin nanosheets for efficient hydrogen evolution. *Journal of the American Chemical Society*. 135: 17881–17888.

[16]. Lin, S.H.; Kuo, J.L. 2015. Activating and tuning basal planes of MoO_2, MoS_2, and $MoSe_2$ for hydrogen evolution reaction. *Physical Chemistry Chemical Physics*. 17: 29305–29310.

[17]. Li, Y.G.; Wang, H.L.; Xie, L.M.; Liang, Y.Y.; Hong, G.S.; Dai, H.J. 2011. MoS_2 nanoparticles grown on graphene: An advanced catalyst for the hydrogen evolution reaction. *Journal of the American Chemical Society*. 133: 7296–7299.

[18]. Liao, L.; Zhu, J.; Bian, X.J.; Zhu, L.N.; Scanlon, M.D.; Girault, H.H.; Liu, B.H. 2013. MoS_2 formed on mesoporous graphene as a highly active catalyst for hydrogen evolution. *Advanced Functional Materials*. 23: 5326–5333.

[19]. Staszak-Jirkovsky, J.; Malliakas, C.D.; Lopes, P.P.; Danilovic, N.; Kota, S.S.; Chang, K.C.; Genorio, B.; Strmcnik, D.; Stamenkovic, V.R.; Kanatzidis, M.G.; Markovic, N.M. 2016. Design of active and stable Co-Mo-S_x chalcogels as pH universal catalysts for the hydrogen evolution reaction. *Nature Materials*. 15: 197–203.

[20]. Ji, Z.; Trickett, C.; Pei, X.; Yaghi, O.M. 2018. Linking molybdenum–sulfur clusters for electrocatalytic hydrogen evolution. *Journal of the American Chemical Society*. 140: 13618–13622.

[21]. Sorribes, I.; Liu, L.; Doménech-Carbó, A.; Corma, A. 2018. Nanolayered cobalt–molybdenum sulfides as highly chemo- and regioselective catalysts for the hydrogenation of quinoline derivatives. *ACS Catalysis*. 8: 4545–4557.

[22]. Solozhenko, V.L.; Lazarenko, A.G.; Petitet, J.P.; Kanaev, A.V. 2001. Bandgap energy of graphite-like hexagonal boron nitride. *Journal of Physics and Chemistry of Solids*. 62: 1331–1334.

[23]. Zeng, H.B.; Zhi, C.Y.; Zhang, Z.H.; Wei, X.L.; Wang, X.B.; Guo, W.L.; Bando, Y.; Golberg, D. 2010. "White graphenes": Boron nitride nanoribbons via boron nitride nanotube unwrapping. *Nano Letters*. 10: 5049–5055.

[24]. Uosaki, K.; Elumalai, G.; Noguchi, H.; Masuda, T.; Lyalin, A.; Nakayama, A.; Taketsugu, T. 2014. Boron nitride nanosheet on gold as an electrocatalyst for oxygen reduction reaction: Theoretical suggestion and experimental proof. *Journal of the American Chemical Society*. 136: 6542–6545.

[25]. Zheng, Y.; Jiao, Y.; Ge, L.; Jaroniec, M.; Qiao, S.Z. 2013. Two-step boron and nitrogen doping in graphene for enhanced synergistic catalysis. *Angewandte Chemie International Edition*. 52: 3110–3116.

[26]. Karnik, R.N. 2014. Materials science: Breakthrough for protons. *Nature*. 516: 173–175.

[27]. Patil, I.M.; Lokanathan, M.; Ganesan, B.; Swami, A.; Kakade, B. 2017. Carbon nanotube/boron nitride nanocomposite as a significant bifunctional electrocatalyst for oxygen reduction and oxygen evolution reactions. *Chemistry A European Journal*. 23: 676–683.

[28]. Chen, X.; Chang, J.; Yan, H.; Xia, D. 2016. Boron nitride nanocages as high activity electrocatalysts for oxygen reduction reaction: Synergistic catalysis by dual active sites. *Journal of Physical Chemistry C*. 120: 28912–28916.

[29]. Zhang, P.; Hou, X.; Li, S.; Liu, D.; Dong, M. 2015. Curvature effect of O_2 adsorption and dissociation on SiC nanotubes and nanosheet. *Chemical Physics Letters*. 619: 92–96.

[30]. Seifert, G.; Fowler, P.W.; Mitchell, D.; Porezag, D.; Frauenheim, Th. 1997. Boron-nitrogen analogues of the fullerenes: Electronic and structural properties. *Chemical Physics Letters*. 268: 352–358.

[31]. Häglund, J.; Guillermet, A.F.; Grimvall, G.; Körling, M. 1993. Theory of bonding in transition metal carbides and nitrides. *Physical Review B*. 48: Article no. 11685.

[32]. Rasaki, S.A.; Zhang, B.; Anbalgam, K.; Thomas, T.; Yang, M. 2018. Synthesis and applications of nano-structured metal nitrides and carbides: A review. *Progress in Solid State Chemistry*. 50: 1–15.

[33]. Shi, J.; Jiang, B.; Li, C.; Yan, F.; Wang, D.; Yang, C.; Wan, J. 2020. Review of transition metal nitrides and transition metal nitrides/carbon nanocomposites for supercapacitor electrodes. *Materials Chemistry and Physics*. 245: Article no. 122533.

[34]. Yang, M.; Cui, Z.; DiSalvo, F.J. 2012. Mesoporous vanadium nitride as a high performance catalyst support for formic acid electrooxidation. *Chemical Communications*. 48: 10502–10504.

[35]. Pande, P.; Deb, A.; Sleightholme, A.E.S.; Djire, A.; Rasmussen, P.G.; Penner-Hahn, J.; Thompson, L.T. 2015. Pseudocapacitive charge storage via hydrogen insertion for molybdenum nitrides. *Journal of Power Sources*. 289: 154–159.

[36]. Xiao, P.; Chen, W.; Wang, X.A. 2015. Review of phosphide-based materials for electrocatalytic hydrogen evolution. *Advanced Energy Materials*. 5: Article no. 1500985.

[37]. Clark, P.; Wang, X.; Oyama, S.T. 2002. Characterization of silica-supported molybdenum and tungsten phosphide hydroprocessing catalysts by ^{31}P nuclear magnetic resonance spectroscopy. *Journal of Catalysis*. 207: 256–265.

[38]. Liu, P.; Rodriguez, J.A. 2005. Catalysts for hydrogen evolution from the [NiFe] hydrogenase to the $Ni_2P(001)$ surface: The importance of the ensemble effect. *Journal of the American Chemical Society*. 127: 14871–14878.

[39]. Ryu, J.; Jung, N.; Jang, J.H.; Kim, H.-J.; Yoo, S.J. 2015. In situ transformation of hydrogen-evolving CoP nanoparticles: Toward efficient oxygen evolution catalysts bearing dispersed morphologies with Co-oxo/hydroxo molecular units. *ACS Catalysis*. 5: 4066–4074.

[40]. Wang, Y.; Kong, B.; Zhao, D.; Wang, H.; Selomulya, C. 2017. Strategies for developing transition metal phosphides as heterogeneous electrocatalysts for water splitting. *Nano Today*. 15: 26–55.

[41]. Kibsgaard, J.; Jaramillo, T.F. 2014. Molybdenum phosphosulfide: An active, acid-stable, earth-abundant catalyst for the hydrogen evolution reaction. *Angewandte Chemie International Edition*. 53: 14433–14437.

[42]. Fu, S.; Zhu, C.; Song, J.; Engelhard, M.H.; Li, X.; Du, D.; Lin, Y. 2016. Highly ordered mesoporous bimetallic phosphides as efficient oxygen evolution electrocatalysts. *ACS Energy Letters*. 1: 792–796.

[43]. Li, Y.; Zhang, H.; Xu, T.; Lu, Z.; Wu, X.; Wan, P.; Sun, X.; Jiang, L. 2015. Under-water super-aerophobic pine-shaped Pt nanoarray electrode for ultrahigh-performance hydrogen evolution. *Advanced Functional Materials*. 25: 1737–1744.

[44]. Chen J.G. 1996. Carbide and nitride overlayers on early transition metal surfaces: Preparation, characterization, and reactivities. *Chemical Reviews*. 96: 1477–1498.

[45]. Jansen, S.A.; Hoffmann, R. 1988. Surface chemistry of transition metal carbides: A theoretical analysis. *Surface Science*. 197: 474–508.

[46]. Sun, J.; Liang, B.; Huang, Y.; Wang, X. 2016. Synthesis of nanostructured tungsten carbonitride (WN_xC_y) by carbothermal ammonia reduction on activated carbon and its application in hydrazine decomposition. *Catalysis Today*. 274: 123–128.

[47]. Chen, W.-F.; Muckerman, J.T.; Fujita, E. 2013. Recent developments in transition metal carbides and nitrides as hydrogen evolution electrocatalysts. *Chemical Communications*. 49: 8896–8909.

[48]. Weidman, M.C.; Esposito, D.V.; Hsu, I.J.; Chen, J.G. 2010. Electrochemical stability of tungsten and tungsten monocarbide (WC) over wide pH and potential ranges. *Journal of the Electrochemical Society*. 157: F179–F188.

[49]. Weidman, M.C.; Esposito, D.V.; Hsu, I.J.; Chen, J.G. 2012. Comparison of electrochemical stability of transition metal carbides (WC, W_2C, Mo_2C) over a wide pH range. *Journal of Power Sources*. 202: 11–17.

[50]. Messner, M.; Waleczyk, D.J.; Palazzo, B.G.; Norris, Z.A.; Taylor, G.; Carroll, J.; Pham, T.X.; Hettinger, J.D.; Yu, L. 2018. Electrochemical oxidation of metal carbides in aqueous solutions. *Journal of the Electrochemical Society*. 165: H3107–H3114.

[51]. Wang, L.; Soler, Z.; Pumera, M. 2015. Voltammetry of layered black phosphorous: Electrochemistry of multilayer phosphorene. *ChemElectroChem*. 2: 324–327.

[52]. Lloret, V.; Rivero-Crespo, M.A.; Vidal-Moya, J.A.; Wild, S.; Doménech-Carbó, A.; Heller, B.S.J.; Shin, S.; Steinrück, H.-P.; Maier, F.; Hauke, F.; Varela, M.; Hirsch, A.; Leyva-Pérez, A.; Abellán, G. 2019. Few layer 2D pnictogens catalyze the alkylation of soft nucleophiles with esters. *Nature Communications*. 10: Article no. 509.

[53]. Davies, T.J.; Hyde, M.E.; Compton, R.G. 2005. Nanotrench arrays reveal insight into graphite electrochemistry. *Angewandte Chemie International Edition*. 44: 5121 –5126.

[54]. Hermes, M.; Scholz, F. 2000. The electrochemical oxidation of white phosphorus at a three-phase junction. *Electrochemistry Communications*. 2: 845–852.
[55]. Martínez-Periñán, E.; Down, M.P.; Gibaja, C.; Lorenzo, E.; Zamora, F.; Banks, C.E. 2017. Antimonene: A novel 2D nanomaterial for supercapacitor applications. *Advanced Energy Materials*. Article no. 1702606.
[56]. Tian, W.; Zhang, S.; Huo, C.; Zhu, D.; Li, Q.; Wang, L.; Ren, X.; Xie, L.; Guo, S.; Chu, P.K.; Zeng, H.; Huo, K. 2018. Few-layer antimonene: Anisotropic expansion and reversible crystalline-phase evolution enable large-capacity and long-life Na-ion batteries. *ACS Nano*. 12: 1887–1893.

10 Electrochemistry of Porous Carbon-Based Materials

10.1 CARBONS AS ELECTROCHEMICAL MATERIALS

Elemental carbon is usually handled in three forms: graphite, diamond, and amorphous carbon. Graphite and amorphous carbon have been extensively used in electrochemistry because of their high electrical conductivity, chemical stability, versatility, and low cost. For electrochemical applications, such materials can be manufactured in bars, powders, and fibers or even form conducting composites using appropriate binders. A number of carbon-based materials—namely, pyrolytic carbon, carbon blacks, activated carbons, graphite fibers, whiskers, glassy carbon, etc.—have been used in electrochemistry for decades [1].

Recently, a lot of new carbon materials have been introduced in the electrochemistry world—from boron-doped diamond electrodes to fullerenes; from carbon black, porous carbons, and carbon fiber electrodes to carbon nanotubes (CNTs) [2]. Glassy carbon (or vitreous carbon) is routinely used in electrochemistry. It is produced by thermal degradation of selected organic polymers followed by carbonization at temperatures ca. 1800 °C and constitutes a widely used working electrode. Graphene—discovered in 2004 [3]—and its derivatives [4] are making a profound impact in science and technology. Several properties of carbons (for precise definitions of different types of carbons, see Ref. [5]) make these materials of particular interest for electrochemical applications; carbon materials are relatively inert in contact with most electrolytes while retaining a high degree of surface activity and admitting different functionalization/derivatization procedures. Novel porous carbon materials have attracted attention for their use as adsorbents, gas storage and catalyst support, and electrode materials in supercapacitors, batteries, fuel cells, and separation techniques based on electro-sorption, etc. [6].

10.2 CARBON ACTIVATION, FUNCTIONALIZATION, AND DOPING

The electrochemical behavior of 'traditional' carbon electrodes (pyrolitic graphite, glassy carbon) is sensitive to the cleaning and conditioning of the surface. It has long been hypothesized that the more electrochemically active sites are edge plane graphite sites, while basal plane sites are non-electroactive or much less active. It has been reported, however, that the heterogeneous electron transfer rates of edge and basal planes of graphite are essentially identical using ruthenium hexaammonium [7], but significantly differ using $Fe(CN)_6^{3-}/Fe(CN)_6^{4-}$ and other couples [8].

Activation of carbon electrodes (i.e. procedures to increase their electrochemical performance) via mechanical, thermal, chemical, and electrochemical treatments not only implies the redistribution of active sites but also the generation of surface functionalities. Thermal treatments generate three main types of functionalities [6]: (a) chemically fixed groups, such as carbonyl, which are electrochemically inactive and can be degassed above 800 °C; (b) carboxylic type groups degassed above 400 °C mostly as CO_2; and (c) surface groups with high electrochemical activity, typically quinone/hydroquinone units. These groups are responsible for voltammetric peaks appearing in contact with aqueous electrolytes and can be generated electrochemically by the application of different potential inputs [9,10].

Among others, adsorption, deposition, covalent attachment, in-site functionalization of self-assembled monolayers, and electropolymerization strategies have been reported for the modification of carbon electrode surfaces [11]. N-derivatization of carbons through reduction of diazonium salts has found application in electrochemical sensing [12].

Another important methodology is doping of carbons with heteroatoms, either leading to the formation of electron excess (*n*-type, e.g., doping with N) of electron-deficient (*p*-type, e.g., doping with B) [13]. This strategy has led to enhanced electrocatalytic activity in CNTs [14] and graphene [5,15]. Co-doping with B and N is another strategy expanded—among other possibilities—by introducing tunable contents of pyridinic N, graphitic N, BC_3, and C-B(N)O [16].

10.3 POROUS CARBONS

The term porous carbons is used to designate the materials prepared by carbonization of organic felts or fibers (cellulose, polyacrylonitrile) under an inert atmosphere or reduced conditions. Complex pyrocarbon structures are formed depending on temperature and other parameters. The pyrolysis eliminates the volatile components, including the heteroatoms, and increases the proportion of conjugated carbon atoms in the sp^2 state, progressively increasing conductivity. The morphology and porosity of amorphous carbons can be modulated via mild burn-off processes using oxidants. This practice leads to high porosity, where the pores may occupy a large portion (until 80%) of the total volume of the material. Chemical and thermal treatments are used for the activation of carbons. Chemical activation, usually by means of KOH, introduces oxygen functionalities in the carbon surface.

Ordered mesoporous carbons (OMCs) possess a uniform, periodic distribution of mesopores with high surface area [17]. Such materials can be prepared by nanocasting ordered mesoporous silicas such as MCM-48 or SBA-15 and mesocellular silica foams [17,18]. During the OMC replication process, the internal porous structure of the silica template is inversely replicated while the external morphology of the primary particles is preserved, the control of particle sizes being of interest. Nongraphitizing carbons are those that cannot be converted to graphite by further high-temperature treatment. Here, the carbon precursors (wood, biomass, etc.) remain in the solid phase during carbonization resulting in rigid amorphous structures [6].

Porous carbons have encountered application in electrochemical sensing, electrical energy storage as supercapacitors, and separation techniques based on electro-adsorption, such as water desalination by capacitive de-ionization (see Chapter 14).

10.4 CARBON NANOTUBES AND NANORIBBONS

Carbon nanotubes CNTs [19–21] and nanoribbons [22] are nanostructured materials having a common fiber-type structure. In carbon nanotubes, sp^2-hybridized carbon atoms are arranged in graphite-type sheets building-up seamless hollow tubes capped by fullerene-type (*vide infra*) hemispheres. Single-walled carbon nanotubes (SWNTs) are constituted by single hollow tubes with a diameter between 0.4 and 2 nm. Multi-walled carbon nanotubes (MWNTs) are formed by concentric tubes with a diameter between 2 and 100 nm. SWNTs can act as semiconductors or metallic conductors, while MWNTs behave as metallic conductors.

As for carbon powders, different activation strategies—typically involving chemical activation protocols with acids or alkalis—can be used. Figure 10.1 shows TEM images of untreated carbon nanotubes (ACNT) and activated with KOH and thermally treated at 800 °C using different KOH/ACNT weight ratios of specimens labeled as 3 (ACNT-3) and 5 (ACNT-5) [23]. Inactivated nanotubes are multiwalled with diameters between 10 and 20 nm with smooth outer and inner walls. After activation, both the morphology and structure of nanotubes change, the tubes become short and distorted and some of the nanotubes become fractured or even powdered. These effects can be attributed to the strong chemical reaction between KOH and carbon nanotubes at high temperatures, the structural alteration being more severe as the KOH/ACNT ratio increases. Interestingly, some of the active sites of the surface of nanotubes are etched with KOH and become pores, increasing the

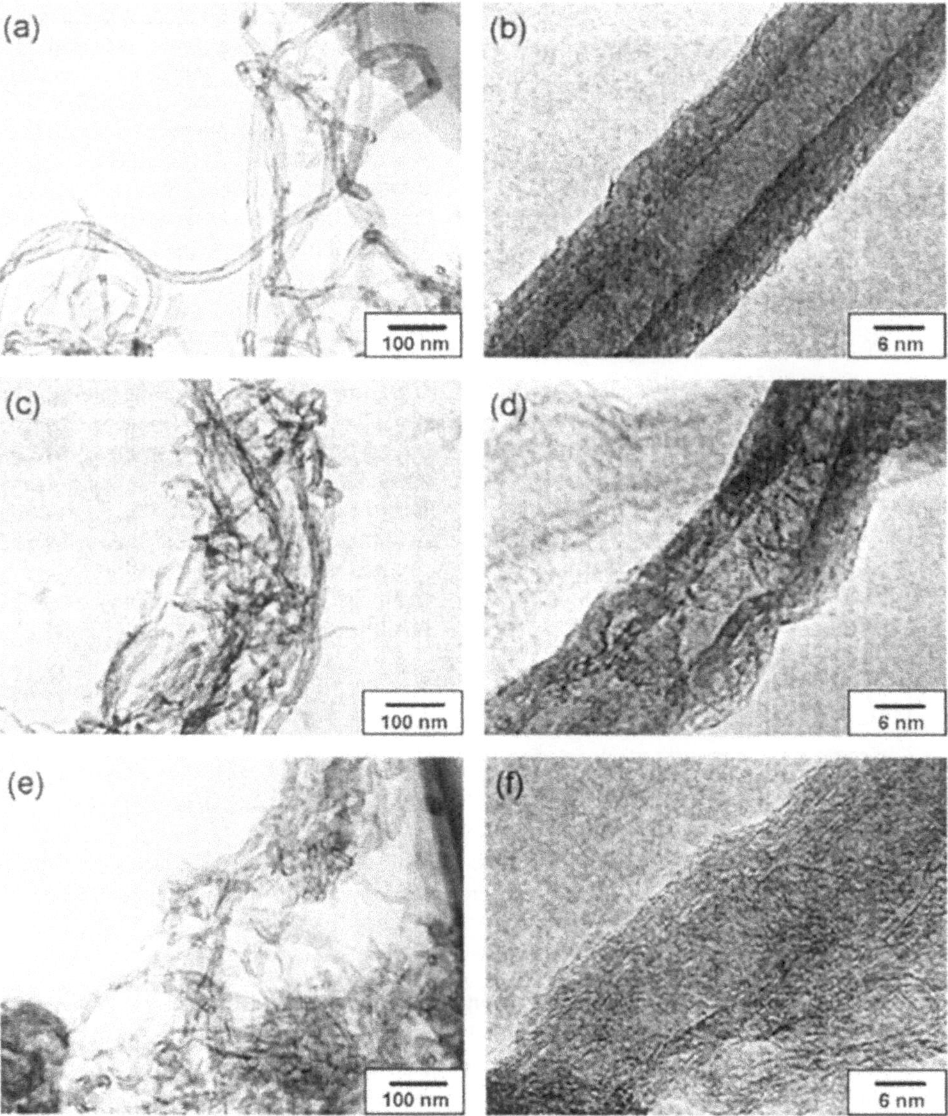

FIGURE 10.1 TEM Images of Untreated Carbon Nanotubes (a,b; ACNT) and Activated with KOH and Thermally Treated at 800 °C Using Different KOH/ACNT Weight Ratios of 3 (c,d; ACNT-3) and 5 (e, f; ACNT-5). Reproduced from Ref. [23] (Xu et al. *Electrochim. Acta.* 2008, 53: 7730–7735), with Permission.

surface area of the material. Concomitantly, the destruction of the graphitic structure of the nanotubes produces a decrease in conductivity.

Figure 10.2 shows CVs of nanotubes in contact with 7 M aqueous KOH at different potential scan rates. At low sweep rates, the voltammogram is close to the ideal rectangular profile corresponding to a purely capacitive behavior. Solely at high scan rates, polarization effects distort the voltammetric profile. In turn, the impedance spectra display a semicircle at high frequency and a nearly vertical line in the low-frequency range. This last denotes the capacitive behavior of the ACNTs. In these spectra the sloping linear region at high-to-medium frequency, often observed in porous systems, is absent. The absence of the feature, related to diffusion resistance, suggests that,

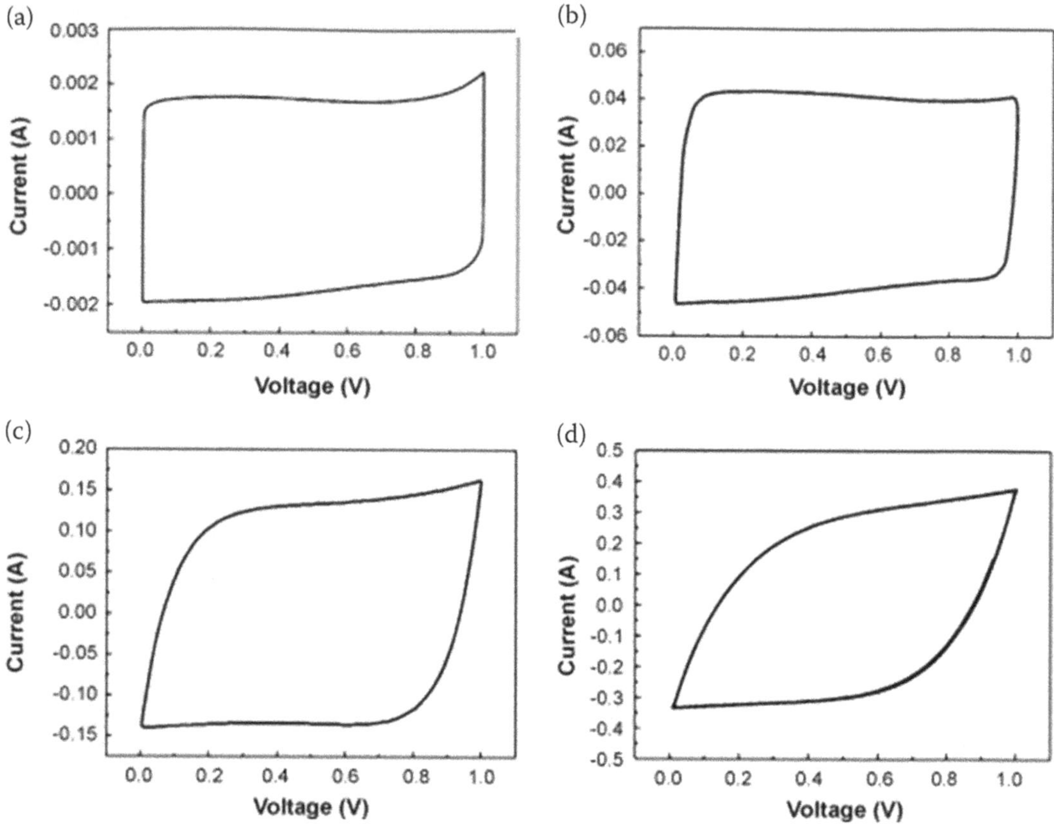

FIGURE 10.2 CVs of ACNT3-Based EDLC in Contact with 7 M KOH Aqueous Electrolyte at the Scan Rate of (a) 2, (b) 50, (c) 200, and (d) 500 mV s^{-1}. Reproduced from Ref. [23] (Xu et al. *Electrochim. Acta.* 2008, 53: 7730–7735), with Permission.

as far as the pore size of the ACNTs is large enough for the access of ions in KOH aqueous electrolyte, the diffusion is not the control factor in the kinetics of the electrode process [23].

Carbon nanoribbons (CNRs) can be described as quasi-one-dimensional systems consisting of elongated stripes of single-layered graphene with a finite width. Graphene nanoribbons can either be metallic or semiconducting depending on the crystallographic direction of the ribbon axis, the main difference between CNRs and CNTs is the existence of sharp edges in CNRs, so edge effects are currently under intensive research [24].

The electrochemical response of carbon nanotubes is dominated by double-layer charging with only a small contribution of Faradaic pseudocapacitance due to surface oxides, so Faradaic processes are essentially absent. In contact with aqueous electrolytes, however, weak peaks may eventually appear. As indicated in Section 10.2, these peaks are presumably due to Faradaic pseudocapacitance associated with oxygen-containing surface functionalities on CNTs and/or carbonaceous impurities. Such processes were found to be pH-dependent and can generically be represented as [25]:

$$=\mathrm{C}=\mathrm{O} + \mathrm{H}^{+} + \mathrm{e}^{-} \rightarrow \ \equiv\mathrm{C}-\mathrm{OH} \tag{10.1}$$

The appearance of weak voltammetric peaks in contact with nonaqueous electrolytes can be attributed to water traces or Li^{+} insertion [26].

As previously noted, doping with heteroatoms is a common strategy to enhance the electrocatalytic performance of CNTs. Depending on the type, position, and distribution of the heteroatoms, various types of nanostructured junctions with different electronic properties would be prepared. This involves double coaxial CNTs of doped and undoped multiwalls or double coaxial CNTs composed of N- and B-doped multiwalls. Doping CNTs with N or B modifies the electronic structure and conducting properties of the materials. The increase in conductivity with respect to that of undoped CNTs can be attributed to the increase in hopping frequency between the conduction and valence bands [27]. This kind of system may exhibit *p-n* junction characteristics of interest for obtaining nanoelectronic devices.

Apart from 'atomic doping' [14,15], 'intercalation doping' of both *n*- and *p*-types can be obtained in electrochemical turnovers, for instance, in Bu_4NBF_4/MeCN electrolytes [28,29]:

$$\{CNT\} + zBu_4N^{+}_{solv} + ze^{-} \rightarrow \{CNT^{z-} \cdots (Bu_4N^{+})_z\} \quad (10.2)$$

$$\{CNT\} + zBF_4{}^{-}{}_{solv} \rightarrow \{CNT^{z+} \cdots (BF_4{}^{-})_z\} + ze^{-} \quad (10.3)$$

About their application in lithium batteries (see Chapter 14), lithium insertion into SWNTs exhibits fast kinetics and absence of staging and Li^+ ions are accommodated in nanochannels occurring inside the nanotube bundles and ropes [30]. However, the practical application of CNTs is made difficult due to the appearance of hysteresis and irreversible capacity. These features can be associated with their mesoporous character that determines the ease evacuation of Li^+ responsible for hysteresis, as well as the easy access of solvated ions to the active surface, determining irreversible capacity [23] (Figure 10.3).

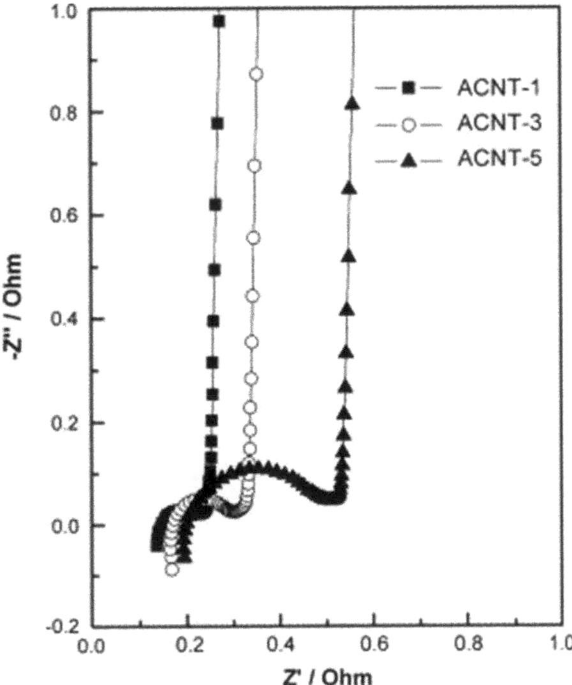

FIGURE 10.3 Nyquist Plots of ACNT Prepared with Different KOH/CNT Ratios in Contact with 7 M KOH Aqueous Electrolyte. Reproduced from Ref. [23] (Xu et al. *Electrochim. Acta.* 2008, 53: 7730–7735), with Permission.

10.5 GRAPHENE(S)

According to the IUPAC definition [31], graphene is "a single carbon layer of graphite structure, describing its nature by analogy to a polycyclic aromatic hydrocarbon of quasi-infinite size." The term graphene, however, is used in a wide sense, including multilayer materials, chemically modified materials (*vide infra*), and damaged graphene sheets containing sp^3-hybridized carbon defects and/or adatoms [32]. Four types of directly related materials can be distinguished [4,5]:

a. Graphene constituted by two dimensional sp^2 bonded carbon layer in a hexagonal arrangement with single-atom thickness.
b. Graphite oxide (GPO) resulting from the oxidation of graphite to provide oxygen functional groups on the basal planes and increase interlayer spacing.
c. Graphene oxide (GO), resulting from the controlled oxidation of graphene, consists of a single layer of graphite oxide usually produced by the chemical treatment of graphite through oxidation, with subsequent dispersion and exfoliation.
d. Reduced graphene oxide (rGO) is a product of the reduction (chemical, electrochemical, thermal, solvothermal, etc.) of graphene oxide.

GPO and GO can be easily prepared [33,34] and are characterized by the presence of oxygen functionalities in the form of hydroxyl and epoxy groups on the basal plane, accompanied by carboxy, carbonyl, phenol, lactone, and quinone at the sheet edges. The conductivity of these materials depends strongly on their chemical and atomic structures on the degree of structural disorder associated with the presence of a sp^3 carbon fraction and sp^3 C–O bonding. GO films are typically insulating but their reduction to rGO increases the conductivity by several orders of magnitude and transform the material into a semiconductor. However, reduction procedures cannot completely remove the oxygen functional groups from GO, so residual oxygen functional groups remain as well as the high density of defects generated during the oxidation process. This means that rGO differs substantially from the graphene [32]. Figure 10.4 shows a scheme for the preparation of GO and rGO [35].

These materials can be electrochemically discriminated using EIS [35,36]. Figure 10.5 compares the Bode plots of the (minus) phase angle vs. log(frequency) for deposits of powders of graphite oxide; GO and rGO prepared by electrochemical and thermal reduction of GO onto GCE, as well as the bare GCE, are all in contact with 0.1 M KCl [36]. At high frequencies, all materials display a similar response corresponding to a purely resistive behavior, as expected considering

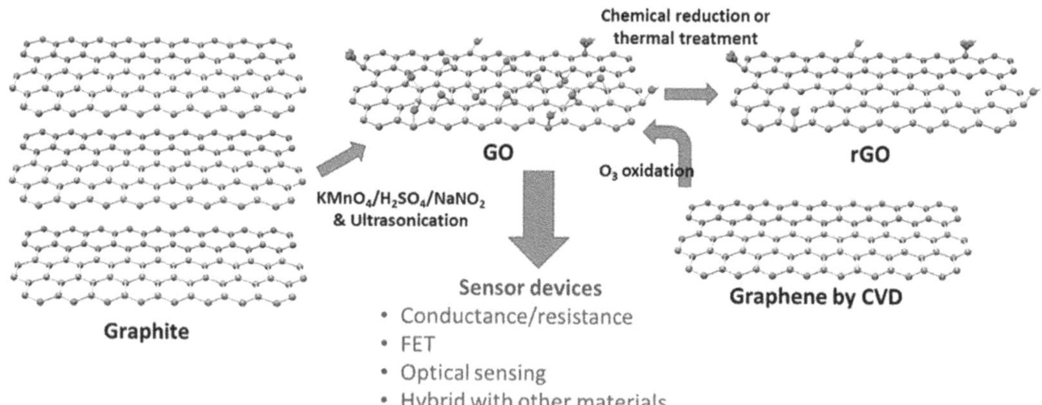

FIGURE 10.4 Scheme for the Preparation and Applications of GO and rGO. Reproduced from Ref. [35] (Toda et al. *Anal. Chim. Acta.* 2015, 878: 43–53), with Permission.

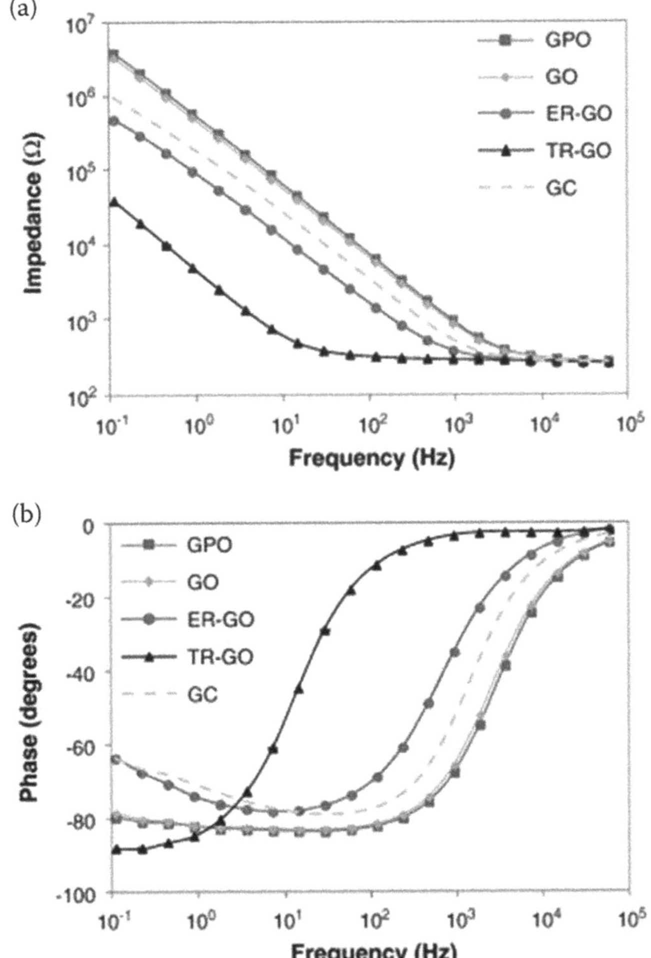

FIGURE 10.5 Bode Plots of Log(Total Impedance) (a) and (Minus) Phase Angle (b) vs. Log(Frequency) of EIS Recorded in 0.1 M KCl for GPO: Graphite Oxide; GO: Graphene Oxide; ER-GO: Electrochemically Reduced Graphene Oxide; TR-GO: Thermally Reduced Graphene Oxide; GC: Bare Glassy Carbon Electrode. Reproduced from Ref. [36] (Bonanni and Pumera, *Electrochem. Commun.* 2013, 26: 52–54), with Permission.

that the impedance is due to the resistance of the solution. At medium and lower frequencies, the (minus) phase angle increase reaching values near 90°, corresponding to a predominantly capacitive behavior; however, there are significant differences between the rGOs (in particular, that prepared by thermal reduction) and the other materials [37].

Figure 10.6 depicts impedance data corresponding to EIS recorded in 0.1 M PBS (pH = 7), 0.1 M KCl plus 10 mM $K_3Fe(CN)_6$/10 mM $K_4Fe(CN)_6$ redox probe for rGO prepared by different methods (by electrochemical, thermal, and photocatalytic reduction of GO) [37]. Experimental data reveal significant differences between these materials, including the fitting of impedance spectra to different equivalent circuits, denoting the significant sensitivity of the electrochemical properties to the different proportion of defect sites, type, and distribution of oxygen functionalities, etc.

An enormous variety of composites using graphene, GO, and rGO have been investigated for electrocatalysis and sensing. These are accompanied by a variety of functionalization and doping

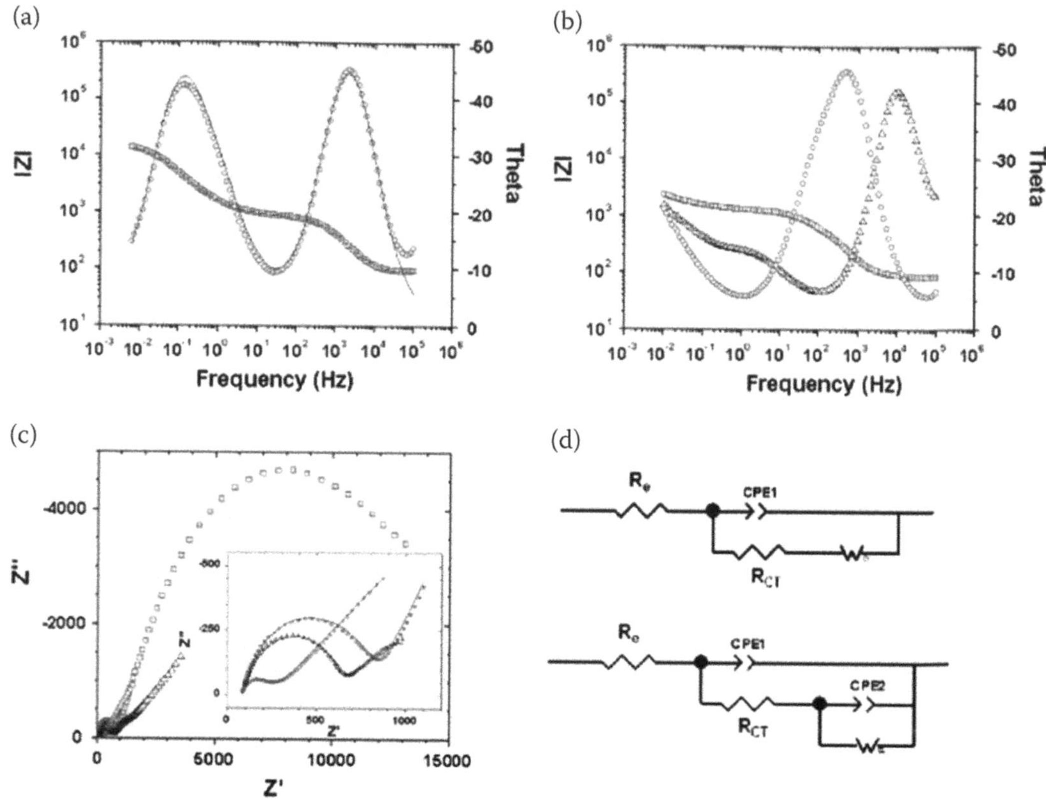

FIGURE 10.6 Bode Plots of Impedance Modulus and (Minus) Phase Angle vs. Log(Frequency) of EIS Recorded in 0.1 M PBS (pH = 7), 0.1 M KCl Containing 10 mM $K_3Fe(CN)_6$/10 mM $K_4Fe(CN)_6$. Impedance Data for Graphite Oxide, Graphene Oxide; Electrochemically Reduced Graphene Oxide, Thermally Reduced Graphene Oxide, and Bare Glassy Carbon Electrode. Impedance Modulus and Phase Angle Diagrams vs. Frequency of Bode Plots for GO (a), Electrochemically and Thermally Reduced GO (b) GO/GCE. (b) Nyquist Plots GO (□), Electrochemically Reduced GO (○) and Photocatalytically Reduced GO (△). Solid Line (—) Correspond to the Fitting of the Experimental Data. Inset; High-Frequency Interval. (d) Electrical Equivalent Circuits. Reproduced from Ref. [37] (Casero et al. *Electrochem. Commun.* 2012, 20: 63–66), with Permission.

strategies [4–6]. Hydrogenation of graphene leads to graphane, a material with interesting properties [32].

Several aspects can be underlined from the electrochemical point of view [32]. Graphene behaves—to some extent—similarly to graphite; the exfoliation of graphite into graphene sheets increases the amount of basal plane per volume without significant change in the amount of available edge plane. Accordingly (see Section 10.2), graphene does not provide relevant advantages to study compounds electrochemically active at the edge plane but inactive at the basal plane of graphene. In fact, higher loading of graphene on the electrode surfaces led to slower electron transfer rates due to the blockage of electroactive edges, while graphitic islands remaining in the surface of graphene materials play an active role in their electrochemistry.

In agreement with the above considerations, Figure 10.7 depicts the SECM image of a graphene electrode using 2 mM $K_3Fe(CN)_6$ as the redox mediator [38]. The graphene electrode was biased at 0.8 V vs. Ag/AgCl, while the tip electrode was biased at −0.1 V. Under these conditions, the current passed through the tip electrode reflects the differences in redox conductivity of the surface so that the basal plane graphene appears in green/blue and the defect sites displaying electroactivity

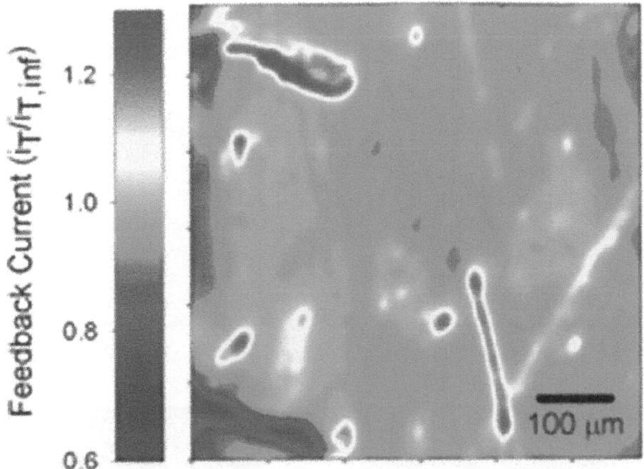

FIGURE 10.7 Intermediate-Negative Feedback SECM Image for a Graphene Electrode in Contact with 2 mM $K_3Fe(CN)_6$ (the Bulk Graphene Appears in Green/Blue and Defects of Higher Activity Appear in Orange). The Graphene Electrode Was Biased at 0.8 V and the Tip Potential Was Fixed at −0.1 V vs. Ag/AgCl. Reproduced from Ref. [38] (Tan et al. *ACS Nano*. 2012, 6: 3070–3079), with Permission.

appear as small circles in orange. The large features correspond to regions where graphene was deliberately removed by scratching the surface with a glass tip.

The electrochemical activity of GOs and rGOs is strongly conditioned by the density of defects and the number and type of oxygen functionalities [32]. These are responsible for the observed voltammetric features, as illustrated in Figure 10.8, for colloidal GO solutions in 50 mM PBS for GO obtained from vein graphite and graphite nanofibers [39]. The voltammograms show cathodic and anodic waves attributed to aldehyde, epoxide, peroxide, and hydroxyl groups [32] but much of this electrochemistry needs clarification.

The electrochemistry of functionalized graphene is also influenced by the electrode morphology, so apparent reaction kinetics varies with minute amounts of electrode porosity [40]. For reversible and quasi-reversible redox couples, the voltamperometric response constitutes a superposition of thin-film diffusion-related effects within the porous electrode and a standard thick film response at a flat electrode, as illustrated in Figure 10.9a,b. In the case of electrochemically irreversible redox processes (Figure 10.9c,d), porous electrodes show large peak shifts compared to flat electrodes. Here, because of the increasing number of accessible sites, the electrode porosity increases the reaction rate defining a morphology-related electrocatalytic effect [40].

10.6 FULLERENES

The term fullerenes is used to design a family of carbon allotropes whose structure is based on fused five- and six-membered rings forming spherical or elongated closed-cage structures [41]. The most popular structure is that of C_{60}, constituted by spherical polyhedra containing 12 pentagons and 20 hexagons where carbon atoms are located at the vertices of the resulting truncated icosahedron. Carbon atoms in fullerenes possess a sp^{2}-type electronic configuration, in contrast with graphitic carbons where additional reactions at the double bond junction between six-membered rings are allowed. This is due to the pyramidalization of sp^2 carbons because of the curvature of the structure. Accordingly, a variety of functionalized fullerenes can be prepared [42]. Additionally, fullerenes can encapsulate anions and neutral molecules, forming the so-called endohedral fullerenes and several types of polymeric chains [43].

Fullerenes have high electron affinity, so electrochemical reduction processes are easily observed. The solution-phase electrochemical response of C_{60} and C_{70} fullerenes in MeCN/toluene

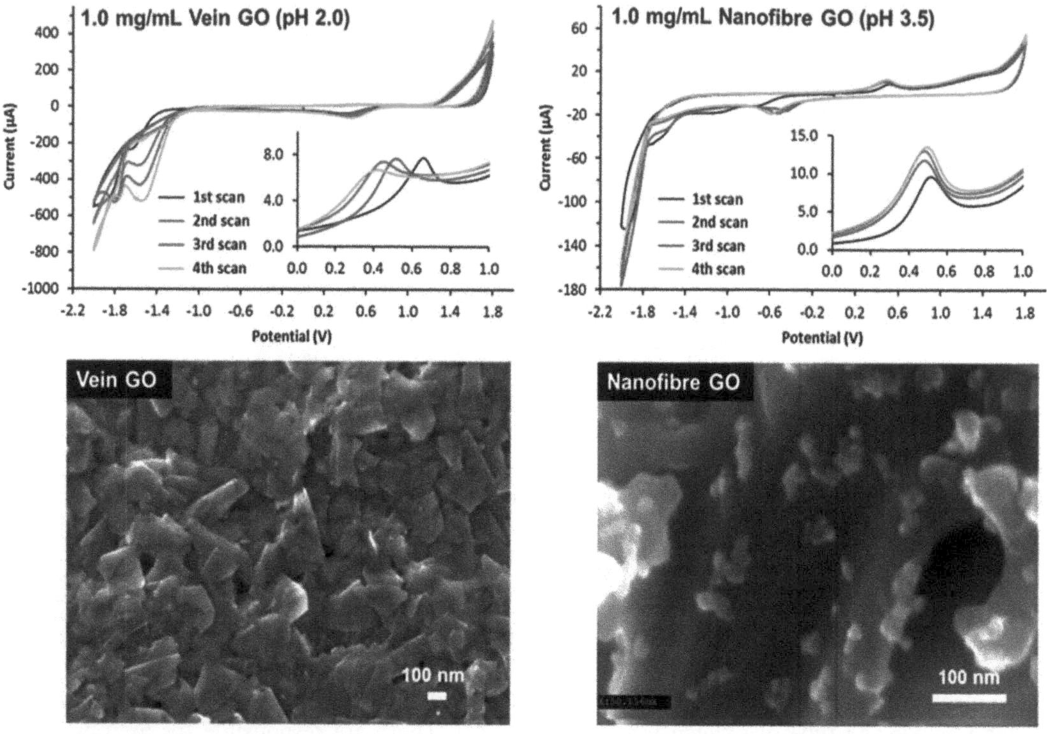

FIGURE 10.8 CVs of Colloidal GO Solutions in 50 mM PBS for GO Obtained from Vein Graphite (Left, Conc. 1.0 mg mL^{-1}, pH 2.0 Buffer), and GO Obtained from Graphite Nanofibers (Right, Conc. 1.0 mg mL^{-1}, pH 3.5 Buffer). Inset Diagrams of Oxidation Peaks Are Shown on the Right. Potential Scan Rate 100 mV s^{-1}. Also Shown Are Scanning Electron Micrographs of Vein GO and Nanofiber GO Particles. Reproduced from Ref. [39] (Sheng and Pumera, *Electrochem. Commun.* 2014, 43: 87–90), with Permission.

mixtures at low temperature consisted of six reversible one-electron reduction at potentials between –0.97 to –3.26 V vs. Fc/Fc^+ [44]. This can be represented as successive one-electron transfer processes:

$$C_{60} \xrightarrow{e^-} C_{60}^{-} \xrightarrow{e^-} C_{60}^{2-} \xrightarrow{e^-} C_{60}^{3-} \xrightarrow{e^-} C_{60}^{4-} \xrightarrow{e^-} C_{60}^{5-} \xrightarrow{e^-} C_{60}^{6-} \quad (10.4)$$

However, higher fullerenes and endohedral fullerenes provide a more complicated electrochemistry.

The electrochemistry of solid fullerenes in contact with aqueous [45] and nonaqueous electrolytes has also been studied. Following Bond et al. [46], microcrystalline C_{60} adhered to gold, glassy carbon, and platinum electrodes in contact with Bu_4NClO_4/MeCN and can be reduced by 3–5 electrons to species in solution, while 1- and 2-electron reduction steps only lead to a minor level of dissolution of C_{60}^{-} and C_{60}^{2-}. The electrochemical pattern is complicated by the adsorption processes of electrochemically generated ions, the adsorption strength decreasing by increasing the charge of the fullerene anionic species. Solid-state electron transfers are limited to the processes:

$$\{C_{60}\} \xrightarrow{Bu_4N^+,\, e^-} \{C_{60}^{-} \cdots Bu_4N^+\} \xrightarrow{Bu_4N^+,\, e^-} \{C_{60}^{2-} \cdots (Bu_4N^+)_2\} \quad (10.5)$$

EQCM experiments confirmed that C_{60}^{3-} species can be oxidatively precipitated on gold, the corresponding process being represented as:

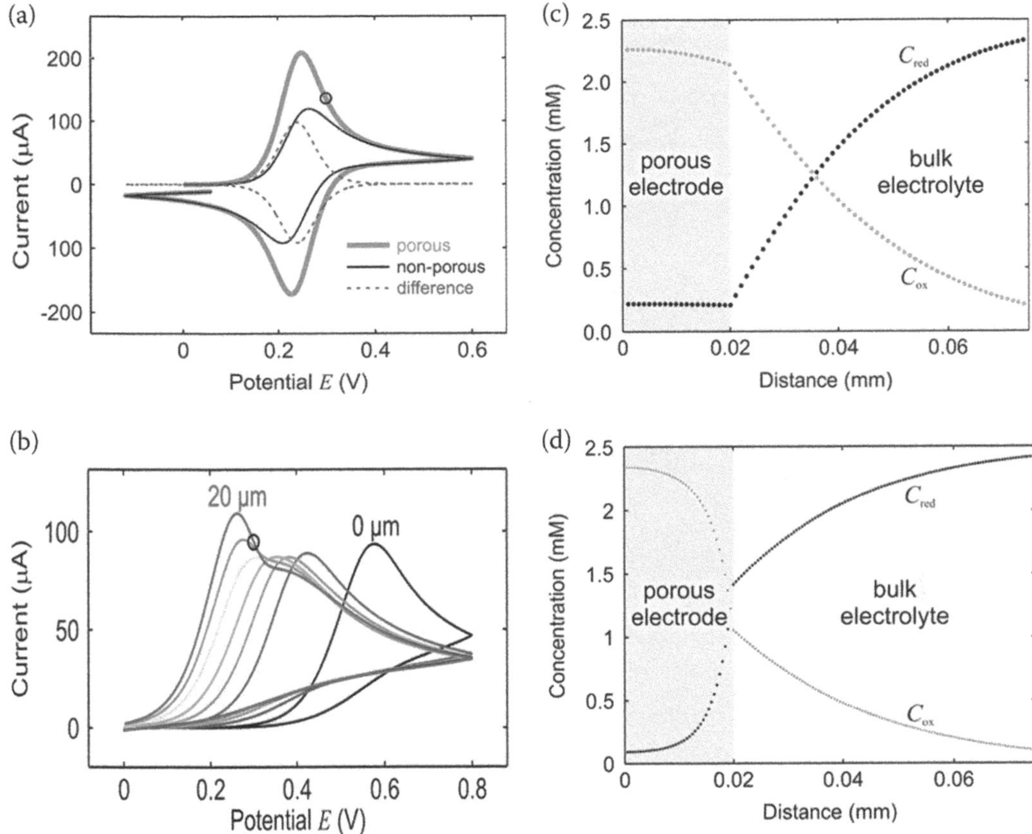

FIGURE 10.9 (a) Numerical Simulations for Quasi-reversible Processes Taking $k° = 0.05$ cm s^{-1}, $E° = 240$ mV, and $A° = 0.22$ cm^2; Theoretical CVs for a Flat Electrode, a Porous Electrode with $d = 20$ μm, and the Difference Between Them (Dotted Line). (b) Concentration Profiles for Reduced and Oxidized Species Within at a Potential of 300 mV. (c) Numerical Simulations for Irreversible Process with $k° = 10^{-7}$ cm s^{-1}, $E°$ = 0 mV, and $A° = 0.22$ cm^2; Theoretical CVs for Thicknesses of the Porous Film of 0, 1, 2, 5, 10, 15, and 20 μm. (d) Concentration Profiles for a 20 μm Thick Electrode at 300 mV. From Ref. [40] (Punkt et al. *J. Phys. Chem. C.* 2013, 117: 16076–16086), with Permission.

$$C_{60}{}^{3-}{}_{solv} + 2Bu_4N^+_{solv} \rightarrow \{C_{60}{}^{2-} \cdots (Bu_4N^+)_2\} + e^- \quad (10.6)$$

Fullerenes are versatile components able to form different macro- and supramolecular structures by their ability to entrap small molecules and their capacity to be embedded into large macrocyclic molecules (cyclodextrins, calixarenes, porphyrins, crown ethers), and derivatize with dendrimer and polymer structures [47]. Fullerenes can form homopolymers but can also be incorporated into polymeric chains, forming a 'pearl necklace' structure or acting as pendant substituents in the side chains of polymers, forming the so-called 'charm bracelet' polymers [42]. Fullerene-epoxide-based polymers containing an epoxy ring connecting two adjacent fullerene cages can be used as precursors for a large group of epoxy resins, which can be synthesized from fullerene epoxides such as $C_{60}O$. Fullerene-epoxide polymerization can be electrochemically promoted in O_2-containing aprotic solvents. Electropolymerization is initiated by the well-known one-electron reversible reduction of O_2 to $O_2{\cdot}^-$, followed by nucleophilic addition of the radical superoxide to the fullerene:

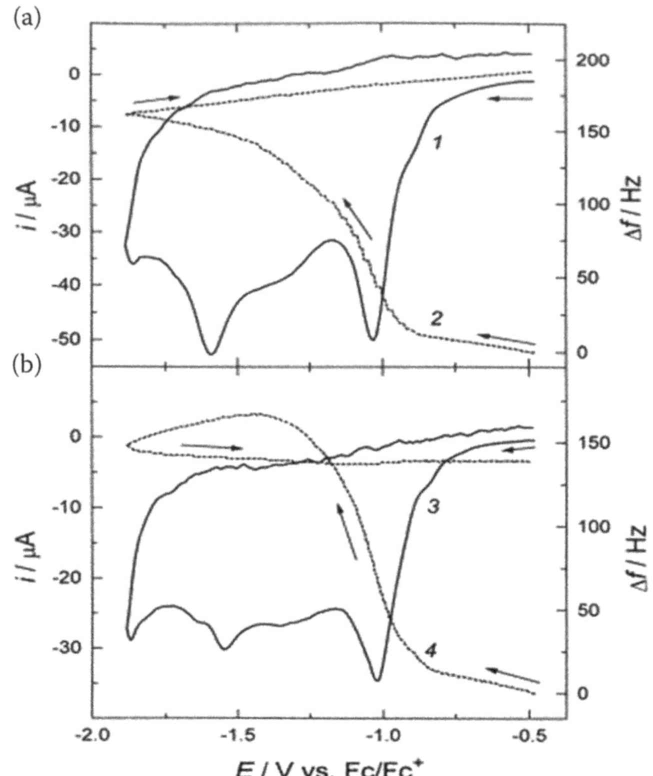

FIGURE 10.10 Simultaneous CVs (Curves 1 and 3) and Frequency Change vs. Potential Traces (Curves 2 and 4) for C_{78} Film, Drop-Coated onto the 5 MHz Au-Quartz Electrode, in 0.1 M KPF_6/MeCN. Potential Scan Rate (a) 100 and (b) 50 mV s^{-1}. Reproduced from Ref. [49] (Kutner et al. *Electrochim. Acta*. 2002, 47: 2371–2380), with Permission.

$$O_2 \xrightarrow{e^-} O_2^{\cdot -} \xrightarrow{C_{60}} C_{60} - O - O^- \xrightarrow{C_{60}} C_{60}O^{\cdot -} + C_{60}O \qquad (10.7)$$

$C_{60}O\cdot^-$ and $C_{60}O$ can induce further polymerization of the fullerene.

Polymers of fullerenes with transition metal complexes exhibit electrochemical activity due to the reduction of fullerene moieties. This process is accompanied by the transport of cations from the supporting electrolyte into the film [48]. Figure 10.10 shows the monitoring of microgravimetric changes by EQCM accompanying the voltammetry of a C_{78} film drop-coated onto the 5 MHz Au-quartz electrode immersed into 0.1 M KPF_6/MeCN [49]. This methodology permits us to obtain information on the solid-state or solution phase nature of the involved electrochemical processes. In the negative potential scan, three cathodic peaks are recorded, but only the former is accompanied by a sharp variation in the frequency. This suggests that the film dissolves at potentials where the first cathodic peak is formed without mass changes so the two latter peaks correspond to the redox processes in solution.

Polymerization of fullerenes and functionalized fullerenes has been also widely treated. In particular, $C_{60}M$ films can be electrochemically reduced due to the electroactivity of fullerene moieties. A typical process is [47]:

$$(Pd - C_{60} - Pd)_n + nTAA^+ + ne^- \rightarrow (Pd - C_{60}^- - Pd\cdots TAA^+)_n \qquad (10.8)$$

where TAA^+ denotes a tetraalkylammonium ion from the electrolyte. Remarkably, the electrochemical reduction of the polymeric film results in an increase in conductivity [50].

Electropolymerizations in the presence of a high concentration of metal give rise to the formation of metal nanoparticles.

A variety of hybrid materials involving fullerene building blocks has been described. 'Double cables' are polymers of transition metal cations and fullerene derivatives, including the so-called fullerene peapods, which can be described as supramolecular complexes formed by filling SWNTs by fullerenes [51].

10.7 DIRECT ELECTROCHEMICAL SYNTHESIS OF CARBON MATERIALS

Electrochemical carbons are synthetic solids mainly consisting of atoms of elemental carbon that can be prepared electrochemically from suitable precursors [52]. Electrochemical carbonization is characterized by three specific features: (a) ability to obtain unstable carbon chains, (b) easy template of carbon nanostructured materials by the precursors, and (c) defined kinetics of certain reactions, controlling film thickness.

Electrochemical carbonization leads to carbines and polyynes, materials that can also be prepared by the action of laser or synchrotron radiations on graphitic carbons or from methods like cluster beam deposition or hot filament-assisted sputtering. As a result, polymeric structures constituted by sp-hybridized carbon chains of the type $-(C{\equiv}C)_n-$ and $=(C{=}C)_n=$ can be obtained. Electrosynthesis of carbon materials started from poly(tetrafluoroethylene), originally discovered by Jansta and Dousek [53], using reduction with a liquid amalgam of alkali metal. Here, the reaction yields a polyyne stabilized by interspersed nanocrystalline alkali metal fluoride which, in turn, protects the carbenoid skeleton against cross-linking and/or chemical attack from the reactive environment. This can be represented as:

$$-((CF_2-CF_2)_n- + 4nM^+(Hg) + 4ne^- \rightarrow (-C{\equiv}C-)_n + 4nMF \tag{10.9}$$

The resulting carbon possesses a small superstoichiometric over-reduction, equivalent to n-doping; the extra negative charge is compensated by the corresponding cations of alkali metals. Actual reaction products, however, are mesoscopic sp^2-hybridized carbons with short oligoyne sequences ($n < 8$) whereas polyyne is spontaneously cross-linked to graphene. Direct synthesis of fullerenes and nanotubes can be performed by electrochemical dehalogenation of perhalogenated hydrocarbons.

10.8 APPLICATIONS

Porous carbons, nanotubes, and the different graphenes have attracted considerable attention on practical issues of electrocatalysis, hydrogen storage, lithium batteries, and supercapacitors. Except for fullerenes that have a rich Faradaic electrochemistry, the electrochemical behavior of porous carbons and carbon nanotubes is dominated by double-layer charging processes with little to no contribution of Faradaic pseudocapacitance of surface oxide functionalities previously commented.

An example of this double layer response can be seen in Figure 10.2 [23], where the voltammograms present the characteristic box-like shape of an ideal capacitor. Values of specific capacitance of nanotubes are significantly enhanced after modifications—such as electrodeposition of a thin layer of conducting polymers—due to the contribution of pseudo-Faradaic properties of the polymer. Electrochemical capacitors based on carbon materials can be divided into double-layer capacitors, where only a pure electrostatic interaction between ions and charged electrode surfaces occur, and supercapacitors based on the occurrence of Faradaic pseudocapacitance reactions (studied in more detail in Chapter 14). It is generally assumed that the charge in porous carbons is mainly stored in the double layer at the electrode/electrolyte interface. Remarkably, several studies indicate that no clear relationship between capacitance and specific surface area exists [54]. In principle, the specific capacitance increases as the specific surface area of the carbon increases, but, as far as this quantity depends on the distance between polarized carbon surface and solvated ions, the presence of mesopores becomes determinant for ion transport.

The capacitance of graphite and carbon nanotubes is closely related to the proportion of sp^3 defects in the sp^2 lattice, increasing with the sp^2 edge planes and defects [32]. In fact, the theoretical surface area of graphene (2630 $m^2\ g^{-1}$) is twice that of single-walled carbon nanotubes with ca. a hundred times larger than graphite powder. Capacitance effects are also notably influenced by N- and S-doping, although these features are not well understood. Pseudocapacitance effects are promoted by association of the carbon support to metal oxide nanoparticles such as NiO or MnO_2, and the preparation of composites with conducting polymers.

Carbon materials are widely used in fuel cells and batteries. Here, carbons (mainly graphene) act as a conductive support for metal or metal oxide catalysts. Pt/C catalysts constitute the reference commercial materials, but their use is obviously conditioned by the high cost and limited availability of platinum. Accordingly, intensive research is devoted to the development of noble metal-free electrocatalysts. Among other possibilities, an association of carbons to nanoparticles of transition-metals, hybrid nanoarchitectures with metal oxides, sulfides, etc. (see Chapter 9), and heteroatom-doped graphene, are under study.

Much research is also focused on the building of carbon-based anodes for lithium batteries. These include graphene-type materials such as porous 3D graphene, graphene doped with carbon nanotubes and/or fullerenes, and heteroatom-, metalloid-, metal-, metal oxide-, or metal sulfide-doped graphene [32]. It is believed that both large interlayer d-spacing and active defects (vacancies and edges) can enhance lithium storage capability and facilitate Li^+ diffusion (see Chapter 14). Carbon materials are also used as cathodes in these batteries to hinder the agglomeration of active electrode components ($LiCoO_2$, etc.) and increase conductivity.

Carbon materials are extensively used in electrochemical sensing. Solid-contact ion-selective electrodes (ISEs) can employ different carbon materials as transducers between the selective membrane and the electrode surface (see Chapter 12). Carbon materials (mainly nanotubes and graphenes) materials have been extensively used in a tremendous variety of composites to prepare sensing voltammetric and amperometric (also impedimetric) electrodes. The large conductivity and high specific surface area of carbon materials contributes to enhance the electrochemical responses of analytes, such as heavy metal ions (Cd^{2+}, Cu^{2+}, Hg^{2+}, Pb^{2+}, Zn^{2+}, etc.) relative to conventional graphite and glassy carbon electrodes. Formation of composites with other components such as metal, metal oxide, sulfide, nanoparticles, conducting polymers, chitosan, etc. is routinely investigated for analytical purposes. Bioanalytes such as glucose, ascorbic acid, dopamine, uric acid, NADH, etc. are repeatedly studied and electrode modification strategies include immobilization of biomolecules and aptamers as well as the formation of molecularly imprinted polymers. As will be discussed in Chapter 12, however, a better knowledge of the role of the components of such composites and the elucidation of the electrochemical pathways and possible synergistic effects between the composite components is needed.

10.9 INFLUENCE OF IMPURITIES ON CARBON ELECTROCHEMISTRY

Manufacturing carbon materials for electrochemical applications implies a series of operations on raw materials. As a result, different impurities can be present in the final carbon product either from the raw material and/or as a result of the preparative process. In particular, the presence of amorphous carbon and surfactants can also influence the electrochemical response of the material. Surfactants are used to prepare suspensions of carbon materials as stabilizers and, although generally non-electroactive, can modify the impedance response of the material.

The main sources for the preparation of carbon materials are natural graphite and several other natural carbonaceous materials whose graphitization at temperatures >2500 °C yields synthetic graphite. Natural graphite has purity levels between 80 and 98% and contains significant proportions of cobalt, nickel, and iron. Milling treatments to manufacture commercial graphite flakes also incorporates metallic impurities. The influence of these residual metallic impurities in the electrocatalytic performance of carbon materials is currently a matter of debate, but it appears that a significant fraction of reported electrocatalytic effects can be associated with metallic impurities accompanying CNTs and graphenes [32]. Current research is working to elucidate this matter, of obvious relevance in respect to catalytic applications of carbon materials.

REFERENCES

[1]. Yoshimura, S.; Chang, R.P.H. Eds. 1998. *Supercarbon. Synthesis, Properties and Applications.* Springer, Berlin.

[2]. Echegoyen, L.E.; Herranz, M.A.; Echegoyen, L. 2006. Carbon, fullerenes. In Bard, A.J.; Stratmann, M.; Schoilz, F.; Pickett, C.J. Eds. *Inorganic Electrochemistry. Encyclopedia of Electrochemisttry*, vol. 7. Wiley-VCH, Weinheim, pp. 143–201.

[3]. Novoselov, K.S.; Geim, A.K.; Morozov, S.V.; Jiang, D.; Zhang, Y.; Dubonos, S.V.; Grigorieva, I.V.; Firsov, A.A. 2004. Electric field effect in atomically thin carbon films. *Science.* 306: 666–669.

[4]. Chen, D.; Feng, H.; Li, J. 2012. Graphene oxide: Preparation, functionalization, and electrochemical applications. *Chemical Reviews.* 112: 6027–6053.

[5]. Higgins, D.; Zamani, P.; Yu, A.; Chen, Z. 2016. The application of graphene and its composites in oxygen reduction electrocatalysis: a perspective and review of recent progress. *Energy & Environmental Science.* 9: 357–390.

[6]. Noked, M.; Soffer, A.; Aurbach, D. 2011. The electrochemistry of activated carbonaceous materials: Past, present, and future. *Journal of Solid State Electrochemistry.* 15: 1563–1578.

[7]. Chen, P.; McCreery, R.L. 1996. Control of electron transfer kinetics at glassy carbon electrodes by specific surface modification. *Analytical Chemistry.* 68: 3958–3965.

[8]. Davies, T.J.; Hyde, M.E.; Compton, R.G. 2005. Nanotrench arrays reveal insight into graphite electrochemistry. *Angewandte Chemie International Edition.* 44: 5121–5126.

[9]. Barisci, J.N.; Wallace, G.; Chattopadhyay, D.; Papadimitrakopoulos, F.; Baughman, R.H. 2003. Electrochemcal properties of single-wall carbon nanotube electrodes. *Journal of the Electrochemical Society.* 150: E409–E415.

[10]. Barisci, J.N.; Wallace, G.G.; Baughman, R.H. 2000. Electrochemical studies of single-wall carbon nanotubes in aqueous solutions. *Journal of Electroanalytical Chemistry.* 488: 92–98.

[11]. Barrière, F.; Downard, A.J. 2008. Covalent modification of graphitic carbon substrates by non-electrochemical methods. *Journal of Solid State Electrochemistry.* 12: 1231–1244.

[12]. Wildgoose, G.G.; Leventis, H.C.; Davies, I.J.; Crossley, A.; Lawrence, N.S.; Jiang, L.; Jones, T.G.J.; Compton, R.G. 2005. Graphite powder derivatised with poly-L-cysteine using "building-block" chemistry—a novel material for the extraction of heavy metal ions. *Journal of Materials Chemistry.* 15: 2375–2382.

[13]. Saito, S. 1997. Carbon nanotubes for next-generation electronics devices. *Science.* 278: 77–78.

[14]. Yadav, R.M.; Wu, J.; Kochandra, R.; Ma, L.; Tiwary, C.S.; Ge, L.; Ye, G.; Vajtai, R.; Lou, J.; Ajayan, P.M. 2015. Carbon nitrogen nanotubes as efficient bifunctional electrocatalysts for oxygen reduction and evolution reactions. *Applied Materials & Interfaces.* 7: 11991–12000.

[15]. Zhao, Y.; Yang, L.; Chen, S.; Wang, X.; Ma, Y.; Wu, Q.; Jiang, Y.; Qian, W.; Hu, Z. 2013. Can boron and nitrogen co-doping improve oxygen reduction reaction activity of carbon nanotubes? *Journal of the American Chemical Society.* 135: 1201–1204.

[16]. Qin, L.; Wang, L.; Yang, X.; Ding, R.; Zheng, Z.; Chen, X.; Lv, B. 2018. Synergistic enhancement of oxygen reduction reaction with BC_3 and graphitic-N in boron- and nitrogen-codoped porous graphene. *Journal of Catalysis.* 359: 242–250.

[17]. Ryoo, R.; Joo, S.H.; Kruk, M.; Jaroniec, M. 2001. Ordered mesoporous carbons. *Advanced Materials.* 13: 677–680.

[18]. Corma, A. 1997. From microporous to mesoporous molecular sieve materials and their use in catalysis. *Chemical Reviews.* 97: 2373–2420.

[19]. Oberlin, A.; Endo, M. 1976. Filamentous growth of carbon through benzene decomposition. *Journal of Crystals Growth.* 32: 335–349.

[20]. Wiles, P.G.; Abrahamson, J. 1978. Carbon fibre layers on arc electrodes—I: Their properties and cool-down behaviour. *Carbon*. 16: 341–349.
[21]. Iijima, S.; 1991. Helical microtubules of graphitic carbón. *Nature*. 354: 56–57.
[22]. Novoselov, K.S.; Geim, A.K.; Morozov, S.V.; Jiang, D.; Zhang, Y.; Dubonos, S.V.; Grigorieva, I.V.; Firsov, A.A. 2004. Electric field effect in atomically thin carbon films. *Science*. 306: 666–669.
[23]. Xu, B.; Wu, F.; Su, Y.; Xao, G.; Chen, S.; Zhou, Z.; Yang, Y. 2008. Competitive effect of KOH activation on the electrochemical performances of carbon nanotubes for EDLC: Balance between porosity and conductivity. *Electrochimica Acta*. 53: 7730–7735.
[24]. Barone, V.; Hod, O.; Scuseria, G.E. 2006. Electronic structure and stability of semiconducting graphene nanoribbons. *Nano Letters*. 6: 2748–2754.
[25]. Barisci, J.N.; Wallace, G.G.; Baughman, R.H. 2000. Electrochemical studies of single-wall carbon nanotubes in aqueous solutions. *Journal of Electroanalytical Chemistry*. 488: 92–98.
[26]. Barisci, J.N.; Wallace, G.G.; Baughman, R.H. 2000. Electrochemical quartz crystal microbalance studies of single-wall carbon nanotubes in aqueous and non-aqueous solutions. *Electrochimica Acta*. 46: 509–517.
[27]. Yang, Q.-H.; Xu, W.; Tomita, A.; Kyotani, T. 2005. The template synthesis of double coaxial carbon nanotubes with nitrogen-doped and boron-doped multiwalls. *Journal of the American Chemical Society*. 127: 8956–8957.
[28]. Frackowiak, E.; Beguin, F. 2002. Electrochemical storage of energy in carbon nanotubes and nanostructured carbons. *Carbon*. 40: 1775–1787.
[29]. Kavan, L.; Rapta, P.; Dunsch, L.; Bronikowski, M.J.; Willis, P.; Smalley, R.E. 2001. Electrochemical tuning of electronic structure of single-walled carbon nanotubes: In-situ Raman and vis-NIR study. *Journal of Physical Chemistry B*. 105: 10764–10771.
[30]. Claye, A.S.; Nemes, N.M.; Janossy, A.; Fischer, J.E. 2000. Structure and electronic properties of potassium-doped single-wall carbon nanotubes. *Physical Review B: Condensed Matter and Materials Physics*. 62: R4845–R4848.
[31]. Fitzer, E.; Kochling, K.H.; Boehm, H.P.; Marsh, H. 1995. Recommended terminology for the description of carbon as a solid (IUPAC Recommendations 1995). *Pure and Applied Chemistry*. 67: 473–506.
[32]. Ambrosi, A.; Chua, C.K.; Bonnani, A.; Pumera, M. 2014. Electrochemistry of graphene and related materials. *Chemical Reviews*. 114: 7150–7188.
[33]. Dreyer, D.R.; Park, S.; Bielawski, C.W.; Ruoff, R.S. 2010. The chemistry of graphene oxide. *Chemical Society Reviews*. 39: 228–240.
[34]. Hummers, W.S.; Offeman, R.E. 1958. Preparation of graphitic oxide. *Journal of the American Chemical Society*. 80: 1339.
[35]. Toda, K.; Furue, R.; Hayami, S. 2015. Recent progress in applications of graphene oxide for gas sensing: A review. *Analytica Chimica Acta*. 878: 43–53.
[36]. Bonanni, A.; Pumera, M. 2013. High-ressolution impedance spectroscopy for graphene characterization. *Electrochemistry Communications*. 26: 52–54.
[37]. Casero, E.; Parra-Alfambra, A.M.; Petit-Domínguez, M.D.; Pariente, F.; Lorenzo, E.; Alonso, C. 2012. Differentiation between graphene oxide and reduced graphene by electrochemical impedance spectroscopy (EIS). *Electrochemistry Communications*. 20: 63–66.
[38]. Tan, C.; Rodríguez-López, J.; Parks, J.J.; Ritzert, N.L.; Ralph, D.C.; Abruña, H.D. 2012. Reactivity of monolayer chemical vapor deposited graphene imperfections studied using scanning electrochemical microscopy. *ACS Nano*. 6: 3070–3079.
[39]. Sheng, A.Y.; Pumera, M. 2014. Direct voltammetry of colloidal grapheme oxides. *Electrochemistry Communications*. 43: 87–90.
[40]. Punkt, C.; Pope, M.A.; Aksay, I.A. 2013. On the electrochemical response of porous functionalized graphene electrodes. *Journal of Physical Chemistry C*. 117: 16076–16086.
[41]. Kroto, H.W.; Heath, J.R.; O'Brien, S.C.; Curl, R.F.; Smalley, R.E. 1985. C_{60}: Buckminsterfullerene. *Nature*. 318: 162.
[42]. Hirsch, A. 1993. The chemistry of the fullerenes: An overview. *Angewandte Chemie International Edition*. 32: 1138–1141.
[43]. Guha, S.; Nakamoto, K. 2005. Electronic structures and spectral properties of endohedral fullerenes. *Coordination Chemistry Reviews*. 249: 1111–1132.
[44]. Xie, Q.; Pérez-Cordero, E.; Echogoyen, L. 1992. Electrochemical detection of $C_{60}{}^{6-}$ and $C_{70}{}^{6-}$: Enhanced stability of fullerides in solution. *Journal of the American Chemical Society*. 114: 3978–3980.

[45]. Szucs, A.; Tolgyesi, M.; Nagy, J.B.; Novak, M.; Lamberts, L. 1996. Photoelectrochemical behaviour of C_{60} films in various oxidation states. *Journal of Electroanalytical Chemistry*. 419: 39–46.
[46]. Bond, A.M.; Miao, W.; Raston, C.L. 2000. Identification of processes that occur after reduction and dissolution of C_{60} adhered to gold, glassy carbon, and platinum electrodes placed in acetonitrile (electrolyte) solution. *Journal of Physical Chemistry B*. 104: 2320–2329.
[47]. Winkler, K.; Balch, A.L.; Kutner, W. 2006. Electrochemically formed fullerene-based polymeric films. *Journal of Solid State Electrochemistry*. 10: 761–784.
[48]. Zhang, X.; Jao, K.; Piao, G.; Liu, S.; Li, S. 2009. Voltammetric study of fullerene C_{60} and fullerene C_{60} nanotubes with sandwich method. *Synthetic Metals*. 159: 419–423.
[49]. Kutner, W.; Noworyta, K.; Marczak, R.; D'Souza, F. 2002. Electrochemical quartz microbalance studies of thin-solid films of higher fullerenes: C_{76}, C_{78} and C_{84}. *Electrochimica Acta*. 47: 2371–2380.
[50]. Balch, A.L.; Costa, D.A.; Winkler, K. 1998. Formation of redox-active, two-component films by electrochemical reduction of C_{60} and transition metal complexes. *Journal of the American Chemical Society*. 120: 9614–9620.
[51]. Smith, B.W.; Monthioux, M.; Luzzi, D.E. 1998. Encapsulated C_{60} in carbon nanotubes. *Nature*. 396: 323–324.
[52]. Kavan, L. 1997. Electrochemical carbon. *Chemical Reviews*. 97: 3061–3082.
[53]. Janska, J.; Dousek, F.P. 1973. Electrochemical corrosion of polytetrafluorethylene contacting lithium amalgam. *Electrochimica Acta*. 18: 673–674.
[54]. Kim, Y.J.; Horie, Y.; Ozaki, S.; Matsuzawa, Y.; Suezaki, H.; Kim, C.; Miyashita, N.; Endo, M. 2004. Correlation between the pore and solvated ion size on capacitance uptake of PVDC-based carbons. *Carbon*. 42: 1491–1500.

11 Electrochemistry of Porous Polymers and Hybrid Materials

11.1 INTRODUCTION

Porous polymers received attention as materials that can be used as electrolytes—such as poly (acrylonitrile) or poly(methyl methacrylate)—in high power applications [1]. The scope of available electrochemical applications has been expanded, with the introduction of polymers of intrinsic microporosity (PIMs)—a new class of molecularly rigid and microporous ion-conducting materials [2].

The label of hybrid materials can be assigned to a variety of systems with a common combination of organic and inorganic components blended on the molecular scale. In principle, one can distinguish between organic-inorganic hybrids—*sensu stricto*—when inorganic units are formed in situ by molecular precursors, and nanocomposites when discrete structural units are used in the respective size regime [3].

Most of the development of hybrid materials derives from the use of sol-gel technologies Preparation strategies involve either the combination of building blocks that retain their molecular integrity to a great extent throughout the process of material formation or the transformation of building precursors into a novel material. Hybrid materials can involve relatively weak interactions such as Van der Waals forces or hydrogen bonding (Class I hybrid materials) or covalent bonding between the building blocks (Class II). Most nanocomposites are formed by embedding nanoparticles, nanowires, nanorods, carbon nanotubes, graphene, etc. in an organic polymer.

Over the last decades, a wide variety of composites and hybrid materials have been prepared and electrochemically characterized. In this chapter, only a fraction of these will be presented. In the case of materials described as the result of the attachment of an organic moiety to an inorganic network, one can distinguish between network modification, network building, and network functionalization. In the first case, the organic units are anchored to the inorganic support that only modifies its surface properties. In a functionalized system, a reactive organic group is bonded to the inorganic support, whereas if a mixed network is formed by combining both the organic and inorganic moieties, both moieties are termed as network builders. In this context, the term nanocomposite is used when the size of one of the structural units falls in the 1–100 nm range. Nanoparticles, nanorods and metal-oxo clusters, fullerenes, carbon nanotubes, sphaerosilicates, and oligomeric silsesquioxanes are included in this group. Usually, this term is used if the material is constituted by discrete units in the respective size regime while the systems are designed as hybrid materials when the inorganic units are formed *in situ* by molecular precursors.

Functionalization of polymer chains has been also intensively investigated. This consists of the covalent binding of functional groups to the polymer backbone. Among a myriad of strategies, functionalization with macrocyclic receptors provides a way to complex species in solution for sensing purposes. Here, one can exploit the electroactive character of any of the components (polymer, receptor, and/or guest species) or functionalize the polymer with a second electroactive unit (such as ferrocene or ferrocenyl derivatives). These molecular recognition strategies, falling within the field of supramolecular chemistry, will not be treated here.

In this context, recent developments include inorganic ceramics- and organic polymer-based solid electrolytes. These materials are particularly interesting for energy production.

Here, electrolytes that contain high concentrations of ion charge carriers with minimum polymer concentration are particularly interesting [4].

It should be noted that polymers that participate in the composition of an enormous variety of composites are currently under research using different combinations of units: metal nanoparticles, quantum dots, carbon nanotubes, porous carbons, graphene, silica, zeolites, titania, organic functionalized mesoporous silicas, and metal-organic frameworks, among others [5]. In this chapter, the attention will be focused on conducting organic polymers. This family of materials is receiving increasing attention due to their special electrical and optical properties (similar to metals), mechanical stability, and processability [6,7].

11.2 CONDUCTING POLYMERS

Conducting polymers, discovered in the 1970s [8], are characterized by the exhibition of electronic conductivity and redox conductivity; among the most popular are poly(pyrrole) (PPy), poly(aniline) (PANI), polyacetylene, and poly(3,4-ethylene dioxythiophene) (PEDOT). Electronic conductivity is due to the electronic structure that gives rise to an extensive electron delocalization from π bonds along the polymeric chain and the possibility of inter-chain charge-transfer. As a result, these polymers have semiconducting properties (bandgap around 1.4 eV). This leads to easy excitations and redox reactions, with the presence of different oxidation levels of the polymer [9].

Polyaniline (PANI) has been one of the most remarkable materials of this type of compound attending to its large stability, conductivity, catalytic ability, and electrochemical properties [10]. Figure 11.1 shows a typical experiment of repetitive CV illustrating the electrochemical generation of PANI. In the initial anodic scan, aniline is electrochemically oxidized to a radical cation via the one-electron process. The radical cation rapidly polymerized, giving rise to PANI so that in subsequent potential cycles, a series of coupled cathodic/anodic pairs appear. The height of such peaks increases in successive cycles because of the deposition of new layers of polymer. Two main couples appear, corresponding to the transitions emeraldine/leucoemeraldine and pernigraniline/emeraldine states [11]. A third couple that often appears is attributed to degradation products [12]. The electropolymerization can also be promoted potentiostatically by applying a sufficiently positive constant potential.

Ultimately, PANI can exist in five differently colored forms or 'states'—usually labeled as leucoemeraldine (light yellow), protoemeraldine (green), emeraldine (deep green), nigraniline

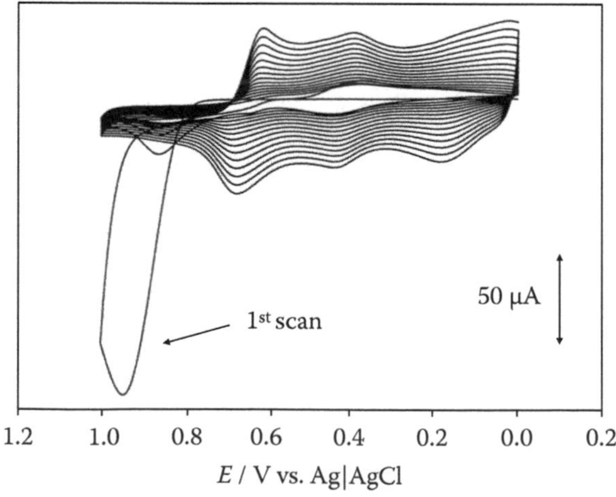

FIGURE 11.1 Repetitive CV Response at Glassy Carbon Electrode for a 10 mM Aqueous Solution of Aniline in 0.10 M H_2SO_4. Potential Scan Rate 50 mV s^{-1}.

(blue), and pernigraniline (violet) [13]—whose interconversion is schematized in Figure 11.2. In the more reduced form, leucoemeraldine, all nitrogen units remain as amine groups, and the protoemeraldine and emeraldine forms incorporate until one-half of nitrogens as protonated imine groups. The nigraniline form possesses amine groups, protonated imine units, and deprotonated imine groups. Finally, the pernigraniline state of the PANI chain corresponds to a fully oxidized form where all imine groups become deprotonated [14].

The voltammetric response of conducting polymers is particularly sensitive to the potential scan rate, as illustrated in Figure 11.3. Here, CVs recorded at different potential scan rates for a poly (3,4-ethylene dioxythiophene) poly-(styrenesulfonate) (PEDOT:PSS) in contact with phosphate buffer saline (PBS) electrolyte are depicted [12]. The voltammetric profile is sensitive to the polymer preparation procedure and the potential scan rate. Peak currents are proportional to potential scan rate, a feature characterizing the voltammetry of surface-confined species. Remarkably, the peaked shape of the voltammograms recorded at slow scan rates becomes narrowed upon increasing scan rate and there is a significant peak potential shift. These features can be attributed to ohmic resistance effects associated with the conducting/nonconducting transition, lowering diffusion rate of ions as a result of the increase of the film thickness, and varying distribution of length of the polymer units, but a complete explanation remains as an open question [13]. Figure 11.4 illustrates the electron transfer processes associated with the interconversion between different PEDOT forms.

The electrochemical processes significantly influence the conductivity of the polymers. In the case of PEDOT:PSS, application of a sufficiently cathodic potential produces cation doping (*n*-doping):

$$\{PEDOT^{+}:PSS^{-}\} + M^{+}_{solv} + e^{-} \rightarrow \{PEDOT^{0}(M^{+})PSS^{-}\} \quad (11.1)$$

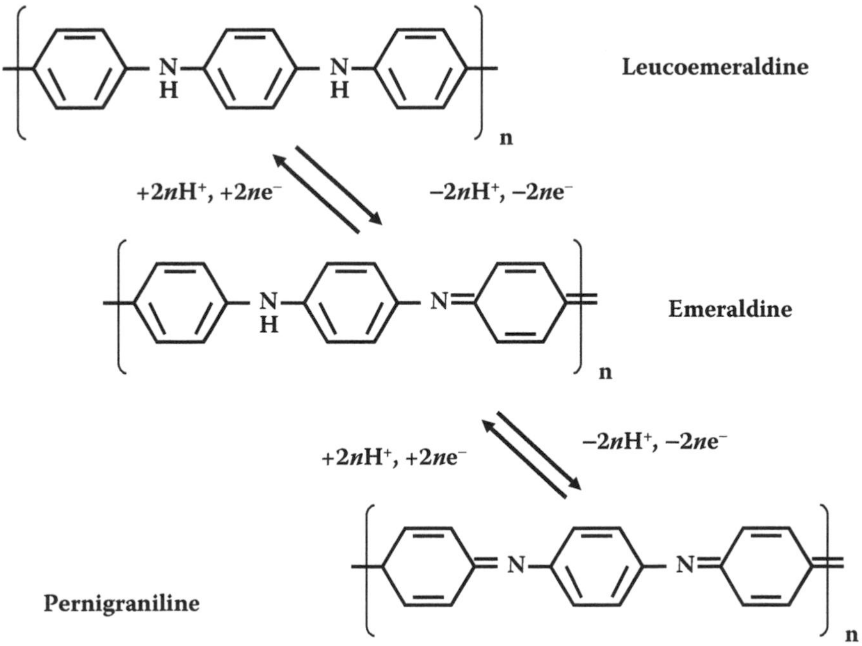

FIGURE 11.2 Schematics for the reductive/oxidative pathways relating the leucoemeraldine, emeraldine and pernigraniline PANI forms.

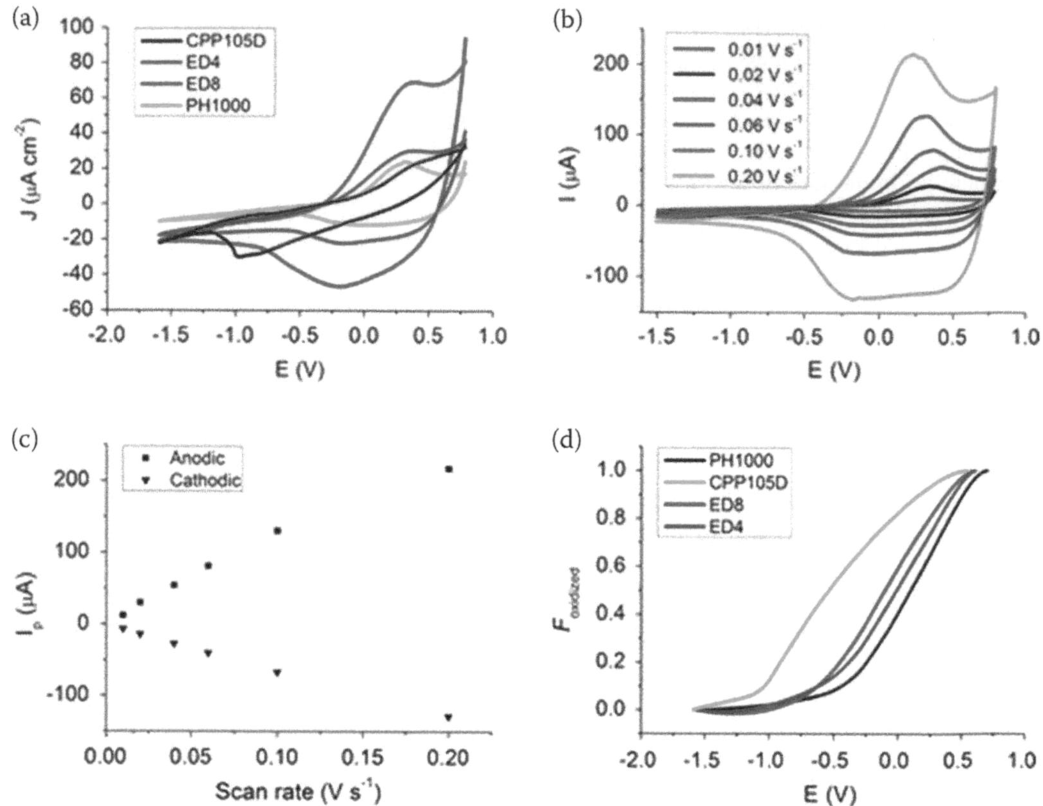

FIGURE 11.3 (a) CVs Recorded for Different PEDOT:PSS Substrates (Potential Scan Rate 20 mV s^{-1}). (b) CVs of PH1000 at Different Scan Rates. (c) Plot of Peak Current vs. Scan Rate. (d) Fraction of Oxidized PEDOT:PSS ($F_{oxidized}$) Calculated for the Different Substrates as a Function of the Applied Potential. Electrolyte: PBS. Reproduced from Ref. [12] (Marzocchi et al. *Appl. Mater. Interfaces*. 2015, 7: 17993–18003), with Permission.

FIGURE 11.4 Scheme for the Electrochemical Interconversion Between PEDOT Forms. Reproduced from Ref. [12] (Marzocchi et al. *Appl. Mater. Interfaces*. 2015, 7: 17993–18003), with Permission.

Then, PEDOT film changes from its oxidized conductive state to its neutral non-conductive state. When an anodic potential input is applied, the partially oxidized polymer moves to a more conductive fully oxidized state. Overoxidation, however, may occur at sufficiently anodic potentials, leading to an insulating state [12].

Figure 11.5 shows experimental data for coupled CV and electrogravimetric experiments on a PEDOT film in contact with 0.1 M $LiClO_4/CH_3CN$ electrolyte [15]. Here, oxidative doping (*p*-doping) occurs:

$$\{PEDOT^0\} + X^-_{solv} \rightarrow \{PEDOT^+(X^-)\} + e^- \qquad (11.2)$$

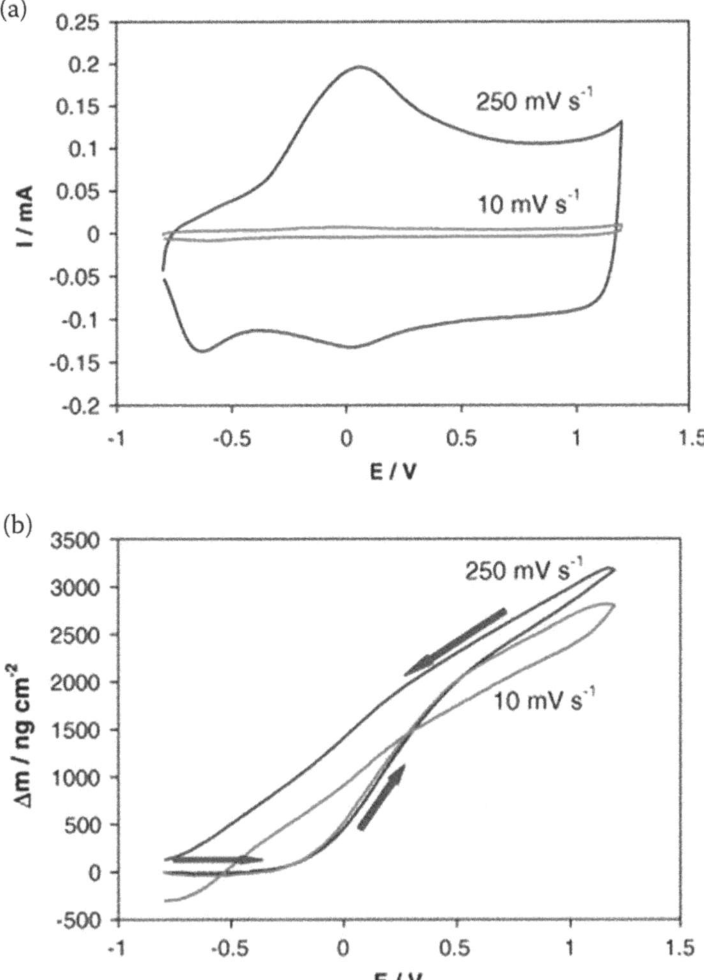

FIGURE 11.5 Experimental Data for Coupled CV (a) and Electrogravimetric (b) Experiments on a PEDOT Film ($\Gamma = 131$ nmol cm^{-2}; of Monomer Units, Coulometrically Assayed on the Basis of a Maximum Doping Level $n = 0.3$). Electrolyte 0.1 M $LiClO_4/CH_3CN$. Film Mass Changes Calculated Using Sauerbrey Equation (Eq. (11.2)). Reproduced from Ref. [15] (Hillman et al. *Electrochem. Commun.* 2007, 9: 1316–1322), with Permission.

The oxidation process yields one single voltammetric peak while the reductive undoping proceeds via two successive voltammetric peaks. This suggests that doping and undoping follow different mechanisms. Film mass changes calculated using Sauerbrey equation (Eq. (1.25)) inform on the fluxes of dopant and solvent and can be interpreted balancing three elemental processes: coupled electron/ion transfer, solvent transfer, and polymer reconfiguration.

Figure 11.6 depicts the fluxes of mobile species as functions of applied potential calculated from EQCM data. Figure 11.7 shows a graphical representation of mechanistic pathways for PEDOT through a cube scheme where the x-coordinate corresponds to coupled electron/anion transfer, the y-coordinate to solvent transfer, and the z-coordinate to polymer reconfiguration. R and O represent the reduced and oxidized PEDOT forms, the subscript S represents the more solvated state of the polymer, and superscripts a and b represent the two limiting polymer configuration states. The arrows represent the different possible processes accounting for the complex relationship between oxidation/doping state, solvation, and polymer reconfiguration.

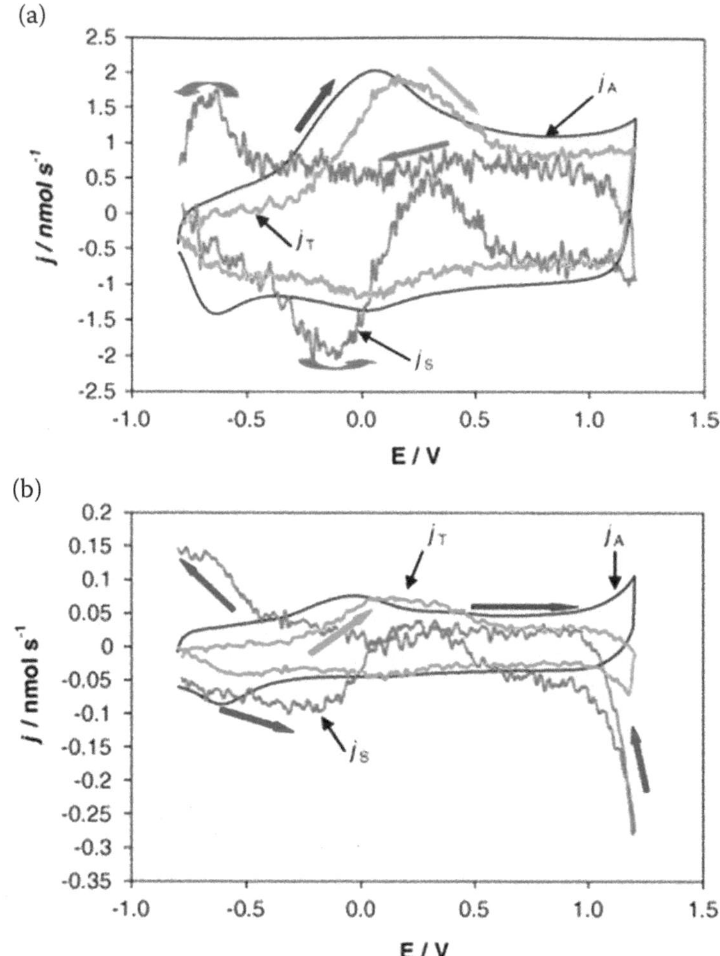

FIGURE 11.6 Fluxes of Mobile Species as Functions of Applied Potential (E) for the Experiment of Figure 11.5 with (a) v = 250 mV s^{-1}; (b) v = 10 mV s^{-1}. j_A: Coupled Anion/Electron Flux; j_T: Combined Fluxes of Anion and Solvent; j_S: Total Mass Flux Minus Anion Flux. Reproduced from Ref. [15] (Hillman et al. *Electrochem. Commun.* 2007, 9: 1316–1322), with Permission.

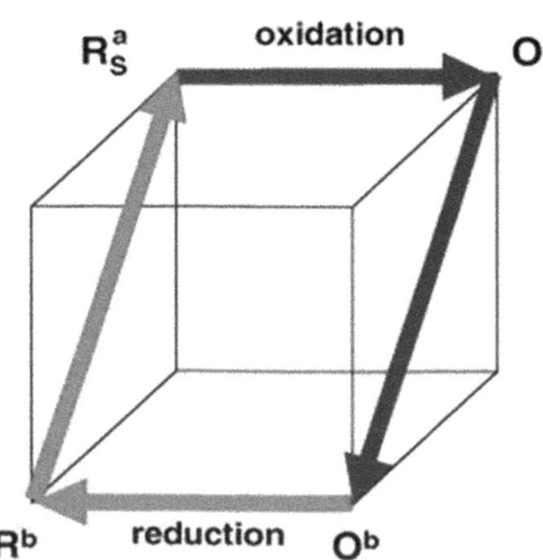

FIGURE 11.7 Graphical Representation of Mechanistic Pathways for PEDOT Redox Switching Defining a Cube Scheme. The x-Coordinate Corresponds to Coupled Electron/Anion Transfer, the y-Coordinate to Solvent Transfer, and the z-Coordinate to Polymer Reconfiguration. Reproduced from Ref. [15] (Hillman et al. *Electrochem. Commun.* 2007, 9: 1316–1322), with Permission.

The impedance behavior of conducting polymers films in contact with aqueous electrolytes is determined by several processes of insertion and transport of charged and non-charged species, namely [16] (a) transfer of electron at the electrode/polymer film interface; (b) transport of ions and water through the film; (c) ingress/issue of ions through the polymer/electrolyte interface; and (d) transport of water and ions through the electrolyte solution. Accordingly, there are electron transfer processes due to the existence of different oxidation states, and simultaneous exchange of electrolyte ions to maintain the film electroneutrality. This situation is parallel to that described in Chapter 2 but incorporating some peculiarities of conducting polymers. Here, delocalized electrons can move through the conjugated systems. This electronic conduction type is accompanied—where allowed—by segmental motion of polymer chains, by exchange reactions consisting of long-distance electron hopping between neighboring redox sites.

Taking polypyrrole as a paradigmatic example, the polymer oxidation generates a localized positive charge in the polymer chain that passes from aromatic to quinoid structure over four pyrrole rings. This leads to the formation of electronic states (polarons) while charge conservation demands the ingress of a negative charge, giving rise to anion doping. Two polarons nearby can generate a bipolaron state. This is energetically favored by the effect of lattice relaxation and polarons, while bipolarons act as charge carriers. Their concentration, which determines the polymer conductivity, in turn, depends on the level of oxidation/doping. The carrier mobility is influenced by conjugation length, determined by the long-range order of the polymer, and the interchain distance.

Figure 11.8 shows the equivalent circuit used to describe this kind of systems [17]. The high-frequency region is modeled by a parallel combination of film resistance R_{bulk} and constant phase element CPE_{bulk} in series with uncompensated solution resistance R_s. The region of middle and low frequencies is modeled by two parallel combinations of charge transfer resistance (R_{ct1} and R_{ct2}) and double layer capacitive (CPE_{dl1} and CPE_{dl2}) elements. These are representative of electron transfer through the electrode-polymer interface and ion transfer through the polymer-electrolyte interface. A Warburg element is added to model ionic and electronic diffusion-migration and can be considered as representative of linear finite restricted diffusion of dopants across the polymer film. This is accompanied in series by a constant phase element (CPE_l) corresponding to the pseudocapacitive charging (*vide infra*) of the film.

These parameters are dependent on temperature, film thickness, applied potential, and types and concentration of electrolytes. There are, however, alternative views of the physical meaning of the different elements of the equivalent circuit. For instance, double-layer capacitive elements have been assigned either to the metal-polymer or the polymer-solution interfaces [16].

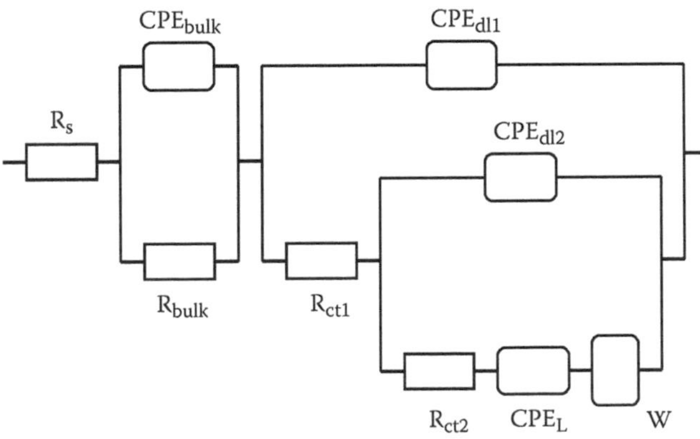

FIGURE 11.8 General Equivalent Circuit to Describe the Impedance Response of Conducting Polymer Coatings on Electrodes.

Similarly, different conductivity models have been proposed, including metallic, film ionic, pores ionic, and variable range charge hopping.

When the polymer film is oxidized, its electronic conductivity can exceed the ionic conductivity due to mobile counter ions. Then, the film behaves as a porous metal with pores of limited diameter and depth. This can be represented by means of Randles-type circuits described in previous chapters. The model circuit includes a charge transfer resistance associated with the electrode/film interface, and a constant phase element representing the charge accumulation that forms the interfacial double-layer [18]. In the low-frequency region, the impedance of the Warburg element, Z_D, can be expressed as a function of a diffusion pseudo-resistance, R_D, a diffusion pseudo-capacitance, C_D, and a diffusion time constant, τ_D, related by:

$$Z_D = \frac{(\tau_D/C_D)\coth(jC_D\tau_D)^{1/2}}{(j\omega C_D)^{1/2}} \tag{11.3}$$

The impedance of a one-dimensional cylindrical pore structure with invariant interfacial impedance Z_o along the wall of the pores can be described in terms of the electrolyte resistance, the impedance for a single pore, Z_p, the impedance of the pore wall, Z_{wall}, the pore radius, r_p, and the pore length, l_p, using the relationships [18]:

$$Z_p = (R_{electrol}Z_{wall})^{1/2}\coth(R_{electrol}/Z_{wall})^{1/2} \tag{11.4}$$

$$Z_{wall} = \frac{Z_o}{2\pi r_p l_p} \tag{11.5}$$

$$R_{electrol} = \frac{l_p}{\sigma\pi r_p} \tag{11.6}$$

Here, σ represents the Warburg coefficient, which can be estimated from the slope of the log Z_{real} vs. log f representation in the region of intermediate frequencies, where:

$$Z_{real} = R_u + R_{ct} + \sigma/\omega^{1/2} \tag{11.7}$$

The film thickness, δ, can be estimated from:

$$\delta = \frac{qM}{zFA\rho} \tag{11.8}$$

Here, q represents the charge necessary to switch from one form to another, M the molecular mass of the monomer, A the electrode area, ρ the density, and z the number of charges of the monomer in polymer chains [18,19].

11.3 HYBRID MATERIALS BASED ON CONDUCTING ORGANIC POLYMERS

Polymers present a characteristic size and dimensionality in between those of molecular and extended species. Two distinct classes of polymer-based, hybrid organic-inorganic materials can be distinguished: 1) those where the polymers constitute the matrix where inorganic species are inserted or integrated, or 2) those where the organic polymer in inserted into a more extended inorganic matrix, although there are obvious borderline materials challenging this classification

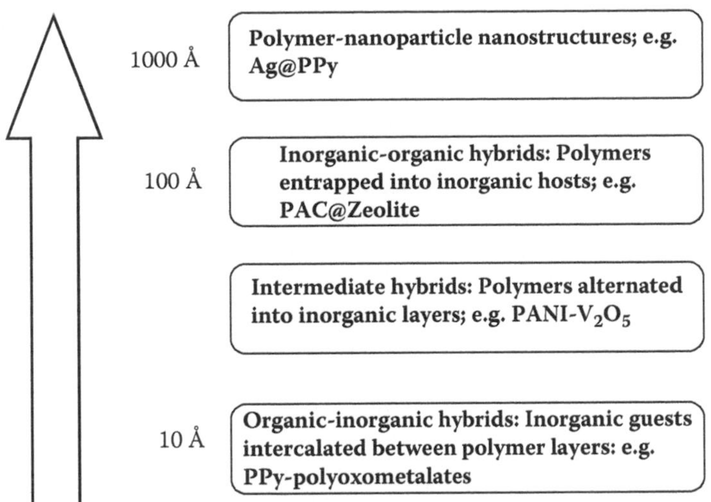

FIGURE 11.9 Types of Hybrids Between Conducting Polymers and Inorganic Species Ordered According to the Scale Length of Their Interactions. Inspired in Refs. [18,20].

like V_2O_5-polyaniline [20]. In the last decades, an enormous variety of hybrid materials incorporating porous polymers have been prepared. These include extended phases to nanoparticles and different nanoarchitecture of metals, metal oxides, etc., metal clusters, and molecular species. Figure 11.9 shows a diagram of the main types of hybrids as a function of the length scale of interaction [18,20]. These include metal nanoparticles coated by polymers and embedded into a polymeric matrix, passing to polymers entrapped to inorganic matrices, to systems where an organic guest is intercalated into an inorganic host (inorganic-organic hybrids), and finally to systems where the inorganic guest becomes intercalated into an organic host (organic-inorganic hybrids).

The insertion of molecular inorganic (or organic) species into a conducting organic polymer network can be obtained by different synthetic routes, including copolymerization, sol-gel template synthesis, electropolymerization, etc. Doping strategies involve the attachment of charged species as charge-balancing ions to the polymer network. Different conducting polymers modified with anionic inorganic species—oxometalates, complexes with organic ligands, metal complexes, macrocyclic compounds, etc.—have been described.

The EIS response of these hybrid materials depends on the film thickness and morphology, applied potential, and the nature of the components of the hybrid system. The hydrophobic nature of the polymer, the level of doping within the film, and the size of ions in contact with the polymer surface are factors to be considered when studying the response of such materials.

Different hybrid materials formed by the insertion of organic polymers into inorganic substrates have been prepared. Layered materials are well-known as host matrices for the incorporation of a large variety of polymeric organic species. Generally, the inorganic part is finely dispersed or exfoliated within the polymer, but alternatively, the polymer can form laminae intercalated in laminar solids. This is the case with layered double hydroxides (LDHs, see Chapter 8), whose structure is closely related to that of the brucite, $Mg(OH)_2$, in which the partial substitution of some of the divalent cations in octahedral positions with trivalent metallic ions results in the generation of positive charge supported by the hydroxyl layer. The substitution of the interlamellar anions by negatively-charged species prepares hybrid materials. This goal can be achieved by different procedures; the delamination/restacking is one of the most versatile [21]. This method consists of the delamination of the LDH system by mechanic shaking in organic solvents and subsequent addition of the anionic substituent molecule to the delaminated material, promoting the flocculation of the intercalated material. Using this methodology, an LDH-supported polyaniline material can be prepared by the intercalation of an emeraldine conductive PANI solution into a ZnAl-NO_3 LDH inorganic host. Attachment of PANI to this inorganic support determines the modulation of the electrochemical properties of the polymer in its different oxidation states as a result of the interaction with the inorganic support. Again, the voltammetric response is electrolyte-dependent [22].

In this context, an alternative way to stabilize reactive charged species resulting from *p*- and *n*-doping of polyacetylenes consists of their anchorage to a nonconducting inorganic matrix. This strategy has been widely used to stabilize organic reactive species into inorganic hosts,—zeolites, in particular (see Chapter 5). This type of material combines the redox activity of the organic guest with the ability of zeolites to stabilize positively-charged species, stabilizing reactive cations, blocking cross-link reactions, and protecting such reactive species from the attack of external reagents, particularly oxygen.

Figure 11.10 compares the CV response of zeolite-associated poly(tiophenylacetylene) (PTA@Y) in contact with Bu_4NPF_6/MeCN, Et_4NClO_4/MeCN, and, $LiClO_4$/MeCN electrolytes [23]. In all electrolytes, a well-defined oxidation peak appears at 1.44 V, followed by an ill-defined wave near 2.4 V. In contact with Bu_4NPF_6/MeCN and Et_4NClO_4/MeCN electrolytes, no coupled cathodic peaks were recorded in the subsequent cathodic scan even at potential scan rates of 1,000 mV s^{-1}, denoting that the main oxidation process corresponds to an irreversible electron transfer or an electron transfer followed by a fast chemical reaction. At more negative potentials, a weak reduction wave near –1.0 V was recorded. In contrast, in contact with $LiClO_4$/MeCN, a well-defined cathodic peak appears at 0.82 V followed by overlapping cathodic signals at –1.05 and –1.51 V.

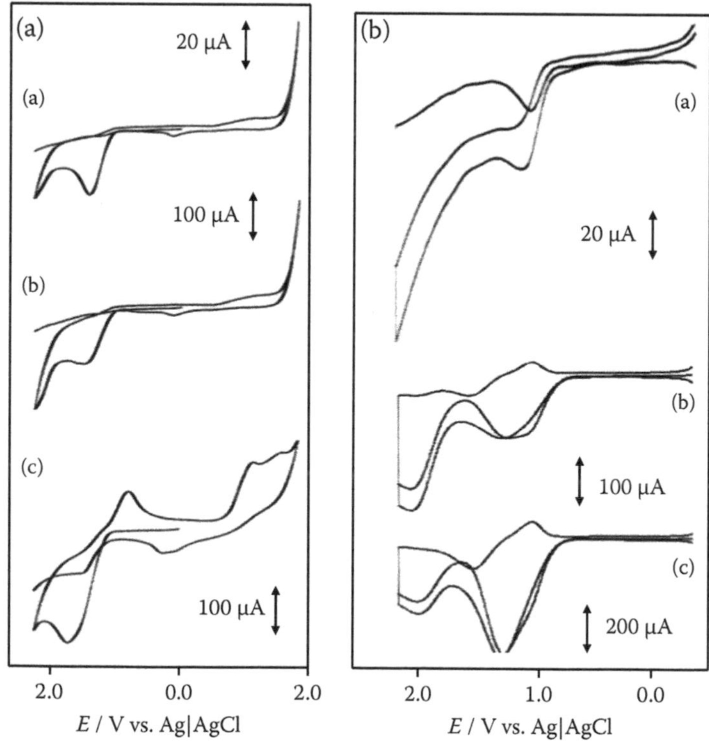

FIGURE 11.10 (a) CVs and (b) Forward, Backward, and Net Currents in SWVs of PTA@Y-Modified Electrodes Immersed into: (a) 0.10 M Bu_4NPF_6/MeCN, (b) 0.10 M Et_4NClO_4/MeCN, and (c) 0.10 M $LiClO_4$/MeCN. CVs: Potential Scan Rate 50 mV s^{-1}. SWVs: Potential Step Increment 4 mV; Square Wave Amplitude 25 mV; Frequency 5 (a,b) and 15 (c) Hz. Reproduced from Ref. [23] (Doménech-Carbó et al. *Electrochem. Commun.* 2006, 8: 1335–1339), with Permission.

Remarkably, peak currents for PTA@Y immersed in $LiClO_4$/MeCN were considerably larger than those recorded for PTA@Y-modified electrodes in contact with Bu_4NPF_6/MeCN and Et_4NClO_4/MeCN electrolytes. In the second and successive scans, peak currents decrease rapidly for PTA@Y immersed in Bu_4NPF_6/MeCN and Et_4NClO_4/MeCN electrolytes, while a slow decay of peak currents was observed for modified electrodes in contact with $LiClO_4$/MeCN. Separation of the cathodic and anodic components of the net current (measured at the end of forward and backward pulses) in SWVs provided only anodic components for PTA@Y electrodes immersed in Bu_4NPF_6/MeCN, as also depicted in Figure 11.10. In contrast, SWVs display well-developed components for PTA@Y electrodes in contact with $LiClO_4$/MeCN. This feature, indicative of reversible electron transfer processes, was found to be more pronounced on decreasing square wave frequency.

The reduction of zeolite-associated poly(tiophenylacetylene) oligomers in solution can be described as a one-electron per monomer electron transfer process, where an anion radical is formed. Similarly, the reduction of zeolite-associated poly(thienylacetylene) can be described in terms of the formation of radical anion species coupled with the ingress of electrolyte cations M^+ (= Bu_4N^+, Et_4N^+, Li^+) into the zeolite framework (equivalent to the *n*-type doping in redox polymers in solution). Now, radical anion species eventually stabilized into the zeolite pore/channel system are formed, as described for zeolite-associated Meisenheimer anions (see Chapter 5). Since the reduction process requires the ingress of electrolyte cations (M^+) into the zeolite system, reduction of zeolite-associated species will be allowed for Li^+ ions while bulky, size-excluded, Bu_4N^+—and to a lesser extent, Et_4N^+ —block reduction processes.

Oxidation of zeolite-associated poly(thienylacetylene) can be described via anion insertion (i.e. the typical *p*-type doping process in redox polymers) or via desorption of charge-balancing Na^+-exchanged cations from the zeolite system to the solution phase. Following the general ideas for the electrochemistry of electroactive species attached to porous solids previously described, the voltammetric response of PTA@Y in contact with Bu_4NPF_6/MeCN and Et_4NClO_4/MeCN electrolytes should be attributed to the population of poly(tienylacetylene) chains located on the particle surface, producing a unique, weak oxidation peak. In contrast, the response of PTA@Y immersed in $LiClO_4$/MeCN results from superposition of externally located poly(tiophenylacetylene) oligomers (displaying an irreversible behavior) and zeolite-associated poly(thienylacetylene) units (displaying an essentially reversible behavior). Consistently, the response in the second and subsequent scans in repetitive voltammetry is rapidly exhausted for PTA@Y immersed in Bu_4NPF_6/MeCN and Et_4NClO_4/MeCN, whereas it is significantly more reversible for $LiClO_4$/MeCN.

In this scheme, zeolite-associated species display a reversible electrochemistry because of the compartmentalization effect due to the encapsulation of poly(thienylacetylene) oligomers into the zeolite framework. It should be noted that the electrochemical oxidation of thiophenes having free 2 or 5 positions can result in dimerization and oligomerization arising from post-electron-transfer reactions. These processes can result frequently in apparent irreversible voltammetric profiles [24]. In the case of thiophene-substituted polymers, such oxidation process may result in the formation of cross-links via α,α' or α,β' coupling between thiophene units [25]. Then, the reversibility of electron transfer processes of PTA@Y in contact with $LiClO_4$/MeCN electrolytes can be attributed to the exigent spatial constraints due to the rigidity of the zeolite framework, stabilizing thiophene substituents by impeding side reactions of polarons to occur.

Polyoxometalate-polymer hybrid materials have been widely studied [26]. The strong oxidizing potential and acid character of acidic forms of oxometalates ('phosphomolybdic acid', for instance) are convenient properties in polymerizing monomers such as aniline, pyrrole, or thiophene to yield the corresponding *p*-doped polymers, where the bulky polyoxometalate species become trapped within the polymeric matrix via oxidative doping. The amount of electroactive guest anions incorporated within the polymer matrix is limited by the doping level attainable for the polymer—the spatial distribution within the polymeric matrix depending on the conditions of preparation of the hybrid material. It should be noted that de-insertion of doping species can occur when the polymer is reduced, resulting in the loss of the desired activity. This is one of the reasons for studying polyoxometalate/conducting polymer hybrid materials because the low diffusion rate of these species relative to that of common doping anions (ClO_4^-, Cl^-) makes it appropriate candidates to remain anchored to the polymer network during oxidative steps. In fact, polyoxometalate-doped polymers can run as cation-inserting redox materials, a property of interest, for instance, for lithium batteries.

11.4 COMPOSITE MATERIALS BASED ON CONDUCTING POLYMERS

An enormous variety of composites including polymers is under study for applications in sensing, optoelectronics, catalysis, etc. Here, the building blocks retain their structural individuality, although frequently, the frontier between hybrid materials and nanocomposites becomes diffuse. For instance, multilayer PANI plus GO materials were prepared by layer-by-layer assembly as schematically depicted in Figure 11.11 [27]. This material presents high specific capacitance superior to that of the separate components. This feature is attributed to a synergistic effect between PANI and GO resulting from an increase of the film porosity that allows a larger charge to be stored by the electrical double-layer mechanism.

This can be illustrated by the values of the specific capacitance of several systems listed in Table 11.1 [7,28–35]. One can see that the value of this quantity for graphene/PANI composite is clearly larger than those of the separate components. Polymer-based composites are also extensively applied in fuel cells as structural supports, coatings on structural supports, catalytic materials, and coating of the catalyst. In batteries these are extensively employed as active masses and additives [28].

The use of porous carbons, carbon nanotubes, and graphene forms in composites, in general, aims to increase electronic conductivity and surface area, whereas conducting polymers provide conductivity and mechanical binding and/or coverage accompanied by their load of intercalated charge. Metal and metal oxide nanoparticles are also included in composites. Nanoparticle stabilization from solutions requires the use of a protecting agent (capping) to prevent aggregation phenomena resulting in peculiar nanoparticle properties. Adsorbed polymer coatings, adsorbed ion layers, monolayers of organic ligands, and dendrimers, among others, have been used for this purpose.

Much of these composites have been applied for electrocatalytic sensing purposes, including films of conducting polymer/polyoxometallate/metal nanoparticles, materials integrating conducting polymers, oxides, and inorganic molecules, or multi-walled CNTs and/or graphene, etc. Recent advances include more control of the structure and composition and green synthesis of the composites. Challenges are open, however, regarding the fundamental electrochemical properties of materials, as far as the detailed knowledge of the catalytic mechanisms.

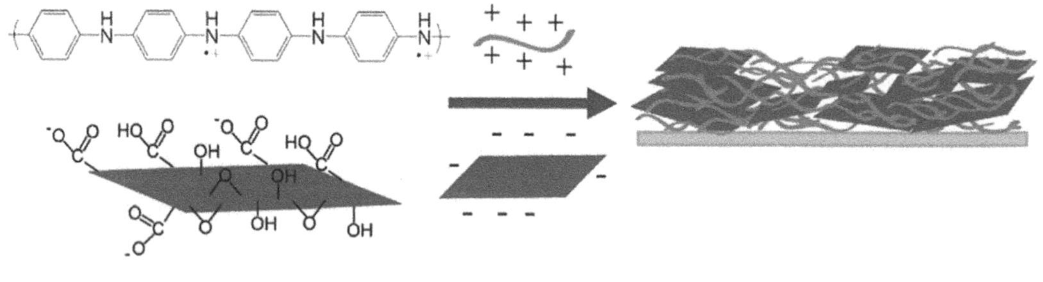

FIGURE 11.11 Scheme of Multilayer Assembly of PANI Nanofiber with rGO. Reproduced from Ref. [27] (Jeon et al. *J. Mater. Chem. A.* 2015, 3: 3757–3767), with Permission.

TABLE 11.1
Specific Capacitance (C_{sp}) of Several Supercapacitor Materials Based on Conducting Polymers

Material	C_{sp}/F g^{-1}
PEDOT (max. teor.) [29]	210
PPy (max. teor.) [29]	620
PANI (max. teor.) [29]	750
Aggregates of defective graphene [31]	100
Cross-linked GO [32]	210
Graphene/PANI [33]	965
Graphene/PANI/Co_3O_4 [34]	1250
PPy/TiN/PANI [35]	1470

Data from Refs. [7,28–35].

11.5 PHOTOELECTROCHEMISTRY

Conducting polymers and their hybrids and composite materials are also used in photoelectrochemical applications. These include organic solar cells (OSCs), electrochromism, and molecular recognition [7].

Organic solar cells are devices that employ organic molecules as light absorbers to generate electron-hole separation to cause a potential difference. This is a particular class of photovoltaic cells (typically constituted by semiconducting materials) where the light absorber is a small molecule or a conducting polymer [36].

Electrochromism is a property widely studied in conducting polymers [37]. First, it is pertinent to note that for many practical purposes, HOMO and LUMO energies (E_{HOMO}, E_{LUMO}, respectively) and energy gap, E_{gap}, can be approximated by the onset potentials (E_{onset}) measured in CVs. These are defined as the potential values obtained from the intersection of two tangents drawn at the rising current and the baseline charging current of the CV curves. In voltammetric experiments in organic solvents, the energies can be calculated as E_{HOMO} (E_{LUMO}) = – [–(E^{onset} – 0.45) – 4.75] eV [38]. In this equation, E^{onset} represents the onset potential vs. Ag^+/Ag, 4.75 eV represents the energy level of ferrocene below the vacuum, and 0.45 V is the value of the ferrocenium/ferrocene couple vs. Ag^+/Ag. Notice that it is assumed that the potential of the ferrocenium/ferrocene couple is solvent-independent.

Many other conducting polymers have been combined with inorganic materials such as Prussian blue and metal hexacyanoferrates, metal nanoparticles, metal oxide nanoparticles, quantum dots, metallic complexes, silica, and other materials to build electrochromic devices. Current research is focused on controlling color tuning and diminishing switching time, in turn depending on the rate of ion diffusion. Among other synthetic strategies, the intercalation of conducting polymer layers into electrochromic materials such as Prussian blue has been reported [39].

11.6 POLYMERS OF INTRINSIC POROSITY

These is a class of amorphous organic microporous materials characterized by the presence of interconnected pores less than 2 nm diameter [40]. This intrinsically microporous system develops from the combination of, for instance, a catechol and an aryl halide, resulting in a rigid polymeric structure. Polymers of intrinsic porosity (PIMs) combine synthetic diversity of

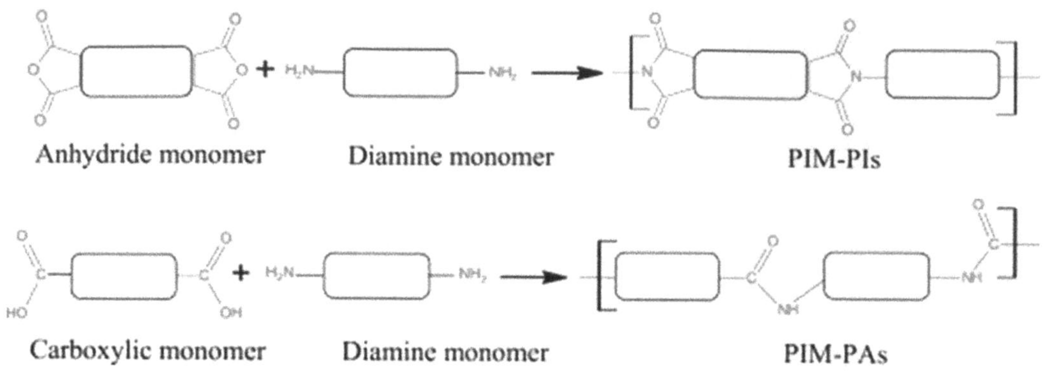

Anhydride monomer Diamine monomer PIM-PIs

Carboxylic monomer Diamine monomer PIM-PAs

FIGURE 11.12 Some Schematic Reactions to Prepare PIMs. From Ref. [41] (Ramimoghadam et al. *Int. J. Hydrogen Energ*. 2016, 41: 16944–16965), with Permission.

organic polymers and the high internal surface area of conventional microporous materials [41]. Figure 11.12 depicts examples of schematic reactions to prepare PIMs [41].

The applications of these materials are concentrated in gas storage, but electrode modification by PIMS is receiving increasing attention [42,43]. PIM-modified electrodes can offer permeability and selectivity towards solution species. In particular, PIMs can immobilize a water-immiscible organic phase to form an organo-gel, allowing the performance of liquid-liquid voltammetry [44], which parallels the three-phase voltammetry of immobilized droplets [45]. Then, typical electrochemical processes can be represented as:

$$\{Ox\}_{org\ phase} + M^{+}_{aq} + e^{-} \rightarrow \{Rd\}_{org\ phase} + M^{+}_{org\ phase} \tag{11.9}$$

$$\{Rd\}_{org\ phase} + X^{-}_{aq} \rightarrow \{Ox\}_{org\ phase} + X^{-}_{org\ phase} + e^{-} \tag{11.10}$$

Electrocatalytic and photoelectrocatalytic effects have been detected at PIM membranes. Figure 11.13 shows an example of this electrochemistry illustrated by the voltammetry of Fe(II)-tetraphenylporphyrinato complex (FeTPP) immobilized into a film of PIM-EA-TB (see Figure 11.12) in the presence of O_2 and H_2O_2 in aqueous phosphate buffer at pH 2 [46]. There is a catalytic effect of FeTPP on both ORR and H_2O_2 reduction processes that significantly depends on the weight ratio of FeTPP:PIM. At the expense of a detailed mechanistic study, it seems likely that charge carrier transport in the microporous catalyst structure plays a crucial role [46]. In the next years, research will enhance the understanding of these electrochemical processes and will expand the applications of PIMs in the electrochemistry field.

11.7 POLYMER ELECTROLYTES

Polymer electrolytes are a family of polymeric materials characterized by its high ionic conductivity [47,48]. One characteristic example is poly(vinyl alcohol)/poly(acrylic acid) [49]. These electrolytes are under investigation for solid-state lithium batteries. Here, inorganic ceramic electrolytes are extensively used because of their mechanical and chemical stability, high ionic conductivity, and high lithium transference number. These electrolytes, however, suffer from several drawbacks: poor contact with the electrodes and high density and reactivity with several cathode materials.

To overcome these problems, much research has been focused on polymer solid electrolytes. These materials, however, offer low ionic conductivities and low lithium transference numbers at

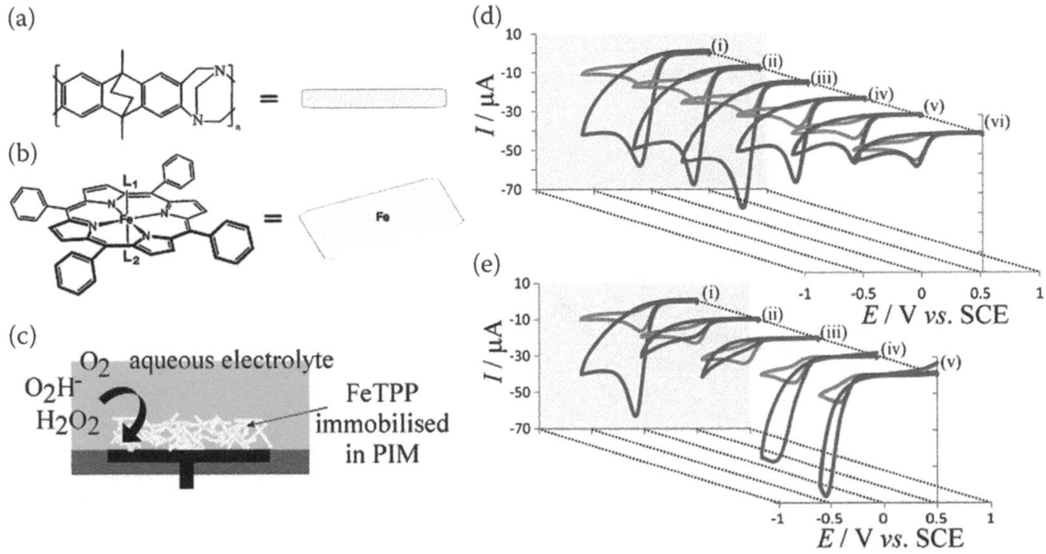

FIGURE 11.13 Molecular Structure of (a) PIM-EA-TB and (b) FeTPP and (c) scheme for the Immobilization of FeTPP into the Rigid Microporous Structure. (d) CVs for the Reduction of Ambient Oxygen (Blue) and 4 mM H_2O_2 (Red) in 0.1 M Phosphate Buffer at pH 2 at Electrodes Coated With (i) 1:1, (ii) 1:2, (iii) 1:5, (iv) 1:10, (v) 1:20, and (vi) 1:40 Weight Ratio FeTPP:PIM. Potential Scan Rate 20 mV s^{-1}. Reproduced from Ref. [46] (Rong et al. *Electrochem. Commun.* 2014, 6: 26–29), with Permission.

room temperature. Additionally, the nature of the ion transport mechanism in these materials is controversial. The more extended view considers that ion conductivity in polymer electrolytes is determined by the local motion of polymer segments associated with the intra- and inter-chain hopping of cations [4].

For these reasons, composites of ceramic materials with polymer electrolytes in both ceramic-in-polymer and polymer-in-ceramic configurations are being explored. Figure 11.14 (upper part) compares the Nyquist plots for a composite of LiI, poly(ethylene oxide) (P(EO)), and $LiAlO_2$ ceramic filler at different temperatures sandwiched between two blocking electrodes [4]. The Nyquist plots define a depressed loop at high frequencies followed by a linear branch at low frequencies. The conductivity of the material can be calculated from the area, thickness, and bulk resistance of the film—the last quantity being determined from the high-frequency intercept on the real axis (σ_1) or from the diameter of the high-frequency loop (σ_2). Figure 11.14 (lower part) shows the Arrhenius plots for these two conductivities for composites using $LiAlO_2$ (LAO:LiI:P(EO)) and $Li_{10}SnP_2S_{12}$ (LSPS:LiI:P(EO)) ceramic materials. Here, different linear patterns can be seen. These plots provide information on the dominant phase and preferred conduction pathway of the material. Factors influencing conductivity are the degree of crystallinity, the dielectric permittivity of the polymer, and coupling ions with the polymer chain. The conductivity of solid polymer electrolytes can be enhanced by using polymer hosts with high dielectric permittivity favoring electrolyte dissociation and low glass transition temperature and/or two-phase polymer blends. Factors such as the grain boundaries between polycrystalline electrolytes and structural defects can also notably influence the conductivity of the materials.

11.8 FINAL REMARKS

The study of porous polymers is an expanding area of research with applications in catalysis, sensing, energy production and storage, gas storage, among others. Apart from specific developments in these domains, one direction of development is the refinement of synthesis post-synthetic

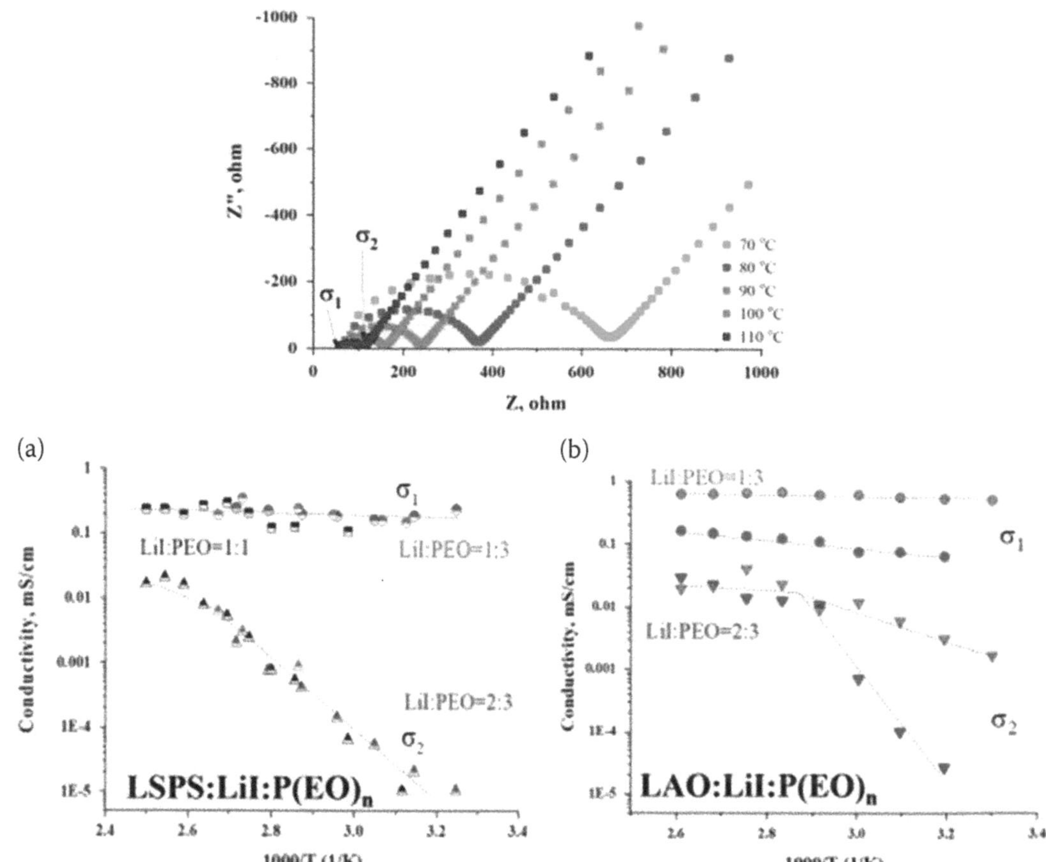

FIGURE 11.14 Upper part: Nyquist Plots of the LiI:P(EO)$_3$ -LiAlO$_2$ Electrolyte at Different Temperatures. Lower part: Arrhenius Plots of Ion Conductivity of LSPS:LiI:P(EO) and LAO:LiI:P(EO) Electrolytes, Containing 75–80% of Ceramics. Reproduced from Ref. [4] (Menkin et al. *Electrochim. Acta.* 2019, 304: 447–455), with Permission.

modification with functional groups to avoid pore-blocking [18,20]. A second line of interest is aimed at the synthesis of multifunctional porous polymers incorporating several groups balancing the good distribution of active sites while maintaining porosity. In parallel, the theoretical description of thermochemical parameters as well as the transport/diffusion kinetics in molecular sieving, heterogeneous catalysis, and energy storage has to be necessarily considered [50].

REFERENCES

[1]. Devaraj, G.; Guruviah, S.; Seshadri, S.K. 1990. Pulse plating. *Materials Chemistry and Physics*. 25: 439–461.

[2]. Xia, F.J.; Pan, M.; Mu, S.C.; Malpass-Evans, R.; Carta, M.; McKeown, N.B.; Attard, G.A.; Brew, A.; Morgan, D.J.; Marken, F. 2014. Polymers of intrinsic microporosity in electrocatalysis: Novel pore rigidity effects and lamellar palladium growth. *Electrochimica Acta*. 128: 3–9.

[3]. Kickelbick, G. Ed. 2007. *Hybrid Materials. Synthesis, Characterization, and Applications*. Wiley-VCH, Weinheim.

[4]. Menkin, S.; Lifshitz, M.; Haimovich, A.; Goor, M.; Blanga, T.; Greenbaum, S.G.; Goldbourt, A.; Golodnitsky, D. 2019. Evaluation of ion-transport in composite polymer-in-ceramic electrolytes. Case study of active and inert ceramics. *Electrochimica Acta*. 304: 447–455.

[5]. Slater, A.G.; Cooper, A.I. 2015. Function-led design of new porous materials. *Science*. 348: artic. 8075.

[6]. Wu, D.; Xu, F.; Sun, B.; Fu, R.; He, H.; Matyjaszewski, K. 2012. Design and preparation of porous polymers. *Chemical Reviews*. 112: 3959–4015.

[7]. Wolfart, F.; Hryniewicz, B.M.; Góes, M.S.; Corrêa, C.M.; Torresi, R.; Minadeo, M.A.O.S.; Córdoba de Torresi, S.I.; Oliveira, R.D.; Marchesi, L.F.; Vidoytti, M. 2017. Conducting polymers revisited: Applications in energy, electrochromism and molecular recognition. *Journal of Solid State Electrochemistry*. 21; 2489–2515.

[8]. Shirakawa, H.; Louis, E.J.; MacDiarmid, A.G.; Chiang, C.K.; Heeger, A.J. 1977. Synthesis of electrically conducting organic polymers: Alojen derivatives of polyacetylene, $(CH)_x$. *Journal of the Chemical Society Chemical Communications*. 16: 578–580.

[9]. Inzelt, G. 2008. *Conducting Polymers—A New Era in Electrochemistry*. In Scholz, F. Ed. Monographs in Electrochemistry Series. Springer, Berlin.

[10]. Scotto, J.; Marmisollé, W.A.; Posadas; D. 2019. About the capacitive currents in conducting polymers: the case of polyaniline. *Journal of Solid State Electrochemistry*. 23: 1947–1965.

[11]. Huang, J.; Virji, S.;Weiller, B.H.; Kaner, R.B. 2003. Polyaniline nanofibers: Facile synthesis and chemical sensors. *Journal of the American Chemical Society*. 125: 314–315.

[12]. Marzocchi, M.; Gualandi, I.; Calienni, M.; Zironi, I.; Scavetta, E.; Castellani, G.; Fraboni, B. 2015. Physical and electrochemical properties of PEDOT:PSS as a tool for controlling cell growth. *Applied Materials & Interfaces*. 7: 17993–18003.

[13]. Holze, R. 2017. From current peaks to waves and capacitive currents—On the origins of capacitor-like electrode behavior. *Journal of Solid State Electrochemistry*. 21: 2601–2607.

[14]. Prakash, R. 2002. Electrochemistry of polyaniline: Study of the pH effect and electrochromism. *Journal of Applied Polymer Science*. 83: 378–385.

[15]. Hillman, A.R.; Daisley, S.J.; Bruckenstein, S. 2007. Kinetics and mechanism of the electrochemical p-doping of PEDOT. *Electrochemistry Communications*. 9: 1316–1322.

[16]. Rubinstein, I.; Sabatony, E.; Rishpon, J. 1987. Electrochemical impedance analysis of polyaniline films on electrodes. *Journal of the Electrochemical Society*. 134: 3078–3083.

[17]. Lvovich, V.F. 2009. A perspective on electrochemical impedance analysis of polyaniline films on electrodes. *Electrochemical Society Interface*. 18: 62–66.

[18]. Galal, A.; Darwish, S.A.; Ahmed, R.A. 2007. Hybrid organic/inorganic films of conducting polymers modified with phtalocyanines. II. EIS studies and characterization. *Journal of Solid State Electrochemistry*. 11: 531–542.

[19]. Duic, L.; Grigic, S. 2001. The effect of polyaniline morphology on hydroquinone/quinone redox reaction. *Electrochimica Acta*. 46: 2795–2803.

[20]. Gómez-Romero, P.; Ayyad, O.; Suárez-Guevara, J.; Muñoz-Rojas, D. 2010. Hybrid organic-inorganic materials: From child's play to energy applications. *Journal of Solid State Electrochemistry*. 14: 1939–1945.

[21]. Liu, Z.P.; Ma, R.Z.; Osada, M.; Iyi, N.; Ebina, Y.; Takada, K.; Sasaki, T. 2006. Synthesis, anion exchange, and delamination of Co–Al layered double hydroxide: Assembly of the exfoliated nanosheet/polyanion composite films and magneto-optical studies. *Journal of the American Chemical Society*. 128: 4872–4880.

[22]. Doménech-Carbó, A.; Coronado, E.; Lardies, N.; Martí, C.; Doménech-Carbó, M.T.; Ribera, A. 2008. Solid-state electrochemistry of LDH-supported polyaniline hybrid inorganic-organic material. *Journal of Electroanalytical Chemistry*. 624: 275–286.

[23]. Doménech-Carbó, A.; Galletero, M.S.; García, H.; Peris, E. 2006. Electrolyte-driven electrochemical amplification by poly(thienylacetylene) encapsulated within zeolite Y. *Electrochemistry Communications*. 8: 1335–1339.

[24]. Garcia, P.; Pernaut, J.M.; Witgens, V.; Valat, P.; Gaunier, F.; Delabouglise, D. 1992. Effect of end substitution on electrochemical and optical properties of oligothiophenes. *Journal of Physical Chemistry*. 97: 513–516.

[25]. Roncali, J. 1992. Conjugated poly(thiophenes): Synthesis, functionalization, and applications. *Chemical Reviews*. 92: 711–738.

[26]. Vaillant, J.; Lira-Cantu, M.; Cuentas-Gallegos, K.; Casañ-Pastor, N.; Gomez-Romero, P. 2006. Chemical synthesis of hybrid materials based on PAni and PEDOT with polyoxometalates for electrochemical supercapacitors. *Progress in Solid State Chemistry*. 34: 147–159.

[27]. Jeon, J.-W.; Kwon, S.R.; Lutkenhaus, J.L. 2015. Polyaniline nanofiber/electrochemically reduced graphene oxide layer-by-layer electrodes for electrochemical energy storage. *Journal of Materials Chemistry A*. 3: 3757–3767.

[28]. Holze, R.; Wu, Y.P. 2014. Intrinsically conducting polymersin electrochemical energy technology: Trends and progress. *Electrochimica Acta*. 122: 93–107.

[29]. Snook, G.A.; Kao, P.; Best, A.S. 2011. Conducting-polymer-based supercapacitor devices and electrodes. *Journal of Power Sources*. 196: 1–12.
[30]. Ambrosi, A.; Chua, C.K.; Bonnani, A.; Pumera, M. 2014. Electrochemistry of graphene and related materials. *Chemical Reviews*. 114: 7150–7188.
[31]. Stoller, M.D.; Park, S.; Zhu, Y.; An, J.; Ruoff, R.S. 2008. Graphene-based ultrcapacitors. *Nano Letters*. 8: 3498–3502.
[32]. Tang, L.A.L.; Lee, W.C.; Shi, H.; Wong, E.Y.L.; Sadovoy, A.; Gorelik, S.; Hobley, J.; Lim, C.T.; Loh, K.P. 2012. Highly wrinkled cross-linked graphene oxide membranes for biological and charge-storage applications. *Small*. 8: 423–431.
[33]. Jin, Y.; Fang, M.; Jia, M. 2014. In situ one-pot synthesis of graphene/polyaniline nanofiber composite for high-performance electrochemical capacitors. *Applied Surface Science*. 308: 333–340.
[34]. Lin, H.; Huang, Q.; Wang, J.; Jiang, J.; Liu, F.; Chen, Y.; Wang, C.; Lu, D.; Han, S. 2016. Self-assembled graphene/polyaniline/Co_3O_4 ternary hybrid aerogels for supercapacitors. *Electrochimica Acta*. 191: 444–451.
[35]. Xie, Y.; Wang, D. 2016. Supercapacitance performance of polypyrrole/titanium nitride/polyaniline coaxial nanotube hybrid. *Journal of Alloys and Compounds*. 665: 323–332.
[36]. Roncali, J. 2009. Molecular bulk heterojunctions: An emerging approach to organic solar cells. *Accounts of Chemical Research*. 42: 1719–1730.
[37]. Malinauskas, A.; Holze, R. 1998. Cyclic UV-Vis spectrovoltammetry of polyaniline. *Synthetic Metals*. 97: 31–36.
[38]. Nguyen, T.; Martini, I.B.; Liu, J.; Schwartz, B. 2000. Controlling interchain interactions in conjugated polymers: The effects of chain morphology on exciton–exciton annihilation and aggregation in MEH–PPV films. *Journal of Physical Chemistry B*. 104: 237–255.
[39]. DeLongchamps, D.M.; Hammond, P.T. 2004. Multiple-color electrochromism from layer-by-layer-assembled polyaniline/Prussian blue nanocomposite thin films. *Chemistry of Materials*. 16: 4799–4805.
[40]. Mason, C.R.; Maynard-Atem, L.; Heard, K.W.J.; Satilmis, B.; Budd, P.M.; Friess, K.; Lanc, M.; Bernardo, P.; Clarizia, G.; Jansen, J.C. 2014. Enhancement of CO_2 affinity in a polymer of intrinsic microporosity by amine modification. *Macromolecules*. 47: 1021–1029.
[41]. Ramimoghadam, D.; MacA Gray, E.; Webb, C.J. 2016. Review of polymers of intrinsic microporosity for hydrogen storage applications. *International Journal of Hydrogen Energy*. 41: 16944–16965.
[42]. Xue, Y.Q.; Zheng, S.S.; Xue, H.G.; Pang, H. 2019. Metal-organic framework composites and their electrochemical applications. *Journal of Materials Chemistry A*. 7: 7301–7327.
[43]. Marken, F.; Madrid, E.; Zhao, Y.; Carta, N.; McKeown, M.B. 2019. Polymers of intrisic microporosity in triphasic electrochemistry: Perspectives. *ChemElectroChem*. 6: 4332–4642.
[44]. Langley, A.R.; Carta, M.; Malpass-Evans, R.; McKeown, N.B.; Dawes, J.H.P.; Murphy, E.; Marken, F. 2018. Linking the Cu(II)/(I) potential to the onset of dynamic phenomena at corroding copper microelectrodes immersed in aqueous 0.5 M NaCl. *Electrochimica Acta*.260: 348–357.
[45]. Scholz, F.; Schröder, U.; Gulabowski, R.; Doménech-Carbó, A. 2014. *Electrochemistry of Immobilized Particles and Droplets*, 2nd edit. Springer, Berlin.
[46]. Rong, Y.Y.; Malpass-Evans, R.; Carta, M.; McKeown, N.B.; Attard, G.A.; Marken, F. 2014. High density heterogeneisation of molecular electrocatalysts in a rigid intrinsically microporous polymer host. *Electrochemistry Communications*. 46: 26–29.
[47]. Di Noto, V.; Lavina, S.; Giffin, G.A.; Negro, E.; Scrosati, B. 2011. Polymer electrolytes: Present, past and future. *Electrochimica Acta*. 57: 4–13.
[48]. Golodnitsky, D.; Strauss, E.; Peled, E.; Greenbaum, S. 2015. Review on order and disorder in polymer electrolytes. *Journal of the Electrochemical Society*. 162: A2551–A2566.
[49]. Thayumanasundaram, S.; Rangasamy, V.S.; Seo, J.W.; Locquet, J.P. 2017. Electrochemical performance of polymer electrolytes based on poly(vinyl alcohol)/poly(acrylic acid) blend and pyrrolidinium ionic liquid for lithium rechargeable batteries. *Electrochimica Acta*. 240: artic. 371e378.
[50]. Wu, J.; Xu, F.; Li, S.;Ma, P.; Zhang, X.; Liu, Q.; Fu, R.; Wu, D. 2019. Porous polymers as multifunctional material platforms toward task-specific applications. *Advanced Materials*. 31: artic. 1802922.

12 Electrochemical Sensing via Porous Materials

12.1 ELECTROCHEMICAL SENSING

In general, sensors can be defined as devices that perceive changes in the physical properties of the environment and turn these into a measurable signal [1]. Electrochemical sensors are based on the measurement of any signal associated with electrochemical phenomena. These include a wide variety of techniques, from conductimetry to electrochemical noise and, of course, classical techniques such as electrogravimetry. Here, we focus the attention on techniques based on electrode modification strategies using porous materials. These techniques are devoted to identifying and/or quantifying selected analyte(s) and, in principle, involve:

a. Monitoring any electrochemical process experienced by electroactive analytes; and
b. Monitoring an auxiliary electrochemical process whose parameters are modified by the presence of the analyte. This approach can correspond to two basic possibilities: (b1) the analyte participates in the electrochemical process via catalysis, complexation, etc., or (b2) the analyte competes with electroactive species blocking or modifying their electrochemical response (competitive methods).

Over the two last decades, the research on electrochemical sensing has experienced an exponential growth centered on three main types of electrochemical sensors: potentiometric, amperometric, and potentiodynamic (voltammetric or voltamperometric). Potentiometric sensors are based on the measurement of the voltage of a cell under equilibrium-like conditions, the measured voltage being a known function of the concentration of the analyte. Potentiometric measurements generally involve Nernstian responses under zero-current conditions (i.e. the measurement of the electromotive force of the electrochemical cell) equivalent to the open circuit potential (OCP). Amperometric sensing is based on the record of the current response of an electrode in contact with the system to be analyzed under the application of a given potential input. Amperometric sensors operate under conditions where mass transport is limiting. Potentiodynamic sensors are based on the measurement of the current response of the working electrode without mass-transport limiting conditions. Potentiodynamic methods typically involve cumulative (or preconcentration) steps, as in stripping voltammetry for analyzing trace metals in solution. This scenario has been expanded by the increasing development of photoelectrochemical techniques, chemiresistive, and impedimetric techniques currently submitted to intensive research.

It should be noted that electrochemical methods to determine a given analyte can be affected by interfering species (whose analytical signal is superimposed or can distort the analyte signal) and matrix effects (species disturbing the signal from the analyte by complexation, adsorption, etc.). The search for highly sensitive and highly selective sensors reluctant to undergo interference and matrix effects focuses most of the work in electroanalytical chemistry.

Social demands (miniaturization for, for instance, monitoring processes *in vivo*) involve improved sensing methods accomplishing so-called analytical properties (accuracy and precision, repeatability and reproducibility, sensibility, selectivity, traceability) but also short response time and long-term stability (suitability), economy (reusability, low power consumption), robustness, safety (for operators and for environment), etc. The development of environmentally friendly sensors is receiving considerable attention.

The first part of this chapter will be devoted to conventional electrochemical sensing of analytes in liquid solution. In the second part, gas sensing—a field where diverse porous materials play an essential role—will be treated.

12.2 POTENTIOMETRIC SENSING

pH measurement is the potentiometric sensing *per excellence*. In this field, casting metals consisting of Sb or Bi covered by a thin hydroxide layer and membranes of transition metal oxide (the so-called metal bronzes in particular) show a relatively high selectivity for hydrogen ions [2]. The OCP of conducting oxides is determined by a redox equilibrium established between the oxide surface and the solution [3]:

$$\{MO_x\}_{surf} + 2\delta H^+_{aq} + 2\delta e^- \rightarrow \{MO_{x-\delta}\}_{surf} + \delta H_2O \tag{12.1}$$

The equilibrium potential (zero current) is reached after an initial transient response characterized by an exchange current. A larger surface area leads to a higher exchange current and, consequently, fast response.

Electrochemical reactions for tungsten and molybdenum bronze electrodes can be represented as:

$$\{Na_xWO_3\}_{surf} + yH^+_{aq} + ye^- \rightarrow \{Na_xH_yWO_3\}_{surf} \tag{12.2}$$

$$\{A_xMo_zO_w\}_{surf} + yH^+_{aq} + ye^- \rightarrow \{A_xH_yMo_zO_w\}_{surf} \tag{12.3}$$

$$\{A_xMo_zO_w\}_{surf} + yH^+_{aq} \rightarrow \{A_{x-y}H_yMo_zO_w\}_{surf} + yA^+_{aq} \tag{12.4}$$

Ionophore-based ion-selective electrodes (IES) have been extensively studied over the last decades for selectively determining the concentration of selected ions in solution. Their practical use, however, has been limited to a relatively narrow range of fields because of the limited concentration interval able to be determined with such devices. In general, Nernstian responses, characterized by a linear variation of the OCP on the pH with a slope of $59/n$ mV decade^{-1}, are desired. Much metal/metal oxide electrodes display non-Nernstian responses, but applications in food control and medicine have been proposed. Lowering detection limits can be obtained, among other strategies, through solid-contact ISEs, prepared by direct contact of an ionophore-doped polymeric membrane with a metallic conductor. To avoid problems associated with poorly defined metal/membrane interfaces and unstable phase boundary potentials, intermediate layers with conducting polymers or localized redox-active units separating the sensing membrane and the metallic conductor can be used. More recently, ordered macroporous carbon contacts and other carbons have been tested. The used carbons consist of a skeleton of glassy carbon surrounding a periodic array of uniform spherical pores that are interconnected in three dimensions. Pore sizes are of a few hundred nanometers while skeletal walls are tens of nanometers thick. As a result of the mutually interconnected pore-wall structure of such materials, filling pores with an electrolyte solution results in a nanostructured material with both ionic and electronic conductivity [4].

Solid-state reference electrodes for potentiometric sensors are currently under research. The main problem faced in developing this kind of electrode is connecting the ionic conducting (usually aqueous) solution with an electronic conductor. Since the reference electrode has to maintain a defined potential, the electrochemical reaction with components of the electrolyte must be avoided. Oxides, mixed oxides, and polyoxometalate salts of transition elements can be proposed in preparing solid-state reference electrodes. Tested compounds include tungsten oxide and molybdenum and tungsten oxides [5].

12.3 VOLTAMMETRIC AND AMPEROMETRIC SENSING

Voltammetric and amperometric sensing of selected species in electrolyte solutions is widely extended with an enormous variety of types of electrode modification. Research in this field is oriented towards the achievement of large sensitivity, low detection limit, and high selectivity. Direct sensing involves the measurement of the signal for the oxidation/reduction of the analyte. Then, the above demands require the combination of large electrocatalytic effects accompanied by size, charge, etc. and heightened selectivity, both imposed by the electrode modifier.

In the last two decades, a plethora of voltammetric and amperometric sensors have been reported. These are particularly concentrated in several analytes—glucose, hydrogen peroxide, dopamine, hydrazine, hydroquinone, and catechol—and consist of more or less sophisticated composites including metal nanoparticles, porous carbons, CNTs, graphene, metal oxide nanoarchitectures, porous polymers, etc. Molecularly imprinted polymers are increasingly used as promoters of highly selective responses.

Much of recent literature, however, overemphasizes the catalytic nature of the reported sensors and generically characterizes its electron transfer properties based on CV and EIS data obtained for a redox probe, typically a $Fe(CN)_6^{3-}/Fe(CN)_6^{4-}$ aqueous solution in KCl electrolyte. These practices are, to some extent, problematic because:

a. As discussed in Section 2.3 [6], there is an apparent electrocatalytic effect at porous electrodes due to an increase of the residence time of electroactive species on the electrode [3,7,8]. Ultimately, electrocatalysis can be viewed as implying redox mediation with a catalyst whose oxidation state changes reversibly during the process.
b. The use of the $Fe(CN)_6^{3-}/Fe(CN)_6^{4-}$ (or another electrochemically reversible) redox probe is not particularly relevant if it is not accompanied by a parallel study of the analyte. As described in Section 10.5, the performance of a given electrode (in terms of the rate of electron transfer, typically expressed in terms of the charge transfer resistance, R_{ct}, determined in EIS experiments with the redox probe) may vary substantially from one electroactive species to another [9,10]. This means that a fast electron transfer for the redox probe does not necessarily imply a fast electron transfer (given by the value of the heterogeneous rate constant, Eq. (1.21)) for the analyte under investigation. The logical practice would be to study, in each case, the EIS response of the corresponding analyte.
c. Frequently, the cathodic-to-anodic peak potential separation in the CV of reversible or essentially reversible redox probes is used as a marker for the more or less faster electron transfer at modified electrodes. Since this separation is also influenced by uncompensated ohmic drops in the cell, as described in Chapter 1, there is a need to separate such effects from the 'true' influence of the electrode modifier.
d. Frequently, the (apparent) electroactive area of the modified electrode is determined from the peak current recorded in CV experiments at the reversible redox probe (whose diffusion coefficient must be known) by applying the Randles–Ševčik equation (Eq. (1.6)). The enhancement of the absolute electroactive area relative to unmodified electrodes cannot be taken as a direct indicator of electrocatalysis and does not imply a 'true' increase in the rate of electron transfer. The R_{ct} values should be corrected, as is customary for all impedances in EIS, by the electrode area so that resistive elements are expressed in $\Omega\ cm^2$ and capacitances in $F\ cm^{-2}$ [11].

In summary, the porous structure of many electrode modifiers may provide a significant increase in the effective area and an augment in the residence time of the analytes on the electrode surface by confinement effect. This means that geometric confinement may contribute to the signal enhancement without a 'true' catalytic effect (i.e. modifying the electrochemical rate constant). This effect is larger for slow redox reactions [12,13].

Figure 12.1 shows an example of the voltammetric response of sensing systems. The modified electrodes were constituted by hemoglobin (Hb) adsorbed in mesoporous molecular sieves (MSU), prepared as layer-by-layer films alternating with poly(diallyldimethylammonium) (PDDA). MSU was prepared from the precursor of zeolite Y using ionic liquids as templates. Layer-by-layer films labeled as {MSU/Hb}$_n$/PDDA—n being the number of layers—were deposited over glassy carbon electrode by alternative adsorption of positively and negatively charged species from their solutions [14]. CVs at the bare glassy carbon electrode and at the modified electrode in contact with blank and H_2O_2-containing solutions buffered at pH 6.9 are superimposed in Figure 12.1. The essentially reversible hemoglobin couple

$$\{HbFe^{III}\} + H^+_{aq} + e^- \rightarrow \{HbFe^{II}\} \tag{12.5}$$

observed at {MSU/Hb}$_5$/PDDA-modified electrode in the blank solution, is replaced by an irreversible, significantly more intense cathodic wave in the presence of H_2O_2. The weakening of the anodic peak clearly suggests that there is a reaction between the electrochemically generated $HbFe^{II}$ with H_2O_2 following the classical scheme for electrocatalysis described in Chapter 4.

The modified electrode was used as an amperometric sensor for H_2O_2. A typical amperometric test is shown in Figure 12.2. The staircase current/time variation depicted in this figure corresponds to successive injections of H_2O_2 into the electrochemical cell. The height of the different steps remains essentially constant, denoting that no significant memory effects appear under the reported experimental conditions.

This is an example of a multilayer modification strategy that can be monitored using EIS. Impedance spectra for electrodes with an increasing number of layers are shown in Figure 12.3 using a $K_4Fe(CN)_6$ solution as a redox probe. The Nyquist plots recorded at the formal potential of

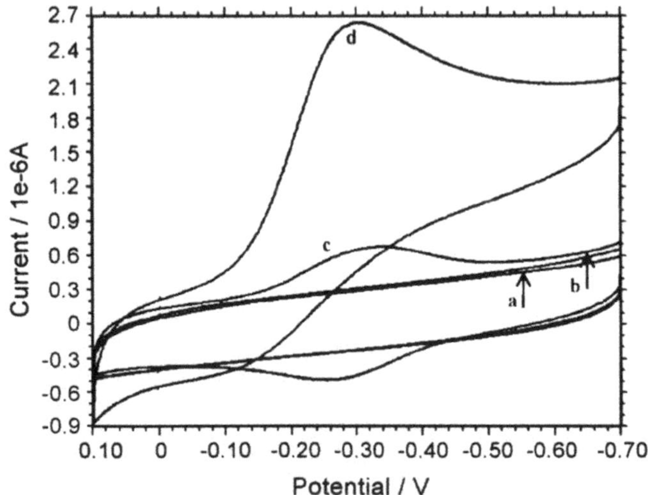

FIGURE 12.1 CVs at: (a,b) Bare Glassy Carbon Electrode; (c,d) {MSU/Hb}$_5$/PDDA for (a,c) a Blank 0.2 M Aqueous Phosphate Buffer (pH 6.9) and, (b,d) 100 μM H_2O_2 Solution in the Above Electrolyte. Potential Scan Rate 100 mV s^{-1}. Reproduced from Ref. [14] (Sun et al. *Talanta*. 2008, 74: 1692–1698), with Permission.

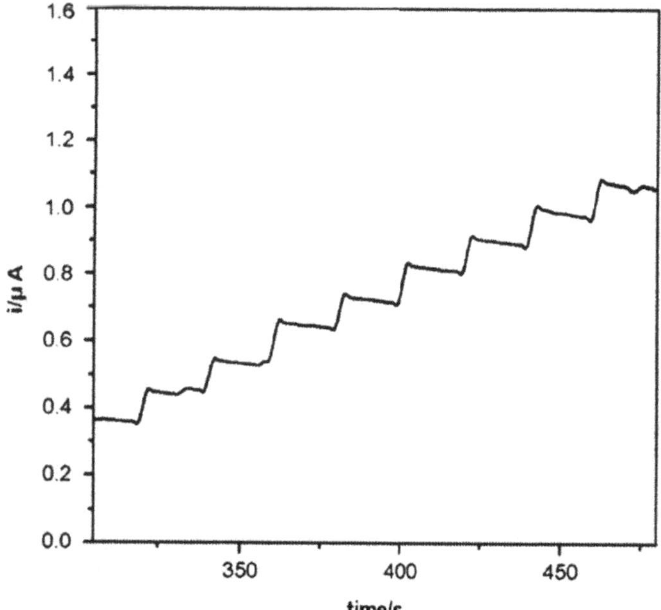

FIGURE 12.2 Amperograms at a Glassy Carbon Electrode Modified with Hemoglobin Adsorbed in Mesoporous Molecular Sieves {MSU/Hb}$_5$/PDDA Upon Successive Additions of 10 µM H_2O_2 to 0.2 M Potassium Phosphate Buffer Solution at pH 6.9. Applied Potential: –0.30 V vs. SCE. Reproduced from Ref. [14] (Sun et al. *Talanta*. 2008, 74: 1692–1698), with Permission.

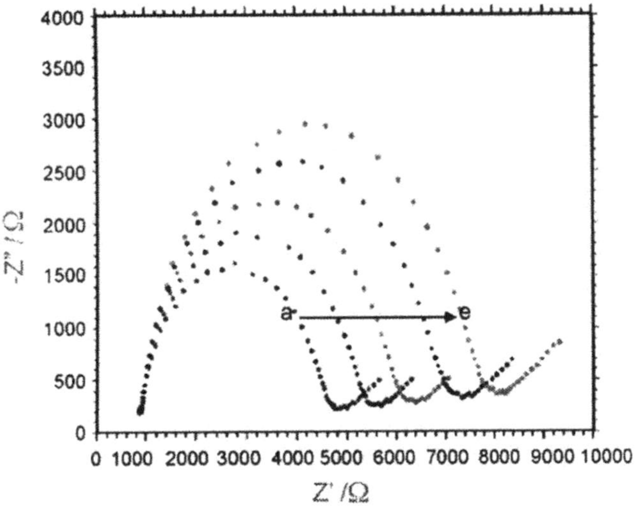

FIGURE 12.3 EIS of Different {MSU/Hb}$_n$/PDDA Electrodes (n varying from 1 to 5) in 5.0 mM Fe $(CN)_6^{3-}/Fe(CN)_6^{4-}$ plus 0.10 M KCl Under Application of a Potential of –0.295 V vs. SCE. Adapted from Ref. [14] (Sun et al. *Talanta*. 2008, 74: 1692–1698), with Permission.

the $Fe(CN)_6^{3-}/Fe(CN)_6^{4-}$ couple (–0.295 V) exhibit a capacitive loop accompanied by a short linear region at low frequencies. This kind of spectra are frequently described assuming that the diameter of the semicircle is representative of the charge transfer resistance (i.e. of the charge transfer rate for the redox probe) but obviously, experimental data must be fitted to a suitable equivalent circuit to properly extract definite conclusions. The loop diameter increases by increasing the number of layers, as expected if the charge-transfer rate for the redox probe is gradually reduced because of the deposition of successive insulating layers on the electrode surface. This means that the optimal sensing conditions must be decided from the compromise between the deposition of stable layers and their insulating effect.

Figure 12.4 depicts two of the basic schemes applied in voltammetric and amperometric sensing of analytes in solution mediated by transduction units immobilized onto electrode surfaces. In the direct approach, (a) the target species binds to the receptor unit of the electrode modifier thus modifying the response of the transductor unit to potential inputs. In the indirect approach, (b) the transducer interacts with a second transducer unit attached to the electrode surface.

A myriad of composite materials is being reported for voltammetric and amperometric sensing. The aim is to attain synergistic combinations of the different components to enhance sensitivity and selectivity. An example relative to gas sensing will be seen in the next chapter, Section 13.7. In most cases, however, there is no detailed knowledge of such effects.

Most applications include electrochemical detection coupled with high-performance liquid chromatography (HPLC) and flux-injection analysis (FIA). In these cases, it generally demands high sensitivity, repeatability, and reproducibility. These properties are conditioned by electrode fouling associated to the formation of solid deposits and adsorbates on the electrode surface. As a result, the electrode experiences "memory effects" with concomitant loss of analytical performance. Correction of such memory effects can often be made by applying electrochemical pre- and post-treatments. These treatments, consisting of the application of successive potential steps, can regenerate the electrode surface before/after measuring.

12.4 SELECTIVITY

As previously noted, potentiodynamic, voltammetric sensing involves the measurement of the current response of the working electrode under no (necessarily) mass-transport limiting

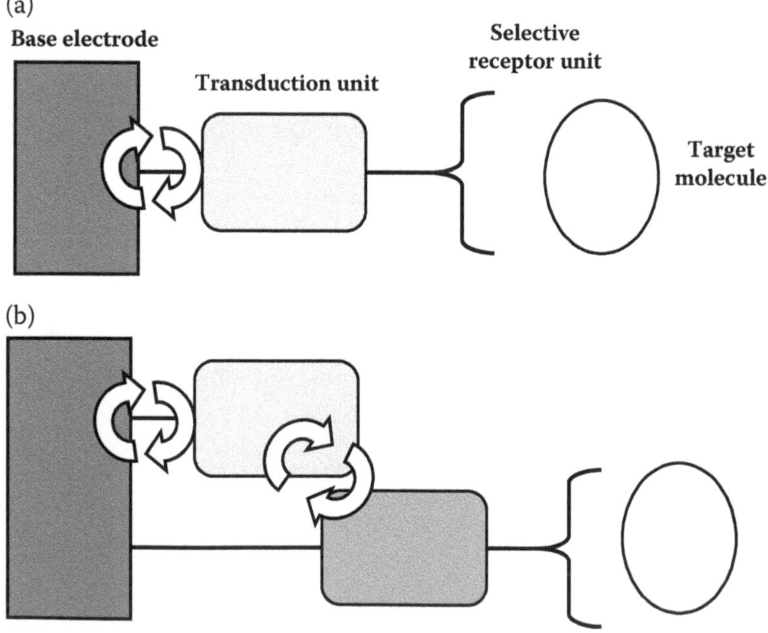

FIGURE 12.4 Scheme for (a) Direct and (b) Indirect Approaches to Detect a Target Analyte Using Transduction Units Immobilized onto Electrode Surfaces. The Arrows Indicate Interfacial Electron Transfer Processes.

conditions. Typical analytical strategies involve the concentration of the analyte in the electrode surface by the application of a reductive (or oxidative) potential step, followed by the application of an oxidative (or reductive) potential scan. In stripping voltammetry devoted to the determination of trace metals in solution, metal ions are reduced to a metal deposit (or an amalgam at Hg electrodes), and their concentration is monitored by measuring the peak current in the oxidative dissolution process recorded upon application of a scan in the positive direction of potentials [15].

Porous materials, used as electrode modifiers, can act as preconcentrating agents to increase the effective concentration of the analyte in the vicinity of the electrode. Large pore size, uniform pore structure, and high loading capacity are general demands for porous materials used for sensing purposes regarding sensitivity. However, for most analytical demands, large sensitivity must also be accompanied by large selectivity.

Voltammetric methods are of particular interest with the selective determination of analytes in the presence of interferents. As a widely studied case, one can mention the determination of dopamine and other neurotransmitter catecholamines in the presence of interfering compounds, namely ascorbic acid and/or uric acid. It is known that dopamine exerts a significant physiological role as an extracellular chemical messenger whose loss in neurons can be associated with serious diseases such as Parkinsonism. Consequently, its determination *in vitro* and *in vivo* is an obvious target in neurochemical studies and has motivated the report of a plethora of voltammetric sensors.

Electrochemical determination of dopamine is made possible by virtue of its well-defined oxidation signal at ca. 0.2 V vs. SCE in aqueous media at physiological pH values. Unfortunately, ascorbic acid accompanies dopamine in large concentration biological samples and is electrochemically oxidized at similar potentials [16,17]. Accordingly, dopamine determination in biological samples necessarily implies high selectivity relative to ascorbic acid (and often other oxidizable components, uric acid, norepinephrine, serotonin, etc.).

Since at physiological pH dopamine is in its ammonium, cationic form ($pK_a = 10.6$) and ascorbic acid is in the form of ascorbate anion ($pK_a = 4.2$), one way to achieve selectivity is by charge-discriminating electrode modifiers. Among other possibilities, zeolites can fulfill appropriate selective conditions because of their recognized affinity for cationic species [18]. This strategy for electrochemically determining dopamine in the presence of excess amounts of ascorbic acid can be reinforced by incorporating catalytic centers into the zeolite support. This is the case with polymer film over glassy carbon electrodes modified with 2,4,6-triphenylpyrylium ion encapsulated into zeolite Y [19].

Here, a variety of molecular recognition strategies based on the functionalization of the electrode modified with molecular groups selectively binding to the analyte can be devised. Additional strategies involve the promotion of shape and size selectivity using inclusion complexes (with β-cyclodextrin and related compounds) [20] or molecularly imprinted polymers (MIPs) [21]. MIPs are prepared from a synthetic polymer matrix forming cavities sculpted around the analyte molecule that acts as a template. Then, the template is removed, and the resulting polymer matrix is used as an electrode modifier able to selectively recognize the target molecule.

An issue to be considered is that the electrochemical oxidation of these analytes involves multi-step pathways. The electrochemical oxidation of dopamine is an apparently irreversible process, due to the occurrence of post-electron transfer reactions. The initial two-proton, two-electron product, dopaminequinone, can undergo a deprotonation accompanied by fast 1,4 (Michael) addition to yield a bicyclic derivative, leucoaminochrome, further oxidized to its quinonic form, aminechrome, either electrochemically or chemically by reaction of dopaminequinone [16,17], as schematized in Figure 12.5 [19]. As a result, the apparent number of electrons consumed during the electrochemical oxidation of dopamine varies from two to four, depending on the time scale of observation. Additionally, electrochemical runs can be accompanied by polymerization processes producing melanin, a process that results in electrode fouling. In turn, ascorbic acid is electrochemically oxidized to a diketolactone, which is rapidly dehydrated to dehydroascorbic acid. This rearranges to another ene-diol, which is further oxidized at higher potentials [22,23].

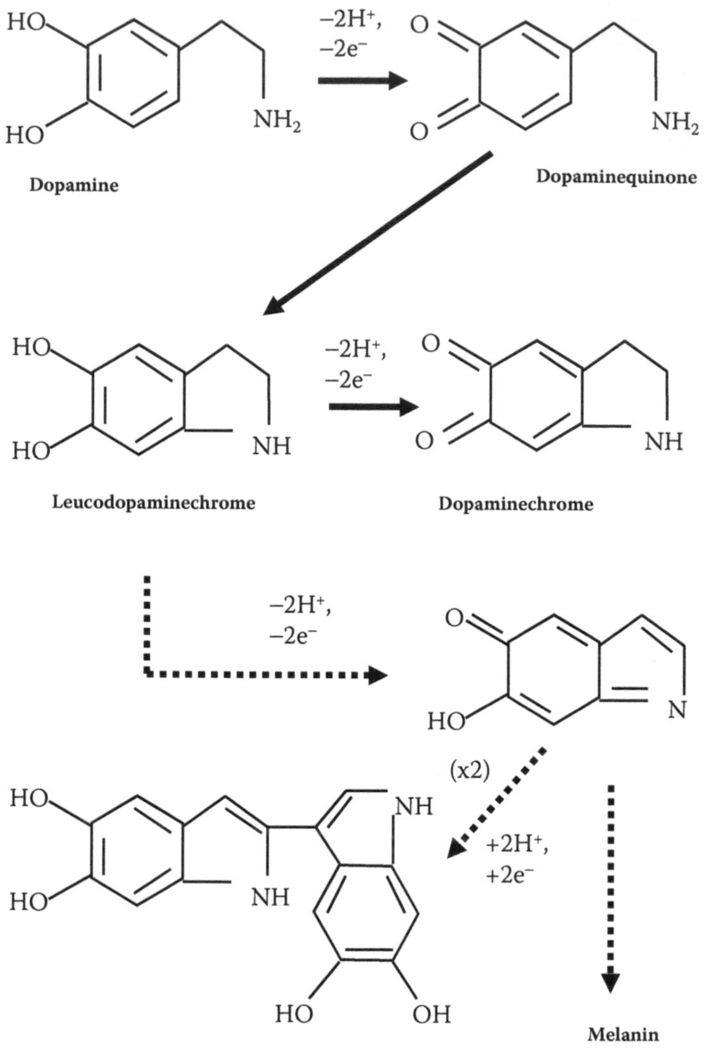

FIGURE 12.5 Simplified scheme for the electrochemical oxidation of dopamine in aqueous media.

The knowledge of these electrochemical pathways, although often ignored in the literature, is of interest in sensing. In the case of dopamine, it is desirable to increase the rate of the initial two-proton, two-electron electron transfer process but also block the subsequent cyclization reaction. The former proceeds via stepped proton-electron transfers, presumably as ECEC or CECE sequences, and the stability of intermediate forms may be of importance to control the rate of the overall process. The second issue increases the apparent reversibility of the electrochemical process and prevents interfering side reactions.

This matter refers to another aspect widely extended in recent literature: the avoidance of a detailed study of the electrochemical pathway involved in sensing. Indeed, it is frequent to find dopamine oxidation (or other multiple-step electrochemical processes such as the oxidation of ascorbic acid and uric acid, the quinone/hydroquinone interconversions, etc.). Frequently, these processes are treated—based on peak current vs. square root of potential scan rate linear representations in a narrow range of sweep rates—as diffusion-controlled processes. In other cases, there is a linear dependence between peak currents and scan rates, and the processes are described as adsorption-controlled. These are approximations valid for analytical purposes under most circumstances, but they must be handled with caution.

Notice that the variation of peak currents with potential scan rate often appears as varying with this last quantity. This feature is generically attributed to a mixed diffusion-adsorption control, but the peculiarities of porous electrodes described in Sections 2.4 and 10.5 should be accounted for. A detailed knowledge of the electrochemical mechanism, including—if existing—the catalytic pathway, is desirable for a proper understanding—and optimization—of the sensing process. In voltammetric and especially in amperometric sensing, calibration graphs divided into different rectilinear sections are reported. The reasons for this behavior are not always discussed and, possibly in most cases, nonlinear calibration curves were pertinent.

12.5 ENANTIOSELECTIVE ELECTROCHEMICAL SENSING

The possibility of chiral electrochemical sensing is another of the interesting analytical applications of the studied materials. In the last decades, a variety of chiral materials with electrochemical activity have been reported [24], including chiral CuO films deposited onto single-crystal Au [25], and epitaxial electrodeposition of ZnO on Au(111) [26].

Chiral polymers can be prepared, as in the case of PANI formation via electropolymerization of aniline in aqueous solution containing (1S)-(+)- or (1R)-(−)-10-camphorsulfonic acid [27]. Formation of inclusion complexes [28] and MIPs [29] are frequent analytical strategies, but there is a possibility of using chiral electroactive doping centers in solids. This is the case with vanadium-doped monoclinic zirconias, where the enantioselective catalytic effect can be attributed to the coordinative arrangement of metal centers. In tetragonal zirconias, each M^{4+} ion (M = Zr, V) is eight-coordinated so that the coordination polyhedra is centrosymmetric and no chiral electrocatalytic effects appear. In monoclinic zirconias, M^{4+} ions possess a non-centrosymmetric seven-coordinated arrangement and there are enantioselective electrocatalytic effects [30].

In most cases, chiral selectivity results in the different effective diffusion coefficient of the electroactive enantiomers. Then, both enantiomers at equal concentrations produce a voltammetric signal at identical potential differing in peak intensity. In this case, the elucidation of the enantiomeric composition (i.e. the ratio between the concentrations of the enantiomers) can only be determined if the total concentration is known.

12.6 CHEMIRESISTIVE AND IMPEDIMETRIC SENSING

Chemiresistive sensing is based on the measurement of changes in the ohmic resistance of a sensor put into contact with a given analyte. These sensors are largely concentrated on gas sensing and will be specifically treated in the next chapter.

Impedimetric (or impedance metric) sensors follow a similar scheme but here, impedance measurements upon application of an alternating potential input are carried out. These sensors, therefore, use the EIS technique as a guide. The most general approach involves the immobilization of a sensing membrane onto a base electrode and the record of the impedance spectra when the system is in contact with the solution containing the analyte [31].

Figure 12.6 schematizes a widely used sensing configuration (so-called Faradaic approach [32]) based on the disposal of a porous sensing membrane onto the electrode surface in contact with a solution of the redox probe (typically an aqueous solution of $K_4Fe(CN)_6/K_3Fe(CN)_6$). The sensing membrane is covered by a porous layer that incorporates a receptor group for the analyte. In the absence of this compound, the membrane pores are open, and the redox probe reaches the base electrode. In the presence of the analyte, its binding to the receptor groups blocks the pores and hinders the access of the redox probe to the electrode surface. In terms of a Randles equivalent circuit, the second situation gives rise to an increase of the charge transfer resistance in EIS measurements.

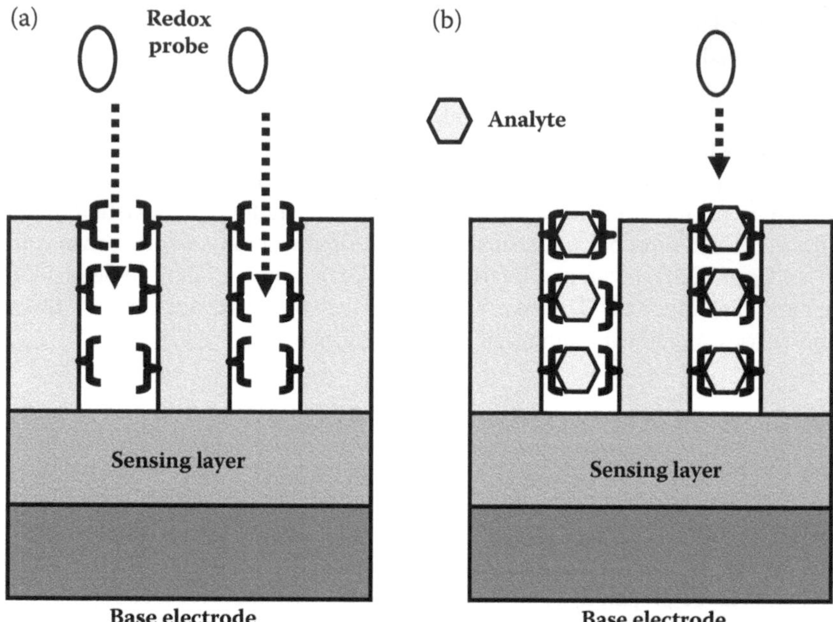

FIGURE 12.6 Typical Configuration of Impedimetric Sensing at Porous Functionalized Electrodes. In the Absence of Analyte (a), the Redox Probe Can Access to the Sensing Layer Deposited Over the Base Electrode. In the Presence of the Analyte (b), the Access of the Redox Probe to the Sensing Layer is Blocked.

A second approach can be made. Here, there is no redox probe and EIS measurements reflect the different impedance responses of a sensing membrane when the analyte is attached to that membrane in the absence of the analyte. The capacitive approach does not need a redox probe in solution, but an electrode tethered one. This is schematized in Figure 12.7 where the basic equivalent circuit used to model this so-called capacitive approach [32] is depicted.

12.7 ELECTROCHEMOLUMINISCENCE AND PHOTOELECTROCHEMICAL SENSING

Electrochemiluminescence (ECL) is a phenomenon consisting of the generation of light as a result of the application of electrochemical inputs. As an analytical technique, it is characterized by its high sensitivity. ECL requires the presence of a molecular species (the luminophore) that can be electrochemically promoted to an excited state that it emits light during the subsequent return to the ground state [33,34]. Typical luminophores utilized in ECL studies are luminol, Ru(II) complexes, quantum dots (QDs), and metal nanoclusters, among others. The luminophore can be directly activated via electrochemistry or by mediation of a redox-active species. This is the case with luminol, which is activated by reactive oxygen species (ROS) generated electrochemically.

Figure 12.8 shows LSVs (a) and electrochemiluminiscent intensity-potential curves (b) of 0.1 mM luminol at a GCE 0.1 M phosphate buffer solution at pH 10.5 in the absence (blue line) and presence (red line) of 10 mM tripropylamine [35]. The voltammogram of luminol displays a main anodic peak that can be attributed to the oxidation of luminol anion (LH^-) to the corresponding radical anion ($L\cdot^-$), subsequently given diazaquinone (L). In the presence of tripropylamine (TPrA), a second anodic peak appears due to the oxidation of this compound to the corresponding radical cation ($TPrA\cdot^+$). The electrochemiluminiscent response was detected by placing the optical window of the electrochemical cell to face the photomultiplier tube that is responsive to photons in the UV/VIS

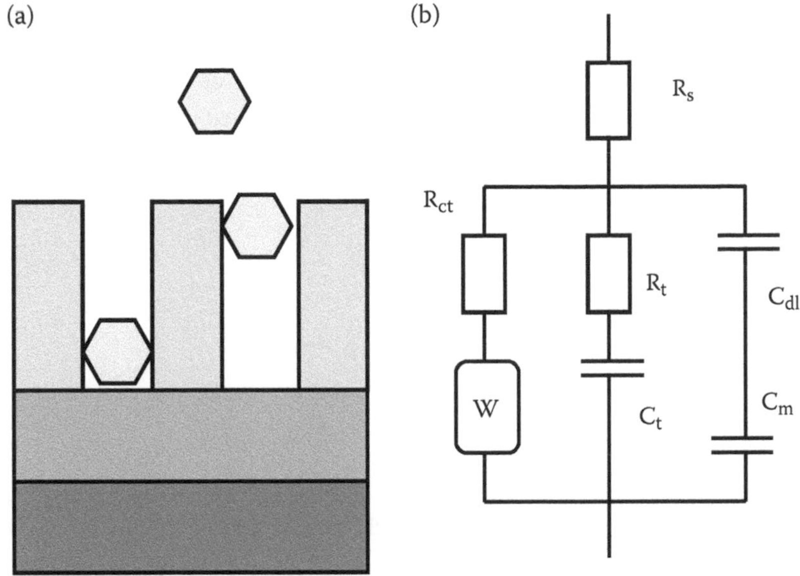

FIGURE 12.7 (a) Schematics and (b) Basic Equivalent Circuit Used to Model Capacitive Impedimetric Sensing Where a Porous Membrane is More or Less Extensively Loaded by the Target Analyte. R_s is the Solution Resistance, C_{dl} the Double Layer Capacitance, C_m the Membrane Capacitance, and R_{ct} the Charge Transfer Resistance. R_t and C_t Represent the Resistive and Capacitive Contributions of the Sensing, Analyte-Loaded, Porous Layer Whose Variations Will be Representative of the Analyte Concentration in the Solution.

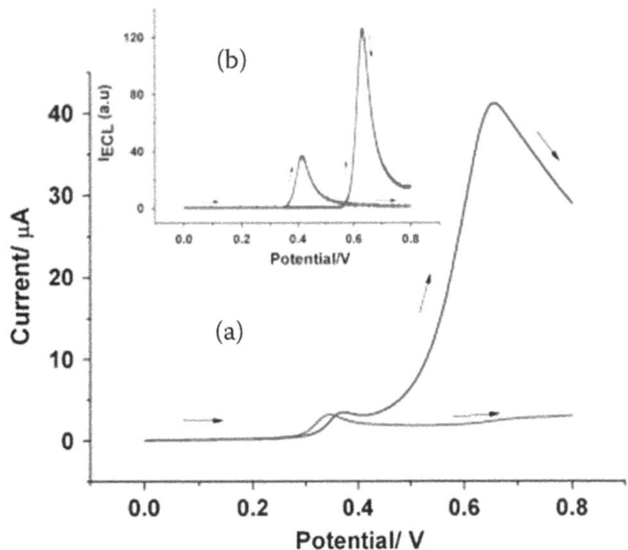

FIGURE 12.8 LSVs (a) and ECL Intensity-Potential Curves (b) of 0.1 mM Luminol at a GCE in 0.1 M Phosphate Buffer Solution at pH 10.5 in the Absence (Blue Line) and Presence (Red Line) of 10 mM TPrA. Potential Scan Rate: 10 mV s^{-1}. PMT Was Set at 700 V. Reproduced from Ref. [35] (Hanif et al. *Electrochim. Acta.* 2016, 196: 245–251), with Permission.

region. The luminol plus TPrA solution produces a strong chemiluminiscent signal whose maximum emission is a 425 nm, corresponding to the characteristic emission of 3-aminophthalate anion (AP^{2-}). The electrochemiluminiscent mechanism can be described in terms of the oxidation of the luminol radical anion by $TPrA\cdot^{+}$ to excited 3-aminophthalate, which emits light when it decays to its ground state. These processes can be represented as:

$$LH^{-} \rightarrow L\cdot^{-} + H^{+} + e^{-} \quad (12.6)$$

$$TPrA \rightarrow TPrA\cdot^{+} + e^{-} \quad (12.7)$$

$$L\cdot^{-} + TPrA\cdot^{+} \rightarrow (AP^{2-})^{*} \quad (12.8)$$

$$(AP^{2-})^{*} \rightarrow AP^{2-} + h\upsilon \quad (12.9)$$

The electrochemiluminiscent effect can be enhanced using PANI/mesoporous carbon and PANI/ graphene composites, among others [36]. Here, the classic electrochemiluminescence pathway involves the oxidation of luminol radical anion by dissolved O_2 with subsequent formation of luminol endoperoxide (LO_2^{2-}) with the generated superoxide radical anion that gives 3-aminophthalate in its excited state:

$$L\cdot^{-} + O_2 \rightarrow O_2\cdot^{-} + L \quad (12.10)$$

$$L\cdot^{-} + O_2\cdot^{-} \rightarrow LO_2^{2-} \quad (12.11)$$

$$LO_2^{2-} \rightarrow (AP^{2-})^{*} + N_2 \quad (12.12)$$

Photoelectrochemical sensing can be seen as the reverse of electrochemiluminiscence analysis. In photoelectrochemical experiments, the current (photocurrent) generated by the analyte, or a sensing species interacting with the analyte after previous light excitation, is detected. The essential idea is that photoelectric species in excited states exchange electrons with the analyte, leading to the change of photocurrent [37].

There are two basic mechanisms of photocurrent generation: (a) the analyte is excited and, in this state, is reduced by a reducing agent in solution; or (b) the excited molecules interchange electrons with the quenching agent so that the resulting oxidized or reduced species obtain or lose electrons from the electrode surface and generate photocurrent. The magnitude of photocurrent depends on the wavelength and intensity of exciting light, the electrode properties, the applied bias voltage, and the composition of electrolytes [38].

Figure 12.9 shows the photocurrent responses of FTO electrodes modified with TiO_2 nanowires, and TiO_2 nanowires with 3-aminopropyltriethoxysilane and 3-aminopropyltriethoxysilane plus N3 dye, bis (4,40-dicarboxy-2,20-bipyridine) dithiocyanato ruthenium (II) before (c) and after (d) incubation with 1 mL of 50 mM Hg^{2+} solution in phosphate buffer at biological pH [39]. 3-Aminopropyltriethoxysilane was used as an anchoring unit to attach the N3 dye to TiO_2 nanowires. Under visible light illumination, TiO_2 nanowires, as well as TiO_2 nanowires plus 3-aminopropyltriethoxysilane, do not produce any significant photocurrent response, as expected due to the large bandgap of the oxide (ca. 3.2. eV). After covalent immobilization of N3 dye, the photocurrent increases dramatically due to the good photoelectric conversion properties of the dye in the visible light region. Upon contact with the Hg^{2+} solution, the photocurrent is reduced drastically. The response of the photosensitizer N3 is attenuated as a result of complexation of Hg^{2+} with the dye, providing a sensitive sensor for such species. Here, the

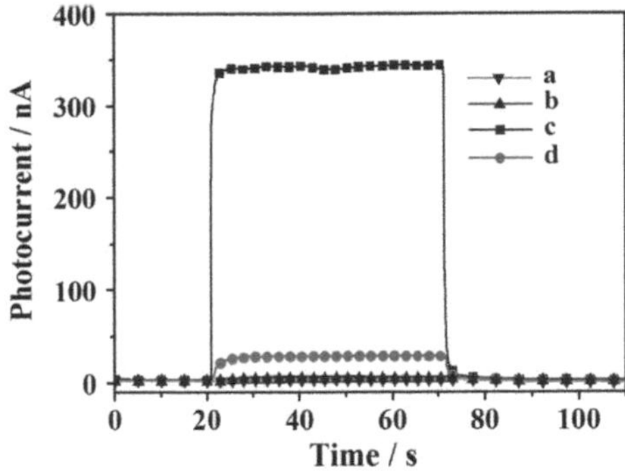

FIGURE 12.9 Photocurrent Responses of FTO Electrodes Modified with TiO_2 Nanowires (a), TiO_2 Nanowires with 3-Aminopropyltriethoxysilane (b) and 3-Aminopropyltriethoxysilane plus N3 Dye, bis(4,40-Dicarboxy-2,20-bipyridine) dithiocyanato ruthenium (II), Before (c) and After (d) Incubation with 1 mL of 50 mM Hg^{2+} Solution. The Measurements Were Performed in Phosphate Buffer Solution at pH 7.4 Under Visible Illumination (540 nm) and Applying a Bias Potential of 0.2 V. Reproduced from Ref. [39] (Hao et al. *Electrochim. Acta.* 2018, 259: 179–187), with Permission.

selectivity relative to other metal cations is provided by the strong complexation of Hg^{2+} ions with the N3 dye.

A variety of materials and architectures are being investigated for photoelectrochemical sensing, including porous anodic metal oxide films [40]. Information on photoelectrochemical activity of modified electrode surfaces can be obtained, apart from voltammetric and impedance techniques, by scanning photoelectrochemical microscopy [41].

12.8 BIOCHEMICAL SENSING

Electrochemical sensing of biochemical products has experienced accelerated growth in the last decades. In a wide sense, biochemical analysis comprises analytical methods that detect species of biological interest using biochemical compounds, tissues, and even cells.

Immobilization of enzymes onto electrode surfaces has been used to assay a range of low molecular weight-specific biomarkers. This methodology has been expanded by the attachment of proteins, antibodies, antibody fragments, or aptamers to the transducer surface [42,43]. Electrochemical immunosensing is highly developed [44] as well as photoelectrochemical immunosensing. Figure 12.10 shows as an example a LED visible-light-driven label-free photoelectrochemical immunosensor based on WO_3/Au/CdS photocatalyst for the sensitive detection of carcinoembryonic antigen.

Figure 12.10a,b shows the photocurrent response of ITO electrodes covered by WO_3, CdS, WO_3 with Au NPs, WO_3/CdS heterojunction, and WO_3/Au/CdS in contact with 0.1 M PBS at pH 7.0 under visible light illumination and a bias potential of 0 V [45]. The modest photocurrent obtained for WO_3 reflects the high hole-electron recombination rate in this semiconductor. The photocurrent is enhanced in the presence of Au NPs as well as CdS, being drastically increased for the ternary WO_3/Au/CdS system. This may be attributed to the well-matched energy levels of WO_3 and CdS heterojunction structure and (see Figure 12.11) the interaction between CdS and WO_3/Au photocatalyst.

Consistently, the ternary WO_3/Au/CdS system displayed lower photoluminescent emission than WO_3 and WO_3/Au, under the excitation of light of 325 nm (Figure 12.10c). The photoluminescent emission depends on the extent of recombination of photogenerated charges [46]. In turn, LSV of the modified electrodes show a clear increase of the WO_3/Au/CdS photocurrent under 430 nm illumination than that in the dark. This photocurrent is higher than WO_3 and WO_3/Au, in agreement with the effective separation of the photoexcited carriers in the ternary system [47]. These features can be associated with the LSPR effect of Au NPs and the heterostructure formation between WO_3 and CdS. The former can be interpreted in terms of three main mechanisms [48]: light scattering, hot-electrons injection, and plasmon resonance energy transfer. In this case, the third effect is dominating [45].

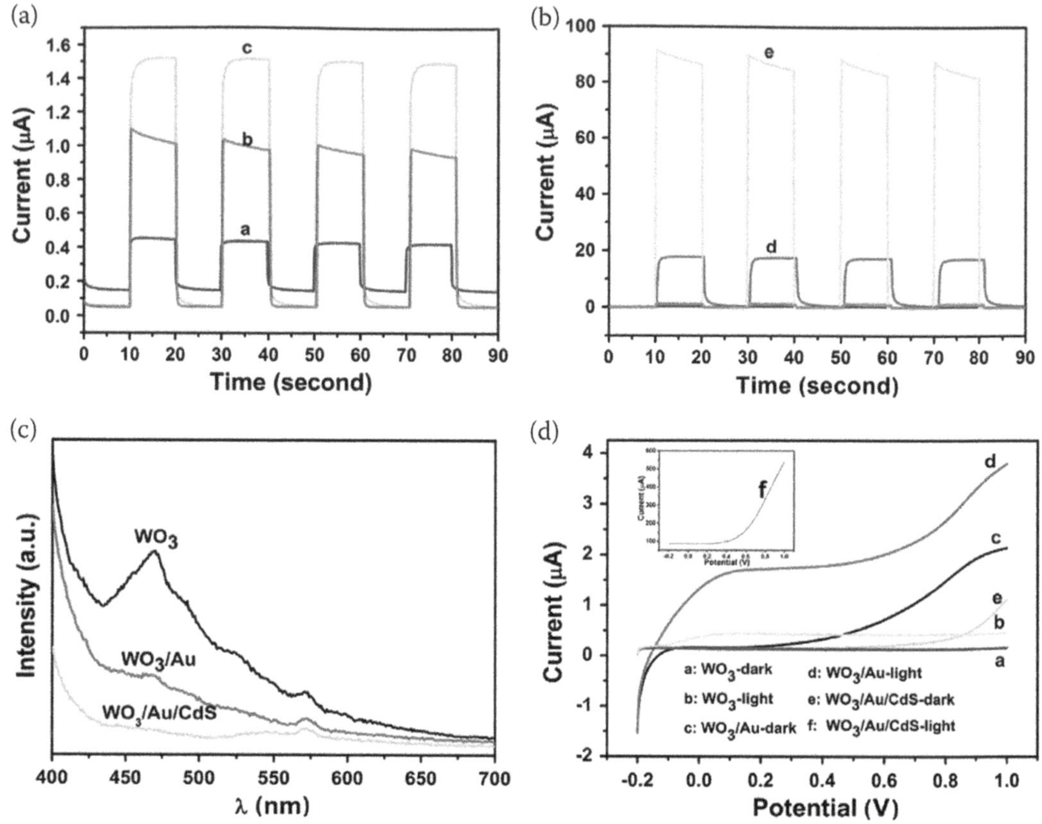

FIGURE 12.10 (a,b) Photocurrent Responses of ITO Electrodes Covered by (a) WO_3, (b) CdS, (c) WO_3/Au NPs, (d) WO_3/CdS, and (e) WO_3/Au/CdS Under Visible Light Illumination and a Bias Potential of 0 V in 0.1 M PBS at pH 7.0 Solution. (c) Photoluminescence (PL) Spectra of WO_3, WO_3/Au, and WO_3/Au/CdS Under the Excitation of 325 nm Light. (d) LSVs of the WO_3, WO_3/Au NPs, and WO_3/Au/CdS Electrodes in 0.1 M PBS at pH 7.0 Solution in the Dark and Under 430 nm Illumination. Reproduced from Ref. [45] (Zhang et al. *Electrochim. Acta.* 2019, 297: 372–380), with Permission.

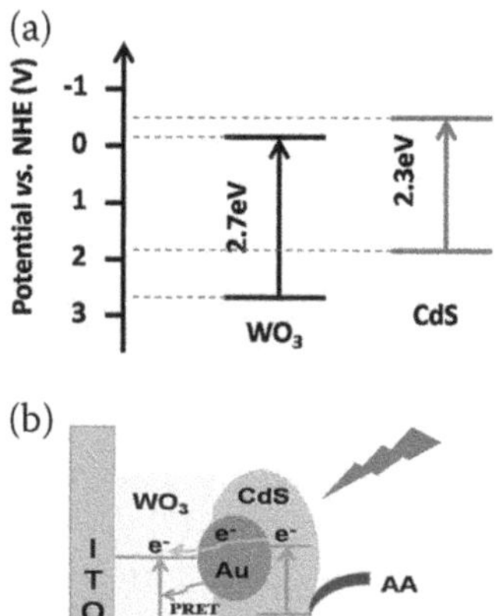

FIGURE 12.11 (a) Schematic Energy Level Diagram of WO_3 and CdS and (b) the Possible Charge Transfer Mechanism at WO_3/Au/CdS Interface in the Presence of Ascorbic Acid as Hole Sacrificial Agent. Reproduced from Ref. [45] (Zhang et al. *Electrochim. Acta.* 2019, 297: 372–380), with Permission.

Figure 12.11 shows the schematic energy level diagram of WO_3 and CdS and the possible charge transfer mechanism at WO_3/Au/CdS electrode in the presence of ascorbic acid as hole sacrificial agent. Under 430 nm light illumination, electrons are excited from the VB to the CB of CdS, and then transferred to CB of WO_3 through Au NPs that act as mediators. This produces an anodic photocurrent while the holes were scavenged by ascorbic acid. As a result of the efficient spatial separation of charge carriers, the photocurrent response of the ternary system was enhanced relative to the separate components.

After immobilization of anti-carcinoembryonic antigen, the described system operates as a sensitive immunosensor to determine the antigen, whose presence decreases the recorded photocurrent [45].

12.9 ELECTROCHEMICAL SWITCH

Modeling of electronic circuitry by means of photochemical and electrochemical systems has received attention in the two last decades. On first examination, a variety of electrochemical systems can be regarded as frequency filters, rectifiers, or amplifiers because of their operating properties parallel to those types of vacuum and semiconducting electronic devices. This is potentially interesting for transduction, sensing, and information storage [49].

In principle, all electrodes displaying electrocatalytic activity can be seen as signal amplifiers. A correlation between electrochemistry and electronics is obtained through field effects sensors and biosensors. In these devices, there is electrostatic modulation of charge carrier mobility in suitably prepared semiconductors. In gate electrodes, there is conductance modulation through the electrostatic environment of the semiconductor surface exposed to the target solution.

The use of a molecular switch in the solution phase has received attention recently [50]. Photochemical logic gates with shown AND, OR, XOR, NOR, and INH functionality have been reported, while electrochemical storage of information by means of electroswitchable systems [51], electrochemical transduction of photonically encoded information or electrochemically encoded information, and electrochemical/photochemical information processing are possible [52,53]. A simple example of a three-state chemical/electrochemical system can be approached from PANI attached to layered double hydroxides (LDH-PANI) using $Fe(CN)_6^{4-}$ redox probe. Here, the voltammetry of redox probe solutions at LDH-PANI-modified electrodes is modulated upon application of different potential inputs, the system operated as an ionic/electronic switch [54].

Figure 12.12a compares CVs for a 2.0 mM solution of $K_4Fe(CN)_6$ in phosphate buffer at (a) unmodified GCE and (b–d) LDH-PANI-modified electrode after conditioning it by applying different constant potential steps between one and five minutes. At unmodified electrodes, a well-defined quasi-reversible couple appears at peak potentials of 0.28 V (anodic) and 0.14 V (cathodic). When no conditioning potential is applied, to the modified electrode, anodic peaks at 0.28 and 0.60 V appear. The second peak corresponds to LDH-PANI oxidation while the first corresponds to the LDH-PANI-mediated oxidation of $Fe(CN)_6^{4-}$, as denoted by the increased cathodic-to-anodic peak potential separation and the significant decrease of the cathodic peak for the $Fe(CN)_6^{4-}$/$Fe(CN)_6^{3-}$ couple with respect to that recorded at bare GCE. If a conditioning potential of –1.25 V is applied, CVs exhibit only the second oxidation peak at 0.60 V. Finally, when a conditioning potential of 0.90 V is applied to pristine LDH-PANI electrodes, the subsequent CV initiated at 0.0 V in the anodic direction shows no oxidation nor reduction peaks.

This behavior suggests that LDH-PANI can be described in terms of a three-state system where they can be accessed by applying the adequate potential inputs from the parent emeraldine form. Under potentials ranging between 0.90 and –0.80 V in contact with aqueous phosphate buffer, the three states are stable (as denoted upon repetitive cycling the potential scan) although there is no entirely reversible switching between all states. Application of such potential inputs can lead to modeling two combinational logic functions, the electrochemical system acting as a switchable

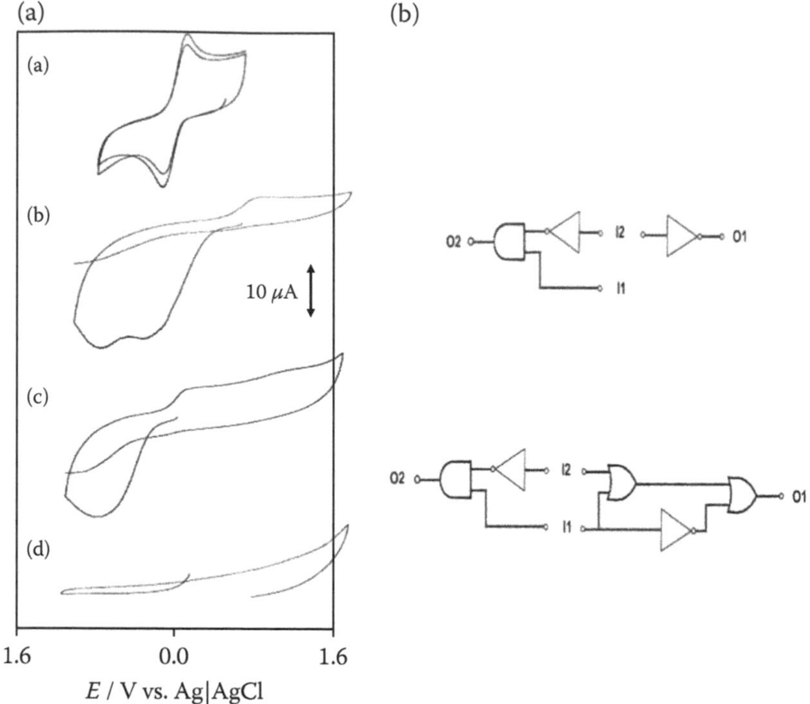

FIGURE 12.12 (a) CVs for a 2.0 mM $K_4Fe(CN)_6$ Solution in 0.50 M Phosphate Buffer at pH 7.0 at: (a) Unmodified GCE, and (b–d) LDH-PANI-Modified GCE After Application of a Constant Potential Step of: (b) 0.0, (c) –1.25, (d) 0.85 V During 2 min. Potential in the CVs Initiated at 0.0 V in the Positive Direction. Potential Scan Rate 100 mV s^{-1}. (b) Schematics for the LDH-PANI-Based Logic Circuits Acting in Modes: (a) I and (b) II (See Text). Adapted from Ref. [54] (Doménech-Carbó et al. *J. Electroanal. Chem.* 2008, 624: 275–286), with Permission.

to transduce one chemical and one electrochemical input (I1, I2) into two electrochemical outputs (O1, O2), corresponding to the oxidation peaks for LDH-PANI and $Fe(CN)_6^{4-}$.

Combinational logic circuits equivalent to the LDH-PANI switch are depicted in Figure 12.12b. A positive logic convention (absence = 0, presence = 1) is used for $Fe(CN)_6^{4-}$ input (I1), as customary. Two alternative encodings can be used for the electrochemical input (I2) by combining preconditioning potentials of (i) 0.0 (off = 0) and +1.0 V (on = 1) and (ii) 0.0 (off = 0) and –1.25 V (on = 1). The corresponding truth tables are shown in Table 12.1. Analysis of the logic behavior of the LDH -PANI switch in mode (i) reveals that the input I1 has no influence on the output O1 that is equivalent to a NOT connected to the input I2. I1 and I2 issues can be combined through NOT and AND gates (INHIBIT gate), producing the output O2. This situation is repeated in mode (ii), whereas the input O1 can be represented as a combination of one NOT and two OR gates.

12.10 PERSPECTIVES

The explosive growth of electrochemical sensing involves an enormous variety of methodologies. Research lines include the development of chemometric methods [55], the application of novel electrochemical techniques [56], and the consideration of new systems as analytes—in particular, solid-state [57] and non-linear systems. A research line of increasingly growing interest is the development of microfluidic devices [58].

TABLE 12.1
Truth Tables for the ZnAl-PANI Switch Operating in Modes I and II

	Mode	I			Mode	II	
I1	I2	O1	O2	I1	I2	O1	O2
0	0	1	0	0	0	1	0
0	1	0	0	0	1	1	0
1	0	1	1	1	0	1	1
1	1	0	0	1	1	1	0

Adapted from ref. [54] (Doménech-Carbó et al. *J. Electroanal. Chem.* 2008, 624: 275–286), with permission.

Future developments should increase the number of accessible analytes as well as disposable strategies. Apart from increased sensitivity and selectivity, a significant increase in robustness is demanded for practical purposes. Importantly, a major implementation of electroanalytical sensing in the practical biomedical, food, environmental, etc. fields is desirable. This is a challenge requiring fluid communication between electrochemistry teams and the biomedical, etc. world. Frequently, electrochemists test the suitability of their sensors by analyte-enriched samples on substrates such as urine, river water, etc. where originally, the analyte is absent or present below the detection limit of the method. This is, of course, a valid practice, but unexpected interference and matrix effects may appear in complex multicomponent systems. Probably, the effective implementation of electrochemical methods into the clinical, etc. practice would take benefit of studies based on 'real' real samples and case studies.

The improvement of the analytical capabilities of electrochemical sensors probably will require the combination of the advances in materials science via new synthetic routes (environmentally friendly green routes in particular) with a better understanding of fundamental detection processes. This last aspect implies to recover attention to molecular electrochemistry to improve the knowledge of the redox mechanisms (including synergistic effects in sensing at composite electrodes) as well as surface chemistry (notably expanded by the use of DFT calculations).

REFERENCES

[1]. Gründler, P. 2007. *Chemical Sensors.* Springer, Berlin-Heidelberg.

[2]. Vonau, W.; Guth, U. 2006. pH monitoring: A review. *Journal of Solid State Electrochemistry.* 10: 746–752.

[3]. Han, J.-H.; Lee, E.; Park, S.; Chang, R.; Chung, T.D. 2010. Effect of nanoporous structure on enhanced electrochemical reaction. *Journal of Physical Chemistry C.* 114: 9546–9553.

[4]. Lai, C.-Z.; Joyer, M.M.; Fierke, M.A.; Petkovich, N.D.; Stein, A.; Bühlmann, P. 2009. Subnanomolar detection limit application of ion-selective electrodes with three-dimensionally ordered macroiporous (3DOM) carbon solid contacts. *Journal of Solid State Electrochemistry.* 13: 123–128.

[5]. Guth, U.; Gerlach, F.; Decker, M.; Oelssner, W.; Vonau, W. 2009. Solid-state reference electrodes for potentiometric sensors. *Journal of Solid State Electrochemistry.* 13: 27–39.

[6]. Ward, K.R.; Compton, R.G. 2014. Quantifying the apparent 'catalytic' effect of porous electrode surfaces. *Journal of Electroanalytical Chemistry.* 724: 43–47.

[7]. Punckt, C.; Pope, M.A.; Liu, J.; Lin, Y.H.; Aksay, I.A. 2010. Electrochemical performance of graphene as effected by electrode porosity and graphene functionalization. *Electroanalysis.* 22: 2834–2841.

[8]. Punkt, C.; Pope, M.A.; Aksay, I.A. 2013. On the electrochemical response of porous functionalized graphene electrodes. *Journal of Physical Chemistry C.* 117: 16076–16086.

[9]. Chen, P.; McCreery, R.L. 1996. Control of electron transfer kinetics at glassy carbon electrodes by specific surface modification. *Analytical Chemistry.* 68: 3958–3965.

[10]. Davies, T.J.; Hyde, M.E.; Compton, R.G. 2005. Nanotrench arrays reveal insight into graphite electrochemistry. *Angewandte Chemie International Edition*. 44: 5121–5126.
[11]. Bai, L.; Gao, L.; Conway, B.E. 1993. Problem of *in situ* real-area determination in evaluation of performance of rough or porous, gas-evolving electrocatalysts Part 2.-Unfolding of the electrochemically accessible surface of rough or porous electrodes: A case-study with an electrodeposited porous Pt electrode. *Journal of the Chemical Society Faraday Transactions*. 89: 243–249.
[12]. Keeley, G.P.; Lyons, M.E.G. 2009. The effects of thin layer diffusion at glassy carbon electrodes modified with porous films of single-walled carbon nanotubes. *International Journal of Electrochemical Sciences*. 4: 794–809.
[13]. Streeter, I.; Wildgoose, G.G.; Shao, L.D.; Compton, R.G. 2008. Cyclic voltammetry on electrode surfaces covered with porous layers: An analysis of electron transfer kinetics at single-walled carbon nanotube modified electrodes. *Sensors and Actuators B*. 133: 462–466.
[14]. Sun, Z.; Li, Y.; Zhou, T.; Liu, Y.; Shi, G.; Jin, L. 2008. Direct electron transfer and electrocatalysis of haemoglobin in layer-by-layer films assembled with Al-MSU-S particles. *Talanta*. 74: 1692–1698.
[15]. Wang, J. 1985. *Stripping Analysis. Principles, Instrumentation and Applications*. VCH, Weinheim-New York.
[16]. Amatore, C.; Savéant, J.-M. 1978. Do ECE mechanisms occur in conditions where they could be characterized by electrochemical kinetic techniques? *Journal of Electroanalytical Chemistry*. 86: 227–232.
[17]. Ciolkowski, E.L.; Maness, K.M.; Cahill, P.S.; Wightman, R.M.; Evans, D.H.; Fosset, B.; Amatore, C. 1994. Disproportionation during electrooxidation of catecholamines at carbon-fiber microelectrodes. *Analytical Chemistry*. 66: 3611–3617.
[18]. Wang, J.; Walcarius, A. 1996. Zeolite-modified carbon paste electrode for selective monitoring of dopamine. *Journal of Electroanalytical Chemistry*. 407: 183–187.
[19]. Doménech-Carbó, A.; García, H.; Doménech, M.T.; Galletero, M.S. 2002. 2,4,6-Triphenylpyrylium ion encapsulated into zeolite Y as a selective electrode for the electrochemical determination of dopamine in the presence of ascorbic acid. *Analytical Chemistry*. 74: 562–569.
[20]. Atta, N.F.; Galal, A.; Ali, S.M.; El-Said, D.M. 2014. Improved host–guest electrochemical sensing of dopamine in the presence of ascorbic and uric acids in a β cyclodextrin/Nafion®/polymer nanocomposite. *Analytical Methods*. 6: 5962–5971.
[21]. Song, W.; Chen, Y.; Xu, J.; Yang, X.; Tian, D. 2010. Dopamine sensor based on molecularly imprinted electrosynthesized polymers. *Journal of Solid State Electrochemistry*. 14: 1909–1914.
[22]. Rueda, M.; Aldaz, A.; Sanchez-Burgos, F. 1978. Oxidation of L-ascorbic acid on a gold electrode. *Electrochimica Acta*. 1978, 23, 419–424.
[23]. Hu, I.F.; Kuwana, T. 1986. Oxidative mechanism of ascorbic acid at glassy carbon electrodes. *Analytical Chemistry*. 58: 3235–3239.
[24]. Scholz, F.; Gulaboski, R.; Mirceski, V.; Langer, P. 2002. Quantification of the chiral recognition in electrochemically driven ion transfer across the interface water/chiral liquid. *Electrochemistry Communications*. 4: 659–662.
[25]. Kothari, H.M.; Kulp, E.A.; Boonsalee, S.; Nikiforov, M.P.; Bohannan, E.W.; Poizot, P.; Nakanishi, S.; Switzer, J.A. 2004. Enantiospecific Electrodeposition of chiral CuO films from copper(II) complexes of tartaric and amino acids on single-crystal Au(001). *Chemistry of Materials*. 16: 4232–4244.
[26]. Limmer, S.J.; Kulp, E.A.; Switzer, J.A. 2006. Epitaxial electrodeposition of ZnO on Au(111) from alkaline solution: Exploiting amphoterism in Zn(II). *Langmuir*. 22: 10535–10539.
[27]. Majidi, M.R.; Kane-maguire, L.A.P.; Wallace, G.G. 1994. Enantioselective electropolymerization of aniline in the presence of (+) - or (−)-camphorsulfonate ion: A facile route to conducting polymers with preferred one-screw-sense helicity. *Polymer*. 35: 3113–3115.
[28]. Ou, J.; Zhu, Y.H.; Kong, Y.; Ma, J.F. 2015. Graphene quantum dots/β-cyclodextrin nanocomposites: A novel electrochemical chiral interface for tryptophan isomer recognition. *Electrochemistry Communications*. 60: 60–63.
[29]. Kong, Y.; Ni, J.; Wang, W.; Chen, Z. 2011. Enantioselective recognition of amino acids based on molecularly imprinted polyaniline electrode column. *Electrochimica Acta*. 56: 4070–4074.
[30]. Doménech-Carbó, A.; Alarcón, J. 2007. Microheterogeneous electrocatalytic chiral recognition at monoclinic vanadium-doped zirconias: Enantioselective detection of glucose. *Analytical Chemistry*. 79: 6742–6751.
[31]. Kimmel, D.W.; LeBlanc, G.; Meschievitz, M.E.; Cliffel, D.E. 2012. Electrochemical sensors and biosensors. *Analytical Chemistry*. 84: 685–707.

[32]. Santos, A.; Davis, J.J.; Bueno, P.R. 2014. Fundamentals and applications of impedimetric and redox capacitive biosensors. *Journal of Analytical and Bioanalytical Techniques*. S7: artic. 016.
[33]. Li, L.; Chen, Y.; Zhu, J.-J. 2017. Recent advances in electrochemiluminescence analysis. *Analytical Chemistry*. 89: 358–371.
[34]. Valenti, G.; Fiorani, A.; Li, H.; Sojic, N.; Paolucci, F. 2016. Essential role of electrode materials in electrochemiluminescence applications. *ChemElectroChem*. 3: 1990–1997.
[35]. Hanif, S.; Han, S.; John, P.; Gao, W.; Kitte, S.A.; Xu, G. 2016. Electrochemiluminiscence of luminol-tripropylamine system. *Electrochimica Acta*. 196: 245–251.
[36]. Dong, Y.-P.; Zhang, J.; Ding, Y.; Chu, X.-F.; Chen, J. 2013. Electrogenerated chemiluminiscence of luminal at polyaniline/graphene modified electrode in neutral solution. *Electrochimica Acta*. 91: 240–245.
[37]. Dong, D.; Zheng, D.; Wang, F.Q.; Yang, X.Q.; Wang, N.; Li, Y.G.; Guo, L.H.; Cheng, J. 2004. Quantitative photoelectrochemical detection of biological affinity reaction: Biotin-avidin interaction. *Analytical Chemistry*. 76: 499–501.
[38]. Zhang, Z.-X.; Zao, C.-Z. 2013. Progress of photoelectrochemical analysis and sensors. *Chinese Journal of Analytical Chemistry*. 41: 436–444.
[39]. Hao, Y.; Cui, Y.; Qu, P.; Sun, W.; Liu, S.; Zhang, Y.; Li, D.; Zhang, F.; Xu, M. 2018. A novel strategy for the construction of photoelectrochemical sensing platform based on multifunctional photosensitivizer. *Electrochimica Acta*. 259: 179–187.
[40]. Zaraska, L.; Gawlak, K.; Wiercigroch, E.; Malek, K.; Kocieł, M.; Andrzejcuk, M.; Marzec, M.M.; Jarosz, M.; Brzózka, A.; Sulka, G.D. 2019. The effect of anodizing potential and annealing conditions on morphology, composition and photoelectrochemical activity of porous anodic tin oxide films. *Electrochimica Acta*. 319: 18–30.
[41]. Kimmich, D.; Taffa, D.H.; Dosche, C.; Wark, M.; Wittstock, G. 2016. Photoactivity and scattering behavior of anodically and cathodically deposited hematite photoanodes – A comparison by scanning photoelectrochemical microscopy. *Electrochimica Acta*. 202: 224–230.
[42]. Willner, I.; Zayats, M. 2007. Electronic aptamer-based sensors. *Angewandte Chemie International Edition*. 46: 6408–6418.
[43]. Xiao, Y.; Lubin, A.A.; Heeger, A.J.; Plazco, K.W. 2005. Label-free electronic detection of thrombin in blood serum by using an aptamer-based sensor. *Angewandte Chemie International Edition*. 44: 5456–5459.
[44]. Oliveira Monteiro, T.; Costa dos Santos, C.; Martimiano do Prado, T.; Santos Damos, F.; Silva Luz, R.C.; Fatibello-Filho, O. 2020. Highly sensitive photoelectrochemical immunosensor based on anatase/rutile TiO_2 and Bi_2S_3 for the zero-biased detection of PSA. *Journal of Solid State Electrochemistry*. 24: 1801–1809.
[45]. Zhang, B.; Wang, H.; Zhao, F.; Zeng, B. 2019. LED visible-light driven label-free photoelectrochemical immunosensor based on WO_3/Au/CdS photocatalyst for the sensitive detection of carcinoembryonic antigen. *Electrochimica Acta*. 297: 372–380.
[46]. Yu, S.; Lee, S.; Yeo, J.; Han, J.; Yi, J. 2014. Kinetic and mechanistic insights into the all solid all solid-state Z-schematic system. *Journal of Physical Chemistry C*. 118: 29583–29590.
[47]. Wang, H.; Ye, H.; Zhang, B.; Zhao, F.; Zeng, B. 2017. Electrostatic interaction mechanism based synthesis of a Z-scheme BiOI-CdS photocatalyst for selective and sensitive detection of Cu^{2+}. *Journal of Materials Chemistry A*. 5: 332–341.
[48]. Ingram, D.; Linic, S. 2011. Water splitting on composite plasmonic-metal/semiconductor photoelectrodes: Evidence for selective plasmon-induced formation of charge carriers near the semiconductor surface. *Journal of the American Chemical Society*. 133: 5202–5205.
[49]. Katz, E. 2011. Processing electrochemical signals at both sides of interface: Electronic vs. chemical signal processing. *Journal of Solid State Electrochemistry*. 15: 1471–1480.
[50]. de Silva, A.P.; Gunaratne, H.Q.N.; McCoy, C.P. 1997. Molecular photoionic and logic gates with bright fluorescence and "off–on" digital action. *Journal of the American Chemical Society*. 119: 7891–7892.
[51]. Katz, E.; Lioubashevsky, O.; Willner, I. 2004. Electromechanics of a redox-active rotaxane in a monolayer assembly on an electrode. *Journal of the American Chemical Society*. 126: 15520–15532.
[52]. Willner, I.; Pardo-Yissar, V.; Katz, E.; Ranjit, K.T. 2001. A photoactivated 'molecular train' for optoelectronic applications: light-stimulated translocation of a β-cyclodextrin receptor within a stoppered azobenzene-alkyl chain supramolecular monolayer assembly on a Au-electrode. *Journal of Electroanalytical Chemistry*. 497: 172–177.

[53]. Pérez-Inestrosa, E.; Montenegro, J.-M.; Collado, D.; Suau, R.; Casado, J. 2007. Molecules with multiple light-emissive electronic excited states as a strategy toward molecular reversible logic gates. *Journal of Physical Chemistry C*. 111: 6904–6909.
[54]. Doménech, A.; Coronado, E.; Lardies, N.; Martí, C.; Doménech, M.T.; Ribera, A. 2008. Solid-state electrochemistry of LDH-supported polyaniline hybrid inorganic-organic material. *Journal of Electroanalytical Chemistry*. 624: 275–286.
[55]. Díaz-Cruz, J.M.; Esteban, M.; Ariño, C. 2019. *Chemometrics in Electroanalysis*. In Scholz, F. Ed. Monographs in Electrochemistry Series. Springer Nature, Cham, Switzerland.
[56]. Molina, A.; González, J. 2016. *Pulse Voltammetry in Physical Electrochemistry and Electroanalysis. Theory and Applications*. In Scholz, F. Ed. Monographs in Electrochemistry Series. Springer, Berlin.
[57]. Scholz, F.; Schröder, U.; Gulabowski, R.; Doménech-Carbó, A. 2014. *Electrochemistry of Immobilized Particles and Droplets*, 2nd edit. Springer, Berlin.
[58]. Nge, P.N.; Rogers, C.I.; Woolley, A.T. 2013. Advances in microfluidic materials, functions, integration, and applications. *Chemical Reviews*. 113: 2550–2583.

13 Electrochemical Gas Sensing

13.1 GAS SENSING

Gas sensing has experienced a considerable expansion during the last decades. A chemical gas sensor can be defined as a device that can change one or more of its physical properties when exposed to a gaseous species so that the change can be quantified [1]. In general, these devices include a recognition unit and a transduction element; if there is no recognition element, the system is called a concentration transducer but not a chemical sensor [2].

Solid-state gas sensors require the reversible interaction of the gas with the surface of a solid-state material, monitoring this interaction through the measurement in the change of conductivity, capacitance, work function, mass, optical properties, or any other representative property [3]. Solid electrolyte sensors are applied to sense three types of gaseous systems [4]: (a) gas mixtures in chemical equilibrium, (b) gas mixtures containing the analyte beside inert gases, and (c) gas mixtures of non-equilibrated gases.

Apart from optical (infrared spectroscopy, photoionization detection, and fluorescence) and mass (quartz crystal microbalance) sensors, the main techniques of gas sensing can be divided into chemiresistive and electrochemical [5]. Chemiresistive gas sensors are based on the change in electrical resistance of the sensing material with the presence of target analytes. These can be divided into those involving the adsorption of the gas onto the surface of a metal oxide semiconductor, and those (catalytic) involving the previous reaction of the analyte with a catalytic bead. Relatively recent expansions include impedance metric and field-effect transistor-based sensors [6].

The electrochemical gas sensors operate through electron transfer processes between the target gas and a sensing material and can be divided into potentiometric and amperometric (and catalytic). In its simplest version, a potentiometric gas sensor for CO_2 may operate by monitoring the pH changes occurring in a solution after being permeated by the gas through a suitable membrane [5]. This type of sensor is receiving increasing attention due to the possibility of integrating a variety of electrolytes (ionic liquids in particular), electrode materials, and cell configurations. Here, however, we will discuss solid-state sensors involving porous materials. Figure 13.1 depicts a simplified scheme for these systems comprising sensing electrode separated from the gas sample by a permeable membrane. The electrode is in contact with a suitable electrolyte where the reference and counter electrodes are embedded.

In the case of conventional potentiometric gas sensors, the gas to be detected is converted to the mobile component in a solid electrolyte that separates a reference compartment and a test compartment. The difference of potential established between the two sides of the solid electrolyte is dependent on the difference of thermodynamic activity across the solid electrolyte of the species that will equilibrate with the charge-transporting ions in the solid electrolyte. Here, sensing involves — to any extent — charge transport across the solid, so that the sensing process is extremely sensitive to temperature and the presence of doping species.

13.2 CHEMIRESISTIVE SENSORS

Chemiresistors are based on the change in electrical resistance of the sensing material in the presence of target analytes. The classical systems are semiconducting metal oxides devoted to oxygen sensing. Here, four basic processes can be accounted [3,7]:

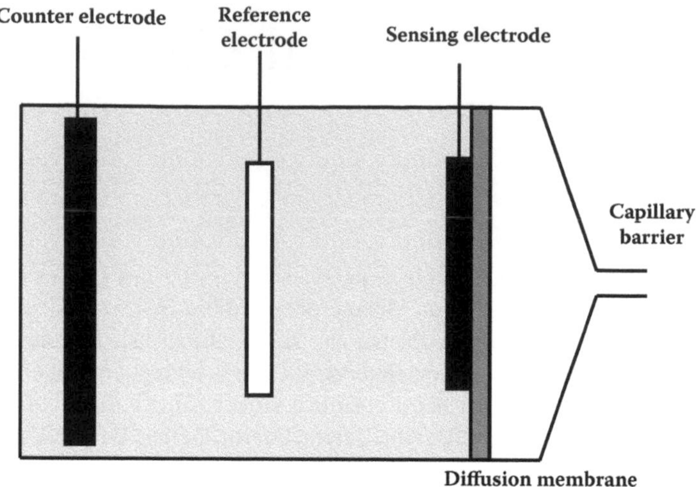

FIGURE 13.1 Schematics for an Electrochemical Gas Sensing.

$$O_{2\,gas} \rightarrow O_{2\,ads} \tag{13.1}$$

$$O_{2\,ads} + e^- \rightarrow O^-_{2\,ads} \tag{13.2}$$

$$O^-_{2\,ads} + e^- \rightarrow 2O^-_{ads} \tag{13.3}$$

$$O^-_{ads} + e^- \rightarrow O^{2-}_{ads} \tag{13.4}$$

These adsorbed oxygen forms are a function of operating temperature. For temperature < 100°C, the O_2^- form predominates, in the temperature range of 100–300 °C, O^- predominates, whereas for temperatures > 300 °C, O^{2-} is the prevailing form [8]. The capture of electrons from the conduction band of the oxide leads to an alteration in the charge carrier's distribution and hence, on the conductivity of the semiconducting metal oxide. The conductivity response depends on the type of semiconductor and the oxidizing/reducing properties of the gas. In the presence of the oxidizing (reducing) gases, the conductivity of *n*-type metal oxides (In_2O_3, SnO_2, TiO_2, V_2O_5, WO_3, ZnO) decreases (increases). In the case of *p*-type semiconductors (CuO, β-MnO_2), reducing (oxidizing) gases decrease (increase) the conductivity of the material. The essential idea is that adsorbed molecules can transfer (or extract) electrons on (from) the conduction band of the metal oxide, modifying the charge carrier distribution and the conductivity of the material. In the case of oxygen, this can be represented as reduction processes described by Eqs. (13.2–13.4). In the case of an H_2S sensor operating at temperatures above 350 °C, the adsorbed oxygen species react with the analyte as [9]:

$$H_2S_{gas} + 3O^{2-}_{ads} \rightarrow SO_{2gas} + H_2O + 6e^- \tag{13.5}$$

The performance of chemiresistive gas sensors depends on three factors: receptor function, transducer function, and utility factor. These are schematized in Figure 13.2 for *n*- and *p*-type semiconductors [10]. The essential factor to be accounted for is the interaction of the sensor constituents with the surrounding atmosphere containing oxygen and target gases. The net amount

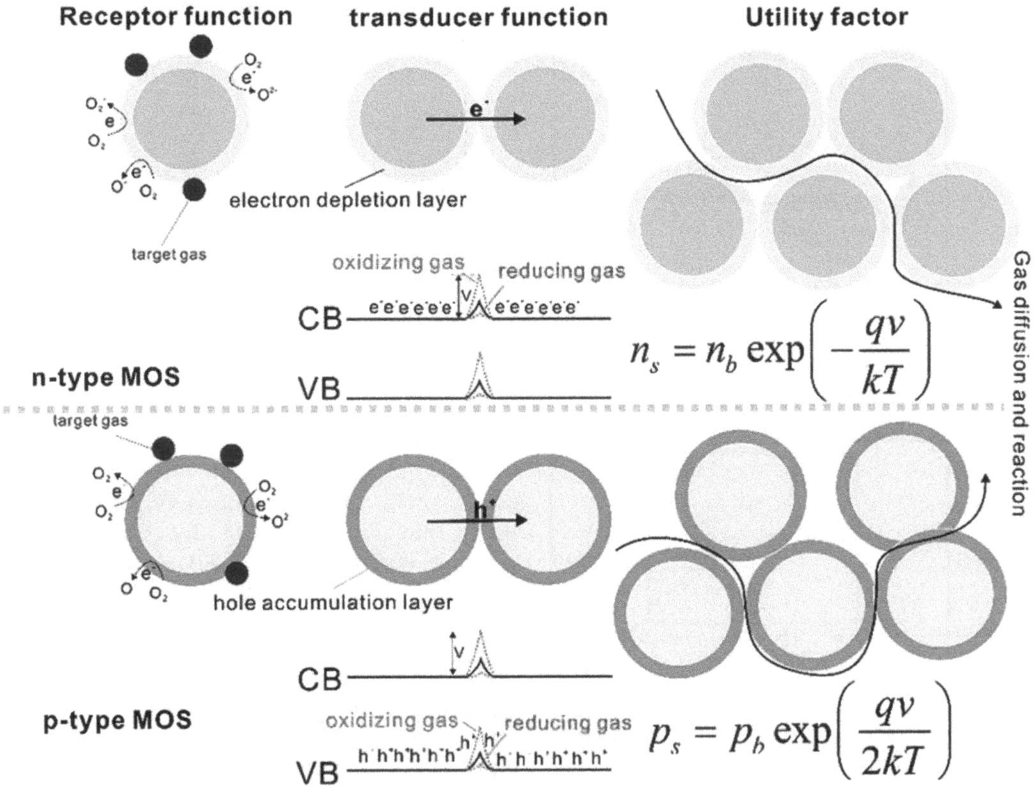

FIGURE 13.2 Scheme for the Gas Sensing Mechanisms Operating in *n*- and *p*-Type Semiconducting Metal Oxides. Reproduced from Ref. [11] (Zhang et al. *PhysChemChemPhys*. 2017, 19, 6313–6329), with Permission.

of oxygen adsorbed decides the sensing properties, which directly depend on the specific surface area of the sensing materials [11].

The sensing electrodes are usually coated as layers or screen printed on insulating substrates (SiO_2, alumina), often on a micro hot plate to control the temperature [6]. The response, however, is sensitive to the presence of other gases and to humidity, in particular. Typical current/voltage responses are illustrated in Figure 13.3, where the current/voltage characteristics of a series of chemiresistive Co_3O_4 thin films — prepared by electron-beam deposition of Co films on (0001) Al_2O_3 substrate followed by oxygen annealing at three different temperatures (500 °, 650 °, and 800 °C) — are depicted [7]. Experimental data approximate linear variations of the logarithm of the conductivity with temperature (see also Figure 13.3). This type of dependence can be extended in most cases to the variation of conductivity (or equivalently, resistivity) with the logarithm of the partial pressure (or concentration) of the target gas at a given temperature.

These variations can be rationalized by considering that the conductivity (σ) of a semi-conducting film can be taken as:

$$\sigma = AT^{3/2}e(\mu_e + \mu_h)e^{-\frac{E_g}{2k_BT}} \tag{13.6}$$

where μ_e, μ_h, represent the mobility of electrons and holes ($m^2\ V^{-1}\ s^{-1}$), E_g the energy gap of the semiconductor, k_B the Boltzmann constant, and A a temperature-independent term, corresponding to the pre-exponential term in an Arrhenius type law for conductivity:

(a)

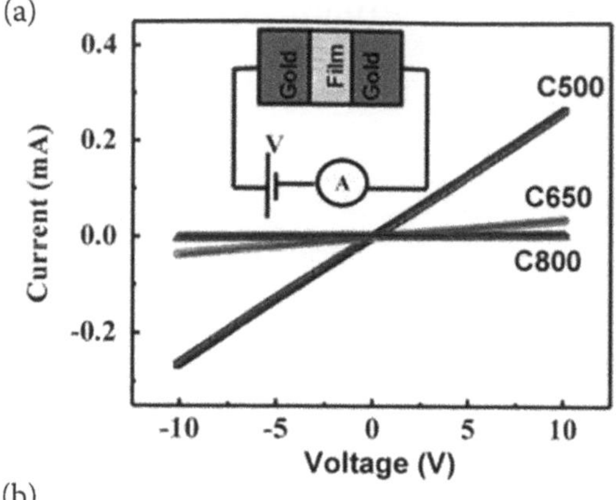

(b)

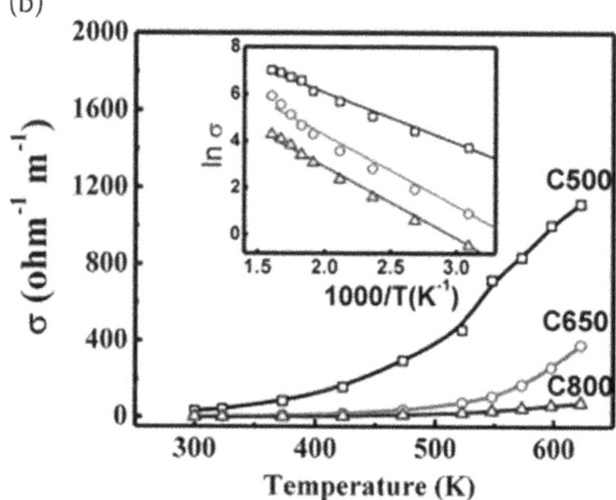

FIGURE 13.3 (a) Current-Voltage Characteristics of Co_3O_4 Thin Films Prepared By Electron-Beam Deposition of Co Films on (0001) Al_2O_3 Substrate Followed by Oxygen Annealing at 500°, 650°, and 800°C (Labeled Respectively as C500, C650, and C800) at Room Temperature. Inset: Schematic Diagram Used for Measurements. (b) Temperature Dependence of Electrical Conductivity for C500, C650, and C800 Films. Inset: Inverse Temperature Dependence of Logarithmic Conductivity. Reproduced from Ref. [7] (Balouria et al. *Sens. Actuators B*. 2013, 176: 38–45), with Permission.

$$\sigma T = Ae^{-\frac{E_a}{k_B T}} \tag{13.7}$$

E_a being the corresponding activation energy.

Doping of metal oxides significantly modulates their resistive response. In the case of cubic zirconia (ZrO_2), moderate doping with rare earth cations such as Y^{3+} or Sc^{3+} increases conductivity, the maximum being reached when the concentration of acceptor-type dopants is close to the minimum necessary to stabilize the cubic phase [12]. For yttria-stabilized zirconia (YSZ), $Y_xZr_{1-x}O_{2-x/2}$ ceramics, the maximum of conductivity is obtained for $x \approx 0.10$. Higher concentrations of dopant cations decrease conductivity by the effect of their progressive association with oxygen vacancies. The working temperature range extends from 400–1600 °C for partial pressures of oxygen between 10^{-20} and 10 bar.

It should be noted that the conductivity in ionic solids involves ion transport mechanism between coordinating sites (site-to-site hopping) and local structural relaxation. In general, the conductivity measured by ac measurements, σ_{ac}, can be expressed as a function of the conductivity from dc measurements, σ_{dc}, and the frequency:

$$\sigma_{ac} = \sigma_{dc} + B\omega^{\alpha} \tag{13.8}$$

where α is the power-law exponent and B a constant. According to the jump relaxation model [13], α represents the quotient between the back-hop rate and the site relaxation time. Then, the total conductivity of an oxide-type material with mixed ionic-electronic conduction, σ, can be expressed as:

$$\sigma = \sigma_{ion} + G(T)p(O_2)^{\pm 1/m} \tag{13.9}$$

where σ_{ion} denotes the ionic conductivity of the material and the subsequent term represents the electronic conductivity of the material, expressed as the product of a temperature-dependent constant, $G(T)$, and the partial pressure of O_2, $p(O_2)$. The sign in front of $1/m$ is positive when the electronic conductivity is primarily due to electron holes (*p*-type conductivity) and positive when it is primarily due to excess electrons (*n*-type conductivity) [14]. One can obtain that, under equilibrium conditions, the concentration of excess electrons and their contribution to electrical conductivity in YSZ is proportional to $p(O_2)^{-1/4}$ or $p(O_2)^{-1/6}$. Similarly, the concentration of defect electrons becomes proportional to $p(O_2)^{1/4}$ or $p(O_2)^{1/6}$.

13.3 IMPEDANCE SENSORS

These are based on the record of the impedance response of the sensing semiconducting material upon adsorption of the target gas. Figure 13.4 depicts two impedance representations of a $La_{0.8}Sr_{0.2}Fe_{0.95}Pd_{0.05}O_{3-\delta}$ sensor operating at 650 °C exposed to various NO_2 concentrations [15]. The equivalent circuit contains three R-CPE parallel couples where the parameter, R1 is the contact resistance and effect of external wires and the impedance elements R2, CPE1, R3, and CPE2 correspond to the effect of the resistance and capacitance of grain bulk and boundary. The resistance R4, representative of the electrochemical process at the three-phase solid-electrolyte-gas boundary, decreases with increasing concentrations of NO_2. Figure 13.5 depicts the response transients of the sensor upon alternately switching from 300 to 700 ppm NO_2 at a frequency of 10 Hz [15]. In general, the grain boundary resistance is particularly sensitive to changes in gas concentration [16]. Adsorbed oxygen plays a role in the sensing process — in the presence of H_2S gas, there is an increase in the injection of electrons from adsorbed oxygen into the metal oxide conduction band. Then, the height of the potential barrier associated with grain-grain contacts decreases, and the conductivity increases [17].

In general, the performance of the sensor at a given angular frequency ω is expressed as the gain $G(\omega)$ defined as [6]:

$$G(\omega) = \frac{|Z(\omega)|_{air}}{|Z(\omega)|_{problem}} \tag{13.10}$$

The phase angle, however, seems to be particularly sensitive to gas concentration changes [18]. There are a variety of materials used for impedance metric gas sensing, from graphene and its derivatives to MOFs and a variety of metal oxide, metal sulfides, etc. nanoarchitectures [19]. There is, however, a variety of impedance models to describe the recorded responses [6].

13.4 POTENTIOMETRIC SENSORS

Potentiometric sensors are devoted to the analysis of a gas whose ions are mobile in a solid electrolyte. For oxygen measurement, these were typically formed by a layer of stabilized zirconia pressed between two porous Pt electrodes [20]. Such electrodes are in contact with the gas sample and a reference gas, respectively.

Gas sensing is not limited to cases where mobile analyte species exist. Other gases can be detected by adding an auxiliary electrode sensitive to other species that can be equilibrated with the mobile ion. Thus, the addition of sulfates or carbonates permits the measurement of SO_2 and CO_2 [21]. A typical example is constituted by O_2 sensors based on yttrium- or scandium-doped zirconias. The overall electrode reaction, occurring at the three-phase metal/electrolyte/gas boundary, can be represented as:

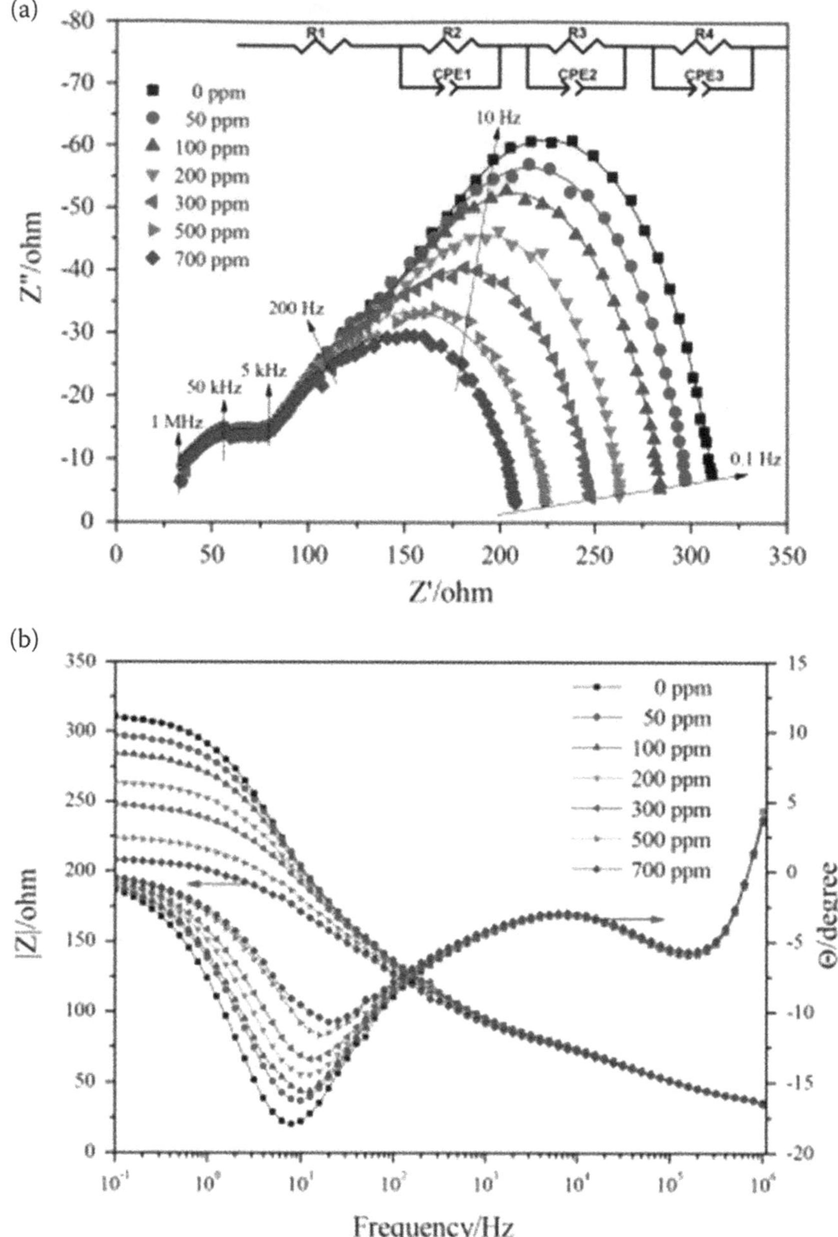

FIGURE 13.4 Impedance Spectra for $La_{0.8}Sr_{0.2}Fe_{0.95}Pd_{0.05}O_{3-\delta}$ Sensor at 650°C Under Various NO_2 Concentrations. Nyquist (a) and Bode (b) Plots. From Ref. [15] (Dai et al. *Electrochim. Acta* 2018, 265: 411–418), with Permission.

$$O_{2\,gas} + 4e^- \rightarrow 2O^{2-}_{electrolyte} \qquad (13.11)$$

To describe the sensing processes, it is convenient to use the Kröger-Vink notation [22] summarized in Table 13.1. Using this approach, the O_2 reduction process can be represented as:

$$O_{2\,gas} + 2\{V_O^{\cdot\cdot}\}_{solid} + 4e^- \rightarrow 2\{O_O^x\}_{solid} \qquad (13.12)$$

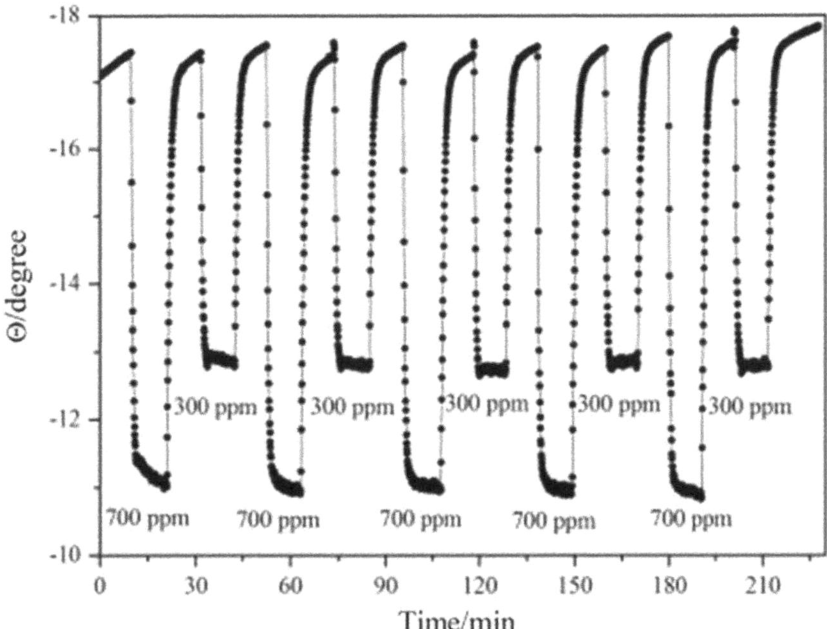

FIGURE 13.5 Transient Response of the $La_{0.8}Sr_{0.2}Fe_{0.95}Pd_{0.05}O_{3-\delta}$ Sensor at 650°C Upon Alternately Switching from 300 to 700 ppm NO_2 at a Frequency of 10 Hz. From Ref. [15] (Dai et al. *Electrochim. Acta* 2018, 265: 411–418), with Permission.

TABLE 13.1

Kröger-Vink Notation for Representing Point Defects in Solids [22]

Point defect	Symbol
Vacancies	*V* (or □ when vanadium is present)
Position occupied in the ideal lattice	Denoted as lower index (M: metal site, X: anion site)
Interstitial position	Lower index i
Formal charge defect with respect to the ideal structure	Superscript n+ or n-
Effective positive charge defect with respect to the ideal structure	Superscript dot (·)
Effective negative charge defect with respect to the ideal structure	Superscript prime (')
Zero effective charge defect with respect to the ideal structure	Superscript cross (×)
Free electrons	e'
Free holes	h·
Polarons	P
Interstitial M^{3+} cation	$M^{\cdots}_i$
M^{2+} cation vacancy	V''_M
X^{2-} anion vacancy	$V^{\cdot\cdot}_X$
X^- anion occupying M^+ cation site	X''_M
Z^- anion occupying X^{2-} anion site	Z'_X
A^{2+} cation at the M^{2+} site	A^x_M
A^{4+} cation at the M^{2+} site	$A^{\cdot\cdot}_M$
Cluster formed by one X^{2-} vacancy and one A^{2+} cation occupying M^{3+} site	$(V^{\cdot\cdot}_X A'_M)$ or $(V_X A_M)$

where $\{V_O^{\cdot\cdot}\}$ denote oxygen vacancies in the YSZ lattice and O^x_O oxide ions occupying anionic lattice positions. The cell can be considered as an oxygen concentration cell where the overall reaction can be described in terms of the transfer of oxygen from one side to the other. The electromotive force of the cell is given by the Nernst equation:

$$E_{eq} = \frac{RT}{4F} \ln \frac{p'_{OX}}{p''_{OX}} \tag{13.13}$$

where p'_{OX} and p''_{OX} are, respectively, the partial pressures of O_2 in the gas sample and the reference gas (usually air with defined humidity). Accordingly, the measurement of open-circuit potentials yields information of the O_2 concentration in the sample relative to the reference gas.

Conductivity in solid oxide-type compounds can be associated to defect formation and oxygen adsorption/desorption processes. The following defect-forming reactions are possible [22]:

$$\text{Null} \rightarrow e' + h^{\cdot} \tag{13.14}$$

$$\text{Null} \rightarrow 2e' + V^{\cdot\cdot}_O + (1/2)O_2 \tag{13.15}$$

where e′ and h· represent, respectively, electrons and holes (i.e. defect electrons) in the Kröger-Vink notation. In its most single version, the oxygen adsorption/absorption process can be represented as a reaction between the oxygen vacancies and gaseous oxygen where holes are generated [23]:

$$V^{\cdot\cdot}_O + (1/2)O_2 \rightarrow O^x_O + 2h^{\cdot} \tag{13.16}$$

Conversely, the pass of oxygen from the oxide phase to the gas phase involves the generation of excess electrons, e′:

$$O^x_O \rightarrow (1/2)O_2 + V^{\cdot\cdot}_O + 2e' \tag{13.17}$$

In this context, the interaction of chlorine gas with metal oxide semiconductors can be represented by the sequence of processes [24]:

$$O''_2 + (1/2)Cl_2 \rightarrow Cl' + (1/2)O_2 + e' \tag{13.18}$$

$$O^x_O + (1/2)Cl_2 \rightarrow Cl^{\cdot}_O + (1/2)O_2 + e' \tag{13.19}$$

$$(1/2)Cl_2 + e' \rightarrow Cl'_{ads} \tag{13.20}$$

$$(1/2)Cl_2 + (V^{\cdot\cdot}_O)_{ads} \rightarrow Cl'_{ads} + (Cl^{\cdot}_O)_{ads} \tag{13.21}$$

High coordination numbers of oxygen, as in fluorite- and perovskite-type structures, are desirable to enable moderate metal ion-oxide ion bonding forces resulting in more interchangeable oxygen sites. Mixed ionic-electronic conductors of the perovskite-type structure provide high oxide ion and electron transport [25]. Doping with isovalent cations of different ionic radii generates mechanical tension in the crystal lattice, whereas heterovalent cations additionally generate acceptor- or donor-type electrical charge defects. Oxide ions are transported through lattice defects — mostly by a vacancy mechanism, rarely by oxygen interstitial sites. The vacancies in the oxygen sublattice are generated by partial reduction of transition metal cations or by heterovalent cationic substitution. Above a certain concentration, oxygen vacancies — which are randomly distributed — play an essential role. Increasing concentrations of ionic and electronic defects in the crystal cause an increase in the transport rate of oxide ions with a parallel decrease in the thermochemical stability of the oxide material. At high defect concentration, the formation of clusters and new crystalline phases can occur, a process that, in general, is accompanied by a decrease in the mobility of oxide ions.

For a substituted perovskite-type material with oxygen vacancy-type defect structure, the overall formula can be written as $A_{1-a}A'_aB_{1-b}B'_bO_{3-x}$, where A, A′ denotes rare earth or earth-alkaline cations, and B, B′ transition metal cations of the fourth row of the periodic system. The equilibrium at a temperature T with a gas phase containing oxygen with a partial pressure $p(O_2)$ can be represented as:

$$\{A_{1-a}A'_aB_{1-b}B'_bO_{3-x-\delta}\}_{solid} + (\delta/2)O_{2\,gas} \rightarrow \{A_{1-a}A'_aB_{1-b}B'_bO_{3-x}\}_{solid} \qquad (13.22)$$

In this equation, $A_{1-a}A'_aB_{1-b}B'_bO_{3-x-\delta}$ represents the oxygen-poorer state and $A_{1-a}A'_aB_{1-b}B'_bO_{3-x}$ the oxygen-richer state. For the overall reaction (13.11), the equilibrium constant will be:

$$K_p = (p(O_2))^{-\delta/2} \qquad (13.23)$$

Related to the reaction of 1 mol of atomic oxygen, usual thermochemical equations yield:

$$\Delta G^{\circ} = (\delta/2)RT\ \ln(p(O_2)) \qquad (13.24)$$

$$\ln(p(O_2)) = -\frac{\Delta H^{\circ}}{RT} + \frac{\Delta S^{\circ}}{R} \qquad (13.25)$$

Thermobalance measurements determine δ values at different temperatures while the concentration of defects is reflected in the unit cells of the crystals, determined from X-ray diffraction (XRD) data. The overall process for $Ba_{0.3}Sr_{0.7}Co_{0.8}M_{0.2}O_{2.5+n-\delta}$ can be represented as [26]:

$$\{Ba_{0.3}Sr_{0.7}Co_{0.8}M_{0.2}O_{2.5+n-\delta}\}_{solid} + (\delta/2)O_{2\,gas} \rightarrow \{Ba_{0.3}Sr_{0.7}Co_{0.8}Fe_{0.2}O_{2.5+n}\}_{solid} \qquad (13.26)$$

The oxygen exchange capacity is directly related to the oxygen/metal stoichiometry. The amount of reversibly exchangeable oxygen in an $A_{1-a}A'_aB_{1-b}B'_bO_{3-x}$ perovskite-type oxide, x, is a function of temperature and gradient of the thermodynamic oxygen activity between the surface and the center of the crystal. Complete occupation of the oxygen sublattice (i.e. $x = 3$) is reached at higher partial pressure of O_2 in the gas phase. At decreasing $p(O_2)$, such materials approach to a level of $p(O_2)$-independent stoichiometry, where $x = a/2$. This situation corresponds to the average oxidation state 3+ of the B cations. At lower partial pressures of O_2, the concentration of oxygen vacancies increases until phase destruction occurs.

A second group of solid-state gas sensors is constituted by those involving equilibration with the mobile component of the solid electrolyte. A typical example is a CO_2 sensor with K_2CO_3 solid electrolyte. Here, K^+ ions are the only mobile ions in the solid electrolyte, so the overall electrochemical process can be represented as [27]:

$$2\{K^+\}_{electrolyte} + CO_{2\,gas} + 1/2O_{2\,gas} + 2e^- \rightarrow \{K_2CO_3\}_{electrolyte} \qquad (13.27)$$

A third group of potentiometric gas sensors involves the use of an additional auxiliary phase. This class can be sub-divided [28] depending on whether the mobile ions of solid electrolyte are the same, different but of the same sign, or different and of different sign than those of the auxiliary phase. An example of this kind of electrode is the CO_2 sensor incorporating Na_2CO_3 and N-alumina.

Although potentiometric gas sensors are based on the equilibration between the gas and the solid electrolyte, these devices can be used in non-equilibrium conditions, provided that steady-state conditions are reached. This situation can be described in terms of mixed potential models such as those applied to describe metal corrosion [27]. The condition to be accomplished is that two separate semi-reactions — oxidation and reduction — occur simultaneously so that the number of electrons produced in the oxidation reaction equals those consumed in the reduction reaction. In the case of metal oxides used in gas sensing, one can write these semi-reactions as:

$$2O^{2-} \rightarrow O_2 + 4e^- \tag{13.28}$$

$$O_2 + 4e^- \rightarrow 2O^{2-} \tag{13.29}$$

In the case of the CO_2/CO system, the semi-reactions can be expressed as:

$$2CO + 2O^{2-} \rightarrow 2CO_2 + 4e^- \tag{13.30}$$

$$2CO_2 + 4e^- \rightarrow 2O^{2-} + 2CO \tag{13.31}$$

The overall reaction will be:

$$2CO + O_2 \rightarrow 2CO_2 \tag{13.32}$$

Then, the solid-state mixed potentials — corresponding to the currents represented by the processes described in Eqs. (13.28) and (13.31) — become equal; i.e., when the electrons released in the former reaction are consumed by the second one [27]. The necessary reference electrode is provided by the solid electrolyte in which ions are mobile. Figure 13.6 depicts the transient responses recorded for electrochemical CO sensors based on SnO_2 electrodes loaded with Pt nanoparticles and an anion-conducting polymer electrolyte [28]. The electrode potential resulting from mixed potential from electrochemical O_2 reduction:

$$(1/2)O_2 + H_2O + 2e^- \rightarrow 2OH^- \tag{13.33}$$

and CO (or H_2) oxidation, represented as [29]:

$$CO + 2OH^- \rightarrow CO_2 + H_2O + 2e^- \tag{13.34}$$

$$H_2 + 2OH^- \rightarrow 2H_2O + 2e^- \tag{13.35}$$

Here, the sensing characteristics appear to be decisively influenced by the state of the surface of Pt nanoparticles.

13.5 AMPEROMETRIC SENSING

Amperometric sensing of gases is based on solid ion-conducting materials, as described for potentiometric gas sensors. Solid-state amperometric gas sensors measure the limiting current (i_L)

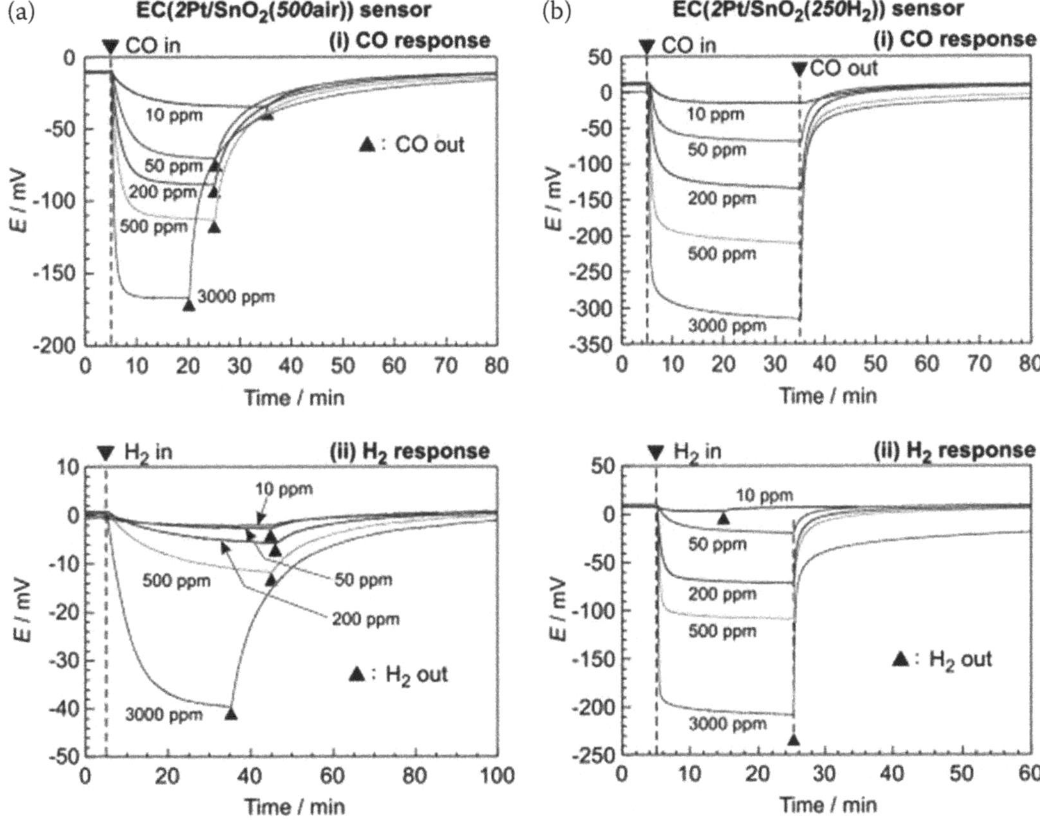

FIGURE 13.6 Response Transients of Two Different CO and H_2 Sensors Based on Pt-Loaded SnO_2 Electrodes and an Anion-Conducting Polymer Electrolyte -Labeled as EC(2Pt/SnO_2(*500*air)) and EC(2Pt/SnO_2(*250*H_2))- in Wet Synthetic Air at 30°C (57% Relative Humidity). Reproduced from Ref. [28] (Hyodo et al. *Sensors and Actuators B*. 2019, 300: artic. 127041), with Permission.

flowing across the electrochemical cell upon application of a fixed voltage, so that the rate of electrode reaction is controlled by the gas transport across the cell. The diffusion barrier consists of small-hole porous ceramics. The limiting current for an oxygen sensor satisfies the relationship:

$$I_L = \frac{4FAD_{OX}T}{RTL} p_{OX} \tag{13.36}$$

here, A represents the area of holes, L their length, and D_{OX} the coefficient of diffusion of O_2 [30]. Figure 13.7 shows an image of YSZ-based amperometric sensor for NO_2 using porous spinel oxide ($NiFe_2O_4$ plus Fe_2O_3) on YSZ as a sensing electrode [31]. The amperometric measurements were carried out at an operating temperature of 500 °C and with an applied potential of −100 mV vs. Pt/air reference electrode providing an excellent linearity (see Figure 13.7) under optimized conditions.

In general, the sensing performance can be improved as a result of catalytic effects [32]. However, it has to be taken into account that the solid catalyst not only catalyzes the sensing electrochemical reaction(s) but can also catalyze gas-phase reactions. For instance, in this case, summarized in Figure 13.7, the sensing reaction:

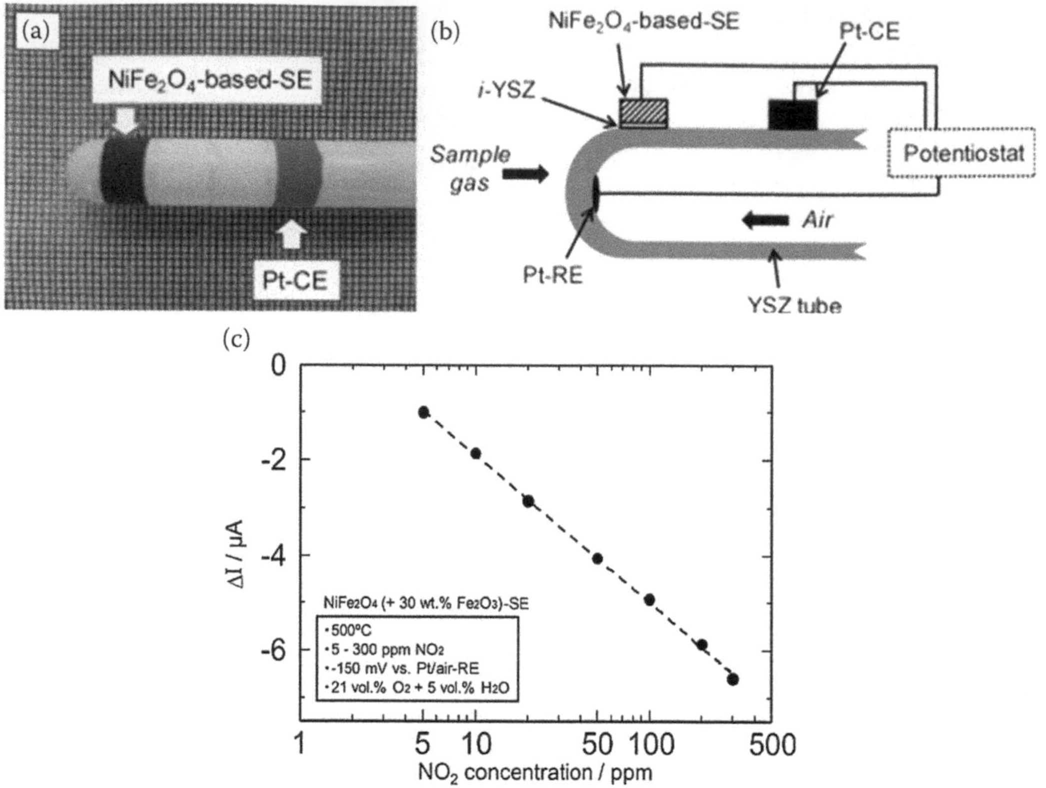

FIGURE 13.7 Exterior View (a) and Schematic Illustration (b) of YSZ-Based Amperometric Sensor Using $NiFe_2O_4$-Based Sensor Unit (SE). (c) Variation of the Currents Measured at 500°C Upon Application of a Potential of –100 mV vs. Pt/Air Reference Electrode with the Concentration of NO_2. Adapted from Ref. [31] (Anggraini et al. *Sensors and Actuators B*. 2018, 259: 30–35), with Permission.

$$NO_2 + 2e^- \rightarrow NO + O^{2-} \tag{13.37}$$

is accompanied by the process:

$$NO_2 \rightarrow NO + (1/2)O_2 \tag{13.38}$$

occurring at the porous sensing electrode/YSZ interface also catalyzed by the electrode material. This means that the recorded current will be governed by the catalytic activity of the catalyst toward gas-phase and electrochemical reactions. Accordingly, there is a need to combine the catalytic activity to gas-phase reactions and the catalytic activity to electrochemical reactions of the target gas to optimize the sensing performance.

13.6 FIELD EFFECT TRANSISTORS

The principle of operation is the modulation of the work function and the conductivity of semiconductors as a result of the adsorption of target gas molecules [33]. Field-effect transistors (FET) consist of a sensing layer placed between two electrodes (source and drain). In classical FETs, a dielectric thin layer is intercalated so that the changes in the drain-source current are due to alterations in the conductivity of the sensing unit. A variety of materials and architectures have been proposed, including graphene, organic semiconductors, as well as interposing an absorbent ionic liquid.

Figure 13.8 shows an example [34] of current/voltage curves recorded at FET sensors. This case corresponds to an ionic-liquid/graphene field-effect transistor used as an NH_3 gas sensor. In the linear region, the relationship between the drain current, I_D, the gate voltage V_G, the difference of potential established between drain and source, V_D, and the threshold potential, V_T, can be written as:

$$I_D = \frac{\mu_m CW}{L}\left(V_G - V_T - \frac{V_D}{2}\right) \tag{13.39}$$

In this equation, μ_m is the mobility of the minority charge carriers, W the width of the channel, C the gate capacitance, and L the channel length.

The threshold potential corresponds to the difference between the work functions of the base semiconductor (usually Si) and the semiconductor constituting the sensing (gate) unit, ϕ_G, ϕ_{DS}:

$$V_T = \frac{\varphi_G - \varphi_{DS}}{C} \tag{13.40}$$

When the FET works in the saturation regime, the drain current becomes:

(a)

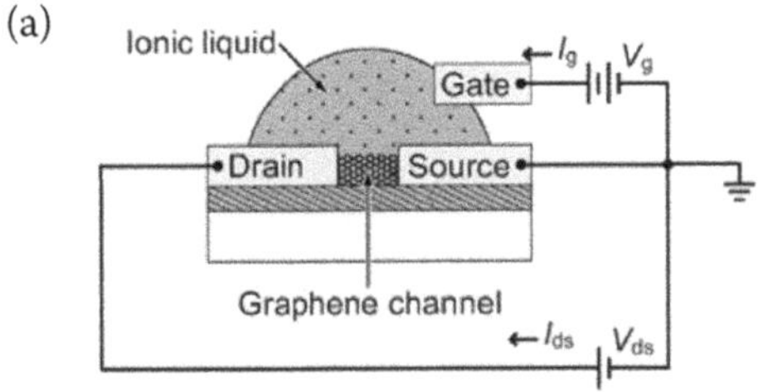

(b)

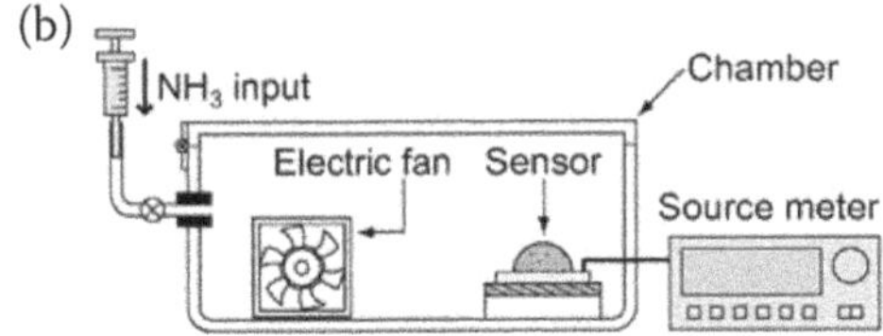

(c)

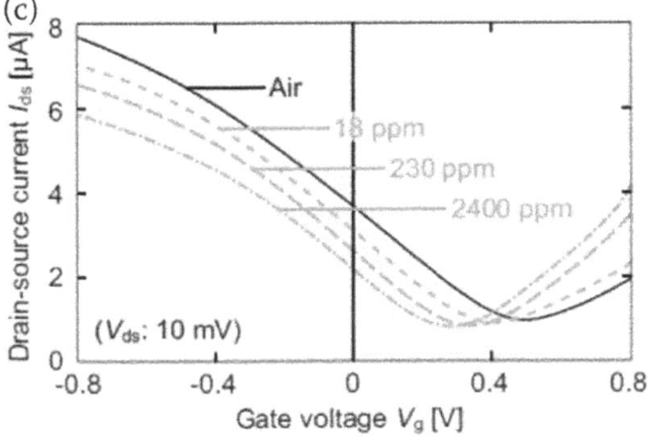

FIGURE 13.8 (a) Scheme of the Ionic-Liquid/Graphene FET Used as NH_3 Gas Sensor with the Corresponding Driving Circuit. (b) Experimental Setup. (c) Current/Voltage Response of a FET Under Air and Exposed to Different NH_3 Concentrations. From Ref. [34] (Inaba et al. *Sensors and Actuators B*. 2014, 195: 15–21), with Permission.

$$I_D = \frac{\mu_m CW}{2L}(V_G - V_T)^2 \tag{13.41}$$

Accordingly, the changes in the gate voltage can be related to the changes in the work function of the sensing unit, in turn depending on the interaction with the target gas. In the case of a mixture of gases where the partial pressure of the target gas is p_G, the following equation applies [33]:

$$V_G = \left(\frac{2LI_D}{\mu_m CW}\right)^{1/2} + \left(p_G + \sum K_i p_i\right) \tag{13.42}$$

where p_i denotes the partial pressure of the gas i and K_i the selectivity coefficient whose meaning is similar to that of the Nicholsky-Eisenman equation in potentiometry (see Chapter 12).

Recent developments include the arrangement as field-modulated chemiresistors [33,34] Here, there is a potential-dependent contact resistance in the vicinity of the drain electrode associated to the space charge created by depletion of the concentration of charge carriers. This effect can be compared to the depletion of the diffusion layer in the vicinity of the electrode occurring in voltammetric experiments in the absence of supporting electrolyte. A variety of materials have been applied for gas sensing via field-effect transistors. Figure 13.9 depicts a scheme of a carbon nanotube network connecting source and drain electrodes [35] and the representation source-drain conductance (G_{SD}) as a function of gate voltage (V_g) before and after thermal evaporation of discontinuous layer of gold.

(a)

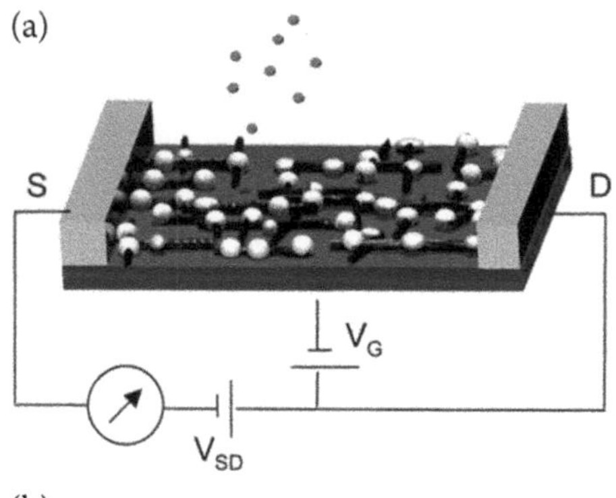

(b)

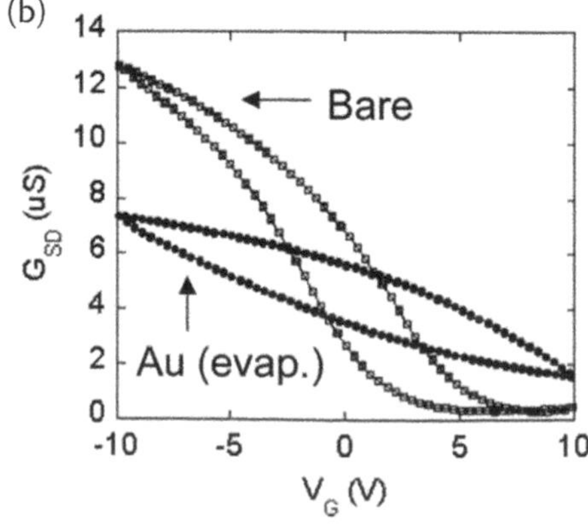

FIGURE 13.9 (a) Scheme of a Carbon Nanotube Network Connecting Source and Drain Electrodes of a Field-Effect Transistor. SWNTs Are Decorated with Metal Nanoparticles (Silver Bullets) for Selective Detection of Analyte Gases (Red Dots). (b) Electronic Measurements, Such as Source-Drain Conductance (G_{SD}), as a Function of Gate Voltage (V_g) Before (Bare) and After Thermal Evaporation of Discontinuous Layer of Gold. Adapted from Ref. [35] (Star et al. *J. Phys. Chem. B.* 2006, 110: 21014–21020), with Permission.

13.7 CONCLUDING REMARKS

The development of novel porous materials is enhancing the number of disposable electrodes, membranes, and electrolytes for gas sensing. Research is logically focused on low-temperature sensing of humidity and environmentally significant gases — NO_x, SO_x, H_2S, NH_3, and H_2 — but acetone, ethanol, and methanol are also under intensive investigation. Sensing materials include doped metal oxides [36], graphene [37] and graphene-based composites [38], and MOFs [39], among others.

In the case of graphene, the binding energy of the gas molecules plays an essential role for sensing purposes. This energy varies depending on the graphene binding site, so the sensing performance is sensitive to the preparation of the material (with expected differences between graphene, GO, and rGO forms). Table 13.2 summarizes some representative binding energies [38,40–43]. Applications

TABLE 13.2
Binding Energies Between Gas Molecules and Graphene Sites [38,40–43]

Gas molecule	Graphene adsorption site	Binding energy/eV	Ref.
O_2	Pristine	–0.006	[40]
	–COOH	0.174	
N_2	Pristine	0.006	[40]
	–COOH	0.027	
NO (N)	Pristine	1.158	[40]
	–COOH	1.187	
NO (O)	Pristine	1.171	[40]
	–COOH	1.157	
H_2O	Pristine	0.044	[41]
	–O–	0.201	
	–OH	0.259	
NH_3	Pristine	0.112	[42]
SO_2	Pristine	0.248	[43]

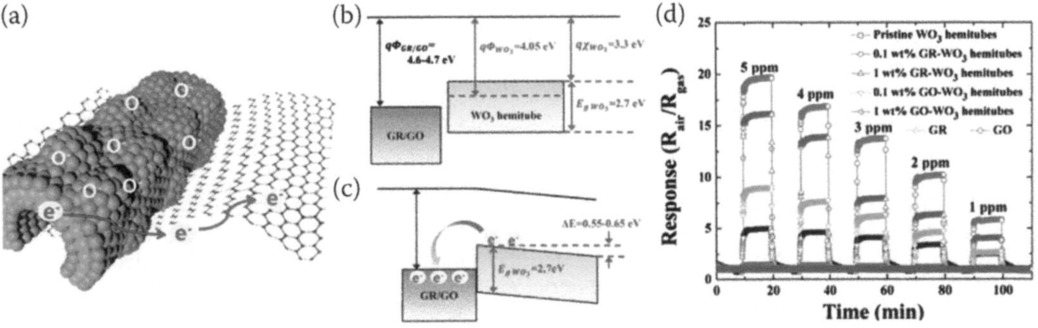

FIGURE 13.10 (a) Scheme of Graphene (GR) and Graphene Oxide (GO) Heterojunction with WO_3 Hemitubes. (b) Band Structure Model for GR/GO and WO_3 Hemitubes (Energies Relative to the Vacuum Level). (c) Schematics for Charge Transfer and Band Bending. (d) Dynamic Sensing Characteristics of H_2S in the Gas Concentration Range of 1–5 ppm at 300°C in the First Cyclic Test. Adapted from Ref. [44] (Choi et al. *ACS Appl. Mater. Interfaces*. 2014, 6: 9061–9170), with Permission.

for medical diagnosis based on breath exhaled gases are under research. Figure 13.10 shows a scheme for an H_2S sensor in breath based on hemitube structured WO_3 functionalized with GO [44]. There is electron transfer from WO_3 to GO causing a band bending of WO_3 of $\Delta E = 0.55 - 0.65$ eV. Then, the electrical resistance of the material decreased in the presence of reducing gases such as H_2S. Importantly, the resistance remains stable with temperature, in contrast with WO_3-graphite hybrids where the resistance is significantly temperature-dependent.

REFERENCES

[1]. Eranna, G. 2012. *Metal Oxide Nanostructures as Gas Sensing Devices*. CRC Press, Boca Raton, p. 326.

[2]. Banica, F.-G. 2012. In *Chemical Sensors and Biosensors: Fundamentals and Applications*. Wiley, Chichester, p. 541.

[3]. Masikini, M. 2020. Review—Metal oxides: Application in exhaled breath acetone chemiresistive sensors. *Journal of the Electrochemical Society*. 167: artic. 037537.

[4]. Bard, A.J.; Inzelt, G.; Scholz, F. Eds. 2008. *Electrochemical Dictionary*. Springer, Berlin-Heidelberg.

[5]. Xiong, L.; Compton, R.G. 2014. Amperometric gas detection: A review. *International Journal of Electrochemical Science*. 9: 7152–7181.

[6]. Balasubramani, V.; Chandraleka, S.; Subba Rao, T.; Sasikumar, R.; Kuppusamy, R.S.; Sridhar, T.M. 2020. Review—Recent advances in electrochemical impedance spectroscopy based toxic gas sensors using semiconducting metal oxides. *Journal of the Electrochemical Society*. 167: artic. 037572.

[7]. Balouria, V.; Samanta, S.; Singh, A.; Debnath, A.K.; Mahajan, A.; Bedi, R.K.; Aswal, D.K.; Gupta, S.K. 2013. Chemiresistive gas sensing properties of nanocrystalline Co_3O_4 thin films. *Sensors and Actuators B*. 176: 38–45.

[8]. Gopel, W.; Schierbaum, K.D. 1995. SnO_2 sensors: Current status and future prospects. *Sensors and Actuators B*. 26: 1–12.

[9]. Zhang, F.; Zhu, A.; Luo, Y.; Tian, Y.; Yang, J.; Qin, Y. 2010. CuO nanosheets for sensitive and selective determination of H_2S with high recovery ability. *Journal of Physical Chemistry C*. 114: 19214–19219.

[10]. Zhu, D.; Fu, Y.; Zang, W.; Zhao, Y.; Xing, L.; Xue, X. 2016. Room-temperature self-powered ethanol sensor based on the piezo-surface coupling effect of heterostructured α-Fe_2O_3/ZnO nanowires. *Materials Letters*. 166: 288–291.

[11]. Zhang, J.; Qin, Z.; Zeng, D.; Xie, C. 2017. Metal-oxide-semiconductor based gas sensors: Screening, preparation, and integration. *Physical Chemistry Chemical Physics*. 19: 6313–6329.

[12]. Kharton, V.V.; Marques, F.M.B.; Atkinson, A. 2004. Transport properties of solid oxide electrolyte ceramics: A brief review. *Solid State Ionics*. 174: 135–149.

[13]. Funke, K.; Roling, B.; Lange, M. 1998. Dynamics of mobile ions in crystals, glasses and melts. *Solid State Ionics*. 105: 195–208.

[14]. Larose, S.; Akbar, S.A. 2006. Synthesis and electrical properties of dense $Bi_2Al_4O_9$. *Journal of Solid State Electrochemistry*. 10: 488–498.

[15]. Dai, L.; Ma, L.; Meng, W.; Li, Y.; He, Z.; Wang, L. 2018. Impedancemetric NO_2 sensor based on Pd doped perovskite oxide sensing electrode conjunction with phase angle response. *Electrochimica Acta*. 265: 411–418.

[16]. Sureshkumar, S.; Venkatachalapathy, B.; Sridhar, T.M. 2019. Enhanced H_2S gas sensing properties of Mn doped ZnO nanoparticles—An impedance spectroscopic investigation. *Materials Research Express*. 6: artic. 075009.

[17]. Balasubramani, V.; Sureshkumar, S.; Rao, T.S.; Sridhar, T.M. 2019. Impedance spectroscopy-based reduced graphene oxide-incorporated ZnO composite sensor for H_2S investigations. *ACS Omega*. 4: 9976–9982.

[18]. Martin, L.P.; Woo, L.Y.; Glass, R.S. 2007. Impedancemetric NO_x sensing using YSZ electrolyte and YSZ/Cr_2O_3 composite electrodes. *Journal of the Electrochemical Society*. 154: J97–J104.

[19]. Khan, Md.A.H.; Rao, M.V.; Li, Q. 2019. Recent advances in electrochemical sensors for detecting toxic gases: NO_2, SO_2 and H_2S. *Sensors*. 19: artic. 905.

[20]. Pasierb, P.; Rekas, M. 2009. Solid-state potentiometric gas sensors-current status and future trends. *Journal of Solid State Electrochemistry*. 13: 3–25.

[21]. Fergus, J.W. 2008. Electrolyte and electrode materials for high temperature electrochemical CO_2 and SO_2 gas sensors. *Sensors and Actuators B*. 134: 1034–1041.

[22]. Kröger, F.A.; Vink, H.J. 1956. Relations between the concentrations of imperfections in crystalline solids. *Solid State Physics*. 3: 307–435.
[23]. Logothetis, E.M. 1980. Resistive-type exhaust gas sensors. *Ceramic Engineering Science Proceedings*. 1: 281–301.
[24]. Tamaki, J.; Naruo, C.; Yamamoto, Y.; Matsuoka, M. 2002. Sensing properties to dilute chlorine gas of indium oxide based thin film sensors prepared by electron beam evaporation. *Sensors and Actuators B*. 83: 190–194.
[25]. Teraoka, Y.; Zhang, H.M.; Furukawa, S.; Yamazoe, M. 1985. Oxygen permeation through perovskite-type oxides. *Chemical Letters*. 14: 1743–1746.
[26]. Girdauskaite, E.; Ullmann, H.; Al Daroukh, M.; Vashook, V.; Bülow, M.; Guth, U. 2007. Oxygen stoiciometry, unit cell volumen, and thermodynamic quantities of perovskite-type oxides. *Journal of Solid State Electrochemistry*. 11: 469–477.
[27]. Fergus, J.W. 2011. Sensing mechanism of non-equilibrium solid-electrolyte-based chemical sensors. *Journal of Solid State Electrochemistry*. 15: 971–984.
[28]. Hyodo, T.; Goto, T.; Takamori, M.; Ueda, T.; Shimizu, Y. 2019. Effects of Pt loading onto SnO_2 on CO-sensing properties and mechanism of potentiometric gas sensors utilizing an anion-conducting polymer electrolyte. *Sensors and Actuators B*. 300: artic. 127041.
[29]. Hyodo, T.; Goto, T.; Ueda, T.; Kaneyasu, K.; Shimizu, Y. 2016. Potentiometric carbon monoxide sensors using an anion-conducting polymer electrolyte and Au-loaded SnO_2 electrodes. *Journal of the Electrochemical Society*. 163: B300–B308.
[30]. Tsipis, E.V.; Kharton, V.V. 2008. Electrode materials and reaction mechanisms in solid oxide fuel cells: A brief review. *Journal of. Solid State Electrochemistry*. 12: 1367–1391.
[31]. Anggraini, S.A.; Yoshida, S.; Ikeda, H.; Miura, N. 2018. Selective NO_2 detection using YSZ-based amperometric sensor attached with $NiFe_2O_4(+Fe_2O_3)$ sensing electrode. *Sensors and Actuators B*. 259: 30–35.
[32]. Ono, T.; Hasei, M.; Kunimoto, A.; Miura, N. 2004. Improvement of sensing performances of zirconia-based total NO_x sensor by attachment of oxidation-catalyst electrode. *Solid State Ionics*. 175: 503–506.
[33]. Janata, J.; Josowicz, M. 2009. Organic semiconductors in potentiometric gas sensors. *Journal of Solid State Electrochemistry*. 13: 41–49.
[34]. Inaba, A.; Yoo, K.; Takei, Y.; Matsumoto, K.; Shimoyama, I. 2014. Ammonia gas sensing using a graphene field-effect transistor gated by ionic liquid. *Sensors and Actuators B*. 195: 15–21.
[35]. Star, A.; Joshi, V.; Skarupo, S.; Thomas, D.; Gabriel, J.C.P. 2006. Gas sensor array based on metal-decorated carbon nanotubes. *Journal of Physical Chemistry B*. 110: 21014–21020.
[36]. Wang, Y.; Liu, B.; Xiao, S.;Wang, X.; Sun, L.; Li, H.; Xie,W.; Li, Q.; Zhang, Q.;Wang, T. 2016. Low-temperature H_2S detection with hierarchical Cr-doped WO_3 microspheres. *ACS Applied Materials & Interfaces*. 8: 9674–9683.
[37]. Yang, M.; Zhang, X.; Cheng, X.; Xu, Y.; Gao, S.; Zhao, H.; Huo, L. 2017. Hierarchical NiO cube/nitrogen-doped reduced graphene oxide composite with enhanced H_2S sensing properties at low temperature. *ACS Applied Materials & Interfaces*. 9: 26293–26303.
[38]. Toda, K.; Furue, R.; Hayami, S. 2015. Recent progress in applications of graphene oxide for gas sensing: A review. *Analytica Chimica Acta*. 878: 43–53.
[39]. Koo, W.-T.; Jang, J.-S.; Kim, I.-D. 2019. Metal organic frameworks for chemiresistive sensors. *Chem*. 5: 1938–1963.
[40]. Omidvar, A.; Mohajeri, A. 2014. Edge-functionalized graphene nanoflakes as selective gas sensors. *Sensors and Actuators B*. 202: 622–630.
[41]. Guo, L.; Jiang, H.-B.; Shao, R.-Q.; Zhang, Y.-L.; Xie, S.-Y.; Wang, J.-N.; Li, X.-B.; Jiang, F.; Chen, Q.-D.; Zhang, T.; Sun, H.-B. 2012. Two-beam-laser interference mediated reduction, patterning and nanostructuring of graphene oxide for the production of a flexible humidity sensing device. *Carbon*. 50: 1667–1673.
[42]. Peng, Y.; Li, J. 2013. Ammonia adsorption on graphene and graphene oxide: A first-principles study. *Frontiers in Environmental Science and Engineering*. 7: 403–411.
[43]. Liu, X.-Y.; Zhang, J.-M.; Xu, K.-W.; Ji, V. 2014. Improving SO_2 gas sensing properties of graphene by introducing dopant and defect: A first principles study. *Applied Surface Science*. 313: 405–410.
[44]. Choi, S.-J.; Fuchs, F.; Demadrille, R.; Grevin, B.; Jang, B.-H.; Lee, S.-J.; Lee, J.-H.; Tuller, H.L.; Kim, I.-D. 2014. Fast responding exhaled-breath sensors using WO_3 hemitubes functionalized by graphene-based electronic sensitizer for diagnosis of diseases. *ACS Applied Materials & Interfaces*. 6: 9061–9170.

14 Supercapacitors, Batteries, Fuel Cells, and Related Applications

14.1 ELECTRICAL ENERGY STORAGE AND CONVERSION

The production, storage of electrical energy, and its conversion into other energy forms are obvious demands from current society. Batteries are power sources where chemical energy is converted into electrical energy by spontaneous electrochemical reactions. Batteries contain one or several cells where reduction and oxidation processes occur in two electrodic compartments lined by a solid or liquid electrolyte. Primary batteries are not designed to be recharged while secondary batteries admit repeated cycles of charge/discharge.

Fuel cells can be defined as energy-producing devices where chemical energy from a fuel and an oxidant is directly converted into electricity and heat through an electrode-electrolyte system. In contrast with batteries—closed systems that chemically store electrical energy—fuel cells can be described as open systems that can operate continuously by maintaining the appropriate flow of reactants.

Capacitors are devices for storing electric energy consisting, ideally, of two conducting elements separated by a dielectric. Under the application of a difference of potential between the conducting elements, the capacitor acquires a charge that can be further released through a consumer circuit. Accordingly, capacitors, apart from their use as frequency filters, rectifiers, etc. in electronic circuitry, can be used for energy storage, especially for shorter periods of time.

Remarkably, charge/discharge processes in batteries proceeds by interconversions of electrode materials involving phase changes and a certain level of irreversibility resulting in a limited life cycle. Increasing stability and cyclability in energy storage devices using non-pollutant and inexpensive materials are the focus of considerable research. In this context, the use of porous materials with high surface area plays an essential role.

Batteries, capacitors, and fuel cells differ in their specific energy (energy/mass ratio) and power (power/mass ratio) (*vide infra*). Batteries and fuel cells can store high specific energy (typically between 120 and 200 W h kg^{-1}) with relatively low specific power (between 0.4 and 3 kW kg^{-1}). The specific energy stored by conventional capacitors is low (< 0.1 W h kg^{-1}) but their specific power is high (up to 1,000 kW kg^{-1}). Electrochemical capacitors store low specific energy (between 4 and 8 W h kg^{-1}) with relatively high specific power (between 5 and 55 kW kg^{-1}).

These quantities are usually reflected in the so-called Ragone plots, consisting of a double logarithmic representation of available energy of an energy storage device for fixed power. The Ragone plot in Figure 14.1 [1] shows a comparative scheme of the different types of devices where characteristic times correspond to lines with unity slope.

The essential idea is that the systems of electrical energy production and storage can be represented in terms of a circuit containing an energy storage device connected to an energy supply. The electrical dynamics of that system can be described by the differential equation [1]:

$$L\frac{\mathrm{d}^2Q}{\mathrm{d}t^2} + R\frac{\mathrm{d}Q}{\mathrm{d}t} + V(Q) = -P\left(\frac{\mathrm{d}Q}{\mathrm{d}t}\right)^{-1} \qquad (14.1)$$

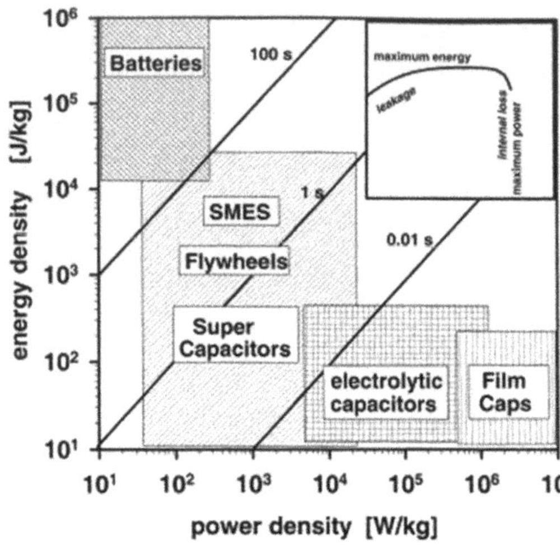

FIGURE 14.1 Ragone Plot Representing the Available Energy of an Energy Storage Device for Fixed Power. Characteristic Times Correspond to Lines with Unity Slope. The Inset Represents the $E(P)$ Curve for Each Individual Energy Storage Device. Reproduced from Ref. [1] (Christen and Carlen, *J. Power Sources*. 2000, 91: 210–216), with Permission.

where Q represents the stored charge at a time t, R an internal series resistor, L an internal inductance, and P the power of energy release. This general equation can be applied to different devices yielding individual representations of the stored energy at a given power $E(P)$, vs. P. These curves present a rapidly descending branch at high P values due to internal dissipation and leakage losses (see inset in Figure 14.1).

14.2 CAPACITORS AND SUPERCAPACITORS

Traditional capacitors use non-polarized dielectrics such as ceramic materials or polymers (polystyrene, polypropylene) intercalated between two metallic electrodes. These devices can be approximated to an ideal plane capacitor whose capacitance C depends on the area of the electrodes A, their separation d, and the dielectric permittivity of the material, ε (i.e. the well-known equation $C = \varepsilon(A/d)$). These devices work typically in the pF to μF capacitance range.

In recent times, new families of electrochemical capacitors, generically called supercapacitors, have been devised. According to the mechanism of charge storage, these devices can be divided into Electric Double-Layer capacitors (EDLs), Faradaic capacitors, and hybrid capacitors [2,3]. The charge storage mechanism of EDLs is based on the electrostatic interaction between electrolyte ions forming an electrical double layer on the electrode surface during charging and discharging. This double layer includes a space charge layer, electrolyte diffusion layer, and compact Helmholtz layer, whose total thickness is about 1 nm [4–6].

Faradaic capacitors store charge by Faradaic redox reaction on the electrode surface also penetrating the electrode. These capacitors can be divided into those involving underpotential deposition [7], redox pseudocapacitance [8], and intercalation pseudocapacitance [9], while hybrid supercapacitors combine both double layer and Faradaic charge storage mechanisms [10].

EDLCs are devices where charge accumulation is produced in the interface between an electronic conductor with a high specific surface area and an ionic conductor (an organic or aqueous electrolyte) [2]. These capacitors exploit the charge accumulation associated to the adsorption in the double layer region formed at the interface between polarizable electrodes and the electrolyte solution, theoretically described with increasing complexity by the models of Helmholtz, Gouy, and Chapman and Stern. In principle, Eq. (14.1) can also be applied to these capacitors but here, d represents the effective thickness of the double layer, typically of the order of few angstroms [11]. In summary, the interface between electrodes and electrolyte solutions can be treated as a capacitor where a layer of adsorbed species is the dielectric separator between the plates [12].

A second type of supercapacitors store charge is based on the pseudocapacitance associated with reversible redox reactions due to specifically adsorbed electrolyte ions at the electrode-electrolyte interface and/or intercalation of electro-active species in the layer lattice [13].

In general, the complete capacitor consists of two interfaces metal/electrolyte, equivalent to two capacitors in series. In a symmetric device, the total capacitance will be half of the capacitance of each of these capacitors. Then, the specific capacitance of the ensemble can be calculated relative to the mass of the combined cell (*m*) as:

$$C_{sp} = \frac{C}{4m} \tag{14.2}$$

This capacitance differs by a factor of 4 from the specific capacitance measured in a conventional three-electrode cell [14]:

$$C_{sp} = \frac{C}{m} \tag{14.3}$$

CV experiments detect the square responses characterizing the purely capacitive behavior (see Figure 10.2 in the corresponding chapter) [15]. Here, the measured current I_{DL} is almost constant and increases linearly on the potential scan rate [16]:

$$I_{DL} = C_{DL}\,v \tag{14.4}$$

The specific capacitance, C_{sp} (F kg^{-1}) can be evaluated from CVs as the quotient between the 'box' current (see Figure 10.2) and the product of the potential scan rate and the mass of the composite layer, *m*:

$$C_{sp} = \frac{I_{DL}}{mv} \tag{14.5}$$

Pseudocapacitance responses are characterized by voltammetric features diverging from the square signals as in Figure 10.2. Pseudocapacitance arises from fast, reversible Faradaic reactions occurring at or near a solid electrode surface over an appropriate range of potential. Such redox reactions can go beyond the surface area and penetrate the bulk of these materials [3,5]. Noble metal oxides such as RuO_2 and IrO_2 provide the most specific capacitance values (around 750 F g^{-1}) but they are toxic and expensive. Then, porous transition metal oxides such as CoO_x, NiO_x, and MnO_2 are currently under intensive research as materials for supercapacitors. The surface Faradaic reactions for NiO are generally described in terms of the processes [17]:

$$NiO + OH^- \rightarrow NiOOH + e^- \tag{14.6}$$

The reversible pseudocapacitance redox reaction might be represented as:

$$NiOOH + H_2O + e^- \rightarrow Ni(OH)_2 + OH^- \tag{14.7}$$

Figure 14.2 shows the voltammetric response of different NiO microstructures obtained from calcinations of $Ni(OH)_2$-based precursors in air at various temperatures accompanied by the SEM image of one of the NiO microstructures [18]. The voltammograms show the typical profile of $NiO/NiO(OH)/Ni(OH)_2$ materials.

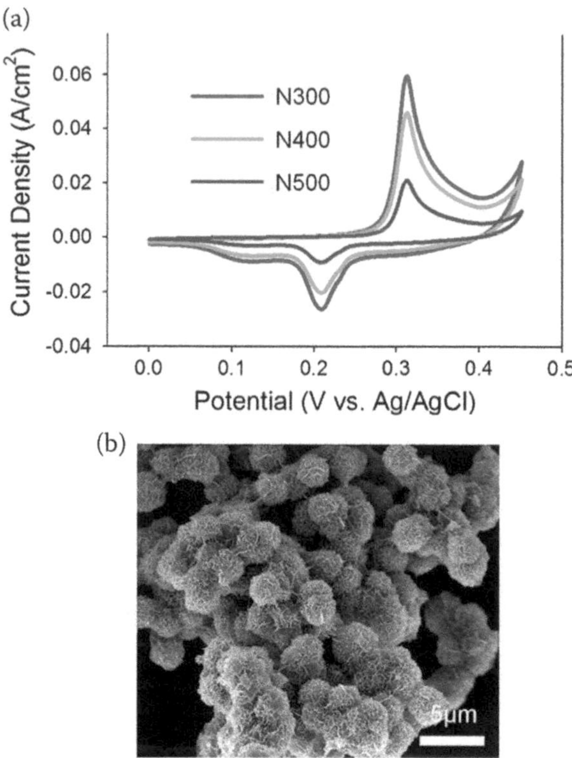

FIGURE 14.2 (a) CVs of Different NiO Microstructures Obtained from Calcinations of $Ni(OH)_2$-Based Precursors in Air at 300°, 400°, and 500 °C for 3 h (labeled as N300, N400, and N500, respectively). (b) SEM Image of a NiO Microstructure. Electrolyte 6 M KOH, Potential Scan Rate 20 mV s^{-1}. Adapted from Ref. [18] (Lee et al. *Electrochim. Acta*. 2011, 56: 4849–4857).

An important test for the performance of capacitors is charge/discharge curves. These correspond to the record of the potential difference vs. time graphs measured upon application of a constant intensity. Typical curves are illustrated in Figure 14.3 for B- and N- co-doped porous carbon composites in contact with 6 M KOH [15]. The specific capacitance can be calculated from charge/discharge curves at a constant current by using the relationship [16]:

$$C_{sp} = \frac{I\Delta t}{m\Delta V_d} \tag{14.8}$$

where I represents the discharge current density, m the mass of the composite, Δt the total time of discharge, and ΔV_d is the potential drop during discharge. The specific capacitance can also be calculated from the imaginary part of impedance at the low-frequency limit at an EIS experiment at open circuit potential as:

$$C_{sp} = \frac{1}{jm\omega Z_{imag}} \tag{14.9}$$

The fraction of metal electroactive sites involved in the Faradaic reaction, f_{ac}, can be calculated from the specific capacitance of the composite using the relationship:

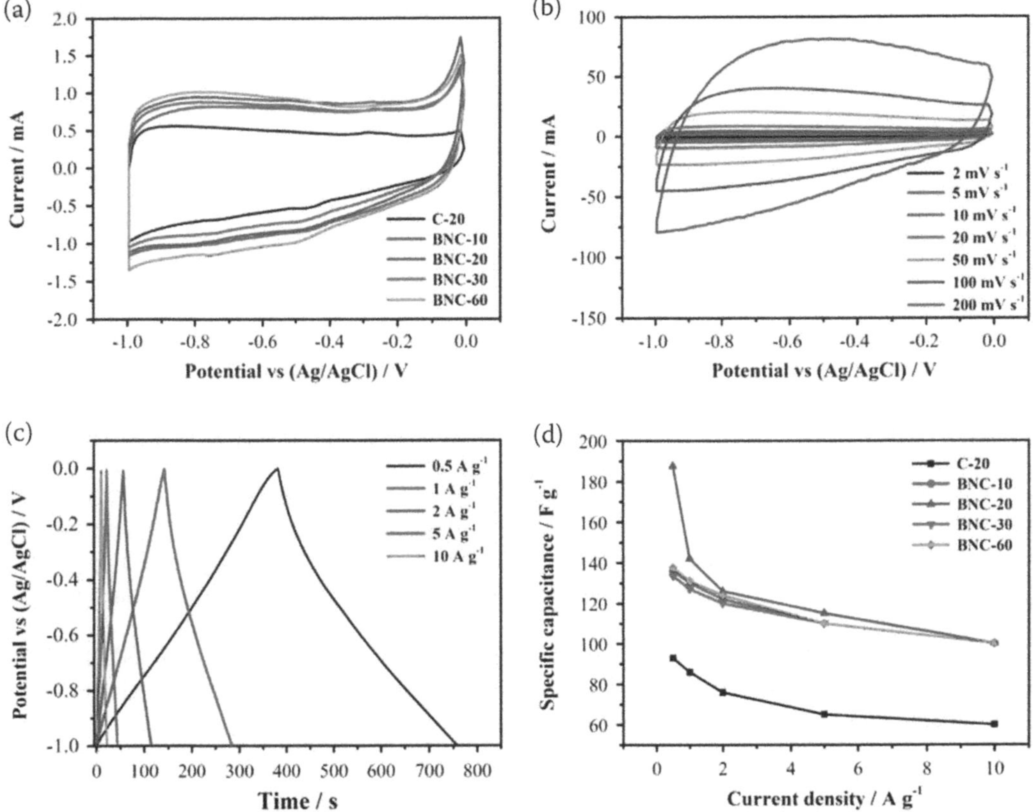

FIGURE 14.3 (a,b) CV Curves for Electrodes Containing C-20, BNC-10, BNC-20, BNC-30, and BNC-60 (a) and BNC-20 (b) B and N Co-doped Porous Carbon Composites in Contact with 6 M KOH Electrolyte; Potential Scan Rate 2 mV s^{-1}. (c) Charge/Discharge Curves of the Electrode Containing BNC-20 at Different Scan Rates and Current Density. (d) Gravimetric Capacitance as Function of the Current Density in the Range of 0.5–10 A g^{-1}. Adapted from Ref. [19] (Luo et al. *Electrochim. Acta.* 2020, 360: artic. 137010), with Permission.

$$f_{ac} = \frac{C_{sp} M \Delta V}{F} \tag{14.10}$$

where ΔV represents the potential window and M the average molecular mass of the material. The specific energy, W_{sp} (W h kg^{-1}), can be estimated by means of the equation:

$$W_{sp} = \frac{CU^2}{2m_{ac}} \tag{14.11}$$

In this equation, C represents the capacitance of the system, U the working voltage, and m_{ac} the total amount of active materials. Typical values of W_{sp} are ca. 50 W h kg^{-1}. The specific power density is given by:

$$P_{sp} = \frac{IU}{4m_{ac}} \tag{14.12}$$

In this equation, I and U represent the constant charge-discharge current and the potential range corrected by ohmic drop. When a constant current is used for charging and discharging processes, coulombic efficiency, η, can be calculated as:

$$\eta(\%) = 100\frac{t_D}{t_C} \tag{14.13}$$

A crucial aspect for practical applications is the stability of the response so that multiple cycling tests (typically 500–1,000 cycles) are routinely performed [20]. A high-rate discharge capability is one of the most important performances of supercapacitors in the application of electrode battery. Recent developments include the use of hybrid electrochemical capacitors, in which intercalations compounds ($Li_4Ti_5O_{12}$) was used as the negative material and activated carbon was used as the positive material [21], but a wide variety of materials including gel electrolytes, ionic liquids, graphene, and different carbon materials, etc. are currently under intensive research (see additional literature).

Figure 14.4 shows the Nyquist plots obtained for NiO microstructure before and after 1,000 cycles in contact with 6 M NaOH [18], data fitting to the equivalent circuit also depicted in the figure. At high frequencies, a small semicircle arc appears, while at low frequencies, the linear region leans more towards the imaginary axis, tending to purely capacitive behavior characteristic of porous electrodes.

With respect to capacitors, supercapacitors take advantages of their virtually unlimited life cycle typically ca. 10^6 cycles coupled with high cycle efficiency (95% or more), low impedance, high rates of charge and discharge with concomitant low charging time (seconds), no danger of overcharge, and relatively low cost/power ratios.

As disadvantages with respect to batteries, one can mention that supercapacitors possess relatively low energy density (from 1/5 to 1/10 than a conventional electrochemical battery), low cell voltages, and high self-discharge rates. Additionally, it should be noted that the linear discharge voltage of capacitors prevents the use of the full energy spectrum, while their use in electric vehicle applications requires electronic control and switching equipment.

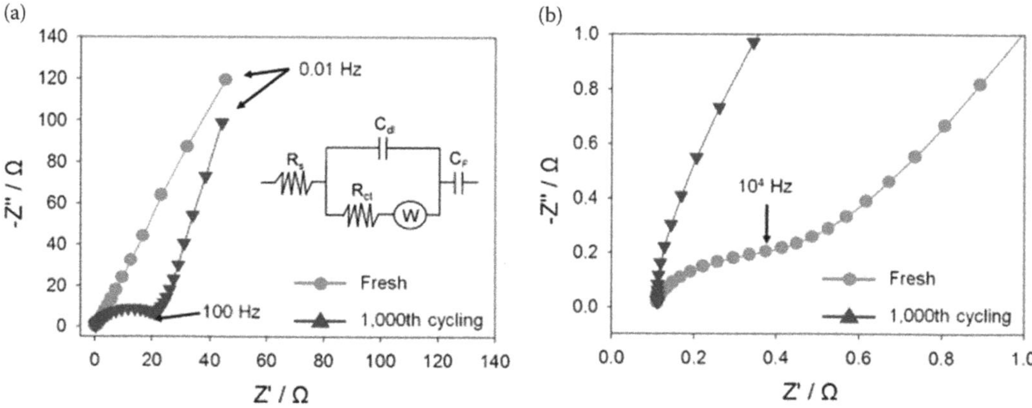

FIGURE 14.4 Nyquist Plots from EIS of a NiO Microstructure Before and After 1,000 Cycles at an Applied Potential of 0.25 V vs. Ag/AgCl in Contact with 6 M NaOH. The Inset in (a) is the Proposed Equivalent Circuit and (b) Indicates Enlarged Nyquist Plots in the High-Frequency Region. Adapted from Ref. [18] (Lee et al. *Electrochim. Acta*. 2011, 56: 4849–4857.), with Permission.

14.3 BATTERIES

Batteries are power sources aimed to maintain a difference of potential between two electrodes—cathode and anode—transforming "chemical energy" into "electrical energy". The characteristics to be balanced to evaluate batteries are:

- Magnitude of the potential and its stability
- Reversibility
- Cyclability (number of charge/discharge cycles)
- Mechanical resistance and weight (energy/mass ratio)
- Rate of energy release (power/mass relationship)
- Maximum current
- Economical and geopolitical accessibility
- Environmentally innocuous

Historically, batteries were created by Volta (1800) and further developed by Daniell (1836) and Leclanché (1868). The former was constituted by a copper electrode immersed into a solution of copper(II) sulfate and a zinc electrode immersed into a solution of zinc sulfate separated by a porous barrier. The second was formed by a solution of NH_4Cl where an electrode of zinc and a graphite electrode covered by MnO_2 paste were immersed. The electrode reactions can be formulated as:

$$\text{Cathodic: } 2MnO_2 + H^+ + 2e^- \rightarrow Mn_2O_3 + OH^- \tag{14.14}$$

$$\text{Anodic: } Zn \rightarrow Zn^{2+} + 2e^- \tag{14.15}$$

These are usually labeled as primary batteries because they cannot be re-charged because the reaction products do not remain adhered to the electrodes. Secondary batteries were first developed by Planté (1859) and Faure (1881), and the representative examples are lead batteries. These are composed of lead cathodes impregnated of PbO_2 and porous lead anodes immersed into a deposit of sulfuric acid solution. The electrode reactions can be written as:

$$\text{Cathodic: } PbO_2 + 2H_2SO_4 + 2e^- \rightarrow 2H_2O + PbSO_4 + SO_4^{2-} \tag{14.16}$$

$$\text{Anodic: } Pb + SO_4^{2-} \rightarrow PbSO_4 + 2e^- \tag{14.17}$$

This is a rechargeable battery, but the number of charge/discharge cycles is limited by the formation of gross crystals of $PbSO_4$.

Non-rechargeable alkaline batteries were introduced in 1960 and can be viewed as an evolution of the Leclanché cell. These batteries contained an anode of amalgamated zinc in contact with KOH and a cathode of graphite plus MnO_2 or Ag_2O. The corresponding reactions can be represented as:

$$\text{Cathodic: } MnO_2 + H_2O + e^- \rightarrow MnO(OH) + OH^- \tag{14.18}$$

$$\text{Anodic: } Zn + 4OH^- \rightarrow Zn(OH)_4^{2-} + 2e^-(\rightarrow ZnO) \tag{14.19}$$

or, alternatively,

$$\text{Cathodic: } Ag_2O + H_2O + 2e^- \rightarrow 2Ag + 2OH^- \quad (14.20)$$

Nickel-cadmium batteries, originally developed by Jungner in 1899, contained electrodes of these metals in contact with a strong alkaline medium (KOH). Formally, the electrode reactions can be written as:

$$\text{Cathodic: } NiO(OH) + H_2O + e^- \rightarrow Ni(OH)_2 + OH^- \quad (14.21)$$

$$\text{Anodic: } Cd + 2OH^- \rightarrow Cd(OH)_2 + 2e^- \quad (14.22)$$

Given the high contaminant character of Cd, their use was prohibited since 2017, being replaced by nickel-metal-hydride batteries described in the next section.

14.4 NICKEL METAL-HYDRIDE BATTERIES

Nickel metal-hydride batteries (NiMH) batteries derived from nickel-cadmium ones replaced cadmium by hydrogen absorbed in a metal alloy as active anodic material, whereas NiOOH remains as the active cathode material. Under charging, the parent $Ni(OH)_2$ is oxidized to NiOOH:

$$xNi(OH)_2 + xOH^- \rightarrow xNiO(OH) + xe^- \quad (14.23)$$

and a metal hydride is generated by water oxidation [22]:

$$M + xH_2O + xe^- \rightarrow MH_x + xOH^- \quad (14.24)$$

The charging process at the metal electrode can be represented in more detail by the following equations, which include the Volmer, Heyrovsky, and Tafel HER processes (see Chapter 4) [23]:

$$xH_2O + xe^- \rightarrow xH_{ads} + xOH^- \quad (14.25)$$

$$2xH_{ads} \rightarrow xH_2 \quad (14.26)$$

$$xH_{ads} + xH_2O + xe^- \rightarrow xH_2 + xOH^- \quad (14.27)$$

$$xH_{ads} \rightarrow xH_{abs} \quad (14.28)$$

resulting in the absorption of hydrogen within the metal electrode. Then, hydrogen diffuses into the bulk of the metal alloy where metal hydride is formed:

$$M + xH_{abs} \rightarrow MH_x \quad (14.29)$$

The hydrogen absorption may vary significantly between different alloys but a representative averaged value is 1.1 hydrogen atoms per metal atom [23]. This process also results in a large expansion of the unit cell volume of the metallic compounds. Figure 14.5 shows a schematic representation of a NiMH battery [24].

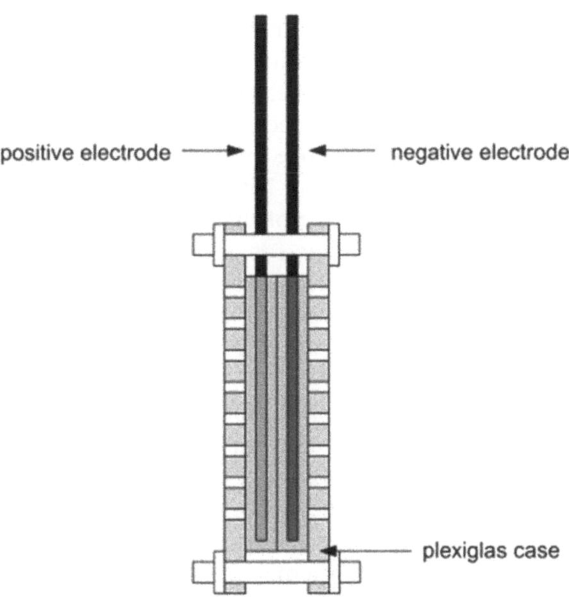

FIGURE 14.5 Scheme of a NIMH Secondary Battery. From Ref. [24] (Slepski et al. *J. Power Sources*. 2013, 241: 121–126), with Permission.

If there is overcharging, metal hydride is formed permanently at the electrode until there is O_2 generation in the $Ni(OH)_2$ electrode through the process:

$$xOH^- \rightarrow (x/4)O_2 + (x/2)H_2O + xe^- \tag{14.30}$$

The generated O_2 can diffuse through the separator and can be reduced to water, preventing the increase of pressure in the hermetic cell:

$$MH_x + (x/4)O_2 \rightarrow M + (x/2)H_2O \tag{14.31}$$

$$xH_2O + xe^- \rightarrow (x/2)H_2 + xOH^- \tag{14.32}$$

Under equilibrium conditions, the amount of O_2 generated through the process described by Eq. (14.30) equals that consumed in the metal electrode via process (14.31), so the internal pressure will be constant. When there is a large hydrogen amount in the metal electrode, the formation of H_2 via HER process (Eqs. (14.25–14.26)) determines a pressure increase in the cell.

Under over-discharge conditions, H_2 can be produced at the $Ni(OH)_2$ electrode (this can also be represented by (14.32)), then diffusing towards the metal electrode where it can react as:

$$M + (x/2)H_2 \rightarrow MH_x \tag{14.33}$$

$$MH_x + xOH^- \rightarrow M + xH_2O + xe^- \tag{14.34}$$

The possibility of gas accumulation has motivated the addition of a resealable safety vent in the top of commercial batteries.

The transport of hydrogen in the metallic electrode has received much attention. Hydride formation begins with a rapid diffusion of a small amount of H into the alloy to produce a dilute solid solution denoted as the α phase [25]. Upon increasing the amount of dissolved hydrogen, a second phase, β hydride, is formed [26]. The diffusion coefficient of hydrogen is critical as limiting in such cases to the reaction rate. Averaged values around 10^{-11} cm^2 s^{-1} are typically obtained from electrochemical methods, but large discrepancies can be found in the literature.

NiMH batteries employ β-$Ni(OH)_2$ as electrode material. This material converts to β-NiOOH during the charging process and this rearranges to γ-NiOOH when overcharged. This last process is accompanied by a significant expansion because of the difference in density between β-NiOOH and γ-NiOOH, which may result in poor electric contact between the current collector and β-Ni$(OH)_2$/β-NiOOH with a concomitant decrease of discharge capacity of the battery. Among others, layered double hydroxides of Ni and other metals—often termed 'stabilized' α- $Ni(OH)_2$ or doped $Ni(OH)_2$—have been tested as electrode materials [27], interlayer anions affecting electrochemical performance [28].

Much research is concentrating on metal alloys for NiMHs. The main conditions to be accomplished are: (a) capacity of hydrogen storage at ambient conditions; (b) oxidation and corrosion resistance in contact with alkaline electrolytes; (c) reversibility of the hydrogenation/dehydrogenation process; and (d) favorable electrochemical kinetics. A number of binary A_xB_y alloys have been studied. These combine one metal whose hydride formation is exothermic with a second metal whose hydride formation is endothermic so that hydrogen can easily be absorbed and desorbed. The more representative groups of binary alloys are AB_5 (prototype $LaNi_5$), AB_2 (prototype ZrV_2), AB (TiFe), A_2B (prototype Mg_2Ni), AB_2 (prototype $TiMn_2$), AB/A_2B (prototype TiNi/Ti_2Ni), and AB_3 (prototype $LaNi_3$), but a wide variety of ternary, etc. alloys are under study.

14.5 LITHIUM BATTERIES

The term lithium battery refers to a family of different devices formally having a common charge/discharge reaction that can be represented as $Li \rightarrow Li^+ + e^-$. Lithium batteries have found application in portable electronic devices because of their high working voltage (up to 4 V), high energy density, excellent cyclability [29], flat discharge characteristics, and long shelf life (up to ten years). These are constituted by a composite anode, a composite cathode, and a porous separator containing the electrolyte phase. Lithiated graphite (Li_xC_6, $0.1 < x < 0.8$) is the most frequently used as anode material while the cathode can b a compound capable of intercalate lithium ions such as Li_yCoO_2 ($0.4 < y < 1.0$) [30].

During discharge intercalated Li leaves anode particles passing to the electrolyte as Li^+ releasing electrons to the external load through a current collector. Li^+ ions migrate/diffuse through the electrolyte toward the cathode. Here, Li^+ receives electrons from the external load, incorporating Li into the cathode particles. The reversible oxidation process at a graphite anode may be represented as [30]:

$$LiC_6 \rightarrow V(C_6) + Li^+_{electrolyte} + e^-_{anode} \quad (14.35)$$

$V(C_6)$ being an intercalation vacancy within the graphite.

The reversible cathodic process at pristine CoO_2 can be described as:

$$V(CoO_2) + Li^+_{electrolyte} + e^-_{cathode} \rightarrow Li(CoO_2) \quad (14.36)$$

where $V(CoO_2)$ is an intercalation vacancy within a cathode particle and $Li(CoO_2)$ is intercalated Li.

There is a possibility, however, of two types of occupation of sites—via intercalation into interstitial sites or via occupying vacancies. In Kröger–Vink notation, these processes can be written as [31]:

$$Li + V_i^x \rightarrow Li^{\cdot}_i + e' \quad (14.37)$$

$$Li + V'_{Li} \rightarrow Li^{x}_{Li} + e' \quad (14.38)$$

respectively.

The use of lithium follows thermochemical considerations due to the negative standard potential of the Li^+/Li pair ($E° = -3.05$ V). While the electromotive force of a cell composed of Cu^{2+}/Cu ($E° = 0.34$ V) and Zn^{2+}/Zn ($E° = -0.71$ V) couples under standard conditions is of 1.1 V, an ideal Li^+/Li plus Au^{3+}/Au ($E° = 1.53$ V) cell would provide a maximum potential difference of 4.5 V. Under this perspective, the lithium-iodine cell, developed in 1986 [31], constituted an antecedent of lithium batteries [32].

The development of lithium batteries has followed a stepped evolution [33]. Given the reactivity of this metal with water, the fabrication of lithium batteries required the use of non-aqueous electrolytes and intercalation electrodes [34]. Lithium was initially used as the anode material. However, it easily reacts with the electrolyte forming a passivated layer on its surface (solid electrolyte interface, SEI) [35]. The next steps were the use of solid polymer electrolytes and the development of batteries using lithiated graphite and lithium cobalt oxide electrodes [36]. These constitute concentration cells where Li^+ ions 'rock' across the electrodes—the so-called lithium rocking chair batteries. Since then, a series of lithium batteries, including lithium-air and lithium-sulfur batteries, have been developed [33].

Cathode materials include lithiated transition metal oxides such as $LiCoO_2$, $LiNiO_2$, $LiMn_2O_4$, and related compounds [37]. Apart from these, different porous materials have been studied for lithium batteries. Thus, V_2O_5 and vanadates (LiV_3O_8, V_6O_{13}) [38], and sol-gel chromium-modified V_6O_{13} [39] have been studied. Such materials possess high cycling capability. A second group of widely studied materials are MnO_2 and related compounds, iron phosphate-based materials [40], and iron sulfides (FeS, FeS_2) [41]. Oxide cathodes can be divided into three groups [42]: (a) layered oxides displaying the so-called O_3 structure with a close-packed array of oxide ions such as $LiCoO_2$, where Li^+ and Co^{3+} ions occupy the (111) planes; (b) spinel oxides such as $LiMn_2O_4$ where Li^+ occupy tetrahedral sites and $Mn^{4+/3+}$ ions occupy octahedral sites in the cubic close-packed array of oxide-ions; and (c) polyanion oxides such as $Li_2Fe_2(MoO_4)_3$ displaying the so-called MASICON structure. The oxides of the first two types offer good electronic conductivity, while the materials of the polyanion oxide class require coating with carbon. The latter, however, offers better safety and thermal stability and provides high charge/discharge rates due to structural integrity. Figure 14.6 shows a scheme of lithium-ion batteries [43].

Figure 14.7 shows a schematic of the time scale at which different phenomena involved in lithium batteries occur accompanied by an equivalent circuit modeling such processes and the corresponding Nyquist plot in EIS measurements [44,45]. At high frequencies, the impedance response corresponds to the ohmic resistance of the electrolyte and contacts. At middle frequencies, two parallel resistance plus CPE units represent the charge transfer during the Li^+ insertion/de-insertion into the cathode and the charge transfer through the anode/SEI boundary. At low frequencies, diffusive effects dominate the impedance response, being represented by a Warburg element.

In the case of polyanion oxides, the central counteraction plays a significant role from the energetic point of view. Figure 14.8 shows the in-vacuum energy diagram comparing the Li^+/Li with the Fe^{3+}/Fe^{2+} couple in Fe_2O_3, $Fe_2(MO_4)_3$, and $Fe_2(SO_4)_3$. The force electromotive of the cell increases as the redox energy of the Fe^{3+}/Fe^{2+} couple decreases [37].

Lithium batteries can operate over a wide temperature range (at least between –40° and 70 °C), the most common solvents being polar organic liquids like acetonitrile, propylene carbonate, methyl formate, etc. Thionyl chloride ($SOCl_2$) and sulfuryl chloride (SO_2Cl_2) are inorganic liquids used as both the solvent and the active cathode material in lithium batteries. $LiClO_4$, $LiBF_4$, and $LiAsF_6$ are the most common salts to support electrolyte dosage. Recent developments include the use of solid electrolytes, namely polyethers and thin films of conducting ceramic materials, among others [33].

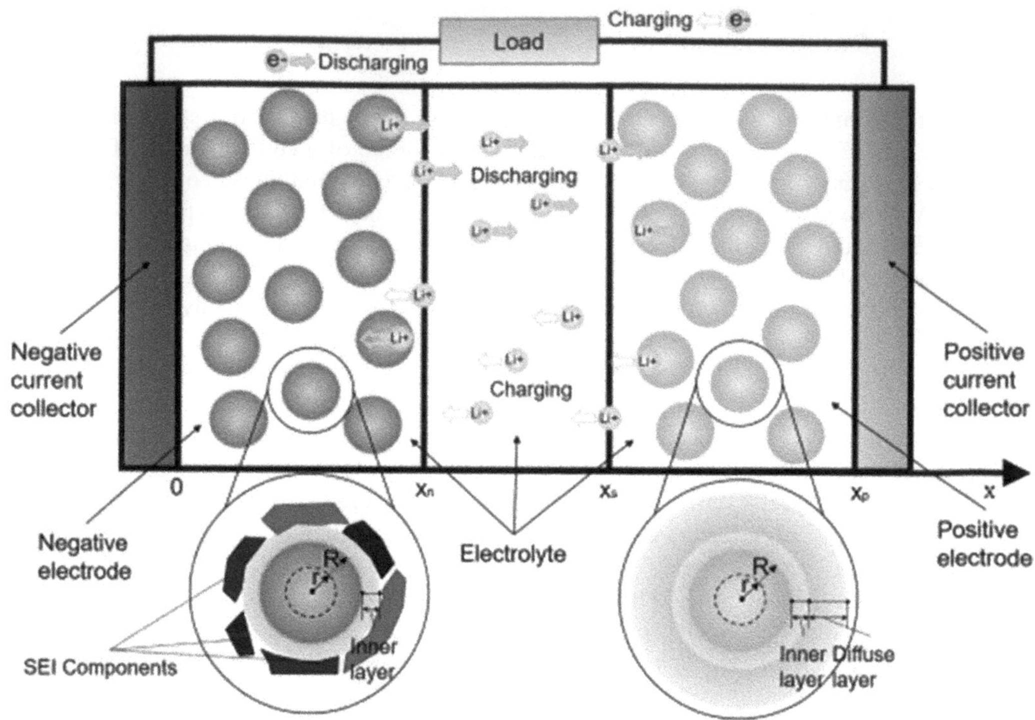

FIGURE 14.6 Schematics of a Lithium-Ion Battery. Reproduced from Ref. [43] (Zhang et al. *Electrochim. Acta.* 2020, 343: artic. 136094), with Permission.

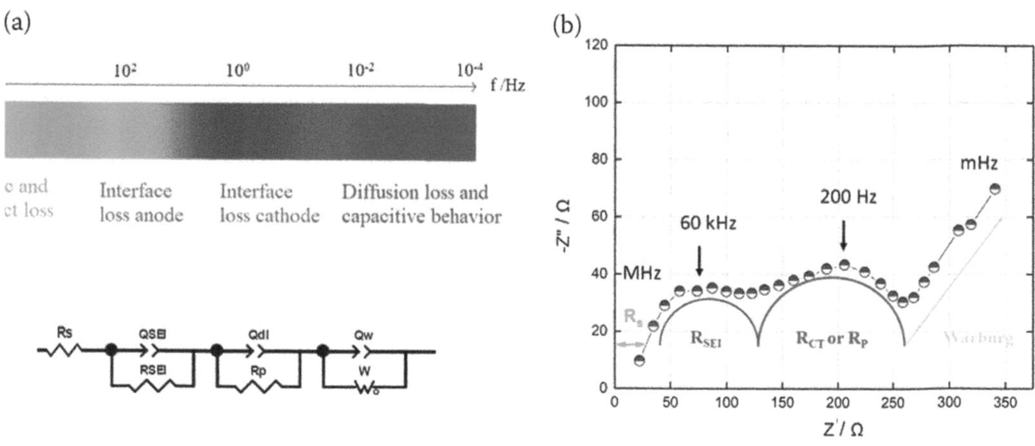

FIGURE 14.7 Equivalent Circuit for Lithium Battery with Polymer Electrolyte. From Ref. [44] (Nangir et al. *J. Electroanal. Chem.* 2020, 873; artic. 114385), with Permission.

Electrochemical performance of cathodes depends significantly on their behavior to lithium diffusion, in turn depending on particle size, surface morphology, homogeneity, and porosity. Porous composite cathodes can be regarded as two contiguous interwoven networks. The cathode-active material network consists of electronically conducting particles, whereas the second network is a polymer electrolyte that fills the pores between the active cathode material particles. The lithium ions participating in cathode reaction and originating from the lithium electrode migrate

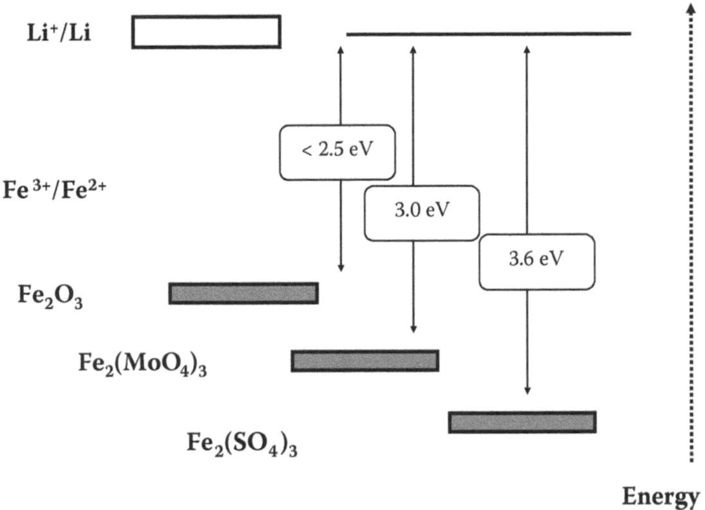

FIGURE 14.8 Energy Diagram for the Redox Energies of the Fe^{3+}/Fe^{2+} Couple for Different Oxo-iron species. From Data in [37].

through the composite into the polymer electrolyte-filled pores of the composite cathode to react with the cathode active material. The mechanism of charge-discharge may be affected by ion transport in the electrolyte network and redox insertion kinetics of the cathode-active material [36]. As general requirements, (i) electrical contact must be ensured between the particles of the cathode-active material, (ii) the electrolyte must completely fill the pores to prevent the separation of cathode-active particles and provide ion conductivity between them, and (iii) the mobility of ions in the polymer-electrolyte phase should be comparable with that in the cathode-active material phase.

The life cycle can be defined as the number of charge/discharge cycles that a battery can go through, at given conditions, before it reaches predefined minimum performance limits. This depends on the rates of charge and discharge, charge and discharge cut-off limits, depth of discharge, self-discharge rate, and temperature of operation. In the case of pyrite batteries, Strauss et al. [41] proposed a multistep mechanism for the charge-discharge mechanism of pyrite in polymer electrolytes in the temperature range 90–135 °C. In the first cycle, the reduction of pyrite occurs in two steps:

$$FeS_2 + 2Li^+ + 2e^- \rightarrow Li_2FeS_2 \quad (14.39)$$

$$Li_2FeS_2 + 2Li^+ + 2e^- \rightarrow Fe + 2Li_2S \quad (14.40)$$

In the second cycle, the process described by Eq. (14.40) is reversed while the insertion/de-insertion of lithium in the Li_2FeS_2 host becomes complicated by phase changes with the formation of pyrrhotite and lithium polysulfides.

The charging and discharging processes of lithium batteries involves the transfer of Li^+ from one ion insertion electrode to another. This process can be regarded as a topotactic intercalation reaction of Li^+ ions into interstitial sites in the crystalline host matrices, eventually accompanied by first-order phase transitions. Electrochemical detection of such phase transitions is conditioned by slow solid-state diffusion of ions into the solid matrix and uncompensated ohmic drops, and can be detected by voltammetric and chronoamperometric experiments. Here, two alternative models can be applied [46]—a quasi-Cottrellian approach and a progressive boundary motion model. In the quasi- Cottrellian approach, it is assumed that the quotient between the charge passed at a time t' after the maximum current, $Q_{t'}$, and the total charge passed at infinite time, Q_∞, are proportional to the square root of time:

$$\frac{Q_{t'}}{Q_\infty} = 4\left(\frac{Dt'}{\pi a}\right)^{1/2} \tag{14.41}$$

If diffusion of lithium ions (with diffusion coefficient D) occurs from two opposite edges of a particle of size a, in the moving boundary model, the $Q_{t'}/Q_\infty$ ratio becomes:

$$\frac{Q_{t'}}{Q_\infty} = 2\gamma\,(D_{phase}\,t)^{1/2} \tag{14.42}$$

D_{phase} being the diffusion coefficient for the propagating phase, and γ a numerical coefficient whose value approaches 0.52. This model leads to interrupted concentration/distance profiles while the quasi-Cottrellian model predicts monotonically decreasing concentration-distance curves to be experimentally tested [46].

Lithium batteries with aqueous electrolytes were introduced in the 1990s [47]. The idea was to occlude lithium inhibiting its reaction with water, allowing the use of water-soluble lithium salts as electrolytes. The overall electrochemical process in the case of $LiMn_2O_4$ can be represented as [48]:

$$Li_{1-x}Mn_2O_4 + xLi^+ + xOH^- \rightarrow LiMn_2O_4 + (x/2)H_2O + (x/4)O_2 \tag{14.43}$$

These batteries offer interesting properties [48]: (a) the ionic conductivity is high, permitting the use of thick electrodes; (b) water-soluble salts ($LiNO_3$, Li_2SO_4) are less expensive than lithium salts soluble in organic solvents (typically $LiPF_6$); and (c) lower cost in separators and cell assembly. On the contrary, the cell potentials are restricted to the range of water splitting (ca. 1.23 V) and the energy density and cycling performance are lower than those in cells using nonaqueous electrolytes.

14.6 OTHER BATTERIES

There is intensive research around new materials for batteries, to a great extent motivated by the fact that lithium is not particularly abundant in the earth's crust and is strongly localized in few sites for extensive extraction. extraction—Among others, ferrate(VI)-based materials are being investigated [49]. The cathodic charge storage of ferrate(VI) can be represented as the reduction processes:

$$FeO_4^{2-} + 3H_2O + 3e^- \rightarrow FeOOH + 5OH^- \tag{14.44}$$

$$FeO_4^{2-} + (5/2)H_2O + 3e^- \rightarrow (1/2)Fe_2O_3 + 5OH^- \tag{14.45}$$

CVs for the reduction of ferrate ions in alkaline solution are shown in Figure 14.9 [50], corresponding to different porous plastic bonded ferrate cathodes in contact with 5 M KOH.

A primary ferrate(VI) battery contains a Fe(VI) cathode and can use a zinc anode and alkaline electrolyte like a conventional alkaline battery. For Ag_2FeO_4 the general discharge reaction would be:

$$Ag_2FeO_4 + (5/2)Zn \rightarrow 2Ag + (1/2)Fe_2O_3 + 5/2\ ZnO \tag{14.46}$$

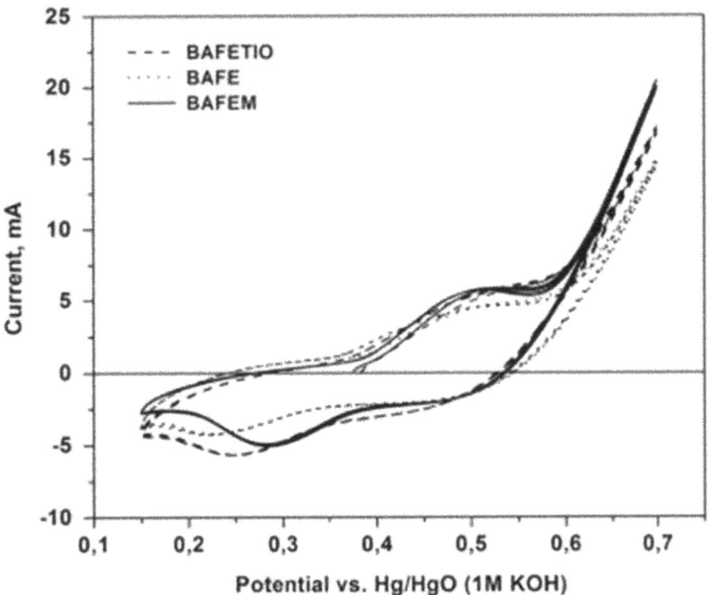

FIGURE 14.9 CV of Different Porous Plastic Bonded Ferrate Cathodes in Contact with 5 M KOH. Potential Scan Rate 100 mV s^{-1}. Reproduced from Ref. [50] (Simičić et al. *Electrochim. Acta.* 2017, 247: 516–523), with Permission.

All-iron batteries are constituted by an iron anode and a precipitated iron salt in the cathode. The basic reactions can be represented as [51]:

$$Fe_2O_3 + 4H^+ + 2e^- \rightarrow 2FeOH^+ + H_2O \quad (14.47)$$

$$Fe \rightarrow Fe^{2+} + 2e^- \quad (14.48)$$

Nickel-iron batteries operate in alkaline media with different iron-based anodes such as FeS and Fe_2O_3 [52]. Within a variety of (aqueous and nonaqueous) metal-air batteries [53], zinc-air batteries are under extensive investigation because of their high theoretical energy density and the abundance of zinc in nature. In these devices, metal oxidation is coupled with O_2 reduction at a suitable cathode. In alkaline media, the anodic process for a zinc battery can be represented as:

$$Zn + 2OH^- \rightarrow ZnO + H_2O + 2e^- \quad (14.49)$$

The theoretical standard potential of this battery is 1.250 V vs. NHE.

14.7 AN OVERVIEW ON ION TRANSFER ELECTROCHEMISTRY

Following Maier [31], we can distinguish three types of ion storage in solids that can be exemplified in Li_xRuO_2 [54]:

- Bulk storage based on dissolution where the guest ion occupies interstitial (intercalation) or vacant (insertion) sites in the solid.
- Storage accompanied by phase changes involving either phase transformation reactions (for instance, $FePO_4$/$LiFePO_4$) or reactions of decomposition (for instance, of $LiCoO_2$ to Co and Li_2O).

- Storage on high-dimensional defects and/or interfaces.

Theoretical modeling of the thermochemistry and kinetics of ion transfer in batteries has received considerable attention [26,55]. Different approaches have been reported [56], the porosity of the electrodes playing a significant role [57,58].

Relevant data can be derived from EIS where diffusion phenomena are modeled by means of Warburg elements. The diffusion coefficient D of the electroactive species can be calculated from the Warburg coefficient, σ_W, as:

$$D = \frac{1}{2}\left(\frac{RT}{AF^2\sigma_W c}\right)^2 \tag{14.50}$$

In this equation, c represents the concentration of the electroactive species and the other symbols have their customary meaning. The charge transfer resistance, R_{ct}, measured in the same EIS exeriments exhibits an Arrhenius-type variation with temperature so that the activation energy, E_a, of the interfacial Li^+ transfer process can be obtained from the slope of the $\ln(R_{ct}/T)$ vs. $1/T$ linear relationship:

$$\frac{\Delta(\ln(R_{ct}/T)}{\Delta(1/T)} = \frac{E_a}{R} \tag{14.51}$$

A more detailed analysis of EIS of lithium-ion batteries distinguishes five parts, schematized in Figure 14.10 [43] based on impedance of ion-insertion electrochemistry [45]. In the extremely high-frequency region (above 10 kHz, labeled as 1), the Nyquist plot is condensed as a point, which is related to the physical internal resistance of lithium ions and electrons passing through the separator, electrolyte, wires, etc. In the high-frequency region, there is a loop representative of the diffusion and transport of lithium ions on the surface of active particles (2). In the region of intermediate frequencies appears a second semicircle that is related to charge transfer through the double layer charge and the solid-electrolyte interface (3). In the low-frequency region, the spectrum displays a diagonal line, which is related to the diffusion of lithium ions (4). Finally, in the extremely low-frequency region (below 10 mHz), the spectrum consists of a loop and a perpendicular line. The semicircle can be related to the crystal structure change of the active particles or the formation of a new phase, whereas the line can be associated with the accumulation and consumption of lithium ions in the active material.

It should be noted that there is the possibility of charge storage in the electrode-electrolyte interface. This space charge region is formed at very low or high potentials and can be viewed as a thin passivation layer that is ion-conducting, but electron-blocking [43] Figure 14.10).

The electromotive force of the battery is determined by the difference between the redox energies of the anode and the cathode (i.e. between the Li^+/Li redox pair and the cathodic pair). In the case of oxide- and sulfide-based cathodes, the latter is limited by the stability of the anionic system that can be evaluated from the in vacuum redox energies [37]. The cell voltage is limited by the top of the S^{2-}:3p band for sulfide cathodes and the top of the O^{2-}:2p band for oxide cathodes. The clearly different positions of such bands in the energy scale limit the voltage of lithium-sulfide and lithium-oxide cells to ca. 2.5 and ca. 4 V, respectively.

14.8 FUEL CELLS

Fuel cells can be defined as energy-producing devices where chemical energy from a fuel and an oxidant is directly converted into electricity and heat without involving combustion cycles, in general defining a steady-state electrode-electrolyte system. In contrast with batteries—closed

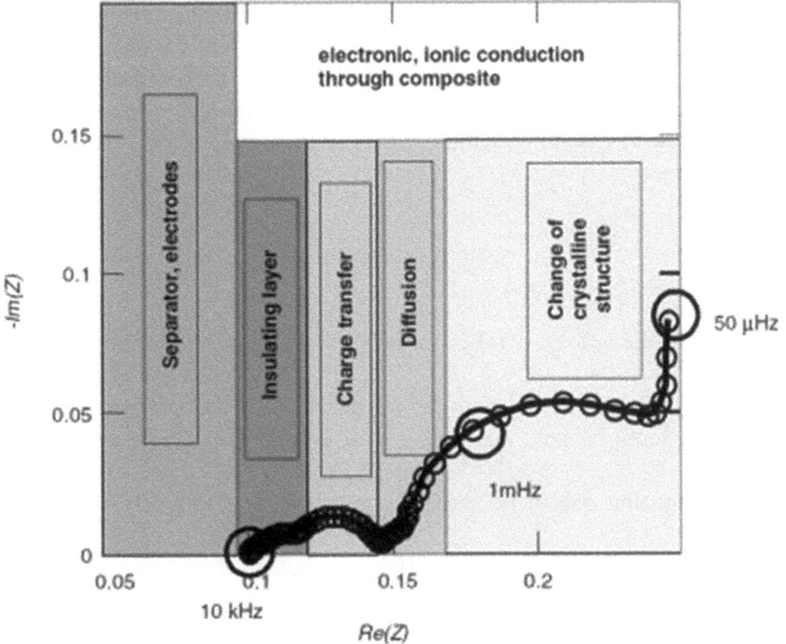

FIGURE 14.10 Typical Impedance Spectrum of Intercalation Electrodes. Reproduced from Ref. [43] (Zhang et al. *Electrochim. Acta.* 2020, 343: artic. 136094), with Permission.

systems that chemically store electrical energy—fuel cells can be described as open systems that can operate continuously by maintaining the appropriate flow of reactants. Here, the electrodes act as catalysts for the overall fuel oxidation reaction—also in contrast with batteries where the electrodes react during charge/discharge cycles [59,60]. Fuel cells provide advantages with respect to other energy conversion systems by virtue of their high efficiency, environmental safety, and possibility to recover exhaust heat.

If the fuel cell system is in thermal and mechanical equilibrium and operates at constant pressure and temperature, through a chemical reaction whose changes of free energy, enthalpy end entropy is, respectively, ΔG, ΔH, and ΔS, the intrinsic maximum efficiency, η_t, is given by:

$$\eta_t = \frac{\Delta G}{\Delta H} = 1 - \frac{T\Delta S}{\Delta H} \tag{14.52}$$

The theoretical efficiency in such systems is close to unity, but, in practice, the practical efficiency is significantly lower. This is due to different mechanical and thermal losses in the system but also to direct chemical reactions between the reactants and secondary electrochemical reactions in the cell. The voltage efficiency of the cell, η_v, is defined as the quotient between the cell voltage, V, at a given cell current, I, and the cell voltage at open circuit, V_r (the maximum value of the cell voltage, equivalent, if the cell is in equilibrium to the reversible cell potential, E_f):

$$\eta_v = \frac{V}{V_r} \tag{14.53}$$

It should be noted that fuel cells may reach stationary open-circuit cell voltages without equilibrium values (the electrochemical equilibrium is not fully attained). When the electrodes are connected through an external circuit, current flows and the cell voltage departs from the stationary open circuit value—a situation generically labeled as polarization of the cell. All phenomena from the departure of the equilibrium potential of the cell are also termed polarization phenomena. The Faradaic efficiency, η_f, is defined as the quotient between the cell current and the theoretical cell current, I_m, calculated by assuming that the reactants are completely consumed and that beside reactions do not take place:

$$\eta_f = \frac{I}{I_m} \tag{14.54}$$

The overall efficiency of the cell, η, is taken as the product between the three above efficiencies (i.e. $\eta = \eta_t\eta_v\eta_f$).

The so-called polarization of a fuel cell can be due to different phenomena, each generally quantified by the overpotential. Three main types of overpotentials can be considered:

- Activation overpotential due to the existence of slow charge transfer steps in the electrochemical reaction.
- Concentration overpotential due to slow diffusion processes in the electrolyte. This arises, in particular, in porous gas diffusion electrodes.
- Resistance overpotential associated with the difference of potential arising between the ends of resistive components of the cell. This can be minimized by using highly conductive electrolytes, electrodes of high area, and a small interelectrode gap.

Typical polarization curves are represented in Figure 14.11, corresponding to Pt/graphene-CNTs H_2/O_2 fuel cell [61]. In the region of low current, activation control predominates whereas, in the central, essentially linear region of the diagram, ohmic control exists. Finally, in the region of almost constant current, mass transfer control prevails. The same figure shows the variation of power density (P, W cm^{-2}) with current density recorded under linear potential scan conditions for the above fuel cell. The maximum performance is obtained in the central current density region.

Considering the different existing activation and resistive limitations, the cell voltage can be expressed as:

$$V = V_r - |\eta_a| - |\eta_c| - IR_{tot} \tag{14.55}$$

here, R_{tot} represents the total resistance of all ohmic components in the cell while η_a, η_c, represent the overpotentials of the anodic and cathodic reactions, respectively.

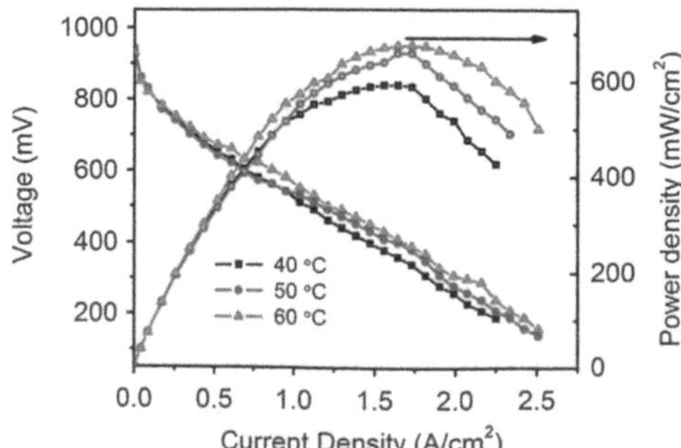

FIGURE 14.11 Polarization Curves and Current Density Curves for PT/Graphene-CNTs Cell at Different Temperatures. Adapted from Ref. [61] (Jyothirmayee Aravind et al. *J. Mater. Chem.* 2011, 21: 18199–18204), with Permission.

TABLE 14.1
Main Types of Fuel Cells

Type of fuel cell	Electrodes	Electrolyte	Primary fuel	Operation temperature (°C)
Alkaline (AFC)	Metal or carbon	NaOH/KOH	Hydrogen	60–90
Polymer electrolyte (PEFC)	Pt on carbon	Polymeric membrane	Reformate hydrogen	80–110
Direct methanol (DMFC)[a]	Pt on carbon	Polymeric membrane	Methanol	80–110
Phosphoric acid (PAFC)	Pt on carbon	H_3PO_4	Reformate hydrogen	160–200
Molten carbonate (MCFC)	Ni + Cr	Li_2CO_3-K_2CO_3	H_2/CO reformate	600–800
Solid oxide (SOFC)	Ni/Y_2O_3-ZrO_2	ZrO_2 doped with Y_2O_3	H_2/CO/CH_4 reformate	800–1,000

[a] Direct carbon, light hydrocarbons, formic acid, and others are also under development.

Fuel cells are usually classified according to their electrolyte and working temperature. The main types are alkaline fuel cells (AFCs), polymer electrolyte fuel cells (PEFCs), direct methanol fuel cells (DMFCs), phosphoric acid fuel cells (PAFCs), molten carbonate fuel cells (MCFCs), and solid oxide fuel cells (SOFCs) whose essential characteristics are listed in Table 14.1. Different fuels have been used—mainly carbon, hydrocarbons, natural gas and biogas, hydrogen, and alcohols. Oxygen from the air is the main oxidant agent, although other oxidants such as CO_2/oxygen/air mixtures and chlorine and chlorine dioxide have been tested.

In proton exchange membrane fuel cells (PEMFCs)—perhaps the most divulgate type of fuel cells—a proton-conducting polymer membrane acts as the electrolyte separating the anode and cathode sides. Among others, porous anodic alumina [62] and mesoporous anastase ceramic membranes [63] have been introduced in this field. Formally, the anodic reaction produces the electrons that flow through the external circuit towards the cathode transferred to the cathodic reactant. The cathode reaction can be written as:

$$O_2 + 2H_2O + 4e^- \rightarrow 4\ OH^- \tag{14.56}$$

whereas the anode reaction can be represented as:

$$H_2 + 2OH^- \rightarrow 2H_2O + 2e^- \tag{14.57}$$

Solid oxide fuel cells (SOFCs) are based on the use of doped metal oxides. Remarkably, there is the production of chemicals in the fuel cell in the course of its operation, so it can also be considered as an electrosynthetic reactor. Solid oxide direct carbon fuel cells use a perovskite cathode (such as $La_{1-x}Sr_xMnO_3$) and YSZ as a solid electrolyte accompanying a Ni/C anode. The main reaction in the electrolyte/C anode can be represented as [64]:

$$C + 2O^{2-}_{YSZ} \rightarrow CO_2 + 4e^- \tag{14.58}$$

This reaction can be accompanied by secondary ones [65]:

$$C + O^{2-}_{YSZ} \rightarrow CO + 2e^- \tag{14.59}$$

$$CO + O^{2-}_{YSZ} \rightarrow CO_2 + 2e^- \tag{14.60}$$

$$C + CO_2 \rightarrow 2CO \tag{14.61}$$

including [66].

$$O_2 + 4e^- \rightarrow 2O^{2-}_{YSZ} \tag{14.62}$$

Thermal and mechanical stability and oxygen ionic conductivity are general demands for materials to be used as electrode materials in SOFCs. Increasing stability and reducing costs requires lowering of the operating temperature. However, the latter is accompanied by an increase of electrode polarization that may be critical for the overall performance [60]. This is because of the apparent activation energies for the interfacial processes are generally higher than those for oxygen ionic transport in solid electrolytes [67]. The reduction of the working temperature results in a lower oxygen vacancy concentration with a concomitant increase of the role of ionic conductivity of electrode material.

As cathode materials, one can mention perovskite-type manganites, $(Ln,A)(Mn)O_{3\pm\delta}$ (Ln = La-Yb or Y; A = Ca, Sr, Ba, Pb), ferrites, $(Ln, A)FeO_{3\pm\delta}$, cobaltites, $(Ln,A)CoO_{3\pm\delta}$, nickelates, and cuprates. Perovskite-type manganites show high electrical conductivity, substantial electrocatalytic activity towards oxygen reduction at temperatures above 1,000–1,100 K, moderate thermal expansion coefficients, predominant electronic conductivity and low oxygen ion diffusivity, and transport properties and electrochemical activity dependent on oxygen non-stoichiometry. Here, electrocatalytic activity under high cathodic polarization is usually correlated with the generation of oxygen vacancies in the electrode surface. Doping with acceptor species in general leads to an increase in the *p*-type conductivity and electrochemical activity of $Ln_{1-x}A_xMnO_{3\pm\delta}$ manganites. Ferrites exhibit higher electronic or mixed conductivity than manganites but provide high thermal expansion coefficients due to oxygen losses at elevated temperatures, and may lead to thermochemical incompatibility with common solid electrolytes. Oxygen diffusivity is enhanced by extensive acceptor doping, but this is usually accompanied by increasing thermal and chemical expansion due to the weakening of metal-oxygen bonds and rising the atomic vibration inharmonicity [60].

Perovskite-related cobaltites show cathodic and transport properties better than those of manganites and ferrites, but also higher chemical and thermal expansion. It is accepted that the performance of ferrite and cobaltite electrodes is governed by the exchange processes at the electrode and electrolyte surfaces and by ion transfer across the cathode/electrolyte interface; the bulk ionic transport in the electrode is important but less critical than the above processes. Nickelates and cuprates are also considered interesting materials, although thermomechanical stability and compatibility with solid oxide electrolyte ceramics are, in general, lower with respect to the aforementioned materials.

Manganite-, cobaltite-, and ferrite-based perovskites, however, cannot be thermodynamically stable under anodic conditions. The most common materials for anodes consist of ceramic-metal composites (cermets) containing yttria-stabilized zirconia (YSZ) and Ni or other metals. In such systems, the metallic phase acts as an electronic conductor and catalyst. Here, composite layers should provide percolation paths for electrons, oxygen ions, and gas so that the electrochemical performance of the anode material is strongly dependent on its microstructure. The YSZ fraction in the composite, its porosity, and the sizes of the Ni and YSZ particles play a crucial role in determining the properties of the electrode. Microstructural differences are important in this context. It should be noted that, at elevated temperatures, Ni-YSZ cermets with high metal content experience fast degradation due to the coarsening of metal particles. Problems due to microstructural reconstruction, strains, and failure, accompanied by rising overpotentials during redox cycling, are also associated to high metal contents in these composites.

Doped ceria have also been extensively used as anodes in SOFC cells. The main advantage of such materials is the high catalytic ability of ceria with regard to combustion reactions involving oxygen—particularly, those using hydrocarbons and biogas as fuels. Reduced $CeO_{2-\delta}$ materials possess a significant mixed oxygen ionic and *n*-type electronic conductivity, properties that can be enhanced by acceptor-type doping. An additional metal must be added, because the electronic conduction in ceria is lower than necessary to avoid critical ohmic losses and/or current constriction effects on the overall anodic polarization. Other materials—namely, spinels, fluorites, pyrochlores, and molybdates, among others—have been proposed for SOFC anodes.

Composite cathode materials, surface modification of solid electrolytes and electrodes and insertion of nanosized particles of catalysts, and the application of oxide-activating agents onto the electrode surface are strategies to improve the overall electrochemical performance of fuel cells. There is a growing interest in preparing porous anodes with gas permeability, chemical stability, mechanical resistance, high electrical conductivity, and high thermal shock resistance.

A relevant aspect to emphasize is that the operation of SOFCs and other fuel cells involves electrochemical processes at the three-phase electrode/electrolyte/gas phase boundary. This ultimately refers to the models of ion-insertion solids described in Chapter 3 and has received considerable attention to model the involved electrochemical processes [68].

Direct methanol fuel cells are based on the electrochemical oxidation of methanol at an anode. The main problem is the disposal of a suitable catalyst for the complete oxidation of methanol to CO_2 and H_2O:

$$CH_3OH + H_2O \rightarrow CO_2 + 6H^+ + 6e^- \tag{14.63}$$

Intensive research is currently carried out to obtain efficient catalysts for this reaction. Ethanol [69] and formic acid [70], among others, are alternative fuels to methanol. This has originated intensive research on materials with electrocatalytic properties, Pt, Pd, and their alloys, in particular. Here, the detailed knowledge of the electrochemical pathway is a necessary but complex task. For instance, methanol oxidation at Pt electrodes in alkaline media a multi-step scheme consisting of the following processes has been proposed [71]:

$$Pt + OH^- \rightarrow PT\text{-}(OH)_{ads} + e^- \tag{14.64}$$

$$2Pt + CH_3OH \rightarrow Pt\text{-}H + Pt\text{-}(CH_3O)_{ads} \tag{14.65}$$

$$Pt\text{-}(OH)_{ads} + Pt\text{-}(CH_3O)_{ads} \rightarrow Pt_2\text{-}(CH_2O)_{ads} + H_2O \tag{14.66}$$

$$Pt\text{-}(OH)_{ads} + Pt_2\text{-}(CH_2O)_{ads} \rightarrow Pt_3\text{-}(CHO)_{ads} + H_2O \tag{14.67}$$

$$Pt\text{-}(OH)_{ads} + Pt_3\text{-}(CHO)_{ads} \rightarrow Pt_2\text{-}(CO)_{ads} + 2Pt + H_2O \tag{14.68}$$

$$Pt\text{-}(OH)_{ads} + Pt_2\text{-}(CO)_{ads} \rightarrow Pt\text{-}(COOH)_{ads} + 2Pt \tag{14.69}$$

$$Pt\text{-}(OH)_{ads} + Pt\text{-}(COOH)_{ads} \rightarrow 2Pt + CO_2 + H_2O \tag{14.70}$$

14.9 FUEL CELLS AND ELECTROCOGENERATION

There is a continuously growing interest in the development of cogenerative fuel cells—fuel cells involving the generation of useful chemicals. The cogeneration processes require favorable thermodynamic and kinetic conditions. Such systems involve a conventional fuel cell, an external circuit for using electrical energy, and a system for recovering the useful chemicals produced. Ideally, this methodology can be extended to portable devices.

The use of porous materials focuses a significant part of research because of their ability to provide electrodes with high specific area and their capacity to store reaction products—gaseous products, in particular. A remarkable aspect is that, frequently, a variety of products result from electrochemical reactions in the fuel cell, so selectivity for a given product is another factor to be considered in cell design. For instance, SOFC cells using bismuth oxide-based catalysts can produce the oxidative dehydrogenation of propylene to 1,5-hexadiene and benzene via the following reactions [72]:

$$\text{Anode: } C_3H_6 + O^{2-} \rightarrow C_3 \text{ dimers} + 2e^- \quad (14.71)$$

$$\text{Cathode: } O_2 + 4e^- \rightarrow 2O^{2-} \quad (14.72)$$

A variety of cogeneration processes has been described in recent literature. In particular, this can be achieved in AFCs consisting of a carbon cathode covered with a layer of catalyst mixed with carbon black and polytetrafluoroethylene (PTFE), KOH electrolyte, and a carbon cloth covered with a Pt-catalyst anode. In these cells, the cathode and anode reaction can be represented as [73]:

$$\text{Cathode: } O_2 + H_2O + 2e^- \rightarrow HO_2^- + OH^- \quad (14.73)$$

$$\text{Anode: } H_2 + 2OH^- \rightarrow 2H_2O + 2e^- \quad (14.74)$$

The aim is to yield electrical energy and HO_2^-, ultimately releasing H_2O_2. The global reaction in the cell is [74]:

$$H_2 + O_2 + OH^- \rightarrow H_2O + HO_2^- \quad (14.75)$$

Another interesting example is the oxidation of short-chain to long-chain hydrocarbons in PEMFC cells [74]. For instance, ethane oxidation to hexane can be represented by the sequence of reactions:

$$C_2H_6 \rightarrow C_2H_5\cdot + H^+ + e^- \quad (14.76)$$

$$2C_2H_5\cdot \rightarrow C_4H_{10} \quad (14.77)$$

$$C_4H_{10} \rightarrow C_4H_9\cdot + H^+ + e^- \quad (14.78)$$

$$C_2H_5\cdot + C_4H_9\cdot \rightarrow C_6H_{14} \quad (14.79)$$

TABLE 14.2
Some Examples of Cogenerative Processes in Fuel Cells Involving the Use of Porous Materials [74–80]

Electrode	Anode reaction	Cathode reaction	Type of fuel cell/ source
Porous Pt	$O_2 + 4\ H^+ + 4e^- \rightarrow 2\ H_2O$	$C_4H_8 + H_2O \rightarrow CH_3COC_2H_5 + 2H^+ + 2e^-$	AFC [75]
Porous Pt	$H_2 \rightarrow 2H^+ + 2e^-$	$C_6H_6 + 6H^+ + 6e^- \rightarrow C_6H_{12}$	AFC [76]
Active carbon	$H_2 \rightarrow 2H^+ + 2e^-$	$O_2 + 2H^+ + 2e^- \rightarrow H_2O_2$	AFC [77]
$(Bi_2O_3)_{0.85}(La_2O_3)_{0.15}$	$2C_3H_6 + 3O^{2-} \rightarrow C_6H_6 + 3H_2O + 6e^-$	$O_2 + 4e^- \rightarrow 2O^{2-}$	SOFC [78]
$ZrO_2(Y_2O_3)$	$PhCH_2CH_3 + O^{2-} \rightarrow PhCH{=}CH_2 + H_2O + 2e^-$	$O_2 + 4e^- \rightarrow 2O^{2-}$	SOFC [79]

A brief list of electric generation processes is shown in Table 14.2 [75–80], where representative examples involving porous materials and different types of fuel cells are included. In most cases, electric generation allows simultaneous removal of organic pollutants and flue gas emissions. Examples of this possibility will be treated in Chapter 16.

REFERENCES

[1]. Christen, T.; Carlen, M.W. 2000. Theory of Ragone plots. *Journal of Power Sources*. 91: 210–216.

[2]. Kotz, R.; Carlen, M. 2000. Principles and applications of electrochemical capacitors. *Electrochimica Acta*. 45: 2483–2498.

[3]. Shi, J.; Jiang, B.; Li, C.; Yan, F.; Wang, D.; Yang, C.; Wan, J. 2020. Review of transition metal nitrides and transition metal nitrides/carbon nanocomposites for supercapacitor electrodes. *Materials Chemistry and Physics*. 245: artic. 122533.

[4]. Conway, B.E. 1991. Transition from "supercapacitor" to "battery" behavior in electrochemical energy storage. *Journal of the Electrochemical Society*. 138: 1539–1548.

[5]. Sarangapani, S.; Lessner, P.; Forchione, J.; Griffith A.; Laconti, A.B. 1990. Advanced double layer capacitors. *Journal of Power Sources*. 29: 355–364.

[6]. Hu, C.C.; Tsou, T.W. 2002. Ideal capacitive behavior of hydrous manganese oxide prepared by anodic deposition. *Electrochemistry Communications*. 4: 105–109.

[7]. Herrero, E.; Buller, L.J.; Abruna, H.D. 2001. Underpotential deposition at single crystal surfaces of Au, Pt, Ag and other materials. *Chemical Reviews*. 101: 1897–1930.

[8]. Sugimoto, W.; Iwata, H.; Yokoshima, K.; Murakami, Y.; Takasu, Y. 2015. Proton and electron conductivity in hydrous ruthenium oxides evaluated by electrochemical impedance spectroscopy: The origin of large capacitance. *Journal of Physical Chemistry B*. 109: 7330–7338.

[9]. Ali, G.A.M.; Osama, A.F.; Makhlouf, S.A.; Yusoff, M.M.; Chong, K.F. 2014. Co_3O_4/SiO_2 nanocomposites for supercapacitor application. *Journal of Solid State Electrochemistry*. 18: 2505–2512.

[10]. Hu, L.; Chen, W.; Xie, X.; Liu, N.; Yang, Y.; Wu, H. 2011. Symmetrical MnO_2-carbon nanotube-textile nanostructures for wearable pseudocapacitors with high mass loading. *ACS Nano*. 5: 8904–8913.

[11]. Frackowiak, E.; Beguin, F. 2001. Carbon materials for the electrochemical storage of energy in capacitors. *Carbon*. 39: 937–950.

[12]. Noked, M.; Soffer, A.; Aurbach, D. 2011. The electrochemistry of activated carbonaceous materials: Past, present, and future. *Journal of Solid State Electrochemistry*. 15: 1563–1578.

[13]. Shukla, A.K.; Sampath, S.; Vijayamohanan, K. 2000. Electrochemical supercapacitors. Energy storage beyond batteries. *Current Science*. 79, 1656–1671.

[14]. Stoller, M.D.; Ruoff, R.S. 2010. *Energy & Environmental Science*. 3: 1294–1301.

[15]. Zheng, D.; Zhao, F.; Li, Y.; Qin, C.; Zhu, J.; Hu, Q.; Wang, Z.; Inoue, A. 2019. Flexible NiO microrods/nanoporous Ni/metallic glass electrode with sandwich structure for high performance supercapacitors. *Electrochimica Acta*. 297: 767–777.
[16]. Holze, R. 2017. From current peaks to waves and capacitive currents—On the origins of capacitor-like electrode behavior. *Journal of Solid State Electrochemistry*. 21: 2601–2607.
[17]. Xing, W.; Li, F.; Yan, Z.F.; Lu, G.Q. 2004. Synthesis and electrochemical properties of mesoporous nickel oxide. *Journal of Power Sources*. 134: 324–330.
[18]. Lee, J.W.; Ahn, T.; Kim, J.H.; Ko, J.M.; Kim, J.-D. 2011. Nanosheets based mesoporous NiO microspherical structures via facile and template-free method for high performance supercapacitors. *Electrochimica Acta*. 56: 4849–4857.
[19]. Luo, L.; Zhou, Y.; Yan, W.; Wu, X.; Wang, S.; Zhao, W. 2020. Two-step synthesis of B and N co-doped porous carbon composites by microwave-assisted hydrothermal and pyrolysis process for supercapacitor application. *Electrochimica Acta*. 360: artic. 137010.
[20]. Liang, Y.-Y.; Bao, S.-J.; Li, H.-L. 2007. Nanocrystalline nickel cobalt hydroxides/ultrastable Y zeolite composite for electrochemical capacitors. *Journal of Solid State Electrochemistry*. 11: 571–576.
[21]. Amatucci, G.G.; Badway, F.; Du pasquier, A.; Zheng, T. 2001. An asymmetric hybrid nonaqueous energy storage cell. *Journal of the Electrochemical Society*. 148: A930–A939.
[22]. Saldan, I. 2010. Primary estimation of metal hydride electrode performance. *Journal of Solid State Electrochemistry*. 14: 1339–1350.
[23]. Tliha, M.; Khaldi, C.; Boussami, S.; Fenineche, N.; El-Kedim, O.; Mathlouthi, H.; Lamloumi, J. 2014. Kinetic and thermodynamic studies of hydrogen storage alloys as negative electrode materials for Ni/MH batteries: A review. *Journal of Solid State Electrochemistry*. 18: 577–593.
[24]. Slepski, P.; Darowicki, K.; Janicka, E.; Sierczynska, A. 2013. Application of electrochemical impedance spectroscopy to monitoring discharge process of nickel/metal hydride batteries. *Journal of Power Sources*. 241: 121–126.
[25]. Wang, X.L.; Suda, S. 1992. Kinetics of the hydriding-dehydriding reactions of the hydrogen-metal hydride system. *International Journal of Hydrogen Energy*. 17: 139–147.
[26]. Zhang, W.; Srinivasan, S.; Ploehn, H.J. 1996. Analysis of transient hydrogen uptake by metal alloy particles. *Journal of the Electrochemical Society*. 143: 4039–4047.
[27]. Bernard, M.C.; Cortes, R.; Keddam, M.; Takenouti, H.; Bernard, P.; Senyarich, S. 1996. Structural defects and electrochemical reactivity of β-$Ni(OH)_2$. *Journal of Power Sources*. 63: 247–254.
[28]. Lei, L.; Hu, M.; Gao, X.; Sun, Y. 2008. The effect of the interlayer anions on the electrochemical performance of layered double hydroxide electrode materials. *Electrochimica Acta*. 54: 671–676.
[29]. Nazri, G.A.; Pistoia, G. 2004. *Lithium Batteries: Science and Technology*. Kluwer, Boston.
[30]. Colclasure, A.M.; Kee, R.J. 2010. Thermodynamically consistent modeling of elementary electrochemistry in lithium-ion batteries. *Electrochimica Acta*. 55: 8960–8973.
[31]. Maier, J. 2013. Thermodynamics of electrochemical lithium storage. *Angewandte Chemie International Edition*. 52: 4998–5026.
[32]. Phipps, J.P.; Hayes, T.G.; Skarstad, P.M.; Untereker, D. 1986. In-situ formation of solid/liquid composite electrolyte in Li/I_2 batteries. *Solid-State Ionics*. 18–19: 1073–1077.
[33]. Scrosati, B. 2011. History of lithium batteries. *Journal of Solid State Electrochemistry*. 15: 1623–1630.
[34]. Whittingham, M.S. 1978. Chemistry of intercalation compounds: Metal guests in chalcogenide hosts. *Progress in Solid State Chemistry*. 12: 1–99.
[35]. Peled, E.; Golodnitsky, D.; Ardel, G.; Eshkenazy, V. 1995. The sei model –Application to lithium-polymer electrolyte batteries. *Electrochimica Acta*. 14: 2197–2204.
[36]. Mizushima, K.; Jones, P.C.; Wiseman, P.J.; Goodenough, J.B. 1981. Li_xCoO_2 ($0<x\leq1$): A new cathode material for batteries of high energy density. *Solid-State Ionics*. 3–4: 171–174.
[37]. Ammunsdsen, B.; Paulsen, J. 2001. Novel lithium-ion cathode materials based on layered manganese oxides. *Advanced Materials*. 13: 943–956.
[38]. Abraham, K.M.; Goldman, J.L.; Holleck, G.L. 1981. Vanadium oxides as cathodes for secondary lithium cells. *Journal of the Electrochemical Society*. 128: 271–281.
[39]. Leger, C.; Bach, S.; Pereira-Ramos, J.P. 2001. The sol–gel chromium-modified V_6O_{13} as a cathodic material for lithium batteries. *Journal of Solid State Electrochemistry*. 11: 71–76.
[40]. Pahdi, A.K.; Nanjundaswamy, K.S.; Goodenough, J.B. 1997. Mapping of transition metal redox energies in phosphates with NASICON structure by lithium intercalation. *Journal of the Electrochemical Society*. 144: 2581–2586.
[41]. Strauss, E.; Golodnitsky, D.; Peled, E. 2002. Elucidation of the charge-discharge mechanism of lithium/polymer electrolyte/pyrite batteries. *Journal of Solid State Electrochemistry*. 6: 468–474.

[42]. Manthiram, A. 2020. A reflection on lithium-ion battery cathode chemistry. *Nature Communications.* 11: artic. 1550.
[43]. Zhang, Q.; Wang, D.; Yang, B.; Cui, X.; Li, X. 2020. Electrochemical model of lithium-ion battery for wide frequency range applications. *Electrochimica Acta.* 343: artic. 136094.
[44]. Nangir, M.; Massoudi, A.; Tayebifard, S.S. 2020. Investigation of the lithium-ion depletion in the silicon-silicon carbide anode/electrolyte interface in lithium-ion battery via electrochemical impedance spectroscopy. *Journal of Electroanalytical Chemi*stry. 873; artic. 114385.
[45]. Barsoukov, E.; Macdonald, J.R. Eds. 2005. *Impedance Spectroscopy: Theory, Experiment, and Applications*, 2nd edit. Wiley, New York.
[46]. Levi, M.D.; Aurbach, D. 2007. The application of electroanalytical methods to the analysis of phase transitions during intercalation of ions into electrodes. *Journal of Solid State Electrochemistry.* 11: 1031–1042.
[47]. Li, W.; Dahn, J.R. 1995. Lithium-ion cells with aqueous electrolytes. *Journal of the Electrochemical Society.* 142: 1742–1746.
[48]. Manjunatha, H.; Shures, G.S.; Venkatesha, T.V. 2011. Electrode materials for aqueous rechargeable lithium batteries. *Journal of Solid State Electrochemistry.* 15: 431–445.
[49]. Licht, S.; Naschitz, V.; Lin, L.; Chen, J.; Ghosh, S.; Liu, B. 2001. Analysis of ferrate (VI) compounds and super-iron Fe(VI) battery cathodes: FTIR, ICP, titrimetric, XRD, UV/VIS, and electrochemical characterization. *Journal of Power Sources.* 101: 167–176.
[50]. Simičić, M.V.; Čekerevac, M.I.; Nicolić-Bujanović, N.L.; Veljković, I.Z.; Zdravković, M.Z.; Tomić, M.M. 2017. Influence of non-stoichiometry binary titanium oxides addition on the electrochemical properties of the barium ferrate plastic-bonded cathode for super-iron battery. *Electrochimica Acta.* 247: 516–523.
[51]. Yensen, N.; Allen, P.B. 2019. Open source all-iron battery for renewable energy storage. *HardwareX.* 6: e00072.
[52]. Kao, C.-Y.; Chou, K.-S. 2010. Iron/carbon-black composite nanoparticles as an iron electrode material in a paste type rechargeable alkaline battery. *Journal of Power Sources.* 195: 2399–2404.
[53]. Mainar, A.R.; Iruin, E.; Colmenares, L.C.; Kvasha, A.; de Meatza, I.; Bengoechea, M.; Leonet, O.; Boyano, I.; Zhang, Z.; Blazquez, J.A. 2018. An overview of progress in electrolytes for secondary zinc-air batteries and other storage systems based on zinc. *Journal of Energy Storage.* 15: 304–328.
[54]. Balaya, P.; Li, H.; Kienle, L.; Maier, J. 2003. Full reversible homogeneous and heterogeneous Li storage in RuO_2 with high capacity. *Advanced Functional Materials.* 13: 621–625.
[55]. Miranda, D.; Costa, C.M.; Lanceros-Mendez, S. 2015. Lithium ion rechargeable batteries: State of the art and future needs of microscopic theoretical models and simulations. *Journal of Electroanalytical Chemistry.* 739: 97–110. Figs. 4, 9.
[56]. Karthikeyan, D.K.; Sikha, G.; White, R.E. 2008. Thermodynamic model development for lithium intercalation electrodes. *Journal of Power Sources.* 185: 1398–1407.
[57]. Lai, W.; Ciucc, F. 2010. Thermodynamics and kinetics of phase transformation in intercalation battery electrodes – Phenomenological modeling. *Electrochimica Acta.* 56: 531–542.
[58]. Ogihara, N.; Itou, Y.; Kawauchi, S. 2019. Ion transport in porous electrodes obtained by impedance using a symmetric cell with predictable low-temperature battery performance. *The Journal of Physical Chemistry Letters.* 10: 5013–5018.
[59]. Alcaide, F.; Cabot, P.-L.; Brillas, E. 2006. Fuel cells for chemicals and energy cogeneration. *Journal of Power Sources.* 153: 47–60.
[60]. Tsipis, E.V.; Kharton, V.V. 2008. Electrode materials and reaction mechanisms in solid oxide fuel cells: A brief review. *Journal of. Solid State Electrochemistry.* 12: 1367–1391.
[61]. Jyothirmayee Aravind, S.S.; Imran Jafri, R.; Rajalakshmi, N.; Ramaprabhu, S. 2011. Solar exfoliated graphene-carbon nanotube hybrid nano composites as efficient catalyst supports for proton exchange membrane fuel cells. *Journal of Materials Chemistry.* 21: 18199–18204.
[62]. Bocchetta, P.; Conciauro, F.; Di Quarto, F. 2007. Nanoscale membrane electrode assemblies based on porous anodic alumina for hydrogen-oxygen fuel cell. *Journal of Solid State Electrochemistry.* 11: 1253–1261.
[63]. Colomer, M.T. 2006. Proton conductivity of nanoporous anatase xerogels prepared by a particulate sol–gel method. *Journal of Solid State Electrochemistry.* 10: 54–59.
[64]. Delebeeck, L.; Hansen, K.K. 2014. Hybrid direct carbon fuel cells and their reaction mechanisms—A review. *Journal of Solid State Electrochemistry.* 18: 861–882.

[65]. Giddey, S.; Badwal, S.P.S.; Kulkarni, A.; Munnings, C. 2012. A comprehensive review of direct carbon fuel cell technology. *Progress in Energy Combustion Science*. 38: 360–399.
[66]. Li, C.; Shi, Y.; Cai, N. 2011. Effect of contact type between anode and carbonaceous fuels on direct carbon fuel cell reaction characteristics. *Journal of Power Sources*. 196: 4588–4593.
[67]. Yamamoto, O. 2000. Solid oxide fuel cells: Fundamental aspects and prospects. *Electrochimica Acta*. 45: 2423–2435.
[68]. Adler, S.B. 2004. Factors governing oxygen reduction in solid oxide fuel cell cathodes. *Chemical Reviews*. 104: 4791–4843.
[69]. Ouf, A.M.A.; Abd Elhafeez, A.M.; El-Shafei, A.A. 2008. Ethanol oxidation at metal-zeolite-modified electrodes in alkaline medium. Part-1: Gold-zeolite-modified graphite electrode. *Journal of Solid State Electrochemistry*. 12: 601–608.
[70]. Rees, N.V.; Compton, R.G. 2011. Sustainable energy: A review of formic acid electrochemical fuel cells. *Journal of Solid State Electrochemistry*. 15: 2095–2100.
[71]. Beden, B.; Kadirgan, F.; Lany, C.; Leger, J.M. 1982. Oxidation of methanol on a platinum electrode in alkaline medium: Effect of metal ad-atoms on the electrocatalytic activity. *Journal of Electroanalytical Chemistry*. 142: 171–190.
[72]. Jiang, Y.; Yentekakis, I.V.; Vayenas, C.G. 1994. Methane to ethylene with 85 percent yield in a gas recycle electrocatalytic reactor-separator. *Science*. 264: 1563–1566.
[73]. Alcaide, F.; Brillas, E.; Cabot, P.-L. 2002. A small-scale flow alkaline fuel cell for on-site production of hydrogen peroxide. *Electrochimica Acta*. 48: 331–340.
[74]. Li, W.S.; Lu, D.S.; Luo, J.L.; Chuang, K.T. 2005. Chemicals and energy co-generation from direct hydrocarbons/oxygen proton exchange membrane fuel cell. *Journal of Power Sources*. 145: 376–382.
[75]. Langer, S.H.; Colucci-Rios, J.A. 1985. Chemicals with power. *Chemtech*, 15: 226–233.
[76]. Langer, S.H.; Yurchak, S.J. 1969. Electrochemical reduction of the benzene ring by electrogenerative hydrogenation. *Journal of the Electrochemical Society*. 116: 1228–1229.
[77]. Yamanaka, I.; Hasegawa, S.; Otsuka, K. 2002. Partial oxidation of light alkanes by reductive activated oxygen over the (Pd black + $VO(acac)_2$/VGFC) cathode of H_2-O_2 cell system at 298 K. *Applied Catayisis A*. 226: 305–315.
[78]. Di Cosmo, R.; Burrington, J.D.; Grasselli, R.K. 1986. Oxidative dehydrodimerization of propylene over a Bi_2O_3---La_2O_2 oxide ion-conductive catalyst. *Journal of Catalysis*. 102: 234–239.
[79]. Michaels, J.N.; Vayenas, C.G. 1984. Styrene production from ethylbenzene on platinum in a zirconia electrchemical reactor. *Journal of the Electrochemical Society*. 131: 2544–2550.
[80]. Alcaide, F.; Cabot, P.-L.; Brillas, E. 2004. Cogeneration of useful chemicals and electricity from fuel cells. In Brillas, E.; Cabot, P.-L. Eds. *Trends in Electrochemistry and Corrosion at the Beginning of the 21st Century*. Universitat de Barcelona, Barcelona, pp. 141–165.

15 Magnetoelectrochemistry and Photoelectrochemistry of Porous Materials

15.1 MAGNETOELECTROCHEMISTRY

The term magnetoelectrochemistry refers to the study of electrochemical phenomena under the application of magnetic fields. The influence of magnetic fields on electrochemical reactions is well-known and involves metal electrodeposition to electropolymerization [1]. Conversely, Magnetic fields can be induced by electrochemical reactions [2]. The effect of magnetic fields on electrochemical processes in the solution phase is three-fold: (i) may cause energy level splitting in radicals (Zeeman effect); (ii) may bring about diamagnetic orientation of aromatic molecules; and (iii) may cause convection in the electrolytic solution (magnetohydrodynamic effects, *vide infra*).

From the studies of Fahidy [3], the magnetohydrodynamic effect has been widely considered [4]. This effect is explained in terms of a gradient of paramagnetic ions in the vicinity of an electrode surface (i.e. in terms of magnetoconvection). It has been suggested that the magnetic field will cause a transport of all ions because of the difference in the magnetic susceptibility in the solution at the electrode surface [5]. Lorentz force (F_B) due to moving charges in an electrolyte solution can be expressed as:

$$\overrightarrow{F_B} = (nz_+e\overrightarrow{v_+} + nz_-e\overrightarrow{v_-}) \times \overrightarrow{B} \quad (15.1)$$

where B denotes the externally applied magnetic field z_+, z_- are the formal charge of ions, and v_+, v_-, their respective velocities. Notice that even though the velocities and positive and negative mobile charges point in opposite directions, the corresponding charges have opposite signs. Accordingly, cations and anions experience magnetic forces in the same direction. As a result, electrolyte near to the electrode surface experiences a convective effect that increases limiting currents.

A second convective effect is due to the motion of charges across the double layer in the vicinity of the electrode [6]. Charge motion creates an electrokinetic force that can be expressed as [7]:

$$F = \frac{\sigma E}{\delta} \quad (15.2)$$

where σ is the charge density in the diffusion layer, δ the thickness of this layer, and E the intensity of the induced non-electrostatic electrical field. Additionally, there is a paramagnetic gradient force that causes the motion of paramagnetic ions in the diffusion layer. The magnetic force can be expressed as proportional to the gradient of concentration of the magnetic particles [6]:

$$F = \chi_B \frac{B^2}{2\mu_o} \vec{\nabla} c \quad (15.3)$$

χ_B being the molar magnetic susceptibility of the mobile species and μ_o the permeability of vacuum. The magnitude of these forces is generally so low to produce significant effects on electrochemical responses [8].

In considering a solution of molecular species with magnetic dipolar character, the average low-field magnetic moment (m_B) per magnetic dipole of a given species, j, with spin ½ (including common organic radicals) is given by:

$$\overrightarrow{<m_B>_j} = \left(\frac{g\mu_B}{2}\right)^2 \frac{\overrightarrow{B}}{kT} \tag{15.4}$$

In this equation, g is the Landé factor and μ_B the Bohr magneton. The total paramagnetic force acting on the unit volume element, F_{pm}, is given by:

$$\overrightarrow{F_{pm}} = N_A \left(\frac{gm_B}{4}\right)^2 \left(2c_j \frac{\overrightarrow{B} \cdot \nabla \overrightarrow{B}}{kT} + \frac{|\overrightarrow{B}|^2}{kT} \nabla c_j\right) \tag{15.5}$$

Here, c_j designates the concentration of radical species and N_A is the Avogadro's number. The first term in the above equation describes the field gradient paramagnetic force exerted in areas where the magnetic field gradient $\neq 0$, typically occurring in the vicinity of ferromagnetic particles [9]. The second term in Eq. (15.5) describes the effect due to the presence of gradients in the concentration of radical species. This occurs, for instance, in the diffusion layer during electrochemical experiments and can be obscured by other phenomena like convection.

In the context of the electrochemistry of porous materials, it can be expected that the application of magnetic fields can promote alterations in charge transport through materials, electrolyte, and interfaces, and—eventually—variations in the orientation of guest molecules can potentially be reflected in the electrochemical response of the materials. By the first token, magnetic fields are used in electrolysis processes to prepare metals and alloys [10], composites [11], polymers, and other materials [12]. In the case of metals and alloys, electrodeposition under magnetic fields results in differences in the morphology, roughness, and surface properties depending on the intensity and orientation of the field as well as the diamagnetic, paramagnetic, or ferromagnetic character of the metal [6]. The application of magnetic fields also influences the porosity of anodic oxide films [13] and the orientation of polymers resulting from electropolymerization [14]. The application of variable magnetic effects is also the subject of magnetoelectrochemical studies [6].

The effect of static magnetic fields on the electrochemistry of porous materials has received scarce attention. Ideally, confinement of electroactive guest species into porous hosts may produce a magnetic-field sensitive electrochemistry as a result of at least two effects: magnetoconvective influence on the diffusion of mobile ions through the ion-permeable solid, and orientation effects on entrapped molecules. Figure 15.1 shows reported data for bipyrylium bication [1,4-bis(3,5-diphenyl-4-pyrylium)phenylene] (=BTP^{2+}) associated with zeolite Y and MCM-41 mesoporous aluminosilicate (BTP@Y and BTP@MCM, respectively) [15]. Molecular modeling indicates that the BTP^{2+} bication cannot be accommodated in a single cavity of zeolite Y and should occupy two neighboring supercages. The response of BTP@Y in Bu_4N^+/MeCN is restricted to a unique reduction process near –0.40 V, while in the presence of Et_4N^+, two reduction processes at ca. –0.2 and –0.4 V appear. This response, which can be associated with the presence of different boundary-associated topological redox isomers (see Chapter 5), is modified upon application of moderate magnetic fields (0.02–0.2 T) that produces significant modifications in the voltammetric behavior of BTP@MCM and BTP@Y, with variation in the relative height of the peaks and appearance of minor additional signals. These features,

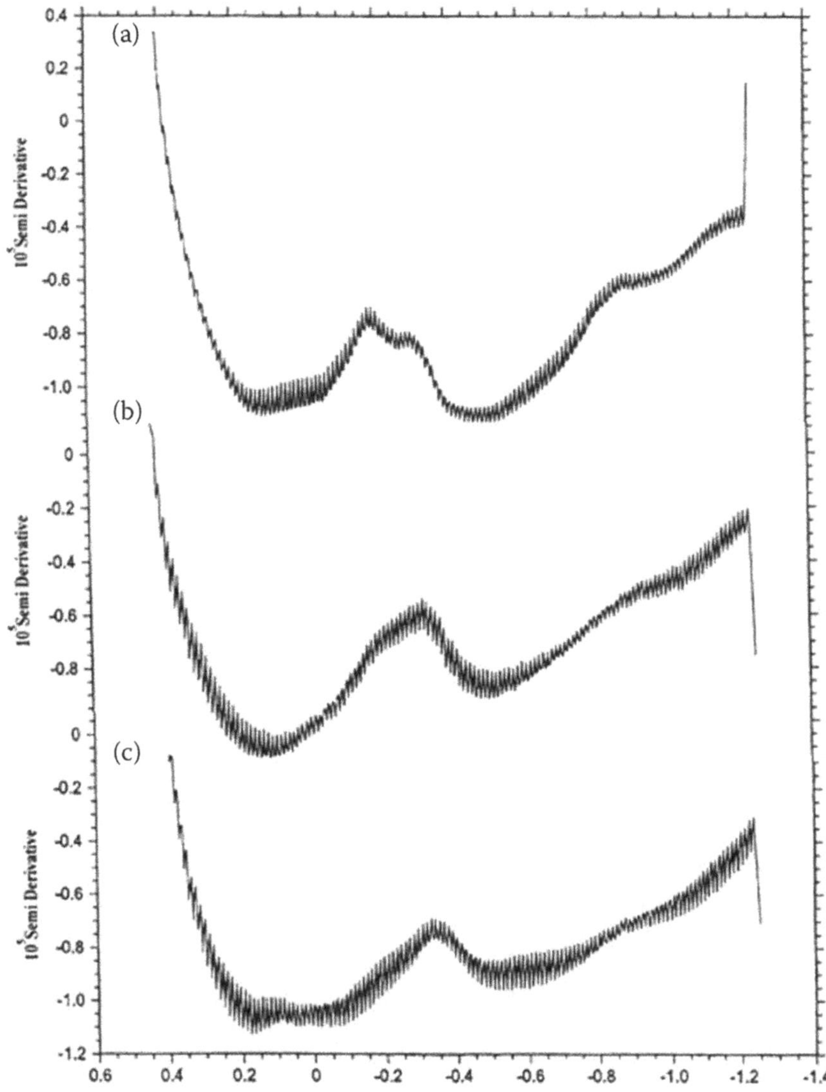

FIGURE 15.1 SWVs for BTP@Y Immersed into (0.05 M Et_4NclO_4 + 0.10 M Bu_4NPF_6)/MeCN Under the Application of a Magnetic Field of 0.2 T Forming Angles of (a) –60°, (b) 0°, (c) 60° with the Vertical Electrode Axis. Potential Scan Initiated at 0.45 V in the Negative Direction. Potential Step Increment 4 mV; Square Wave Amplitude 25 mV; Frequency 15 Hz. Adapted from Ref. [15] (Doménech-Carbó et al. *J. Electroanal. Chem.* 2005, 577: 249–262), with Permission.

depending on the angle between the applied magnetic field and the electrode axis, can be attributed to the superposition of the aforementioned magnetoconvective effect associated ion motion, paramagnetic effect associated with concentration gradients, and orientation effects induced by the applied magnetic field. Future developments should consider magnetically-assisted electrochemical formation of porous layers and the effect of non-stationary magnetic fields because relaxation effects associated with encapsulation would be possibly detected.

15.2 PHOTOELECTROCHEMISTRY

The term photoelectrochemistry is in general applied to all phenomena where photon absorption is accompanied by electrochemical processes. The usual requirements for photoelectrochemical activity are: (a) the semiconductor character of the electrode material, (b) the existence of an electrolyte concentration high enough to significantly exceed the density of charge carriers in the semiconductor, and (c) the semiconductor should be reverse biased with respect to the solution [16].

A significant part of such systems is addressed to produce hydrogen from water, pioneered by Honda and Fujishima [17] who discovered that water can be decomposed into H_2 and O_2 by using a photocell comprising a TiO_2 electrode under light irradiation. Due to their wide band gaps, the majority of the photocatalysts are only active under UV light irradiation, limiting their application in hydrogen production from water and sunlight. For all types of cells, the position of the valence and conduction bands of the semiconductor relative to the redox energy levels of the redox couple is an important parameter to estimate the efficiency of the cell in energetic terms.

As previously noted, photoelectrochemical cells involve semiconductor/electrolyte junctions. To describe this system, it should be kept in mind that for an intrinsic semiconductor at room temperature, the Fermi level (the energy value, E_F, where the probability of a level being occupied by an electron is equal to the probability of being occupied by a vacant) is at the midpoint in the bandgap region between valence and conduction bands. For a doped material, the location of the Fermi level depends on the doping level and dopant concentration. For doped *n*-type materials, E_F lies just below the conduction band edge, whereas for doped *p*-type materials the Fermi level is located slightly above the valence band edge. E_F can be determined from measurements of work functions or electroaffinities and is usually expressed with respect to the vacuum scale of energies (i.e. taking as zero the energy for a free electron in vacuum). It is pertinent to note, however, that most oxides, sulfides, etc. are *n*- or *p*-type semiconductors without need for doping because defects (cationic, vacancies, anionic vacancies, etc.) provide the trap energy levels producing the semiconducting behavior.

When equilibrium is established between a semiconductor and an electrolyte solution, the Fermi level in both phases is equalized via charge transfer between phases. In the case of an *n*-type semiconductor, this implies the passing of electrons from the semiconductor to the solution phase, so that the semiconductor becomes positively-charged and the electrolyte, negatively-charged. The excess charge in the semiconductor resides (contrary to a metal, where excess of charge resides at the surface) in a space charge region. This region can be considered analogous to the diffuse double layer (see Chapter 1) formed in the vicinity of the electrolyte/semiconductor junction. Then, the surface region of the semiconductor (50–2,000 Å-sized) becomes depleted of majority carriers and a depletion layer is formed [18]. This situation is represented in energy diagrams by bending of the bands, as shown in Figure 15.2 for *p*-type and *n*-type semiconductors.

The band bending depends on the applied potential. At the potential where no charge excess exists in the semiconductor, there is no band bending; this is the potential of zero charge or flat band potential, E_{FB}. For *n*-type semiconductors, at potentials negative (positive) with respect to E_{FB}, there is an excess of electrons (i.e. an accumulation (depletion) in the space charge region). For *p*-type semiconductors, at potentials positive (negative) relative to E_{FB} there is an excess of majority charge carriers (i.e. there is accumulation (depletion) in the space charge region [19]).

If there is an accumulation layer, there is disposal of majority charge carriers and the semiconductor behaves similarly to a metallic electrode. If there is a depletion layer, electron transfer processes are blocked to a great extent. Accordingly, an *n*-type semiconductor in the dark behaves like an insulating material at potentials positive to the E_{FB}. At potentials negative to E_{FB}, the electrons move to the semiconductor bulk while the holes move to the semiconductor/electrolyte interface so that the semiconductor acts as a cathode (i.e. an electron acceptor).

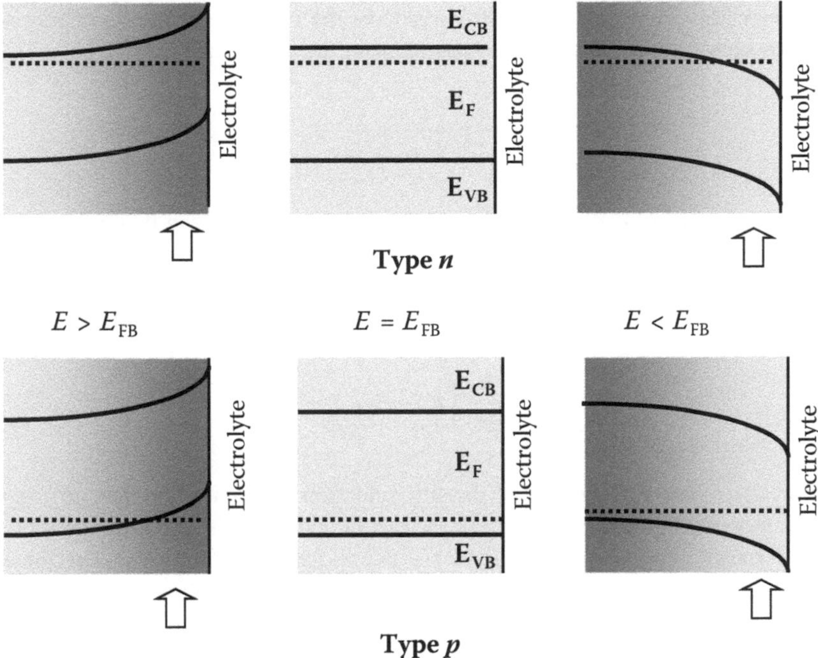

FIGURE 15.2 Energy Diagrams for the Semiconductor/Electrolyte Junction Under Dark Conditions for *n*-Type and *p*-Type Semiconductors Under Application of Different Potentials Relative to the Flat Band Potential. The Arrows Mark the Space Charge Region.

Semiconductor exposition to radiation of sufficiently high energy promotes electrons from the valence to the conduction band. In the conductor bulk, this process is followed by electron-hole recombination and heat-producing, non-radiative decay. In the space charge, however, the electric field in this region favors charge separation. As a result, the behavior of a *n*-type semiconductor at potentials negative of the EFB equals that in the dark (dark cathode), but at potentials positive of the flat band potential, the material acts as an anode (photoanode). Following similar considerations, *p*-type semiconductors act as dark anodes and photocathodes [19].

Accordingly, not only the illumination conditions but also the applied potential must be accounted for to describe the semiconductor photoelectrochemistry [20]. It should be noted, however, that this ideal, reversible behavior of the semiconductor/electrolyte junction is rarely found, deviations from reversibility arising by inherently slow charge transfer across the semiconductor/electrolyte interface, surface oxide formation, corrosion, etc. [18], motivating the development of elaborated models on potential distribution of porous crystalline semiconductor films [21].

There are three basic methods to determine the flat band potential of a semiconductor:

a. Measuring the open circuit potential of the cell (obviously relative to a reference electrode) under irradiation with different intensities. Then, the plot of the photo potential vs. the intensity of the radiation will reach a plateau at the E_{FB}.
b. Recording the photocurrent passing through the cell, varying the potential under constant irradiation. The onset of the current (when the dark current equals the photocurrent) will correspond to E_{FB}.
c. From impedance values at different potentials under depletion conditions. Here, the total

capacitance results from the series combination of the capacitance due to charge separation in the space charge (C_{SC}) and charge separation in the electrolyte double layer (C_H). Then:

$$\frac{1}{C_{exp}} = \frac{1}{C_{SC}} + \frac{1}{C_H} \tag{15.6}$$

Under ordinary conditions, $C_{SC} \ll C_H$, so that the contribution of C_H can be neglected. Then, the system can be treated as an RC circuit evaluating the capacitance as the imaginary component of the measured impedance at high frequencies (usually kHz).

According to the Mott-Schottky model, the space charge capacitance verifies the relationship:

$$\frac{1}{C_{SC}^2} = \frac{2}{Ne\varepsilon\varepsilon_o}\left(E - E_{FB} - \frac{k_B T}{e}\right) \tag{15.7}$$

In this equation, N represents the donor density (electron concentration for an n-type semiconductor or hole concentration for a p-type semiconductor). This equation predicts a linear variation of $1/C^2$ on the applied potential E whose extrapolation to $1/C^2 \rightarrow 0$ determines the flat band potential. Interestingly, the slope of the linear representation discerns between p-type and n-type semiconductors. Figure 15.3 shows Mott-Schottky plot recorded at a hematite photoanode under dark conditions from capacitance measurements in contact with 0.1 M phosphate buffer at pH 7 [22]. Potentials were originally measured at 298 K vs. Ag/AgCl (3 M NaCl) reference electrode and converted to the RHE scale as E (vs. RHE) = E (vs. Ag/AgCl) + 0.0591 pH + $E°$ (Ag/AgCl vs. RHE). This last quantity represents the standard potential of the Ag/AgCl in 3 M NaCl reference electrode vs. RHE, equal to 0.209 V.

15.3 CAPACITIVE VOLTAMMETRY OF SEMICONDUCTORS

The voltammetric response of electroactive semiconducting materials such as MnO_2 or V_2O_5 experiencing proton-assisted reduction processes was treated in Chapter 8. In a number of cases like TiO_2, the voltammetry of metal oxides in contact with aqueous electrolytes does not produce this type of process (at least to a significant extent). Then, the observed voltammetry can be modeled combining two elements—a capacitance related to the double-layer or redox process and a resistance-associated to charge transport. The essential idea is that nanoporous semiconductor solids permeated with a conductive phase accumulate injected electron charges. There is a displacement of the Fermi level toward the semiconductor conduction band that results in an increase of the electron concentration in the film. This yields an intrinsic differential capacitance varying

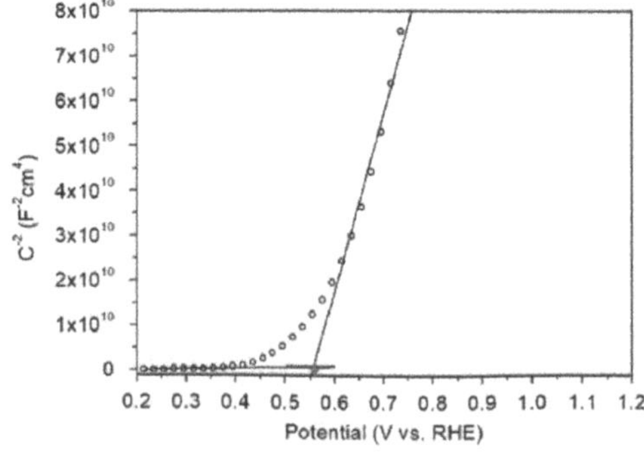

FIGURE 15.3 Mott-Schottky Plot of a Hematite Photoanode Under Dark Conditions. Capacitance Measurements in Contact with 0.1 M Phosphate Buffer at pH 7. Adapted from Ref. [22] (Chae et al. *Electrochim. Acta*. 2019, 297: 784–793), with Permission.

(a)

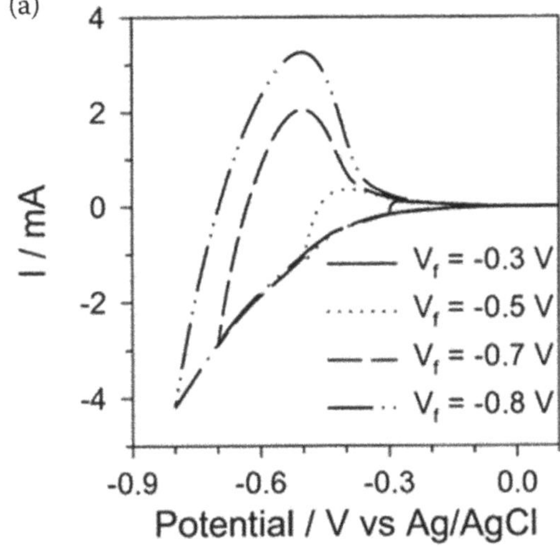

(b)

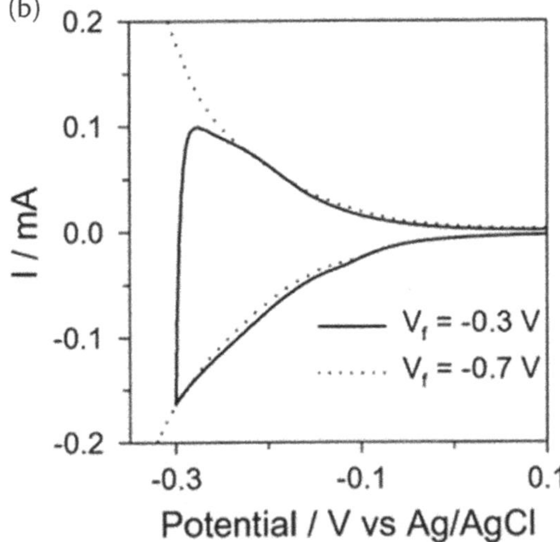

FIGURE 15.4 CVs Recorded for a TiO_2-Modified FTO Electrode Immersed into Aqueous H_2SO_4 Solution at pH 2; Potential Scan Rate 50 mV s^{-1} with Different Return Potentials (a) and zoom of panel for selected voltammograms (b). Reproduced from Ref. [23] (Fabregat-Santiago et al. *J. Phys. Chem. B.* 2003, 107: 758–768), with Permission.

with the applied potential. Figure 15.4 depicts CVs recorded at different return potentials at a TiO_2-modified electrode forming a film in contact with H_2SO_4 solution at pH 2 [23].

One can see that the cathodic current increases exponentially and that, in the anodic region, a broad peak appears. At low potential scan rates, however, the anodic peak vanishes while the cathodic region increases more slowly.

This behavior has been described assuming that the distribution of electronic states in the semiconductor can be divided into two contributions. One is due to extended states at energy E_C and density N_C (i.e. the conduction band) and the other corresponds to monoelectronic band gap localized states. The capacitance associated with the first states, C_{cb}, can be expressed as [23]:

$$C_{cb} = \frac{Le^2 n_o}{k_B T} \exp\left(-\frac{eV_C}{k_B T}\right) = C_{cb0} \exp\left(-\frac{eV_C}{k_B T}\right) \tag{15.8}$$

where L is the film thickness, n_o the concentration of states at zero bias potential, and V_C is determined by the difference between Fermi levels of the macroscopic metal contacts (i.e. the conducting substrate of the semiconductor film and the counter electrode). These parameters can be summarized in a capacitance at zero bias potential, C_{cb0}.

The capacitance associated to trap states will be [23]:

$$C_{trap} = C_{trap0} \exp\left(-\frac{\alpha_c e V_C}{k_B T}\right) \tag{15.9}$$

where α_c is a parameter defined as the ratio between temperature and the temperature representative of tail shape (broadening) of the exponential distribution of states. The typical values of α_c are of ca. 0.5. C_{trap0} represents the capacitance of trap states at zero bias potential.

The model is complemented with a series resistance, representative of charge transport in the solution and through the film, and a parallel resistance representative of the interfacial charge transfer. The current associated with this charge transfer will have a component associated with charging the capacitor, and others associated to electrons jumping to electrolyte acceptor levels [23]. Then, a Faradaic charge transfer resistance, R_F can be taken as:

$$R_F = R_{Fb} + R_{Fa} \exp\left(\frac{\alpha_F e V_C}{k_B T}\right) \tag{15.10}$$

In this equation, α_F represents the electron transfer coefficient of the Faradaic reaction, and R_{Fa}, R_{Fb}, two constants. Theoretical CVs can be obtained by combining the above equations in a Randles-type equivalent circuit and applying a potential linearly variable with time. Numerical integration of the corresponding differential equations yields CV curves agreeing with experimental data such as in Figure 15.4. Qualitatively, this voltammetry is dominated by the capacitive component of the system so that the capacitor stores charge during the cathodic scan of the voltammogram. This process continues when the potential scan is reversed while the current is negative (cathodic). By continuing the anodic scan until the current on anodic becomes positive, the capacitor returns the accumulated charge originating an anodic peak. Interestingly, the maximum peak current in the CV varies dramatically with the potential scan rate (v). For low solution resistance, the maximum current tends to be proportional to v, while tends to be v-independent at large resistance values. At intermediate resistance values, the maximum current becomes proportional to $v^{1/2}$. It should be emphasized, however, that this dependence does not mean diffusion control [23].

15.4 PHOTON ENERGY AND REDOX PROCESSES

When the semiconductor surface in contact with a redox couple Ox/Red is irradiated with light of sufficiently low wavelength (i.e. when the photon energy, $h\nu$, is larger than the bandgap energy), photons are absorbed and electron-hole pairs are created. Although some electron-hole pairs are recombined, the space charge field favors their separation and, in the case of an n-type (p-type) semiconductor, promotes the oxidation (reduction) process for the redox pair in solution when the electrode potential is larger (lower) than the flat band potential. As a result, the oxidation (reduction) process in solution takes place at electrode potentials less positive (negative) than those required to develop this process at inert metal-conducting electrodes.

It should be noted that these effects are generally not observed at potentials for redox couples located at potentials negative (n-type semiconductors) or positive (p-type semiconductors) of the flat band potential. In these cases, the majority carriers (electrons for n-type semiconductors, holes for p-type ones) tend to accumulate near the semiconductor/electrolyte interface and the semiconductor behavior approaches that of a metal electrode.

From the energetic point of view, not only the band gap but also the energetic position of the valence and conduction band edges is important. For instance, for water splitting (*vide infra*), the potentials of the O_2/H_2O and H_2O/H_2 couples must lie between the valence and conduction band edges of the semiconductor [24]. For the case of metal oxides and related compounds using potentials vs. SCE, the position of the valence band edge with respect to the vacuum level, E_{VB}, can be expressed in eV as:

$$E_{VB} = V_{FB} + 4.74 + E\Delta_F + V_H \tag{15.11}$$

In this equation, the term 4.74 (eV) corresponds to the position of the SCE with respect to the vacuum level and ΔE_F represents the difference between the Fermi level and the valence band edge ($\Delta E_F = E_F - E_{VB}$), a parameter that can be estimated from thermoelectric power measurements. V_H is the Helmholtz potential between the electrode and the solution originated from the preferential adsorption of either OH^- or H + ions from the solution on the metal oxide surface. Finally, V_H depends on the pH according to:

$$V_H = 0.059\,(pH_{zc} - pH) \tag{15.12}$$

where pH_{zc} is the pH of zero charge, a parameter that can be estimated with an accuracy of 1–2 pH units from the electronegativity considerations [25]. Figure 15.5 shows the energy band scheme calculated for TiO_2, SnO_2, and RuO_2. It must be underlined that the energy levels and the bandgap are sensitive to structural modifications, presence, type of defects and impurities, and chemiadsorped species [26]. For instance, the bandgap of rutile TiO_2 (3.02 eV) and anatase TiO_2 (3.2 eV) are different and the gap can be significantly reduced depending on the synthetic route.

The above factors can significantly influence the photoelectrochemical response. Thus, oxygen vacancies in the surface of TiO_2 originate Ti^{3+} centers, creating trapping levels for electrons that can

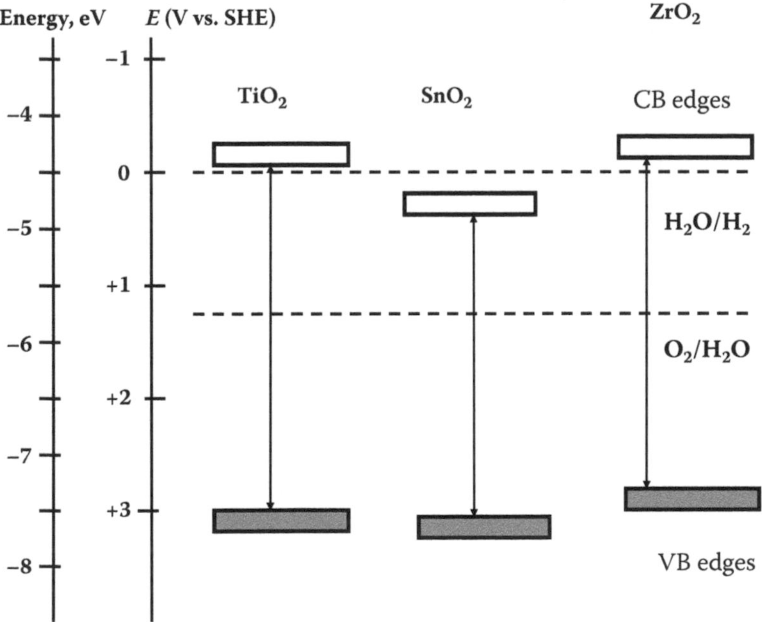

FIGURE 15.5 Energies of the Edges of the Valence and Conduction Bands of TiO_2, SnO_2, and ZrO_2 Compared with the Electrode Potentials of the H_2O/H_2 and O_2/H_2O Couples.

also inhibit electron-hole recombination on illumination. It has been reported that Fe^{3+} and Cu^{2+} doping of TiO_2 inhibits $e^- - h^+$ recombination, whereas Cr^{3+} increases $e^- - h^+$ recombination [27].

Correlation between electrochemical and spectroscopic data usually focuses on the determination of bandgap energy and estimation of the position of the upper edge of the valence band (HOMO energy in the case of molecular systems) and the lower edge of the conduction bad (LUMO energy in the case of molecular systems). Formal electrode potentials are correlated to the vacuum level as previously indicated. In cases such as polymeric systems, the E_{HOMO}, E_{LUMO} energies, and the corresponding electrochemical bandgap can be approached by the voltammetric onset potentials from extreme oxidation and reduction voltammetric peaks [28]. There, are, however, discrepancies between bandgap energies calculated from electrochemical and spectral data [29].

15.5 PHOTOCHEMICAL WATER SPLITTING

Water splitting into H_2 and O_2 is a widely studied process due to the possibility of using H_2 as a fuel. In thermodynamic terms, electrochemical water splitting requires the application of a difference of potential of 1.23 V, corresponding to the application of 2.46 eV per water molecule under standard conditions. Unfortunately, both HER and OER electrode reactions, $2H^+ + 2e^- \rightarrow H_2$ and $2H_2O \rightarrow O_2 + 4\ H^+ + 4e^-$ suffer from significant kinetic constraints, so higher potential differences must be applied in practice. Accordingly, there is considerable research aimed to obtain electrodes that can act catalytically on these processes, as also discussed in Chapters 4 and 7.

Porous semiconductor materials are of application because of the possibility of using solar light to promote their electrocatalytic activity. The general conditions are bandgap lower than 2.2 eV (to acquire reasonable solar efficiency), overlapping of band edges with the potentials of the H_2O/H_2 and O_2/H_2O couples, and fast electron transfer. Of course, chemical and electrochemical stability and environmental safety are necessary demands. As a result, water splitting can be achieved under illumination by applying lower voltages.

Under illumination conditions, *n*-type semiconductors (TiO_2, WO_3, Fe_2O_3) can act as oxidation catalysts and *p*-type semiconductors (CuO) can act as reduction catalysts. The catalytic OER process can be represented as (VB: valence band; CB: conduction band):

$$4OH^- + 4h^+(VB) \rightarrow O_2 + 2H_2O \tag{15.13}$$

whereas the catalytic HER process can be summarized as:

$$2H^+ + 2e^-(CB) \rightarrow H_2 \tag{15.14}$$

These processes are schematized in Figure 15.6. The illumination of the photoanode with a photon of sufficient energy ($h\upsilon$) promotes an electron in the space charge from the valence band of the *n*-type semiconductor to its conduction band. Here, the electron moves towards the bulk of the material, whereas the positive hole generated in the valence band remains near the surface and reacts with water to give O_2 (Eq. (15.10)). In turn, illumination of the surface of the *p*-type semiconductor with a sufficiently energetic photon ($h\upsilon'$) produces a similar electron promotion. However, the electron in the conduction band remains in the surface producing H_2 via the process represented by Eq. (15.14).

Several processes, such as electron or hole trapping in the bulk material, can decrease the catalytic performance—electrons back transfer via diffusion, recombination of back transfer electrons and holes in the valence band, and recombination of electrons in the conduction band with holes at the semiconductor/electrolyte interface. These undesired effects can be minimized by a variety of strategies—for instance, by means of the deposition of an ultrathin layer of compact

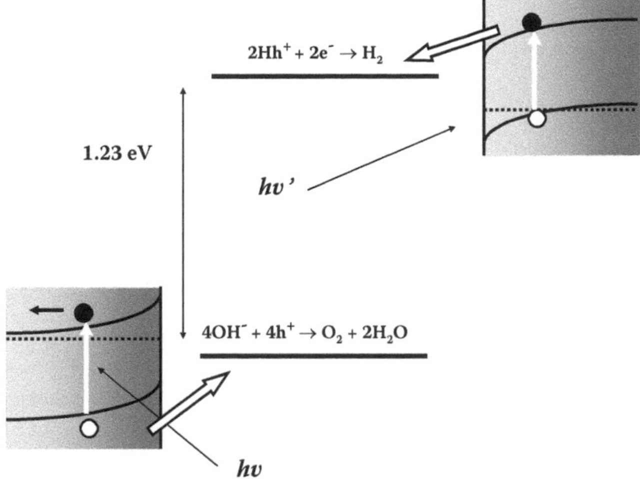

FIGURE 15.6 Scheme of a Photoelectrolysis Cell Based on Two Semiconductor–Liquid Junctions. The *n*-Type Semiconductor Acts as a Photoanode (Water Oxidation into O_2) While the *p*-Type Semiconductor Acts as a Photocathode (H^+ Reduction into H_2).

hematite interposed between the base indium-doped tin oxide electrode (ITO) and an external porous layer of granular hematite [30]. Figure 15.7 depicts the photocurrent-potential curves, photon-to-electron efficiency plots, applied bias potential to photon efficiency plots, and Mott-Schottky plots of hematite and Ti-doped hematite photoanodes in contact with 0.1 M KOH solution under illumination [31].

The above description, based on the formation of space charge regions, is appropriate for compact semiconductor films [32]. An alternative approach [33–35], based on electron and hole trapping, may apply for mesoporous semiconductor films where depletion is extensive.

A variety of materials for photochemical water splitting is under intensive research, including combinations of semiconductors, metal/semiconductor composites, and sensitized semiconductors (*vide infra*) [36]. Another strategy is the use of sacrificial agents; H_2 generation is facilitated by the presence of sacrificial electron donors, whereas O_2 production requires the use of sacrificial electron acceptors. This permits the use of the same semiconducting material as anode and cathode.

15.6 PHOTOELECTROCHEMICAL CELLS

Photoelectrochemical cells are formed by a semiconductor electrode and a suitable counter-electrode immersed into the electrolyte. Three types of photoelectrochemical cells can be distinguished. In photovoltaic cells, the reaction at the counter electrode is simply the reverse of the photoassisted reaction occurring at the semiconductor electrode so that the cell converts radiant energy into electricity without change in the composition of the solution or electrodes.

In photoelectrosynthetic cells, the Ox′/Red′ reaction at the counter electrode is different than that occurring at the semiconductor electrode, Ox/Red. As a result, the overall reaction occurring in the cell is driven by light in a direction contrary to that thermodynamically spontaneous. Then, light energy is converted into chemical energy stored in the resulting products of the electrochemical reaction. For an *n*-type semiconductor, this requires the potential of the Ox/Red couple to lie above the valence band edge while the potential of the Ox′/Red′ couple should be below the flat band potential.

Finally, in photocatalytic cells, the reaction—which is kinetically hindered—is conducted in its thermodynamically spontaneous direction by applying an external potential bias to the cell, the

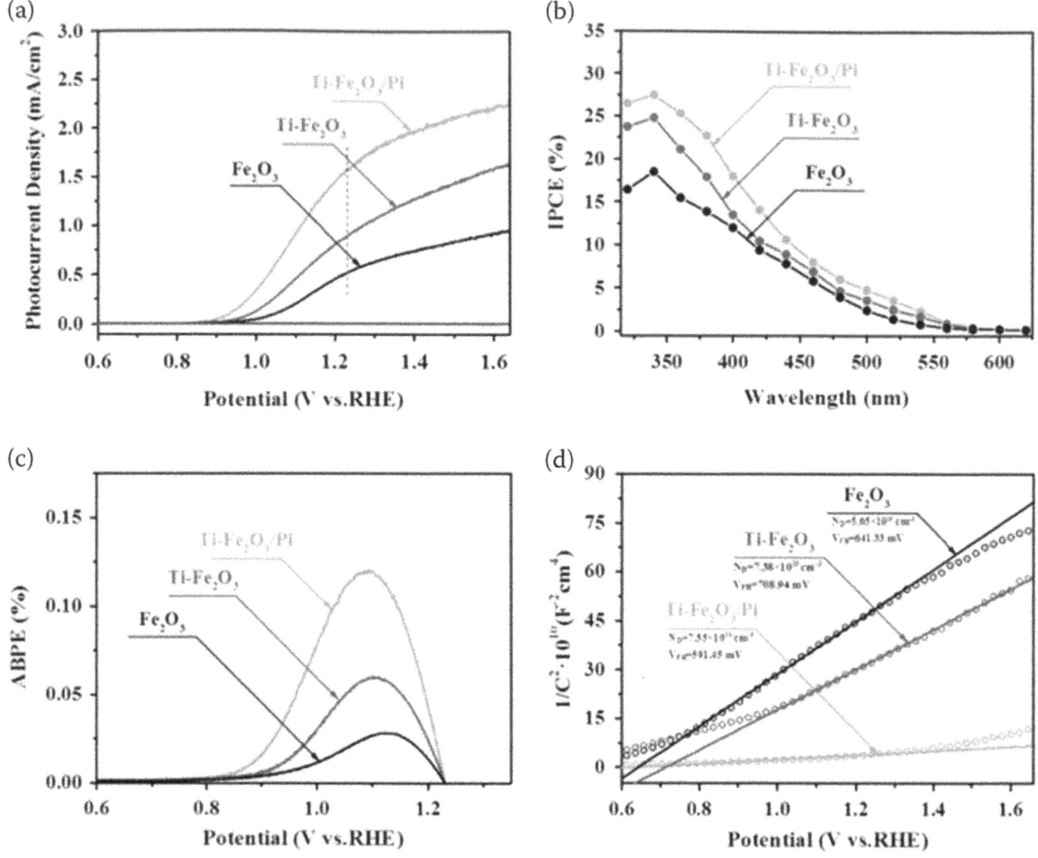

FIGURE 15.7 (a) Photocurrent-Potential Curves, (b) Photon-to-Electron Efficiency Plots, (c) Applied Bias Potential to Photon Efficiency Plots, and (d) Mott-Schottky Plots of Fe_2O_3, Ti-Fe_2O_3, and Ti-Fe_2O_3/Pi Photoanodes Obtained in 0.1 M KOH Solution Under AM 1.5 G Light Irradiation (100 mW cm^{-2}). Reproduced from Ref. [31] (Liu et al. *Electrochim. Acta.* 2019, 307: 197–205), with Permission.

light providing the necessary activation energy for the process. Here, the location of the potentials of the Ox/Red and Ox′/Red′ couples is inverse to the case of photoelectrosynthetic cells. Photocatalytic cells are of application for electrosynthetic purposes as well as electrochemical degradation of contaminants. These matters will be described in Chapter 16.

Because the light used for photoelectrochemical processes must have greater photon energy than the bandgap of the semiconductor, UV irradiation is needed in most cases. A strategy in using visible light is the sensitization of the semiconductor with a layer of a photo-mediator (typically, dyes) that can absorb visible light. The photonically excited mediator injects an electron in the semiconductor band, thus becoming oxidized. If a suitable redox couple exists in the adjacent electrolyte solution, the original mediator form is regenerated via electron exchange with species in solution. This can be described as the sequence of reactions:

$$\{Med\} + h\nu \rightarrow \{Med^{+}\} + e^{-} \tag{15.15}$$

$$\{Med^{+}\} + Red \rightarrow \{Med\} + Ox \tag{15.16}$$

Here, { } denote the layer of mediator deposited over the semiconductor surface. Dye-sensitized solar cells will be described in Section 15.8.

Apart from oxides such as TiO_2, ZnO, or ZrO_2, sulfides (typically CdS), niobates, tantalates, titanates, and other compounds have been tested as photocatalytic materials for splitting water. Composites [36] and porous materials [37] have been introduced as electrode materials in photovoltaic cells in the last years.

It must be considered, however, that electrocatalytic and photoelectrocatalytic processes can involve more complicated pathways. For instance, in the case of O_2 generation at hematite surfaces, Fe(IV) and even Fe(V) oxidation states may be involved via the formation of Fe-oxo surface species [31].

15.7 PHOTOELECTROCATALYSTS ATTACHED TO POROUS SOLIDS

Modulation of the electrocatalytic response by electrochemically- or photochemically-induced changes in the electroactive centers anchored to porous solids is another possibility in photoelectrochemistry. This can be illustrated by the voltammetric behavior of spiropyrans attached to zeolite Y, SP@Y. Spiropyrans are composed of an indoline and a chromene moiety that are linked by a spirocarbon atom. In solution, irradiation with near-UV light ($\lambda < 350$ nm) induces heterolytic cleavage of the spiro-carbon-oxygen bond of the original closed spiropyran to produce a ring-opened species whose structures are depicted in Figure 15.8 [38]. In the absence of UV irradiation, this undergoes ring closure to reform the parent spiropyran. However, upon attachment to aluminosilicates, spiropyrans show a reverse photochemical behavior in which, under illumination with visible light ($\lambda > 450$ nm), the predominant open form undergoes ring closure to the closed spiropyran one. In addition, a third form consisting of protonated open spiropyran is also detected in the solids. Thus, there is a distribution of three different species (closed, unprotonated open, and protonated open forms) whose population varies depending on the polarity and acidity of the solid, the open protonated form being the predominant species in the solids.

Due to the frozen orthogonal configuration of the indolic and phenolate moieties, the closed spiropyran form is a rigid molecule. In contrast, the open forms are significantly more flexible due to the allowed rotation through the single C–C bond connecting the indole and chromene units. In the interior of cages of zeolite Y, in which the spiropyran molecule must occupy a little more space than that available in a single cavity, the conformational freedom should be, at least partially,

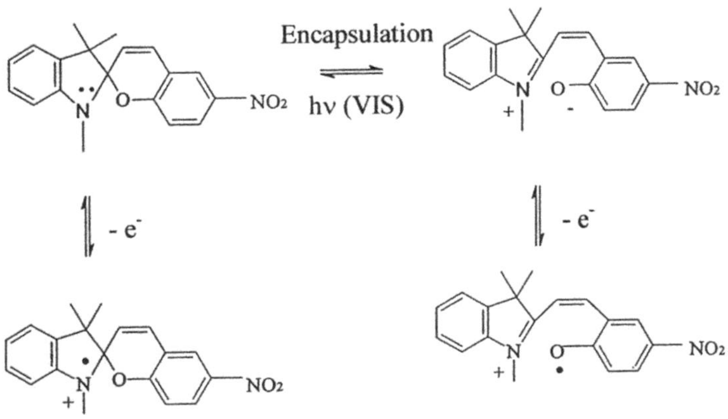

FIGURE 15.8 Main Photoelectrochemical Processes Involved in the Voltammetric Response of SP@Y Systems. From Ref. [38] (Doménech-Carbó et al. *J. Phys. Chem. B.* 2004, 108: 20064–20075), with Permission.

hindered so that the cis configuration with respect to the central C–C bond of the open-chain is retained, as illustrated in Figure 15.8.

Electrochemical data for SP@Y probes were consistent with these above considerations [38]. For our purposes, the relevant point to emphasize is that a light-sensitive effect on the voltammetric response for benzidine oxidation in nonaqueous media at SP@Y-modified electrodes. This can be seen in Figure 15.9, where CVs at paraffin-impregnated graphite electrodes modified with bleached, partially bleached, and pristine SP@Y specimens in contact with a MeCN solution of N,N,N′,N′-tetramethylbenzidine, are shown [38]. The two-peak response of benzidine recorded for pristine SP@Y specimens is identical to that recorded at unmodified graphite electrodes, but peak currents are clearly enhanced upon bleaching the SP@Y specimen. These features are attributable to an electrocatalytic effect that is associated with the closed spiropyran form attached to the zeolite boundary. Comparable results can be obtained for electrochemically active doped materials such as chromium-doped sphene, where the doping metal centers act as a catalytic site (see additional literature).

Electrochromism is the property of change or bleaching of color as a result of either an electron-transfer (redox) process or application of a sufficient electrochemical potential [39]. Apart from electrochromic species in the solution phase (typically viologen and methyl viologen), a series of metal oxides exhibit electrochromism in solid-state, being those of tungsten, molybdenum, iridium, and nickel the most representative. Spectral features determining electrochromism come from intervalence charge-transfer optical transitions where different oxidation states are available.

Electrochromic materials are of interest for displays, smart windows, sunroofs, antiglare car mirrors, electronic strips, etc., and are characterized by the reversible change in their color upon application of light or electrical inputs. The electrochromic effect is generally associated to the ingress/issue of electrons and metal cations. Typical electrochromic materials are oxides that can be colored anodically (IrO_2, Rh_2O_3, NiO_x) or cathodically (MoO_3, WO_3, TiO_2). A typical example is that of tungsten oxide. This compound is transparent when forming thin films. Upon electrochemical reduction, W(V) sites are generated to give an electrochromic material:

$$\{W^{VI}O_3\}_{(colorless)} + xM^+ + xe^- \rightarrow \{M_xW_x^VW_{1-x}^{VI}O_3\}_{(deep\ blue)} \tag{15.17}$$

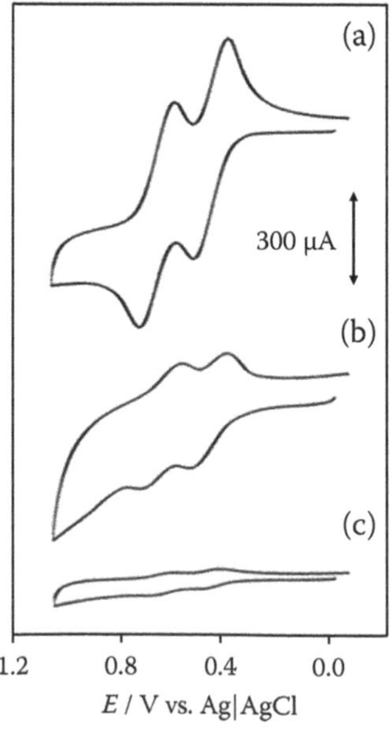

FIGURE 15.9 CVs at Paraffin-Impregnated Graphite Electrodes Modified by (a) Bleached SP@Y; (b) Partly Bleached SP@Y; (c) Pristine SP@Y, Immersed into 3.0 mM N,N,N′,N′-Tetramehylbenzidine Solution in 0.10 M Et_4NClO_4/MeCN. Potential Scan Rate 100 mV s^{-1}. From Ref. [38] (Doménech-Carbó et al. *J. Phys. Chem. B.* 2004, 108: 20064–20075), with Permission.

At low x values, the films acquire a blue coloration caused by intervalence charge transfer between adjacent W^{V} and W^{VI} sites. At higher x values, irreversible insertion occurs, yielding a solid with intense luster. This motivated the coining of the misleading denomination of 'tungsten bronzes.'

There are, however, different proposals to interpret the electrochromism underlying the role of defects, such as oxygen vacancies, polaronic absorption, interband excitations, photo-effected intervalence charge transfer between adjacent W(V) and W(VI) sites, and transitions from the valence band to split off the W(V) state [39]. The response is sensitive to the applied potential and it is better for porous films than for continuous ones [40]. Figure 15.10 shows the transmittance spectra and in situ optical responses of WO_3, prepared over fluor-doped tin oxide (FTO) and indium-doped tin oxide coated with polyethylene terephthalate (PET), for the colored and bleached states [40].

Vanadium (V) oxide is another example of electrochromism where the more intense absorbing state is generated by electrochemical reduction:

$$\{V_2^{V}O_5\}_{(\text{brown-yellow})} + xM^+ + xe^- \rightarrow \{M_xV_x^{IV}V_{1-x}^{V}O_5\}_{(\text{pale blue})} \tag{15.18}$$

where M^+ is typically Li^+ or H^+. In other cases, this state is produced via oxidation; for instance:

$$\{Ir(OH)_3\}_{(\text{colorless})} \rightarrow \{IrO_2{\cdot}H_2O\}_{(\text{blue grey})} + H^+ + e^- \tag{15.19}$$

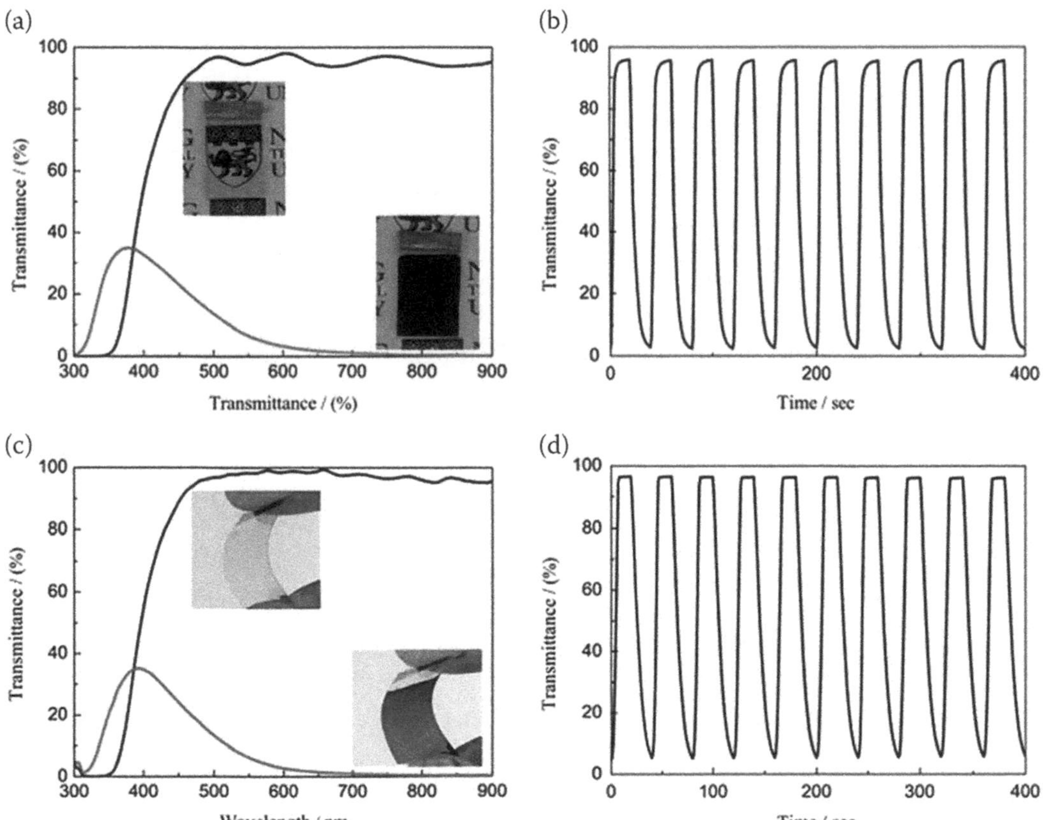

FIGURE 15.10 Transmittance Spectra And In Situ Optical Responses in the Colored and Bleached States of WO_3 Films on Different Substrates. (a,b) FTO; (c,d) ITO-PET. From Ref. [40] (Cai et al. *Chem. Sci.* 2016, 7: 1373–1382), with Permission.

$$\{Ir(OH)_3\}_{(colorless)} + OH^- \rightarrow \{IrO_2{\cdot}H_2O\}_{(blue\text{-}black)} + H_2O + e^- \quad (15.20)$$

Prussian blue is another widely studied electrochromic material. Electrodeposition of Prussian blue films is achieved by the reduction of the brown-yellow soluble complex, iron(III) hexacyanoferrate(III):

$$[Fe^{III}Fe^{III}(CN)_6]_{aq} + M^+_{aq} + e^- \rightarrow [MFe^{III}Fe^{II}(CN)_6]_{solid} \quad (15.21)$$

Oxidation of the $[Fe^{III}Fe^{II}(CN)_6]^-$ blue chromophore yields successively Berlin green, a solid solution $\{Fe^{III}[Fe^{II}(CN)_6]^-\}_x\{Fe^{III}[Fe^{III}(CN)_6]\}_{1-x}$, and yellow $[Fe^{III}Fe^{III}(CN)_6]^{2-}$ [41]:

$$[Fe^{III}Fe^{II}(CN)_6]^- \rightarrow \{[Fe^{III}Fe^{III}(CN)_6]_{2/3}[Fe^{III}[Fe^{II}(CN)_6]_{1/3}\}^{1/3-} + (2/3)e^- \quad (15.22)$$

$$[MFe^{III}Fe^{II}(CN)_6]^- \rightarrow [Fe^{III}Fe^{III}(CN)_6] + e^- \quad (15.23)$$

In turn, the electrochemical reduction of the chromophore yields colorless $[Fe^{II}Fe^{II}(CN)_6]^{2-}$:

$$[MFe^{III}Fe^{II}(CN)_6]^- + e^- \rightarrow [Fe^{III}Fe^{III}(CN)_6]^{2-} \quad (15.24)$$

In solid films, naturally, these processes require the ingress/issue of electrolyte counteractions (see Chapter 3). A collateral aspect of interest is the use and deterioration of Prussian blue as pigment [41] and its interaction with binding media [42] in paintings.

Since the electrochromic properties are strongly related to the ability of the base material to incorporate doping ions, the synthesis of porous materials is receiving growing interest. Tested materials include—among a wide variety—anastase films, amorphous TiO_2 [43], and conducting polymer films [44]. Semisolid organic electrolytes (polyelectrolytes and polymer electrolytes) have been also used in electrochromic devices (see Chapter 9).

High-performance devices using two complementary electrochromic materials are under study. An example is constituted by WO_3 as the primary electrode in and porous NiO working in non-aqueous electrolytes such as propylene carbonate (PC). Figure 15.11 shows the voltammetric response of NiO films in contact with 1 M $LiClO_4$/PC electrolyte. The initial cathodic signal (c1) can be assigned to the insertion process [45]:

$$\{NiO\}_{film} + yLi^+_{solv} + ye^- \rightarrow \{Li_yNiO\}_{film} \quad (15.25)$$

which is reversed in the coupled anodic process (a1). Upon repeatedly cycling the potential scan, a second couple (c2/a2) is progressively enhanced, corresponding to the electrochromic intercalation/de-intercalation process:

$$\{Li_yNiO\}_{film} + xLi^+_{solv} + xe^- \rightarrow \{Li_{y+x}NiO\}_{film} \quad (15.26)$$

The electrochromic efficiency at a given wavelength, $\eta(\lambda)$, can be calculated from spectro-chronocoulometric experiments as [46]:

$$\eta(\lambda) = \frac{\log T(\lambda)_{bleach} - \log T(\lambda)_{color}}{\sigma} \quad (15.27)$$

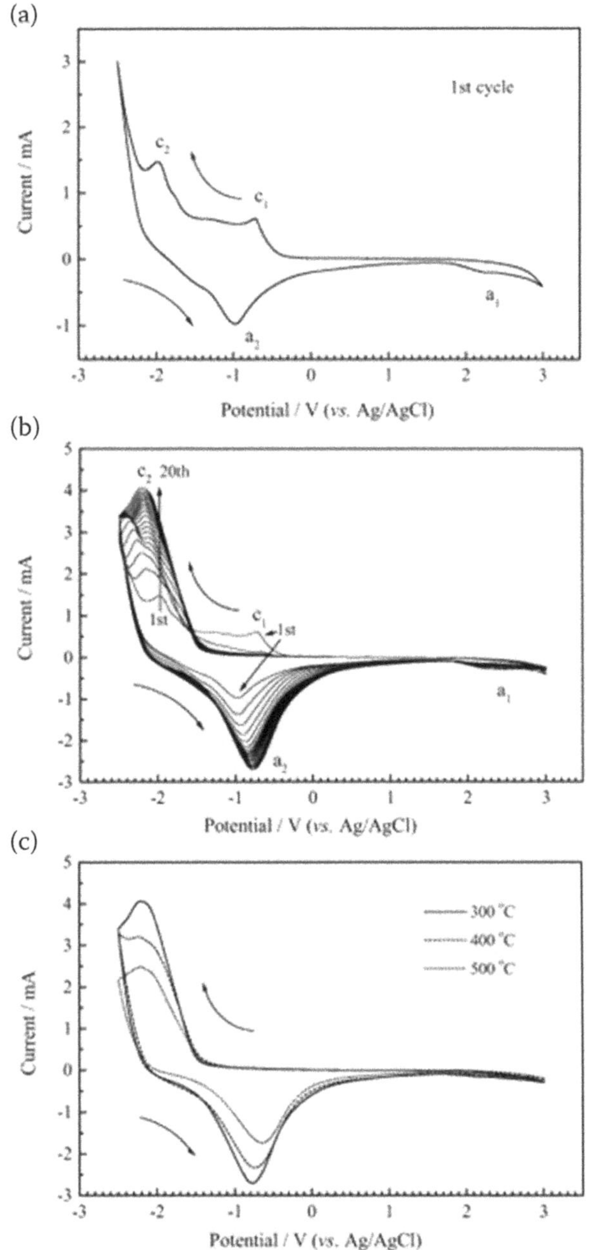

FIGURE 15.11 CVs of NiO Film on ITO Substrate in Contact with 1 M $LiClO_4$/PC. Potential Scan Rate 10 mV s^{-1}. From Ref. [45] (Huang et al. *Electrochim. Acta.* 2011, 56: 4281–4286), with Permission.

In the above equation, $T(\lambda)_{bleach}$ and $T(\lambda)_{color}$ represent, respectively, the transmittance of the bleached and colored forms at the selected wavelength and σ the charge density (C cm^{-2}) necessary to produce the observed chromatic change. Other quantities used to characterize electrochromic films are the coloration efficiency (CE), defined as the change in optical density (ΔOD) per unit of charge (Δq) entered or released from the electrochromic film [40]:

$$CE = \frac{\Delta OD}{\Delta q} \tag{15.28}$$

where the optical density change is expressed as a function of the transmittances of the bleached and colored states, $T_{bleached}$, $T_{colored}$:

$$\Delta OD = \log \frac{T_{bleached}}{T_{colored}} \tag{15.29}$$

Another interesting phenomenon is producing luminescence from electrochemical reactions. This can be achieved by doping crystalline metal oxides and related materials or anodic metal oxide layers [47], or depositing organic molecules onto nanostructured porous materials such as silicon [48].

15.8 DYE-SENSITIZED CELLS

Solar cells are devices aimed to convert sunlight energy into electric energy. Figure 15.12 shows a scheme of solar cells, involving the use of mesoporous TiO_2. Dye-sensitized electrodes, shortly commented in previous chapters, can also be used as photoelectrocatalysts for sensing purposes based on the same principles as solar cells.

Figure 15.12 shows a scheme of dye-sensitized solar cell composed of mesoporous TiO_2 and Pt particles, both deposited onto transparent FTO electrodes with a solid or quasi-solid redox electrolyte containing an electron acceptor (*vide infra*) [49].

The operation of dye-sensitized solar cells involves a set of processes initiated by the photo-excitation of the dye. Then, one electron is injected from the excited sensitizer to the semiconductor so that the dye is transformed in its oxidized state. This process competes with the decay of the excited dye to its ground state. The next step is the transport of electrons through the mesoporous semiconducting film. The back-electron transfer reactions involve the electron transfer from the conduction band to the oxidized sensitizer, or the recombination of electrons with any acceptor species in the electrolyte. To operate in a cyclic way, the oxidized dye must be re-generated. This is achieved with an electron acceptor in the electrolyte that captures electrons from

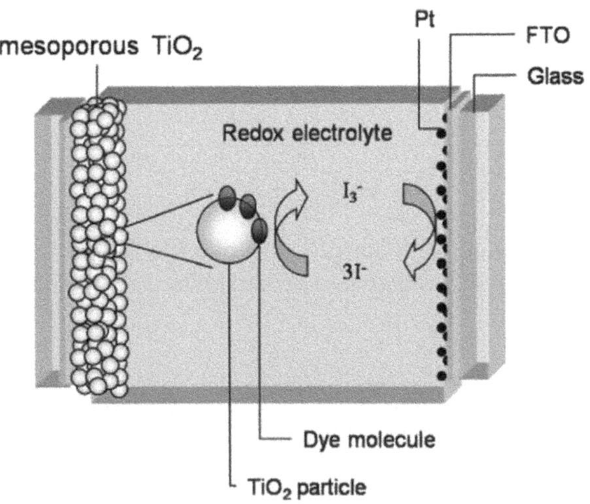

FIGURE 15.12 Scheme of a Dye-Sensitivized Solar Cell. Reproduced from Ref. [49] (Hagfeldt et al. *Chem. Rev.* 2010, 110: 6595–6663), with Permission.

the semiconductor, being subsequently reduced at the counter electrode. In the example illustrated in Figure 15.12, iodide ions act as electron acceptors regenerated at the Pt electrode via the $I_3^- + 2e^- \rightarrow 3I^-$ process.

The main parameters used to describe the performance of solar cells are:

a. Incident photon to current conversion efficiency (IPCE), defined as the ratio between the photocurrent density measured at short-circuit (J_{sc}) under monochromatic illumination to the photon flux that strikes the cell:

$$\mathrm{IPCE} = \frac{J_{sc}(\lambda)}{e\Phi(\lambda)} = 1240\frac{J_{sc} \quad (\mathrm{A\ cm^{-2}})}{\lambda\ (\mathrm{nm})\ P_{in}(\lambda) \quad (\mathrm{W\ cm^{-2}})} \tag{15.30}$$

P_{in} being the incident light power.

b. Fill factor of the cell (FF), defined as the ratio of the maximum power of the solar cell, P_{max}, per unit area to J_{sc} and the open-circuit photovoltage, V_{oc}:

$$\mathrm{FF} = \frac{P_{max}}{J_{sc}V_{oc}} \tag{15.31}$$

c. Overall solar-to electrical energy conversion efficiency, η, defined in terms of the above quantities as:

$$\eta = \mathrm{FF}\frac{J_{sc}V_{oc}}{P_{in}} \tag{15.32}$$

It is worth to note that the behavior of mesoporous electrodes is different compared with their compact analogs because of the lowest conductivity of the film, the built-in electric field in colloidal particles, and the formation of interpenetrating networks between the oxide particles and the electrolyte-containing pores giving rise to phase boundaries of high contact area. Then, porous metal oxide films can be approached by a set of particles through which electrons can percolate by hopping between them [49]. Figure 15.13 shows a scheme of a nanotube array architecture as a photoelectrode in dye-sensitized cells, accompanied by the scanning electron microscopy image of TiO_2 nanotubes defining a porous system [50].

In solid-state dye-sensitized cells, a hole conductor acts as the electrolyte. In typical viscous liquid electrolytes (ionic liquids, for instance), there is high conductivity and ionic strength, so charge transport by migration is negligible. Here, the photocurrent may be limited by the (slow) diffusion of redox components. In the case of the iodide/triiodide electrolyte, an additional conduction pathway (Grotthus mechanism) may occur. This can be represented as:

$$I_3^- + I^- \rightarrow \{I^-\cdots I_2\cdots I^-\} \rightarrow I^- + 3I^- \tag{15.33}$$

TiO_2, ZnO, SnO_2, and a variety of compounds and composite materials have been used as electrodes. These are accompanied by an enormous number of dyes, including ruthenium and osmium complexes, porphyrins, phthalocyanines, and triarylamines, among others. Liquid electrolytes include organic carbonates, ionic liquids, gel, and polymer electrolytes.

The essential demand for dye-sensitized cells is the existence of an efficient electron transfer between the dye in its excited state and the semiconductor. In the case of *n*-type semiconductors, the LUMO level of the photosensitizer should be higher in energy than the conduction band edge of the semiconductor. In the case of *p*-type semiconductors, the HOMO level of the photosensitizer

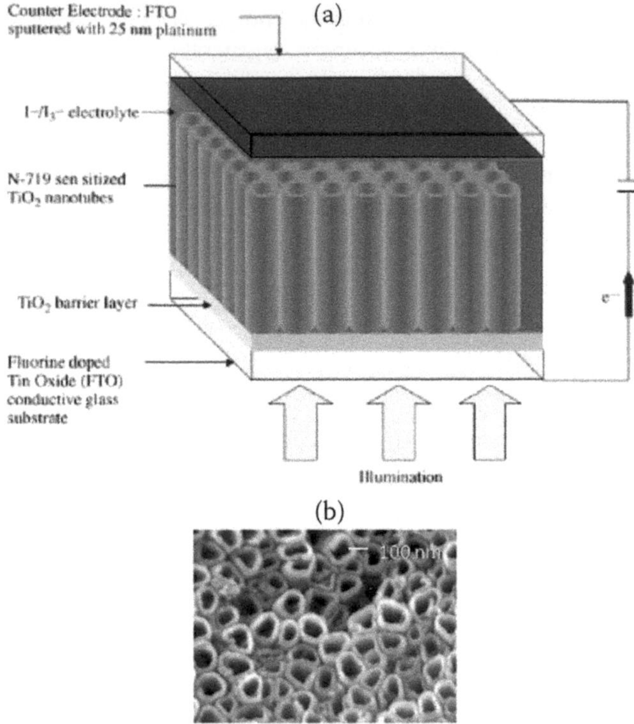

FIGURE 15.13 (a) Schematic Representation of a Nanotube Array Architecture as a Photoelectrode in Dye-Sensitized Cells, and (b) Scanning Electron Microscopy Image (Top View) of TiO_2 Nanotubes. Reproduced from Ref. [50] (Mor et al. *Sol. Energy Mater. Sol. Cells.* 2006, 90, 2011–2075), with Permission.

should be at a more positive potential than the valence band edge of the semiconductor. Figure 15.14 depicts a schematic energy levels diagram corresponding to a TiO_2 cell where a dye S is accompanied by an iodide containing electrolyte [49].

Dye regeneration requires that the oxidized state level of the photosensitizer must be more positive than the redox potential of the electrolyte. Additional requirements are: (a) the photosensitizer

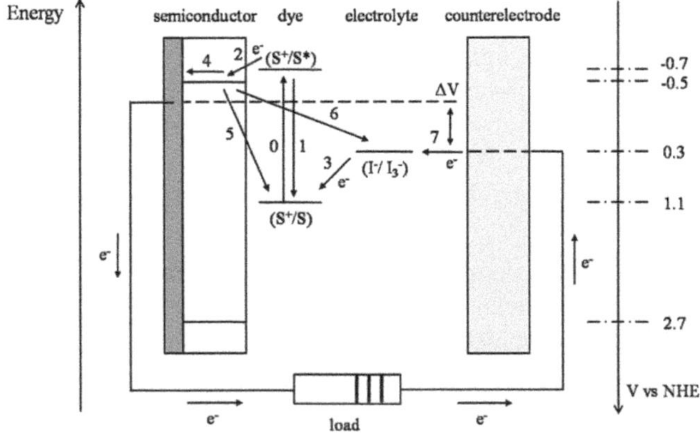

FIGURE 15.14 Energy Level Diagram for a Dye-Sensitized Cell Constituted by an *n*-Type Semiconductor Covered by a Dye in Contact with a Iodide-Containing Electrolyte. Reproduced from Ref. [49] (Hagfeldt et al. *Chem. Rev.* 2010, 110: 6595–6663), with Permission.

should absorb in the whole visible region; (b) the photosensitizer should be thermally, electrochemically, and photochemically stable; (c) the photosensitizer should have anchoring groups (–COOH, $-H_2PO_3$, $-SO_3H$, etc.); and (d) the dye binding to the semiconductor surface can be optimized, avoiding dye aggregation by means of the addition of co-adsorbers.

REFERENCES

[1]. Fahidy, T.Z. 1999. The effect of magnetic fields on electrochemical processes. In Conway, B.E.; Bockris, O.M.; White, R.E. Eds *Modern Aspects of Electrochemistry*, no. 32. Kluwer/Plenum, New York, pp. 333–354.

[2]. Juzeliunas, E. 2007. Advances in detection of magnetic fields induced by electrochemical reactions - A review. *Journal of Solid State Electrochemistry*. 11: 791–798.

[3]. Fahidy, T.Z. 1973. Hydrodynamic models in magnetoelectrolysis. *Electrochimica Acta*. 18: 607–614.

[4]. Ragsdale, S.R.; Grant, K.M.; White, H.S. 1998. Electrochemically generated magnetic forces. Enhanced transport of a paramagnetic redox species in large, nonuniform magnetic fields. *Journal of the American Chemical Society*. 120: 13461–13468.

[5]. Waskaas, M.; Kharkats, Y.I. 1999. Magnetoconvection phenomena: A mechanism for influence of magnetic fields on electrochemical processes. *Journal of Physical Chemistry B*. 103: 4876–4883.

[6]. Kolodziejczyk, K.; Miękoś, E.; Zieliński, M.; Jaksender, M.; Szdzukocki, M.; Czamy, K.; Krawczyk, K. 2018. Influence of constant magnetic field on electrodeposition of metals, alloys, conductive polymers, and organic reactions. *Journal of Solid State Electrochemistry*. 22: 1629–1647.

[7]. Hinds, G.; Spada, F.E.; Coey, J.M.D.; Ní Mhíocháin T.R.; Lyons, M.E.G. 2001. Magnetic field effects on copper electrolysis. *Journal of Physical Chemistry B*. 105: 9487–9502.

[8]. Coey, J.M.D.; Rhen, F.M.F.; Dunne, P.; McMurry, S. 2007. The magnetic concentration gradient force—Is it real? *Journal of Solid State Electrochemistry*. 11: 711–717.

[9]. Leventis, N.; Dass, A.; Chandrasekaran, N. 2007. Mass transfer effects on the electropolymerization current efficiency of 3-methylthiophene in the magnetic field. *Journal of Solid State Electrochemistry*. 11: 727–735.

[10]. Matsushima, H.; Ispas, A.; Bund, A.; Plieth, W.; Fukunaka, Y. 2007. Magnetic field effects on microstructural variation of electrodeposited cobalt films. *Journal of Solid State Electrochemistry*. 11: 737–743.

[11]. Hu, F.; Chan, K.C.; Qu, N.S. 2007. Effect of magnetic field on electrocodeposition behavior of Ni–SiC composites. *Journal of Solid State Electrochemistry*. 11: 267–272.

[12]. Monzon, L.M.A.; Coey, J.M.D. 2014. Magnetic fields in electrochemistry: The Lorentz force. A mini-review. *Electrochemistry Communications*. 42: 38–41.

[13]. Patermarakis, G.; Kapiris, G. 2013. Processes, parameters and mechanisms controlling the normal and abnormal growth of porous anodic alumina films. *Journal of Solid State Electrochemistry*. 17: 1133–1158.

[14]. Cho, M.S.; Yun, Y.Y.; Nam, J.D.; Son, Y.; Lee, Y. 2008. Effect of magnetic field on electrochemical polymerization of EDOT. *Synthetic Metals*. 158: 1043–1046.

[15]. Doménech-Carbó, A.; García, H.; Carbonell, E. 2005. Electro- and magneto-electrochemistry of zeolite Y- and MCM-41-associated bipyrylium ion. *Journal of Electroanalytical Chemistry*. 577: 249–262.

[16]. Bard, A.J.; Inzelt, G.; Scholz, F. Eds. 2008. *Electrochemical Dictionary*. Springer, Berlin-Heidelberg.

[17]. Honda, K.; Fujishima, A. 1972. Electrochemical photolysis of water at a semiconductor electrode. *Nature*. 238: 37–38.

[18]. Bard, A.J.; Faulkner, L.R. 2001. *Electrochemical Methods*. 2nd edit. John Wiley & Sons, New York.

[19]. Bott, A.W. 1998. Electrochemistry of semiconductors. *Current Separations*. 17: 87–91.

[20]. Adán-Mas, A.; Silva, T.M.; Guerlou-Demourgues, L.; Montemor, M.-F. 2018. Application of the Mott-Schottky model to select potentials for EIS studies on electrodes for electrochemical charge storage. *Electrochimica Acta*. 289: 47–55.

[21]. Bisquert, J.; Garcia-Belmonte, G.; Fabregat-Santiago, F. 1999. Modelling the electric potential distribution in the dark in nanoporous semiconductor electrodes. *Journal of Solid State Electrochemistry*. 3: 337–347.

[22]. Chae, S.Y.; Rahman, G.; Joo, O.-S. 2019. Elucidation of the structural and charge separation properties of titanium-doped hematite films deposited by electrospray method for photoelectrochemical water oxidation. *Electrochimica Acta*. 297: 784–793.

[23]. Fabregat-Santiago, F.; Mora-Seró, I.; Garcia-Belmonte, G.; Bisquert, J. 2003. Cyclic voltammetry studies of nanoporous semiconductors. Capacitive and reactive properties of nanocrystalline TiO_2 electrodes in aqueous electrolyte. *Journal of Physical Chemistry B*. 107: 758–768.

[24]. Bard, A.J.; Fox, M.A. 1995. Artificial photosynthesis – Solar splitting of water to hydrogen and oxygen. *Accounts of Chemical Research*. 28: 141–145.
[25]. Koffyberg, F.P.; Benko, F.A. 1982. A photoelectrochemical determination of the position of the conduction band and the valence band edges of *p*-type CuO. *Journal of Applied Physics*. 53: 1173–1177.
[26]. Yang, S.; Prendergast, D.; Neaton, J.B. 2012. Tuning semiconductor band edge energies for solar photocatalysis via surface ligand passivation. *Nano Letters*. 12: 383–388.
[27]. Shaban, Y.A.; Khan, S.U.M. 2009. Carbon-modified (CM)-n-TiO_2 thin films for efficient water splitting to H_2 and O_2 under xenon lamp light and natural sunlight illuminations. *Journal of Solid State Electrochemistry*. 13: 1025–1036.
[28]. Chen, Z.-K.; Huang, W.; Wang, L.-H.; Kang, E.-T.; Chen, B.J.; Lee, C.S.; Lee, S.T. 2000. A family of electroluminiscent silyl-substituted poly(-phenylenevynilene)s: Synthesis, characterization, and structure-property relationships. *Macromolecules*. 33: 9015–9025.
[29]. Haram, S.K.; Quinnn, B.M.; Bard, A.J. 2001. Electrochemistry of CdS nanoparticles: A correlation between optical and electrochemical band gaps. *Journal of the American Chemical Society*. 123: 8860–8861.
[30]. Bouhjar, F.; Bessaïs, B.; Marí, B. 2018. Ultrathin-layer α-Fe_2O_3 deposited under hematite for solar water splitting. *Journal of Solid State Electrochemistry*. 22: 2347–2356.
[31]. Liu, G.; Zhao, Y.; Li, N.; Yao, R.; Wang, M.; Wu, Y.; Zhao, F.; Li, J. 2019. Ti-doped hematite photoanode with surface phosphate ions functionalization for synergistic enhanced photoelectrochemical water oxidation. *Electrochimica Acta*. 307: 197–205.
[32]. Peter, L.M. 2013. Energetics and kinetics of light-driven oxygen evolution at semiconductor electrodes: The example of hematite. *Journal of Solid State Electrochemistry*. 17: 315–326.
[33]. Bertoluzzi, L.; Bisquert, J. 2012. Equivalent circuit of electrons and holes in thin semiconductor films for photoelectrochemical water splitting applications. *Journal of Physical Chemistry Letters*. 3: 2517–2522.
[34]. Morales-Guio, C.G.; Tilley, S.D.; Vrubel, H.; Gratzel, M.; Hu, X.L. 2014. Hydrogen evolution from a copper(I) oxide photocathode coated with an amorphous molybdenum sulphide catalyst. *Nature Communications*. 5: artic. 3059.
[35]. Kida, T.; Guan, G.; Minami, Y.; Ma, T.; Yoshida, A. 2003. Photocatalytic hydrogen production from water over a $LaMnO_3$/CdS composite prepared by the reverse micelle method. *Journal of Materials Chemistry*. 13: 1186–1191.
[36]. Velaski, R.; Muchenski, F.; Mello, R.M.Q.; Micaroni, L.; Roman, L.S.; Hümmelgen, I.A. 2006. Sulfonated polyaniline/poly(3-methylthiophene)-based photovoltaic devices. *Journal of Solid State Electrochemistry*. 10: 24–27.
[37]. Cummings, C.Y.; Marken, F.; Fermi, L.M.; Upul Wijayantha, K.G.; Tahir, A.A. 2011. New insights into water splitting at mesoporous α-Fe_2O_3 films: A study by modulated transmittance and impedance spectroscopies. *Journal of the American Chemical Society*. 134: 1228–1234.
[38]. Doménech-Carbó, A.; García, H.; Casades, I.; Esplá, M. 2004. Electrochemistry of 6-nitro-1′,3′,3′-trimethylspiro[2H-1-benzopyran-2,2′-indoline] associated to zeolite Y and MCM-41 aluminosilicate. Site-selective electrocatalytic effect on N,N,N′,N′-tetramehtylbenzidine oxidation. *Journal of Physical Chemistry B*. 108: 20064–20075.
[39]. Mortimer, R.J. 2011. Electrochromic materials. *Annual Reviews on Materials Research*. 41: 241–268.
[40]. Cai, G.; Cui, M.; Kumar, V.; Darmawan, P.; Wang, J.; Wang, X.; He, L.-S.A.; Qian, K.; Lee, P.S. 2016. Ultra-large optical modulation of electrochromic porous WO_3 film and the local monitoring of redox activity. *Chemical Science*. 7: 1373–1382.
[41]. Samain, L.; Gilbert, B.; Grandjean, F.; Long, G.J.; Strivay, D. 2013. Redox reactions in Prussian blue containing paint layers as a result of light exposure. *Journal of Analytical Atomic Spectrometry*. 28: 524–535.
[42]. Doménech-Carbó, A.; Doménech-Carbó, M.T.; Osete-Cortina, L.; Donnici, M.; Guasch-Ferré, N.; Gasol-Fargas, R.M.; Iglesias-Campos, M.A. 2020. Electrochemical assessment of pigments-binding medium interactions in oil paint deterioration: A case study on indigo and Prussian blue. *Heritage Science*. 8: artic. 71.
[43]. Lin, S.-Y.; Chen, Y.-C.; Wang, C.-M.; Liu, C-C. 2008. Effect of heat treatment on electrochromic properties of TiO_2 thin films. *Journal of Solid State Electrochemistry*. 12: 1481–1486.
[44]. Ma, L.; Li, Y.; Yu, X.; Zhu, N.; Yang, Q.; Noh, C.-H. 2008. Electrochemical preparation of PmeT/TiO_2 nanocomposite electrochromic electrodes with enhanced long-term stability. *Journal of Solid State Electrochemistry*. 12: 1503–1509.
[45]. Huang, H.; Tian, J.; Zhang, W.K.; Gan, Y.P.; Tao, X.Y.; Xia, X.H.; Tu, J.P. 2011. Electrochromic properties of porous NiO thin film as a counter electrode for NiO/WO_3 complementary electrochromic window. *Electrochimica Acta*. 56: 4281–4286.

[46]. Gaupp, C.L.; Welsh, D.M.; Rauh, R.D.; Reynolds, J.R. 2002. Composite coloration efficiency measurements of electrochromic polymers based on 3,4-alkylenedioxythiophenes. *Chemistry of Materials*. 14: 3964–3970.
[47]. Meulenkamo, E.A.; Nelly, J.J.; Blasse, G. 1993. Electrochemically induced characteristic luminescence of metal ions at anodic valve metal oxides. *Journal of the Electrochemical Society*. 140: 84–91.
[48]. Martín, E.; Torres-Costa, V.; Martín-Palmar, J.; Bousoño, C.; Tutor-Sánchez, J.; Martínez-Duart, J.M. 2006. Photoluminiscence of naphthalimide derivatives deposited onto nanostructured porous silicon. *Journal of the Electrochemical Society*. 153: D134–D137.
[49]. Hagfeldt, A.; Boschloo, G.; Sun, L.; Kloo, L.; Pettersson, H. 2010. Dye-sensitivized solar cells. *Chemical Reviews*. 110: 6595–6663.
[50]. Mor, G.K.; Varghese, O.K.; Paulose, M.; Shankar, K.; Grimes, C.A. 2006. A review on highly ordered, vertically oriented TiO_2 nanotube arrays: Fabrication, material properties, and solar energy applications. *Solar Energy Materials and Solar Cells*. 90: 2011–2075.

16 Microporous Materials in Electrosynthesis, Environmental Remediation, and Drug Release

16.1 INTRODUCTION

In this chapter, several fields of application of porous materials will be described. Traditionally, these include electrosynthesis—aimed to prepare new compounds—and electrodegradation of contaminants, both based on electrolysis methods. In the last decades, other applications of porous materials have been developed—like drug release.

Electrosynthetic procedures are those devoted to preparing chemical products using electrochemistry, ordinarily employing controlled current or controlled potential electrolysis in a cell or electrochemical reactor. An electrochemical reactor can be defined as a device where the chemical product is obtained from one or more reagents using electrical energy. Electrolysis cells or electrochemical reactors mainly differ from chemical reactors, using electrical energy as a driving force to promote chemical changes, and by the existence of an interface where the electrochemical reaction takes place [1].

Different classifications of electrochemical reactors can be made depending on their configuration (divided, undivided cathodic and anodic compartments), electrode geometry (bi- and tri-dimensional), and the fluid flow through the reactor (mixing, plug-flow, fluidized baths), among others. There are several demands for electrochemical reactors [2]:

- Security, safety, low cost and ease in operation, and eventually, automatic control;
- Controlled, uniform potential distribution and current density between the electrodes;
- Low cell voltage as possible for minimizing energetic cost;
- Electrode surface/reactor volume ratio must be as higher as possible;
- Low-pressure drop and easy gas evacuation; and
- Control of mass transport and heat transport in the reactor.

The second part of this chapter is devoted to what is generically termed as environmental remediation. Preservation and improvement of the environment is a contemporary necessity that involves all humanity. Terms like sustainable development or green chemistry have become something more than a literary topic, currently converted into widely claimed social demands. Destruction of environmental pollutants is one of the aspects dealing with environmental management—a term where concepts such as environmental preservation, environmental remediation, etc. are included. Here, the attention will be focused in the electrodegradation of contaminants/pollutants. According to the Environmental Protection Agency (EPA), a chemical contaminant is an addition of any species to the environment that produces a deviation of its average composition in any natural media. In turn, a pollutant is an addition of natural or anthropogenic origin whose concentration is high enough to produce undesired effects on living organisms or objects of social interest. Here, we will treat electrochemical methods aimed to convert contaminants/pollutants into innocuous species. In the case of most organic matter, the general aim is to convert (the term mineralization is also used) C into CO_2, H in water, and N into N_2.

In this chapter, the desalination by capacitive deionization at porous materials will also be considered. This technique that is under intensive research is one of the responses to another challenging economic and social problem—the need for affordable clean water, a problem that increased in the last decades due to population growth, climate change, deforestation, extensive groundwater extraction (with the concomitant saltwater ingress in wells and aquifers), etc.

16.2 ELECTROLYTIC SYNTHESIS INVOLVING POROUS ELECTRODES

A variety of porous metallic structures have been used for organic and inorganic electrosynthesis [3]. The methodology used in chemical engineering to describe the performance of chemical reactors can be adapted to the study of electrochemical cells. Electrosynthesis of a variety of industrially relevant products has been studied; for instance, the synthesis of ammonia from natural gas at atmospheric pressure [4,5] or the electrodeposition of metals on porous solids [6]. Porous electrodes are of particular interest because of their high area. There is a possibility, however, of electrode degradation when operating at high temperatures and/or in aggressive electrolytes and/or at large overpotentials. This last problem is of less importance for fuel-cell anodes that operate at relatively low potentials, but it can be of importance for electrochemical reactors. Porous column electrodes are prepared by packing conductive materials (carbon fiber, metal shot) to form a bar. Continuous-flow column electrolytic procedures can provide high efficiencies for electrosynthesis or removal of pollutants in industrial situations.

Industrially relevant products can be prepared by electrosynthesis using microporous materials. Two general approaches can be used: electrosynthesis at conventional electrodes immersed into dispersions of the porous material in the electrolyte, or electrocatalytic synthesis at porous electrodes using a catalyst. This can be an electrode modifier deposited over the basal electrode, but there is the possibility of additional interposing of an electron mediator between the catalyst and the electrode [7]. Different electrode configurations (packed-bed, gas diffusion, trickle bed, rotating porous cylinders, etc.) have been described. As discussed in previous chapters, most of synthetically or energetically relevant electrochemical processes involve high kinetic constraints, represented by cathodic and/or anodic overpotentials. Electrocatalysis allows lowering of such overpotentials and obtaining effective electrosynthetic procedures.

In the case of electron transfer mediators immobilized on electrode surfaces, one can define a specific activity, S_A, as the quotient between catalytic peak current, i_{pcat}, and catalyst deposition charge, Q_{cat} ($S_A = I_{pcat}/Q_{cat}$) [8]. In most systems, one can define apparent activation energy, E_a, satisfying an Arrhenius-type equation:

$$I_p = k\ \exp(-E_a/RT) \tag{16.1}$$

Then, the specific activity should vary with temperature as:

$$S_A = -E_a/RT + \text{constant} \tag{16.2}$$

The encapsulation of catalytically active species into porous solids is one of the possible strategies of particular interest. Microheterogeneous electrocatalysis can be performed with zeolite-based catalysts using suspensions of such materials in organic solvents with low ionic strength by applying a dc voltage to suspensions of zeolite particles, as described by Rolison and Stemple [9] for the Pd^{II}-Cu^{II}-NaY catalyzed oxidation of propene. Electrocatalytic effects in zeolites can be influenced by (a) Lewis base effects associated with framework oxygens [10], (b) 'docking' of the substrate into the supercages [11], or (c) 'boxing effect' due to electronic confinement for the guest molecule entrapped into the zeolite framework [12].

The electrocatalytic properties of MOFs are also under intensive exploitation for electrosynthetic purposes [13]. Much of this research is centered on HER and OER processes aimed at their application in fuel cells. MOFs combine high ionic permeability with the high density of catalytic metal sites. The catalytic performance can be increased by hybridizing MOFs with highly conductive materials (either as MOF supports or encapsulated within the MOFs) and combining MOFs with other active catalysts [14].

16.3 SOLID TO SOLID SYNTHESIS

Although electrosynthetic methods are largely confined in species in solution, there is the possibility of work via solid-to-solid interconversion. This generally implies solids of intrinsic porosity to be permeated by charge-balancing electrolyte ions. For this purpose, the parent solid is immobilized on a basal electrode surface and submitted to the pertinent potential inputs. Solid to solid transformations of 7,7,8,8-tetracyanoquinodimethane in contact with aqueous electrolytes were studied by Bond et al. [15]. Solid to solid phase transformations may be controlled by nucleation and growth of the new crystalline phase, often giving rise to new solids with vast differences in morphology and crystal size. The reaction product can form (a) a solid solution within the parent compound, (b) immiscible inclusions within the parent crystals, and (c) segregated layers from the parent solid [16].

The electrochemical processes may involve intermediate species in solution and/or adsorbed onto surfaces. Electropolymerization processes can be initiated from crystalline solids, as described for the case of diphenylamine attached to platinum and gold electrodes [17]. New electrosynthetic strategies developed by Marken et al. [18] involve redox mediators acting catalytically on solid reagents.

16.4 ELECTROCHEMICAL DEGRADATION OF CONTAMINANTS

Electrochemical methods for degradation of environmental contaminants take advantage of other treatments (incineration, chemical attack) using low operating temperatures and the possibility of an accurate control of variables, influencing the overall process. These methods play a prominent role in the degradation/destruction of environmental pollutants—a strategy addressed to the conversion of pollutants into environmentally innocuous substances. The most divulgated method for the electrochemical degradation of organic contaminants in waters is anodic oxidation, often called electrochemical incineration. This methodology is based on the oxidation of pollutants and can be divided into direct (the pollutants are oxidized at the electrode surface) and indirect when an oxidizing reagent is added [19]. Most oxidative pathways are dominated by the electrochemical generation of adsorbed hydroxyl radical ($OH\cdot$) formed at the surface of the anode as a result of water oxidation. Table 16.1 summarizes the standard redox potentials for selected oxygen and chlorine oxidizing species that can be involved in electrochemical oxidation processes [20].

The hydroxyl radical is a powerful oxidizing agent that can react with organic compounds to give dehydrogenated or hydroxylated derivatives until complete mineralization, yielding water and carbon dioxide. Typically, the contaminated water is treated in the anodic compartment of a divided cell with high-oxygen overvoltage anodes. These include—among others—Pt, boron-doped diamond electrode, RuO_2, PbO_2, SnO_2, or IrO_2, and various doped systems, such as Bi-doped PbO_2 [21].

In principle, it is desirable to obtain a higher current density as possible, because this should lead to higher pollutant removal. There is a need, however, to reach a compromise between energy consumption and mineralization current efficiency (MCE). This can be defined for the total organic content (TOC) removal as [22]:

TABLE 16.1
Electrode Reactions and Standard Potentials for Oxidizing Species Formed in Aqueous Chloride Solutions [20]

Oxidant species	Reaction	$E°$ (V vs. NHE)
Hydroxyl radical	$HO\cdot_{aq} + H^+_{aq} + e^- \rightarrow H_2O$	2.722
Atomic oxygen	$O_g + 2H^+_{aq} + 2e^- \rightarrow H_2O$	2.421
Ozone	$O_{3g} + 2H^+_{aq} + 2e^- \rightarrow O_2 + H_2O$	2.076
Hydrogen peroxide	$H_2O_2 + 2H^+_{aq} + 2e^- \rightarrow 2H_2O$	1.776
Hydrogen peroxyl radical	$HO_2\cdot + 3H^+_{aq} + 3e^- \rightarrow 2H_2O$	1.684
Oxygen	$O_{2g} + 4H^+_{aq} + 4e^- \rightarrow 2\ H_2O$	1.228
Atomic chlorine	$Cl_g + e^- \rightarrow Cl^-_{aq}$	2.410
Chlorine monoxide	$Cl_2O_g + 2H^+_{aq} + 4e^- \rightarrow 2Cl^-_{aq} + H_2O$	2.153
Chlorine dioxide	$ClO_{2g} + 4H^+_{aq} + 5e^- \rightarrow Cl^-_{aq} + 2H_2O$	1.511
Hypochlorous acid	$HClO_{aq} + H^+_{aq} + e^- \rightarrow Cl^-_{aq} + H_2O$	1.494
Chlorine	$Cl_{2g} + 2e^- \rightarrow 2Cl^-_{aq}$	1.395

$$MCE\,(\%) = 100\frac{nFV\,(\mathrm{TOC})}{4.32 \times 10^7 gIt} \tag{16.3}$$

where V is the solution volume (L), (TOC) is the experimental TOC removal, g is the number of carbon atoms of the pollutant, I is the applied current (A), t is the electrolysis time (h), and n is electron stoichiometry of the burning reaction. The numerical coefficient (4.32×10^7) is a conversion factor ($3{,}600\ \mathrm{s\ h^{-1}} \times 12{,}000\ \mathrm{mg\ mol^{-1}}$).

There is the possibility of a direct oxidation of the pollutants and/or the oxidation by highly oxidizing species generated electrochemically. The two main degradation routes consist of (a) the electrochemical generation of physically adsorbed oxygen species (mainly hydroxyl radical), termed as combustion and (b) the generation of active oxygen chemisorbed onto (typically) the metal oxide lattice (termed electrochemical conversion) [23]. If no oxidizable species are present, the active oxygen species continue the pathway directed to produce O_2 (i.e. the OER process described in Chapters 4 and 7). Electrode processes yielding oxygen radical species are:

$$H_2O \rightarrow HO\cdot_{ads} + H^+_{aq} + e^- \tag{16.4}$$

$$H_2O_2 \rightarrow HO_2\cdot_{ads} + H^+_{aq} + e^- \tag{16.5}$$

In the second of the above cases, the oxidation of an organic compound R at a metal oxide (MO_x) electrode can be represented by the sequence of processes [24]:

$$MO_x + H_2O \rightarrow MO_x\text{–}OH + H^+ + e^- \tag{16.6}$$

$$MO_x\text{–}OH \rightarrow MO_{x+1} + H^+ + e^- \tag{16.7}$$

$$MO_x\text{–}OH \rightarrow (1/2)O_2 + MO_x + H^+ + e^- \tag{16.8}$$

$$MO_{x+1} \rightarrow (1/2)O_2 + MO_x + e^- \tag{16.9}$$

$$R + MO_x(OH)_z \rightarrow MO_x + (z/2)CO_2 + zH^+ + ze^- \quad (16.10)$$

$$R + MO_{x+1} \rightarrow MO_x + RO \quad (16.11)$$

Of course, anodic processes involve side reactions. In particular, the OER occurs when the potential reaches the oxygen evolution overpotential, increasing energy consumption. A general requisite for anodes is that they present a large overpotential for OER.

Direct electroreduction methods are typically used for the dechlorination of chlorinated pollutants in waters. The easy removal of Cl from chlorinated organics converts chlorofluorocarbons (CFCs) into hydrochlorofluorocarbons (HCFCs), hydrofluorocarbons (HFCs), and even fluorocarbons (FCs). ECFCs are much less destructive to the atmospheric ozone than CFCs, but HFCs and FCs are harmless to atmospheric ozone, although they may contribute to the greenhouse effect.

The proposed reactions for the reduction of $CFCl_3$ at Hg or Pb cathodes can be represented as the sequence of processes:

$$CFCl_3 \xrightarrow{+2e^-,-Cl^-} CHFCl_2 \xrightarrow{+2e^-,-Cl^-} CH_2FCl \xrightarrow{+2e^-,-Cl^-} CH_3F \quad (16.12)$$

In the last years, hydrocatalytic dehalogenation procedures have been devised. These processes are based on the reductive dehalogenation of adsorbed halogenated organic substrates by adsorbed hydrogen atoms produced electrochemically as in HER processes (see Chapter 4). The sequence of processes can be represented as [24]:

$$M + 2H_2O + 2e^- \rightarrow M(2H_{ads}) + 2OH^- \quad (16.13)$$

$$R\text{–}X + M \rightarrow M(R\text{–}X)_{ads} \quad (16.14)$$

$$M(R\text{–}X)_{ads} + M(2H_{ads}) \rightarrow M(R\text{–}H)_{ads} + HX \quad (16.15)$$

$$M(R\text{–}H)_{ads} \rightarrow M + R\text{–}H \quad (16.16)$$

The electrode material plays an essential role in the above processes. Good catalytic performance to generate reductive atomic hydrogen is required. Pb, Ni, Pd, and their alloys—as well as a variety of modified electrode materials—are under study [24].

Electrodes modified with zero-valent iron nanoparticles can also be used for reductive degradation of contaminants. These can remove aqueous contaminants by reduction to insoluble forms, case of heavy metal ions, and chlorinated species via reductive dechlorination. In particular, zero-valent iron nanoparticles can be used to degrade nitrate and nitrite ions. The overall reactions can be expressed as:

$$5Fe^\circ + 2NO_3^- + 6H_2O \rightarrow 5Fe^{2+} + N_2 + 12OH^- \quad (16.17)$$

$$3Fe^\circ + 2NO_2^- + 4H_2O \rightarrow 3Fe^{2+} + N_2 + 8OH^- \quad (16.18)$$

These reactions, however, compete with the processes:

$$2Fe^\circ + O_2 + 2H_2O \rightarrow 2Fe^{2+}_{aq} + 4OH^- \quad (16.19)$$

$$Fe^{\circ} + 2H_2O \rightarrow Fe_{aq}^{2+} + H_2 + 2OH^- \tag{16.20}$$

As in the case of oxidative treatments, electrode poisoning, undesired or incomplete degradation reactions, and removal of degradation products are concerned.

Porous materials can be used not only as electrode materials but also for ion storage and exchange. Figure 16.1 shows a scheme of an electrochemical-ion exchange reactor for ammonia removal based on zeolites [25]. The reactor incorporates a Cu/Zn cathode and a Ti/IrO_2–Pt anode. The simulated ammonia wastewater is modeled by an $(NH_4)_2SO_4$ solution accompanied by a NaCl solution which acts as a regeneration solution.

Here, chloride is electrochemically oxidized to Cl_2:

$$2Cl^- \rightarrow Cl_2 + 2e^- \tag{16.21}$$

which is hydrolyzed giving hypochlorite which reacts with ammonia.

$$Cl_2 + OH^- \rightarrow ClO^- + H^+ + Cl^- \tag{16.22}$$

$$2ClO^- + 2NH_3 \rightarrow N_2 + 2H_2O + 2H^+ + 2Cl^- \tag{16.23}$$

Nitrate ions can also be produced during the anodic decomposition of ammonia. These ions can be re-reduced to nitrite, nitrogen, and even to ammonia at the cathode, but are again oxidized in the presence of hypochlorite in the reactor. Figure 16.2 shows the representation of the removed concentration of ammonia vs. time, representative of the performance of the electrochemical reactor in Figure 16.1 [25].

The above is an example of indirect methods of electrochemical removal of contaminants. Other indirect methods involve continuous supply of H_2O_2 or other reagents to the contaminated solution [24]. Hydrogen peroxide can be electrochemically obtained in acidic media from:

$$O_2 + 2H^+ + 2e^- \rightarrow H_2O_2 \tag{16.24}$$

In cathodic indirect oxidation with H_2O_2, this reagent is generated in the cathode chamber by reduction of dissolved oxygen [24].

The Fenton procedure (electron-Fenton) involves the addition of Fe^{2+} ions to the solution, resulting in the generation of hydroxyl radicals [2]:

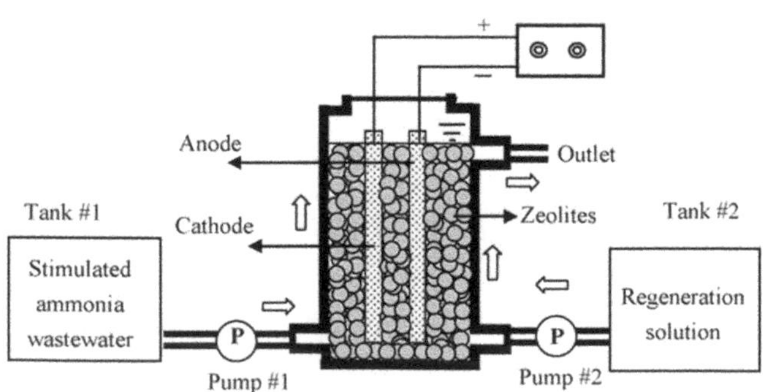

FIGURE 16.1 Scheme of an Electrochemical Reactor for Ammonia Removal Using a Cu/Zn as Cathode and a Ti/IrO_2–Pt Anode and Zeolite as Ion Exchanger. Reproduced from Ref. [25] (Li et al. *Electrochim. Acta.* 2009, 55: 159–164), with Permission.

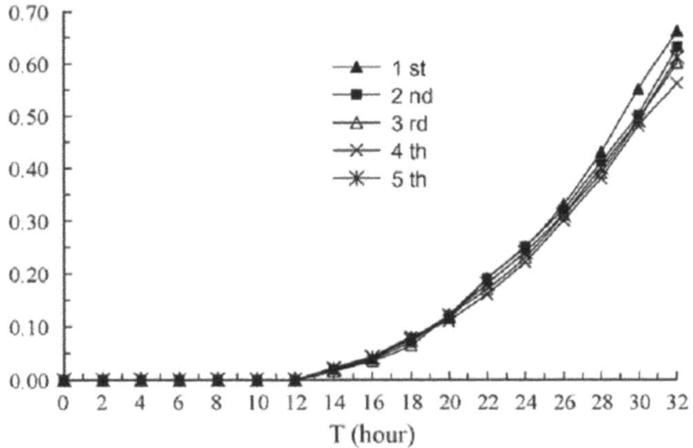

FIGURE 16.2 Performance of the Electrochemical Reactor for Ammonia Removal Zeolite as Ion Exchanger in Terms of the Removed Concentration of Ammonia vs. Time. Reproduced from Ref. [25] (Li et al. *Electrochim. Acta*. 2009, 55: 159–164), with Permission.

$$H_2O_2 + Fe^{2+} \rightarrow Fe^{3+} + HO\cdot + OH^- \quad (16.25)$$

This is a catalytic process that can be propagated via the regeneration of ferrous ions with electrogenerated H_2O_2. This reaction yields hydroperoxyl radical ($HO_2\cdot$), a species that have a much weaker oxidizing power than HO· (see Table 16.1):

$$H_2O_2 + Fe^{3+} \rightarrow Fe^{2+} + HO_2\cdot + H^+ \quad (16.26)$$

Additionally, the reaction of Fe^{3+} with $HO_2\cdot$ and organic radical intermediates is also possible.

16.5 DEGRADATION/GENERATION

Cogeneration strategies were described in Chapter 14 with regard to the simultaneous production of energy and chemicals in fuel cells. A similar concept can be applied in environmental chemistry because electrochemical degradation of most contaminants can yield useful products and, eventually, produce energy [19].

One of the most significant cases is that of CO_2, which can be used as a starting product to generate useful chemicals. The electrochemical reduction of CO_2 to CO ($E^\circ = -0.41$ V vs. SHE at pH = 5) can be homogeneously catalyzed by several species, for instance, by Ni(II)-cyclam complexes. Alternatively, different electrodes can use heterogeneous catalysis. As a result, CO_2 can be used for electrocarboxylations to give a variety of carboxylic acids. Two possible pathways can be considered, having in common the generation of a radical anion of the organic compound:

$$R \rightarrow R\cdot^- + e^- \quad (16.27)$$

The generated radical anion can react as a nucleophile attacking the CO_2, yielding a carboxylated radical anion further electrochemically reduced:

$$R\cdot^- + CO_2 \rightarrow R\text{–}COO\cdot^- \quad (16.28)$$

$$R\text{–}COO\cdot^- + H^+ + e^- \rightarrow RH\text{–}COO^- \quad (16.29)$$

Or, alternatively, can reduce CO_2 giving a $CO_2{\cdot}^-$ radical anion that is subsequently coupled with the parent radical:

$$R{\cdot}^- + CO_2 \rightarrow R + CO_2{\cdot}^- \quad (16.30)$$

$$R{\cdot}^- + CO_2{\cdot}^- + H^+ \rightarrow RH\text{–}COO^- \quad (16.31)$$

Similarly, nitrogen oxides in emissions from automobiles and power plants can be processed in fuel cells using $HClO_4$ as electrolyte, the anodic reaction being hydrogen oxidation to protons [26]. Possible electrochemical reactions are:

$$NO_{2g} + 2H^+_{aq} + 2e^- \rightarrow NO_g + H_2O \quad (16.32)$$

$$2NO_g + 2H^+_{aq} + 2e^- \rightarrow N_2O_g + H_2O \quad (16.33)$$

$$NO_g + 3H^+_{aq} + 3e^- \rightarrow NH_2OH_{aq} \quad (16.34)$$

$$NO_g + 5H^+_{aq} + 5e^- \rightarrow NH_{3\ aq} + H_2O \quad (16.35)$$

Similarly, sulfur dioxide can also be treated to produce sulfuric acid.

Recent research investigates different cell processes; for instance, using gas diffusion electrodes (Figure 16.3) [27] and microfluidic arrangements (Figure 16.4) [28].

16.6 PHOTOELECTROCHEMICAL DEGRADATION

Photo-assisted electrochemical degradation of pollutants is an interesting alternative in principle based on the use of semiconducting electrodes. It is pertinent to underline that in these photo-electrochemical experiments, the point of zero charge (ZCP) of a semiconductor is less important than in chemical photolysis [20]. In these last processes, the ZCP is essential to determine the composition of the adsorption layer. For instance, the ZCP of WO_3 is close to pH 1. Then, under

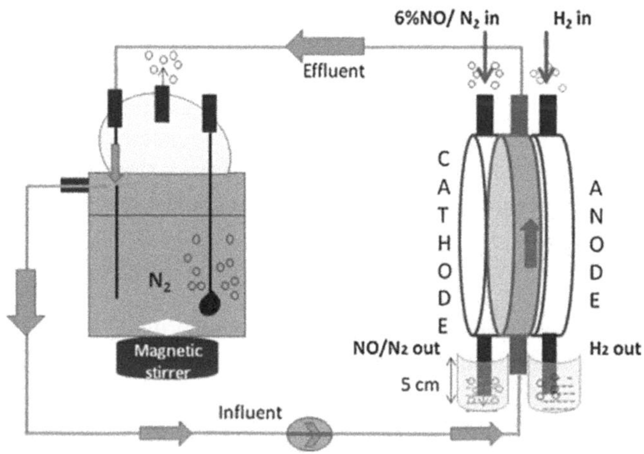

FIGURE 16.3 Schematics of a Gas Diffusion Electrode for Cogeneration of Chemicals and Electricity. Reproduced from Ref. [27] (Alvarez-Gallego et al. *Electrochim. Acta.* 2012, 82: 415–426), with Permission.

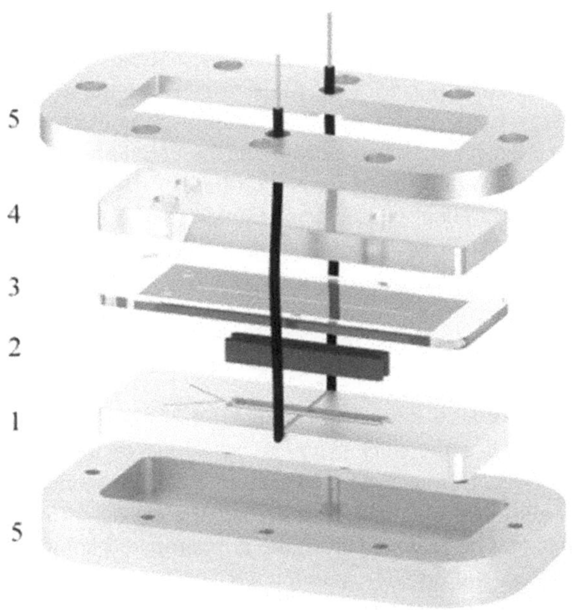

FIGURE 16.4 Scheme of a Microfluidic Reactor for Cogeneration of Chemicals and Electricity. Reproduced from Ref. [28] (Wouters et al. *Electrochim. Acta.* 2016, 210: 337–345), with Permission.

open circuit potential conditions, the surface of the semiconductor is negatively charged at pH > 1, disfavoring the adsorption of anionic species. In photoelectrochemical experiments, however, the surface potential is controlled by the external potential input and the ZCP becomes essentially irrelevant. When the applied potential (see Chapter 15) is more positive than the flat band potential of an *n*-type semiconductor such as WO_3, the surface charge is positive due to the formation of a depletion layer, and anion adsorption is favored.

There are different factors influencing the effectiveness of such processes—intensity and type of irradiation, electrodes, pH, adsorption, and co-adsorption effects, in particular. The pH of the solution influences the photodegradation process in three ways: (i) favoring (or not) the adsorption of ionic species, as previously discussed; (ii) the different pH dependence of the electrode potential of competing processes; and (iii) as a result of the variation of the flat band potential that follows the equation [20]:

$$E_{fb} = E^{o}_{fb} - 0.059\text{pH} \quad (\text{V at } 298 \text{ K}) \tag{16.36}$$

Photocatalytic oxidation of organic pollutants on TiO_2-based materials has been extensively investigated. The catalyst is used in the form of a suspension of fine particles or thin film on robust substrates. In the base solution, water is oxidized at the TiO_2 electrode surface by the photogenerated holes, while organics are oxidized. At low potential bias, the photocurrent increases slowly with the applied potential. This is interpreted as a result of the low separation rates of photogenerated electron-hole pairs under low electric field conditions. At higher bias potentials, there is an almost linear variation of the photocurrent on the potential. This can be rationalized by assuming that the electron transport in the film becomes rate-determining so that the TiO_2 photoelectrode behaves as a constant resistance [29]. By further increasing the potential, the photocurrent levels of and tends to stabilize, thus defining a saturation behavior attributable to scanty holes to facilitate the interface reaction.

TiO_2 has two main drawbacks: the relatively large bandgap requiring ultraviolet (UV) light absorption and its low quantum yields. Improving the photocatalytic efficiency of such systems can be fulfilled by doping and/or using porous TiO_2 films to increase the active area.

A second general strategy is the performance of indirect photoelectrodegradations. The most studied process is the so-called photoelectro-Fenton. In the conventional electro-Fenton method, the formation of complexes of iron ions with some intermediates, such as oxalate ions, can reduce the effectiveness of the process. Then, photoelectro-Fenton methods involve the use of UV irradiation to produce photolysis of such Fe^{3+} complexes and increase the rate of Fe^{2+} regeneration by the photo-Fenton reaction:

$$Fe^{3+} + H_2O + h\nu \rightarrow Fe^{2+} + H^+ + HO\cdot \tag{16.37}$$

which is combined with the processes described by Eqs. (16.24)–(16.26). An alternative is the use of a sacrificial Fe anode that continuously supplies Fe^{2+} ions to the solution by its oxidation process: $Fe \rightarrow Fe^{2+} + 2e^-$. Here, ferrous ions are oxidized by electrogenerated hydrogen peroxide so that an Fe^{3+}-saturated solution is obtained. In this case, the excess of this ion precipitates as Fe $(OH)_3$. Then, the pollutants can be removed from the contaminated solution by their degradation with OH· and coagulation with the $Fe(OH)_3$ precipitate.

16.7 DESALINATION

Apart from distillation, reverse osmosis and electrodialysis are the most known desalination technologies [30]. The capacitive deionization (CDI) of water is based on the electrosorption phenomenon at porous electrodes [31]. In this technique, salt ions are removed from water upon applying a difference of potential between two porous electrodes. The ions are temporarily immobilized into the electrical double layer of the micropores of electrodes so that the ionic charge is locally compensated by the electronic charge acquired by the electrode.

The CDI systems operate cyclically in two steps—electrosorption and discharge. The typical desalination cells consist of two electrodes made of porous carbon. In the electrosorption step, more counterions are adsorbed in the micropores than co-ions are expelled, so the water flowing through the CDI device is desalinated. The duration of this step is typically several minutes and results in a fully charged cell. Then, the difference of potential is suppressed, and ions are released, leading to a flow of water of higher salinity.

Figure 16.5 depicts two common cell architectures consisting of sandwiched carbon electrodes covered by a thin membrane of ion-exchange resin [31]. In the second case (membrane capacitive deionization, MCDI), the cathode is covered by a thin layer of cation-exchange resin and the anode is covered by an anion-exchange resin.

The CDI methodology was pioneered by Blair, Murphy, and their co-workers in the 1960s [32,33]. Initially termed electrochemical demineralization of water, the interest in CDI has increased exponentially, giving rise to a variety of operation modes including stop-flow operation during ion release, salt release at reversed voltage, constant-current operation, and energy recovery from the desalination/release cycle, among others.

The energy consumption in a CDI cell in which a volumetric flux of saline solution, ϕ_{fresh}, gives rise to fluxes of de-mineralized (feed) and concentrated (conc) water can be evaluated as (j = fresh, feed, conc):

$$\Delta G = G_{fresh} + G_{conc} - G_{feed} = RTc_j\varphi_j \ln c_j \tag{16.38}$$

The above equation can be converted into [31]:

(a)

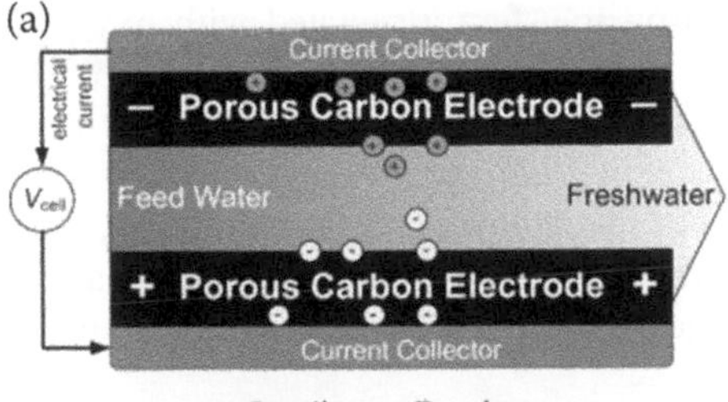

(b)

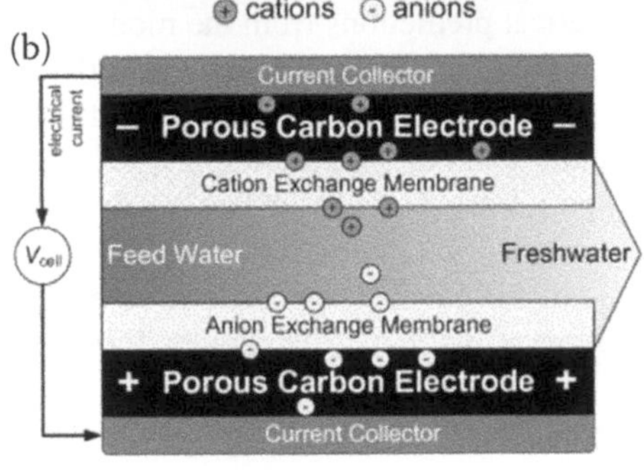

FIGURE 16.5 Scheme of Two Common Water Desalination Cells. (a) Ordinary CDI Two-Electrode Cell; (b) Membrane Capacitive Deionization, MCDI. Here, a Cation-Exchange Membrane is Fixed on the Cathode and an Anion-Exchange Membrane is Attached to the Anode. Reproduced from Ref. [31] (Porada et al. *Progr. Mater. Sci.* 2013, 58: 1388–1442), with Permission.

$$\Delta G = RT\varphi_{\text{fresh}}(c_{\text{feed}} - c_{\text{fresh}})\left[\frac{\ln(c_{\text{feed}}/c_{\text{fresh}})}{1 - c_{\text{feed}}/c_{\text{fresh}}} - \frac{\ln(c_{\text{feed}}/c_{\text{conc}})}{1 - c_{\text{feed}}/c_{\text{conc}}}\right] \tag{16.39}$$

where c_{fresh}, c_{feed}, c_{conc} are, respectively, the saline concentrations in the initial, de-mineralized, and concentrated flows.

As requirements for electrode materials one can cite [31]:

- Large ion-accessible specific surface area
- High chemical and electrochemical stability
- Easy ion mobility within the pore network
- High electronic conductivity
- Low contact resistance between the porous electrode and the current collector
- Good wetting behavior (adequate hydrophilicity)
- Low costs and scalability
- Good processability into film electrodes based on compacted powders, fibers, monoliths, etc.
- Large (natural) abundance and low CO_2 footprinting
- High bio-inertness to avoid bio-fouling

Ultimately, the lower limit to the pore size is the bare ion size or—less restrictive—the hydration ion size (for instance, for Na^+, 1.16 and 3.58 Å, respectively), but larger pore sizes are used. It must be underlined that the capacity for salt electrosorption is related to the surface area, but only a fraction of the same is electrochemically accessible to ions. Additionally, there are limitations to diffusion due to small pores. This requires the control not only of pore shape and size distribution but also of interparticle distances and electrode thickness.

Several models have been developed to interpret electrosorption processes. Initially, the effect of oxygenated functions in carbon (see Chapter 10) was underlined [33], further considering double-layer effects as decisive [34]. Here, the ZPC of the material constituting the electrodes plays a crucial role. The ZPC can be modulated by oxidizing or reducing carbon substrates and using a third reference electrode can be beneficial in this regard [35]. Different electrical double-layer-based models have been proposed [2,36]. As general demands, modeling may account for

Faradaic and non-Faradaic processes. The former are electron transfers associated with oxygen functionalities in carbon surfaces, processes of HER, PER, and ORR, and the electrochemical oxidation of carbons to surface oxidic species and even CO_2. These processes may appear depending on the experimental conditions, applied potential solution pH, etc.

The non-Faradaic processes deal with the characteristics of the double layer in micropores, ion mobility in macropores (spaces between carbon particles), and effects of chemical surface charge mainly associated with carboxylic and amine functionalities existing in carbon surfaces, summarized in Figure 16.6 [2].

Figure 16.7 compares the experimental data with the theoretical predictions from the modified-Donnan model for anion and cation equilibrium adsorption in the micropores of activated carbon electrodes [37]. This figure represents the ion concentration in the micropores as a function of the charge stored in the micropores. These are two experimentally available quantities, the second determined from the integration of current-time curves. Apart from porous carbons (activated carbons and activated carbon cloths, carbon black, carbon aerogels, ordered mesoporous carbons, CNTs, graphene, and its derivatives), other materials are being investigated. These include carbide-derived carbons [38] and MOF-doped carbons [39] among several inorganic membranes [40].

16.8 DRUG DELIVERY

Drug delivery is an active research field built around the controlled release of pharmaceuticals into living organisms. The aim is to increase the efficiency of chemotherapy through a localized release in the body, modulating the biodistribution rate. A variety of active pharmaceutical ingredients have been incorporated into drug release systems, including analgesics, antibiotics, and anticancer drugs [41]. The general idea is to reversibly attach the active pharmaceutical to a substrate so that, when this system is in contact with the desired environment, the drug is released.

For these purposes, a wide variety of nanoarchitectures have been studied—nanoparticles, scaffolds, foams, films, etc. [42]. A significant part of the drug supports are porous systems, including silicon [43] to hydroxyapatite [44], and polymers such as poly(lactic-co-glycolic acid) [42] and poly(g-benzyl-L-glutamate)–poly(glutamic acid) [45]. The drug delivery systems (DDS) involve reaction scheme directly lying with molecular recognition strategies. Among a variety of

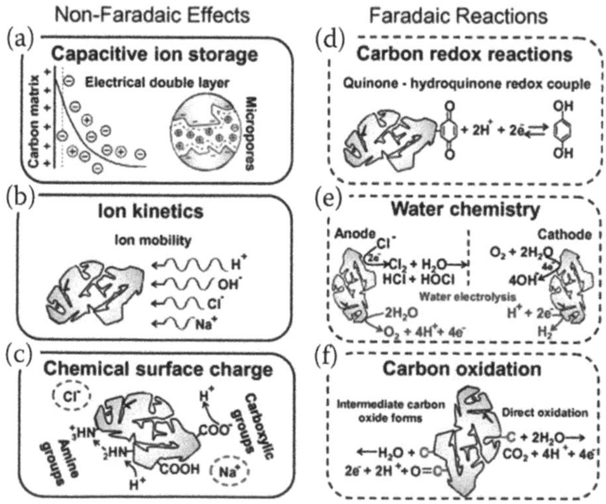

FIGURE 16.6 Pictorial Scheme for the Main Faradaic and Non-Faradaic Processes in at Carbon Electrodes CDI Cells. Reproduced from Ref. [31] (Porada et al. *Progr. Mater. Sci.* 2013, 58: 1388–1442), with Permission.

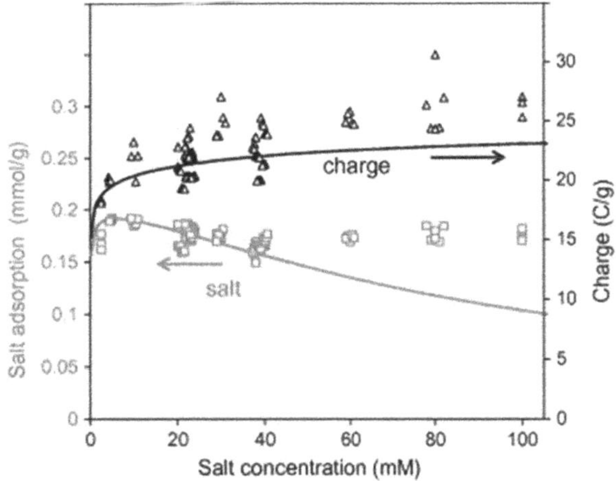

FIGURE 16.7 Anion and Cation Equilibrium Adsorption in the Micropores of Activated Carbon Electrodes. Theoretical Lines from the Donnan-Modified Model and Experimental Data from Ref. [37]. Adapted from Refs. [31] (Porada et al. *Progr. Mater. Sci.* 2013, 58: 1388–1442) and [37] (Biesheuvel et al. *J. Colloid Interface Sci.* 2011, 360: 239–248), with Permission.

possibilities, DDS may consist of a porous substrate, where the drug is attached via encapsulation of binding to a functionalized substrate by means of a specific receptor, the drug, and a sensing unit—bound to the drug or to the substrate. When the environment fulfills a given condition of activation (pH change, for instance) detected by the drug or the sensing unit, the drug release is activated.

In the case of porous polymers, there are four basic pathways of drug release: (a) drug transport through water-filled pores, (b) drug transport through the polymer matrix, (c) osmotic pumping, and (d) polymer dissolution. The first is the most frequent mechanism and usually proceeds by diffusion but can also occur via convection, induced by osmotic pressure. In this case, the polymer that has mobile polymer chains absorbs water, so by increasing the water content, the osmotic pressure also increases, producing swelling and rearrangement of the polymer chains. Transport through the polymer matrix occurs for hydrophobic drugs constituted by small molecules and water-absorbing polymers. Finally, erosion of the polymer causes drug release. This can be accompanied or not by an increase in the rate of drug transport [41]. Figure 16.8 shows a schematic of the basic mechanisms of drug release in polymeric matrices [41].

In principle, the kinetics of drug release can be described by the semiempirical Peppas equation [46] which applies for a diffusion-controlled release without swelling and degrading interferences:

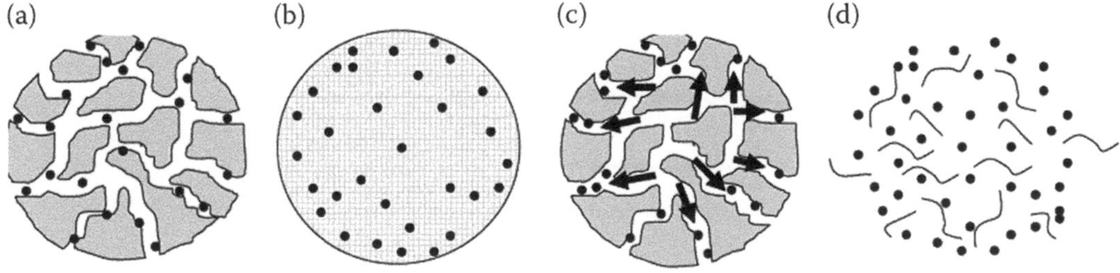

FIGURE 16.8 Schematics of the Basic Mechanisms of Drug Release in Polymeric Matrices: (a) Diffusion Through Water-Filled Pores, (b) Diffusion Through the Polymer, (c) Osmotic Pumping, and (d) Polymer Erosion. Reproduced from Ref. [41] (Fredenberg et al. *Int. J. Pharm.* 2011, 415: 34–52), with Permission.

$$\frac{M_t}{M_\infty} = kt^h \tag{16.40}$$

In this equation M_t is the net amount of drug released at time t, M_∞ is the total amount of drug encapsulated, k is the rate constant h is the release exponent. The value of h is representative of the geometry of the system ($h = 0.43$ for spherical diffusion and $h = 0.50$ for a thin film). This is the simplest rate law for drug release that has been followed by numerous others [41].

Frequently, the released mass/time curves offer two or three steps. In principle, the first phase (I) is associated with the so-called burst release and is attributed to the delivery of surface molecules and/or disintegration of the support particles. This may be followed by a second phase (II) of slow release. This has been attributed to drug transport through the polymer and/or polymer hydration and erosion. The following phase of fast release (phase III or second burst phase) can be assigned to substrate erosion. There are, however, different possibilities. Figure 16.9 compares several curves representing different types of drug release.

The kinetics of drug release is influenced by a number of factors. Regarding diffusion through water-filled pores, it depends on the porous structure of the substrate, which results in a time-dependent diffusion coefficient (D_{eff}) function of the 'true' diffusion coefficient of the drug (D_o), the porosity, and the tortuosity [47,48]:

$$D_{eff} = D_o e^{kt} \tag{16.41}$$

Additionally, pore closure associated with polymer hydration and other factors can also be influential on effective diffusion. On the contrary, diffusion through the polymer is not particularly dependent on the pore structure but can be influenced by the transition between the vitreous to the rubbery states of the polymer as well as the formation of cracks in the polymer.

Apart from the obvious influence of the activation factor initiating drug release, the local environmental conditions also affect the kinetics of the process. Increasing temperature increases chemical reactions, polymer mobility, and diffusion coefficients, whereas high osmolality in the medium decreases the rate of water absorption of the DDS. Other factors are substrate-drug and drug-drug interactions and substrate hydrolysis. Drug release is also influenced by the presence of

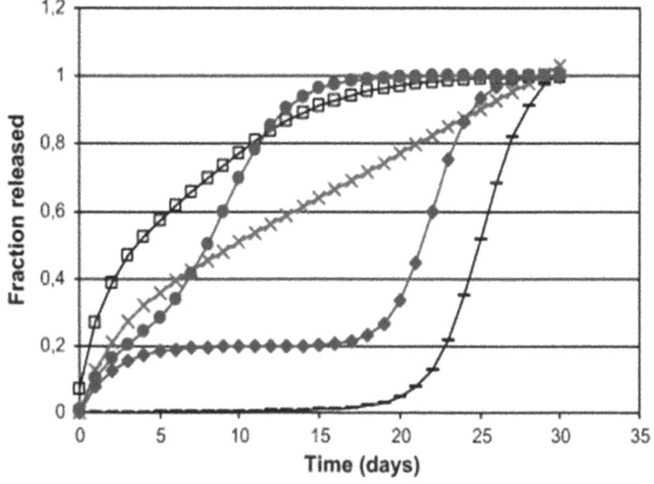

FIGURE 16.9 Typical Drug Release Curves. Open Squares: Burst and a Rapid Phase II. Filled Circles: Tri-phasic Release with a Short Phase II. Crosses: Burst and Zero-Order Release. Filled Diamonds: Tri-phasic Release. Dashes: Bi-phasic Release, Similar to Tri-phasic but Without the Burst Release. Reproduced from Ref. [41] (Fredenberg et al. *Int. J. Pharm.* 2011, 415: 34–52), with Permission.

salts, plasticizing agents, and surfactants in the medium, whereas the pH and the buffering capacity may condition the substrate degradation rate. All these factors draw a complex picture, where encapsulated drugs may be released simultaneously by several of the aforementioned mechanisms and the dominating mechanism may change during the release process [41].

REFERENCES

[1]. Pletcher, D.; Walsh, F.C. 1990. *Industrial Electrochemistry*. Chapman and Hall, New York.

[2]. Brillas, E.; Sirés, I.; Oturan, M.A. 2009. Electro-Fenton process and related electrochemical technologies based on Fenton's reaction chemistry. *Chemical Reviews*. 109: 6570–6631.

[3]. Marken, F.; Atobe, A. Eds. 2018. Solid-to-solid transformations in organic electrosynthesis. In *Modern Electrosynthetic Methods in Organic Chemistry*. CRC Press.

[4]. Marnellos, G.; Zisekas, S.; Stoukide, M. 2001. Synthesis of ammonia at atmospheric pressure with the use of solid state proton conductors. *Journal of Catalysis*. 193: 80–87.

[5]. Wang, B.H.; Wang, J.D.; Liu, R.; Yie, Y.H.; Li, Z.J. 2007. Synthesis of ammonia from natural gas at atmospheric pressure with doped ceria–$Ca_3(PO_4)_2$-K_3PO_4 composite electrolyte and its proton conductivity at intermediate temperature. *Journal of Solid State Electrochemistry*. 11: 27–31.

[6]. Masliy, A.I.; Poddubny, N.P.; Medvedev, A.Zh.; Zherebilov, A.F.; Sukhorukov, D.V. 2008. Electrodeposition on a porous electrode with low initial conductivity: Effect of the oxidant on the dynamics of the cathode deposit mass. *Journal of Electroanalytical Chemistry*. 623: 155–164.

[7]. Zhong, X.; Chen, J.; Liu, B.; Xu, Y.; Kuang, Y. 2007. Neutral red as electron transfer mediator: Enhance electrocatalytic activity of platinum catalyst for metanol electro-oxidation. *Journal of Solid State Electrochemistry*. 11: 463–468.

[8]. Wang, M.Y.; Chen, J.H.; Fan, Z.; Tang, H.; Deng, G.H.; He, D.L.; Kuang, Y.F. 2004. Ethanol electro-oxidation with Pt and Pt-Ru catalysts supported on carbon nanotubes. *Carbon*. 42: 3257–3260.

[9]. Rolison, D.R.; Stemple, J.Z. 1993. Electrified microheterogeneous catalysis in low ionic strength media. *Chemical Communications*. 25–27.

[10]. Teunisson, E.H.; van Santen, R.A.; Jansen, A.P.J.; Van Duijneveldt, F.B. 1993. NH_4^+ in zeolites-coordination and solvation effects. *Journal of Physical Chemistry*. 97: 203–210.

[11]. Derouane, E.G.; André, J.M.; Lucas, A.A. 1988. Surface curvature effects in physisorption and catalysis by microporous solids and molecular sieves. *Journal of Catalysis*. 110: 58–73.

[12]. Zicovich-Wilson, D.M.; Corma, A.; Viruela, P. 1997. Electronic confinement of molecules in microscopic pores. A new concept which contributes to the explanation of the catalytic activity of zeolites. *Journal of Physical Chemistry*. 98: 10863–10870.

[13]. Mueller, U.; Schubert, M.; Teich, F.; Puetter, H.; Schierle-Arndt, K.; Pastré, J. 2006. Metal-organic frameworks-prospective industrial applications. *Journal of Materials Chemistry*. 16: 626–636.

[14]. Zheng, F.; Zhang, Z.; Zhang, C.; Chen, W. 2020. Advanced electrocatalysts based on metal-organic frameworks. *ACS Omega*. 5: 2495–2502.

[15]. Nafady, A.; Al-Qahtani, N.J.; Al-Farhan, K.A.; Bhargava, S.; Bond, A.M. 2014. Synthesis and characterization of microstructured sheets of semiconducting $Ca[TCNQ]_2$ via redox-driven solid-solid phase transformation of TCNQ microcrystals. *Journal of Solid State Electrochemistry*. 18: 851–859.

[16]. Doménech-Carbó, A. 2018. Solid-to-solid transformations in organic electrosynthesis. In Marken, F.; Atobe, A. Eds. *Modern Electrosynthetic Methods in Organic Chemistry*, chap. 8. CRC Press, pp. 167–179.

[17]. Inzelt, G. 2002. Cyclic voltammetry of solid diphenylamine crystals immobilized on an electrode surface and in the presence of an aqueous solution. *Journal of Solid State Electrochemistry*. 6: 265–271.

[18]. Kolodziej, A.; Ahn, S.D.; Carta, M.; Malpass-Evans, R.;. McKeown, N.B.; Chapman, R.S.L.; Bull, S.D.; Marken, F. 2015. Electrocatalytic carbohydrate oxidation with 4-Benzoyloxy-TEMPO heterogenised in a polymer of intrinsic microporosity. *Electrochimica Acta*. 160: 195–201.

[19]. Martinez-Huitle, C.A.; Ferro, S. 2006. Electrochemical oxidation of organic pollutants for the wastewater treatment: direct and indirect processes. *Chemical Society Reviews*. 35: 1324–1340.

[20]. Hepel, M.; Luo, J. 2001. Photoelectrochemical mineralization of textile diazo dye pollutants using nanocrystalline WO_3 electrodes. *Electrochimica Acta*. 47: 729–740.

[21]. Shmychkova, O.; Luk'yanenko, T.; Velichenko, A.; Meda, L.; Amadelli, R. 2013. Bi-doped PbO_2 anodes: Electrodeposition and physico-chemical properties. *Electrochimica Acta*. 111: 332–338.

[22]. Yu, X.; Zhou, M.; Hu, Y.; Groenen Serrano, K.; Yu, F. 2014. Recent updates on electrochemical degradation of bio-refractory organic pollutants using BDD anode: A mini review. *Environmental Science and Pollution Research*. 21: 8417–8431.

[23]. Brillas, E.; Martínez-Huitle, C.A. 2014. Decontamination of wastewaters containing synthetic organic dyes by electrochemical methods. An updated review. *Applied Catalysis B Environmental*. 166–167: 603–643.

[24]. Zhang, M.; Shi, Q.; Song, X.; Wang, H.; Bian, Z. 2019. Recent electrochemical methods in electrochemical degradation of halogenated organics: A review. *Environmental Science and Pollution Research*. 26: 10457–10486.

[25]. Li, M.; Feng, C.; Zhang, Z.; Zhao, R.; Lei, X.; Chen, R.; Sugiura, N. 2009. Application of an electrochemical-ion exchange reactor for ammonia removal. *Electrochimica Acta*. 55: 159–164.

[26]. Langer, S.H.; Pate, K.T. 1980. Electrogenerative reduction of nitric oxide. *Nature*. 284: 434–435.

[27]. Alvarez-Gallego, Y.; Dominguez-Benetton, X.; Pant, D.; Diels, L.; Vanbroekhoven, A.; Genné, I.; Vermeiren, P. 2012. Development of gas diffusion electrodes for cogeneration of chemicals and electricity. *Electrochimica Acta*. 82: 415–426.

[28]. Wouters, B.; Herijgers, J.; De Malsche, W.; Breugelmans, T.; Hubin, A. 2016. Electrochemical characterization of a microfluidic reactor for cogeneration of chemicals and electricity. *Electrochimica Acta*. 210: 337–345.

[29]. Jiang, D.; Zhao, H.; Zhang, S.; John, R. 2003. Characterization of photoelectrocatalytic processes at nanoporous TiO_2 film electrodes: Photocatalytic oxidation of glucose. *Journal of Physical Chemistry B*. 107: 12774–12780.

[30]. Anderson, M.A.; Cudero, A.L.; Palma, J. 2010. Capacitive deionization as an electrochemical means of saving energy and delivering clean water. Comparison to present desalination practices: will it compete? *Electrochimica Acta*. 55: 3845–3856.

[31]. Porada, S.; Zhao, R.; van der Wal, A.; Presser, V.; Biesheuvel, P.M. 2013. Review on the science and technology of water desalination by capacitive deionization. *Progress in Materials Science*. 58: 1388–1442.

[32]. Arnold, B.B.; Murphy, G.W. 1961. Studies on electrochemistry of carbon and chemically modified carbon surfaces. *Journal of Physical Chemistry*. 65: 135–138.

[33]. Murphy, G.W.; Caudle, D.D. 1967. Mathematical theory of electrochemical demineralization in flowing systems. *Electrochimica Acta*. 12: 1655–1664.

[34]. Soffer, A.; Folman, M. 1972. The electrical double layer of high surface porous carbon electrode. *Journal of Electroanalytical Chemistry Interfacial Electrochemistry*. 38: 25–43.

[35]. Cohen, I.; Avraham, E.; Noked, M.; Soffer, A.; Aurbach, D. 2011. Enhanced charge efficiency in capacitive deionization achieved by surface-treated electrodes and by means of a third electrode. *Journal of Physical Chemistry C*. 115: 19856–19863.

[36]. Oren, Y. 2008. Capacitive delonization (CDI) for desalination and water treatment – Past, present and future (a review). *Desalination*. 228: 10–29.

[37]. Biesheuvel, P.M.; Zhao, R.; Porada, S.; van der Wal, A. 2011. Theory of membrane capacitive deionization including the effect of the electrode pore space. *Journal of Colloid and Interface Science*. 360: 239–248.

[38]. Porada, S.; Weinstein, L.; Dash, R.; van der Wal, A.; Bryjak, M.; Gogotsi, Y.; Biesheuvel, P.M. 2012. Water desalination using capacitive deionization with microporous carbon electrodes. *ACS Applied Materials & Interfaces*. 4: 1194–1199.

[39]. Cen, B.; Li, K.; Lv, C.; Yang, R. 2020. A novel asymmetric activated carbon electrode doped with metal-organic frameworks for high desalination performance. *Journal of Solid State Electrochemistry*. 24: 687–697.

[40]. Goh, P.S.; Ismail, A.F. 2018. A review on inorganic membranes for desalination and wastewater treatment. *Desalination*. 434: 60–80.

[41]. Fredenberg, S.; Wahlgren, M.; Reslow, M.; Axelsson, A. 2011. The mechanisms of drug release in poly (lactic-co-glycolic acid)-based drug delivery systems—A review. *International Journal of Pharmaceutics*. 415: 34–52.

[42]. Klose, D.; Siepmann, F.; Elkharraz, K.; Siepmann, J. 2008. PLGA-based drug delivery systems: Importance of the type of drug and device geometry. *International Journal of Pharmaceutics*. 354: 95–103.

[43]. Anglin, E.J.; Cheng, L.; Freeman, W.R.; Sailor, M.J. 2008. Porous silicon in drug delivery devices and materials. *Advances in Drug Delivery Research*. 17: 1266–1277.

[44]. De Miguel, L.; Popa, I.; Noiray, M.; Caudron, E.; Arpinati, L.; Desmaele, D.; Cebrián-Torrejón, D.; Doménech-Carbó, A.; Ponchel, G. 2015. Osteotropic polypeptide nanoparticles with dual hydroxyapatite binding properties and controlled cisplatin delivery. *Pharmaceutical Research*. 32: 1794–1803.

[45]. De Miguel, L.; Cebrián-Torrejón, G.; Caudron, E.; Arpinati, L.; Doménech-Carbó, A.; Ponchel, G. 2015. Bone-targeted cisplatin-complexed poly(g-benzyl-lglutamate)–poly(glutamic acid) block polymer nanoparticles: An electrochemical approach. *Chem Electro Chem.* 2: 748–754.

[46]. Peppas, N.A. 1985. Analysis of Fickian and non-Fickian drug release from polymers. *Pharmaceutica Acta Helvetica.* 60: 110–111.

[47]. Berkland, C.; Kim, K.K.; Pack, D.W. 2003. PLG microsphere size control drug release rate through several competing factors. *Pharmaceutical Research.* 20: 1055–1062.

[48]. Charlier, A.; Leclerc, B.; Couarraze, G. 2000. Release of mifepristone from biodegradable matrices: Experimental and theoretical evaluations. *International Journal of Pharmaceutics.* 200: 115–120.

Additional Literature

CHAPTER 1

Adams, R.N. 1958. Carbon paste electrodes. *Analytical Chemistry*. 30: 1576.

Bond, A.M.; Feldberg, S.W.; Miao, W.; Oldham, K.B.; Raston, C.L. 2001. Modelling of solid-state, dissolution and solution-phase reactions at adhered solid–electrode–solvent (electrolyte) interfaces: Electrochemistry of microcrystals of C_{60} adhered to an electrode in contact with dichloromethane (Bu_4NClO_4). *Journal of Electroanalytical Chemistry*. 501: 22–32.

DuVall, S.; McCreery, R.L. 2000. Self-catalysis by catechols and quinones during heterogeneous electron transfer at carbon electrodes. *Journal of the American Chemical Society*. 122: 6759–6764.

Eftekhari, A. Ed. 2006. *Nanostructured Materials in Electrochemistry*. Wiley-VCH, New York.

Evans, D.H. 2008. One-electron and two-electron transfers in electrochemistry and homogeneous solution reactions. *Chemical Reviews*. 108: 2113–2144.

Grygar, T.; Marken, F.; Schröder, U.; Scholz, F. 2002. Voltammetry of microparticles: A review. *Collection Czechoslovack Chemical Communications*. 67: 163–208.

Hah, H.J.; Kim, J.S.; Jeon, B.J.; Koo, S.M.; Lee, Y.E. 2003. Simple preparation of monodisperse hollow silica particles without using templates. *Chemical Communications*. 1712–1713.

Kuwana, T.; French, W.G. 1964. Electrooxidation or reduction of organic compounds into aqueous solutions using carbon paste electrode. *Analytical Chemistry*. 36: 241–242.

Laviron, E. 1980. A multilayer model for the study of space-distributed redox modified electrodes: Part I. Description and discussion of the model. *Journal of Electroanaytical Chemistry*. 112: 1–9.

Mintova, S.; Olson, N.H.; Valtchev, V.; Bein, T. 1999. Mechanism of zeolite A nanocrystal growth from colloids at room temperature. *Science*. 283: 958–960.

Morales-Saavedra, O.G.; Sánchez-Vergara, M.E.; Ortiz-Rebollo, A.; Ortega-Martínez, R. 2007. Electrical and optical properties of Jäger-nickel (II)-based molecular-material thin films prepared by the vaccum thermal evaporation technique. *Journal of Physics and Chemistry of Solids*. 68: 1571–1582.

Nicholson, R.S. 1965. Theory and application of cyclic voltammetry for measurement of electrode reaction kinetics. *Analytical Chemistry* 37: 1351–1355.

Oldham, K.B. 1991. Steady-state microelectrode voltammetry as a route to homogeneous kinetics. *Journal of Electroanaytical Chemistry*. 313: 3–16.

Plieth, W. 2008. *Electrochemistry for Materials Science*. Elsevier, Amsterdam.

Salgado, L.; Tejo, G.; Meas, Y.; Zayas, T. 2006. Cyclic voltammetry and electrochemical quartz crystal microbalance studies of a rhodized platinum electrode in sulfuric acid solution. *Journal of Solid State Electrochemistry*. 10: 230–235.

Scholz, F.; Lange, B. 1992. Abrasive stripping voltammetry - An electrochemical solid state spectroscopy of wide applicability. *Trends in Analytical Chemistry*. 11: 359–367.

Scholz, F.; Meyer, B. 1994. Electrochemical solid state analysis - State of the art. *Chemical Society Reviews*. 23: 341–347.

Scholz, F.; Meyer, B. 1998. Voltammetry of solid microparticles immobilized on electrode surfaces. *Electroanalytical Chemistry, A Series of Advances*. 20: 1–86.

Schultz, F.A.; Kuwana, T. 1965. Electrochemical studies of organic compounds dissolved in carbon-paste electrodes. *Journal of Electroanalytical Chemistry*. 10: 95–103.

Seeber, R.; Zanardi, C.; Inzelt, G. 2015. Links between electrochemical thermodynamics and kinetics. *Chem. Texts*. 1: artic. 18.

Uchida, H.; Ikeda, N.; Watanabe, M. 1998. Electrochemical quartz crystal microbalance study of copper adatoms on Au(111) electrodes in solutions of perchloric and sulfuric acid. *Journal of Electroanalytical Chemistry*. 452: 97–106.

Yan, H.; Bein, T. 1992. Molecular recognition on acoustic wave devices: Sorption in chemically anchored zeolite monolayers. *Journal of Physical Chemistry*. 96: 9387–9393.

CHAPTER 2

Ban, Z.; Kätelhön, E.; Compton, R.G. 2016. Voltammetry of porous layers: Staircase vs analog voltammetry. *Journal of Electroanalytical Chemistry*. 776: 25–33.

Barcia, O.E.; D'Elia, E.; Frateur, I.; Mattos O.R.; Pebere, N.; Tribollet, B. 2002. Application of the impedance model of de Levie for the characterization of porous electrodes. *Electrochimica Acta*. 47: 2109–2116.

Bayley, H.; Martin, C.R. 2000. Resistive-pulse sensing–From microbes to molecules. *Chemical Re*views. 100: 2575–2594.

Brown, A.P.; Anson, F.C. 1977. Cyclic and differential pulse voltammetric behavior of reactants confined to the electrode surface. *Analytical Chemistry*. 49: 1589–1595.

Chan, H.T.H.; Katelhon, E.; Compton, R.G. 2017. Voltammetry using multiple cycles: Porous electrodes. *Journal of Electroanalytical Chemistry*. 799: 126–133.

Compton, R.G.; Banks, C.E. 2011. *Understanding Voltammetry*, 2nd Edit. Imperial College Press, London.

Friedl, J.; Stimming, U. 2017. Determining electron transfer kinetics at porous electrodes. *Electrochimica Acta*. 227: 235–245.

Godino, N.; Borrisé, X.; Muñoz, F.X.; del Campo, F.J.; Compton, R.G. 2009. Mass transport to nanoelectrode arrays and limitations of the diffusion domain approach: Theory and experiment. *Journal of Physical Chemistry C*. 113: 11119–11125.

Henstridge, M.C.; Dickinson, E.J.F.; Compton, R.G. 2012. Mass transport to and within porous electrodes. Linear sweep voltammetry and the effects of pore size: The prediction of double peaks for a single electrode process. *Russian Journal of Electrochemistry*. 48: 629–635.

Le, T.D.; Zhang, L.; Kuhn, A.; Mano, N.; Vignoles, G.; Lasseux, D. 2019. Upscaled model for diffusion and serial reduction pathways in porous electrodes. *Journal of Electroanalytical Chemistry*. 855: artic. 113325.

Lee, G.-J.; Pyun, S.-I. 2006. Theoretical approach to ion penetration into pores with pore fractal characteristics during double-layer charging/discharging on a porous carbon electrode. *Langmuir*. 22: 10650–10665.

Lee, S.; Zhang, Y.; Harrell, C.; Martin, C.R.; White, H.S. 2004. Electrophoretic capture and detection of nanoparticles at the opening of a membrane pore using scanning electrochemical microscopy. *Analytical Chemistry*. 76: 6108–6115.

Menshykau, D.; Compton, R.G. 2008. The influence of electrode porosity on diffusional cyclic voltammetry. *Electroanalysis*. 20: 2387–2394.

Menshykau, D.; Streeter, I.; Compton, R.G. 2008. Influence of electrode roughness on cyclic voltammetry. *Journal of Physical Chemistry C*. 112: 14428–14438.

Nyikos, L.; Pajkossy, T. 1990. Electrochemistry at fractal interfaces: The coupling of *ac* and *dc* behaviour at irregular electrodes. *Electrochimica Acta*. 35: 1567–1572.

Nyikos, L.; Pajkossy, T.; Borosy, A.P.; Martenyanov, S.A. 1990. Diffusion to fractal surfaces—IV. The case of the rotating disc electrode of fractal surface. *Electrochimica Acta*. 35: 1423–1424.

Sliusarenko, O.; Oleinick, A.; Svir, L.; Amatore, C. 2015. Validating a central approximation in theories of regular electrode electrochemical arrays of various common geometries. *Electroanalysis*. 27: 980–991.

Stromme, M.; Niklasson, G.A.; Granqvist, C.Q. 1996. Fractal dimension of Li insertion electrodes studied by diffusion-controlled voltammetry and impedance spectroscopy. *Physical Review B*. 54: 2968–2971.

Stromme, M.; Niklasson, G.A.; Granqvist, C.Q. 1996. Fractal surface dimension from cyclic *I-V* studies and atomic-force microscopy: Role of noncontiguous reaction sites. *Physical Review B*. 54: 17884–17887.

Zuo, X.B.; Xu, C.S.; Xin, H.W. 1997. Simulation of voltammogram on rough electrode. *Electrochimica Acta*. 42: 2555–2558.

CHAPTER 3

Andrieux, C.P.; Savéant, J.-M. 1982. Kinetics of electrochemical reactions mediated by redox polymer films: Reversible ion-exchange reactions. *Journal of Electroanalytical. Chemistry*. 142: 1–30.

Andrieux, C.P.; Savéant, J.-M. 1984. Kinetics of electrochemical reactions mediated by redox polymer films: Pre-activation (CE) mechanisms. *Journal of Electroanalytical Chemistry*. 11: 65–93.

Andrieux, C.P.; Savéant, J.-M. 1988. Electroneutrality coupling of electron hopping between localized sites with electroinactive counterion displacement. 1. Potential-step plateau currents. *Journal of Physical Chemistry*. 92: 6761–6767.

Doménech-Carbó, A.; Koshevoy, I.O.; Montoya, N.; Pakkanen, T.A.; Doménech-Carbó, M.T. 2012. Solvent-independent electrode potentials of solids undergoing insertion electrochemical reactions: Part II. Experimental data for alkynyl-diphosphine dinuclear Au(I) complexes undergoing electron exchange coupled to anion exchange. *Journal of Physical Chemistry C*. 116: 25984–25992.

Doménech-Carbó, A.; Scholz, F.; Montoya, N. 2012. Solvent-independent electrode potentials of solids undergoing insertion electrochemical reactions: Part III. Experimental data for Prussian blue undergoing electron exchange coupled to cation exchange. *Journal of Physical Chemistry C*. 116: 25993–25999.

Laviron, E. 1984. Electrochemical reactions with protonations at equilibrium: Part X. The kinetics of the *p*-benzoquinone/hydroquinone couple on a platinum electrode. *Journal of Electroanalytical Chemistry*. 164: 213–227.

Lovrić, M.; Hermes, M.; Scholz, F. 2000. Solid state electrochemical reactions in systems with miscibility gaps. *Journal of Solid State Electrochemistry*. 4: 394–401.

Oldham, K.B. 1991. Steady-state microelectrode voltammetry as a route to homogeneous kinetics. *Journal of Electroanaytical Chemistry*. 313: 3–16.

Oldham, K.B.; Myland, J.C. 1994. *Fundamentals of Electrochemical Science*. Academic Press, San Diego.

CHAPTER 4

Araminaitė, R.; Garjonytė, R.; Malinauskas, A. 2010. Electrocatalytic reduction of hydrogen peroxide at Prussian blue modified electrodes: A RDE study. *Journal of Solid State Electrochemistry*. 14: 149–155.

Bieniasz, L.K.; González, J.; Molina, A.; Laborda, E. 2010. Theory of linear sweep/cyclic voltammetry for the electrochemical reaction mechanism involving a redox catalyst couple attached to a spherical electrode. *Electrochimica Acta*. 56: 543–552.

Dharuman, V.; Chandrasekara Pillai, K. 2006. RuO_2 electrode surface effects in electrocatalytic oxidation of glucose. *Journal of Solid State Electrochemistry*. 10: 967–979.

Do, T.Q.N.; Varnicic, M.; Hanke-Rauschenbach, R.; Vidakovic-Koch, T.; Sundmacher, K. 2014. Mathematical modeling of a porous enzymatic electrode with direct electron transfer mechanism. *Electrochimica Acta*. 137: 616–626.

Dubois, N.; Grimaud, A. 2019. The hydrogen evolution reaction: From material to interfacial descriptors. *Chemical Science*. 10: 9165–9181.

González, J.; Soto, C.M.; Molina, A. 2009. Analytical *I–E* response for several multistep potential techniques applied to an electrocatalytic process at mediator modified electrodes. *Electrochimica Acta*. 54: 6154–6160.

Goyal, A.; Marcandalli, G.; Mints, V.A.; Koper, M.T.M. 2020. Competition between CO_2 reduction and hydrogen evolution on a gold electrode under well-defined mass transport conditions. *Journal of the American Chemical Society*. 142: 4154–4161.

Jirkovsky, J.S.; Halasa, M.; Schiffrin, D.J. 2010. Kinetics of electrocatalytic reduction of oxygen and hydrogen peroxide on dispersed gold nanoparticles. *Physical Chemistry Chemical Physics*. 12: 8042–8052.

Karajic, A.; Reculusa, S.; Ravaine, S.; Mano, N.; Kuhn, A. 2016. Miniaturized electrochemical device from assembled cylindrical macroporous gold electrodes. *ChemElectroChem*. 3: 2031–2035.

Koutecky, J.; Brdicka, R. 1947. Fundamental equation for the electrolytic current when depending on the formation rate of the depolarizer and the rate of diffusion; polarographic verification. *Collection of Czechoslovack Chemical Communications*. 12: 337–355.

Kruk, M.; Javoniec, M.; Ko, C.H.; Ryoo, R. 2000. Characterization of the porous structure of SBA-15. *Chemistry of Materials*. 12: 1961–1968.

Laviron, E. 1984. Electrochemical reactions with protonations at equilibrium: Part X. The kinetics of the *p*-benzoquinone/hydroquinone couple on a platinum electrode. *Journal of Electroanalytical Chemistry*. 164: 213–227.

Liardet, L.; Hu, X. 2018. Amorphous cobalt vanadium oxide as highly active electrocatalyst for oxygen evolution. *ACS Catalysis*. 8: 644–650.

Lovrić, M. 2002. Square-wave voltammetry. In Scholz, F. Ed. *Electroanalytical Methods*. Springer, Berlin.

Lu, Q.; Hutchings, G.S.; Yu, W.; Zhou, Y.; Forest, R.V.; Tao, R.; Rosen, J.; Yonemoto, B.T.; Cao, Z.; Zheng,

H.; Xiao, J.Q.; Jiao, F.; Chen, J.G. 2015. Highly porous non-precious bimetallic electrocatalysts for efficient hydrogen evolution. *Nature Communications*. 6: artic. 6567.

Markovic, N.N.; Gasteiger, H.A.; Ross, P.N. 1996. Oxygen reduction on platinum low-index single-crystal surfaces in alkaline solution: Rotating ring disk$_{Pt(hkl)}$ studies. *Journal of Physical Chemistry*. 100: 6715–6721.

Martel, D.; Kuhn, A. 2000. Electrocatalytic reduction of H_2O_2 at $P_2Mo_{18}O_{62}{}^{6-}$ modified glassy carbon. *Electrochimica Acta*. 45: 1829–1836.

McCrum, I.T.; Janik, M.J. 2016. pH and alkali cation effects on the Pt cyclic voltammogram explained using density functional theory. *Journal of Physical Chemistry C*. 120: 457–471.

Narayanan, J.S.; Anjalidevi, G.; Dharuman, V. 2013. Nonenzymatic glucose sensing at ruthenium dioxide–poly(vinyl chloride)–Nafion composite electrode. *Journal of Solid State Electrochemistry*. 17: 937–947.

O'Dea, J.J.; Osteryoung, J.; Osteryoung, R.A. 1981. Theory of square wave voltammetry for kinetic systems. *Analytical Chemistry*. 53: 695–701.

Reculusa, S.; Heim, M.; Gao, F.; Mano, N.; Ravaine, S.; Kuhn, A. 2011. Design of catalytically active cylindrical and macroporous gold microelectrodes. *Advanced Functional Materials*. 21: 691–698.

Ruthstein, S.; Frydman, V.; Kabbaya, S.; Landau, M.; Goldfarb, D. 2003. Study of the formation of the mesoporous material SBA-15 by EPR spectroscopy. *Journal of Physical Chemistry B*. 107: 1739–1748.

Song, J.; Yuan, J.; Li, F.; Han, D.; Song, J.; Niu, L. 2010. Tunable activity in electrochemical reduction of oxygen by gold–polyaniline porous nanocomposites. *Journal of Solid State Electrochemistry*. 14: 1915–1922.

Thakur, B.; Guo, X.; Chang, J. Kron, M.; Chen, J. 2017. Porous carbon and Prussian blue: A highly sensitive electrochemical platform for glucose biosensing. *Sensing and Bio-Sensing Research*. 14: 47–53.

Trasatti, S. 1972. Work function, electronegativity, and electrochemical behavior of metals. 3. Electrolytic hydrogen evolution in acid solutions. *Journal of Electroanalytical Chemistry*. 39: 163–184.

Zeis, R.; Lei, T.; Sieradzki, K.; Snyder, J.; Erlebacher, J. 2008. Catalytic reduction of oxygen and hydrogen peroxide by nanoporous gold. *Journal of Catalysis*. 253: 132–138.

Zheng, Y.; Jiao, Y.; Vasileff, A.; Qiao, S.-Z.Q. 2018. The hydrogen evolution reaction in alkaline solution: From theory, single crystal models, to practical electrocatalysts. *Angewandte Chemie International Edition*. 57: 7568–7579.

CHAPTER 5

Andrieux, C.P.; Hapiot, P.; Savéant, J.-M. 1984. Electron-transfer coupling of diffusional pathways: Theory for potential step chronoamperometry and chronocoulometry. *Journal of Electroanalytical Chemistry*. 117: 49–65.

Baker, M.D.; Senaratne, C.; McBrien, M. 1995. Comment on intrazeolite electron transport mechanism. *Journal of Physical Chemistry*. 99: 12367.

Barrer, R.M.; James, S.D. 1960. Electrochemistry of crystal-polymer membranes. Part I. Resistance measurements. *Journal of Physical Chemistry*. 64: 417–421.

Bedioui, F.; Devynck, J. 1996. Comment on Zeolite-modified electrodes: Intra- versus extrazeolite electron transfer". *Journal of Physical Chemistry*. 100: 8607–8609.

Bessel, C.A.; Rolison, D.R. 1997. Microheterogeneous dispersion electrolysis with nanoscale electrode-modified zeolites. *Journal of Electroanalytical Chemistry*. 439: 97–105.

Chen, H.-Y.; Sachtler, W.M.H. 1998. Activity and durability of Fe/ZSM-5 catalysts for lean burn NO_x reduction in the presence of water vapor. *Catalysis Today*. 42: 73–83.

Doménech-Carbó, A.; Casades, I.; García, H. 1999. Electrochemical evidence for an impeded attack of water to anthracene and thianthrene radical ions located on the outermost layers of zeolites. *Journal of Organic Chemistry*. 64: 3731–3735.

Doménech-Carbó, A.; García, H.; Casades, I.; Esplá, M. 2004. Electrochemistry of 6-nitro-1′,3′,3′-trimethylspiro[2H-1-benzopyran-2,2′-indoline] associated to zeolite Y and MCM-41 aluminosilicate. Site-selective electrocatalytic effect on N,N,N′,N′-tetramethylbenzidine oxidation. *Journal of Physical Chemistry B*. 108: 20064–20075.

Doménech-Carbó, A.; Doménech-Carbó, M.T.; Valle-Algarra, F.M.; Domine, M.E.; Osete-Cortina, L. 2013. On the dehydroindigo contribution to Maya blue. *Journal of Materials Science*. 48: 7171–7183.

Doménech-Carbó, A.; Doménech-Carbó, M.T.; Osete-Cortina, L.; Valle-Algarra, F.M.; Buti, D. 2014.

Isomerization and redox tuning in 'Maya yellow' hybrids from flavonoid dyes plus palygorskite and kaolinite clays. *Microporous and Mesoporous Materials*. 194: 135–145.

Doménech-Carbó, A.; Valle-Algarra, F.M.; Doménech-Carbó, M.T.; Osete-Cortina, L.; Domine, M. 2013. 'Maya chemistry' of organic-inorganic hybrid materials: isomerization, cyclicization and redox tuning of organic dyes attached to porous silicates. *RSC Advances*. 3: 20099–21005.

García, H.; Roth, H.D. 2002. Generation and reactions of organic radical cations in zeolites. *Chemical Reviews*. 102: 3947–4007.

Ghosh, P.K.; Mau, A.W.-H.; Bard, A.J. 1984. Clay-modified electrodes: Part II. Electrocatalysis at bis(2,2′-bipyridyl) (4,4′-dicarboxy-2,2′-bipyridyl)Ru(II)-dispersed ruthenium dioxide—Hectorite layers. *Journal of Electroanalytical Chemistry*. 169: 315–317.

Hashimoto, S. 2011. Optical spectroscopy and microscopy studies on the spatial distribution and reaction dynamics in zeolites. *The Journal of Physical Chemistry Letters*. 2: 509–519.

Hattori, H. 1995. Heterogeneous basic catalysis. *Chemical Reviews*. 95: 537–558.

Herance, J.R.; Concepción, P.; Doménech-Carbó, A.; Bourdelande, J.L.; Marquet, J.; García, H. 2005. Anionic organic guests incorporated within zeolites. Adsorption and reactivity of the Meisenheimer complex in faujasites. *Chemistry A European Journal*. 11: 6491–6502.

Hubbard, B.; Kuang, W.; Moser, A.; Facey, G.A.; Detellier, C. 2003. Structural study of Maya blue: Textural, thermal and solid-state multinuclear magnetic resonance characterization of the palygroskite-indigo and sepiolite-indigo adducts. *Clays and Clay Minerals*. 51: 318–326.

Joyner, R.; Stockenhuber, M. 1999. Preparation, characterization, and performance of Fe–ZSM-5 catalysts. *Journal of Physical Chemistry B*. 103: 5963–5976.

Li, H.-Y.; Anson, F.C. 1985. Electrochemical behavior of cationic complexes incorporated in clay coatings on graphite electrodes. *Journal of Electroanalytical Chemistry*. 184: 411–417.

Long, J.W.; Dunn, B.; Rolison, D.R.; White, H.S. 2004. Three-dimensional battery architectures. *Chemical Reviews*. 104: 4463–4492.

Panov, G.I.; Uriarte, A.K.; Rodkin, M.A.; Sobolev, V.I. 1998. Generation of active oxygen species on solid surfaces. Opportunity for novel oxidation technologies over zeolites. *Catalysis Today*. 41: 365–385.

Rolison, D.R. 1990. Zeolite-modified electrodes and electrode-modified zeolites. *Chemical Reviews*. 90: 867–878.

Rolison, D.R. 1994. The intersection of electrochemistry with zeolite science. In *Advanced Zeolite Science and Applications*, Jansen, J.C.; Stöcker, M.; Karge, H.G.; Weitkamp, J. Eds. *Studies in Surface Science and Catalysis*. 85: 543–587.

Rolison, D.R.; Bessel, C.A.; Baker, M.D.; Senaratne, C.; Zhang, J. 1996. Reply to the comment on "Zeolite-modified electrodes: Intra- versus extrazeolite electron transfer". *Journal of Physical Chemistry*. 100: 8610–8611.

Senaratne, C.; Zhang, J.; Baker, M.D.; Bessel, C.A.; Rolison, D.R. 1996. Zeolite-modified electrodes: Intra- versus extrazeolite electron transfer. *Journal of Physical Chemistry*. 100: 5849–5862.

Shaw B.R.; Kreasy, K.E. 1988. Carbon composite electrodes containing alumina, layered double hydroxides, and zeolites. *Journal of Electroanalytical Chemistry*. 243: 209–217.

CHAPTER 6

Cho, K.; Han, S.-H.; Suh, M.P. 2016. Copper–organic framework fabricated with CuS nanoparticles: Synthesis, electrical conductivity, and electrocatalytic activities for oxygen reduction reaction. *Angewandte Chemie International Edition*. 55 (49): 15301–15305.

Chun, H.; Dybstev, D.N.; Kim, H.; Kim, K. 2005. Synthesis, X-ray crystal structures, and gas sorption properties of pillared square grid nets based on paddle-wheel motifs: Implications for hydrogen storage in porous materials. *Chemistry A European Journal*. 11: 3521–3529.

Férey, G.; Millange, F.; Morcrette, M.; Serre, C.; Doublet, M.-L.; Grenèche, J.-M.; Tarascon, J.-M. 2007. Mixed-valence Li/Fe-based metal-organic frameworks with both reversible redox and sorption properties. *Angewandte Chemie International Edition*. 46: 3259–3263.

Gong, Y.; Shi, H.-F.; Hao, Z.; Sun, J.-L.; Lin, J.-H. 2013. Two novel Co(II) coordination polymers based on 1,4-bis(3-pyridylaminomethyl)benzene as electrocatalysts for oxygen evolution from water. *Dalton Transactions*. 42: 12252–12259.

Jiang, H.-L.; Liu, B.; Lan, Y.-Q.; Kuratani, K.; Akita, T.; Shioyama, H.; Zong, F.; Xu, Q. 2011. From metal-organic framework to nanoporous carbon: toward a very high surface area and hydrogen uptake. *Journal of the American Chemical Society*. 133: 11854–11857.

Liang, Z.; Qu, C.; Xia, D.; Zou, R.; Xu, Q. 2018. Atomically dispersed metal sites in MOF-based materials for electrocatalytic and photocatalytic energy conversion. *Angewandte Chemie International Edition*. 57: 9604–9633.

Liu, D.N.; Lu, K.D.; Poon, C.; Lin, W.B. 2014. Metal–organic frameworks as sensory materials and imaging agents. *Inorganic Chemistry*. 53: 1916–1924.

Maurin, G.; Serre, C; Cooper, A.; Férey, G. 2017. The new age of MOFs and their porous-related solids. *Chemical Society Reviews*. 46: 3104–3107.

Rowsell, L.C.; Yaghi, O.M. 2005. Strategies for hydrogen storage in metal-organic frameworks. *Angewandte Chemie International Edition*. 117: 4670–4679.

Rowsell, J.L.C.; Yaghi, O.M. 2006. Effects of functionalization, catenation, and variation of the metal oxide and organic linking units on the low-pressure hydrogen adsorption properties of metal-organic frameworks. *Journal of the American Chemical Society*. 128: 1304–1315.

Switzer, J.A.; Hung, C.J.; Huang, L.Y.; Switzer, E.R.; Kammler, D.R.; Golden, T.D.; Bohannan, E.W. 1998. Electrochemical self-assembly of copper/cuprous oxide layered nanostructures. *Journal of the American Chemical Society*. 120: 3530–3531.

Usov, P.M.; Huffman, B.; Epley, C.C.; Kessinger, M.C.; Zhu, J.; Maza, W.A.; Morris, A.J. 2017. Study of electrocatalytic properties of metal–organic framework PCN-223 for the oxygen reduction reaction. *ACS Applied Materials & Interfaces*. 9: 33539–33543.

Wang, X.; Lin, H.; Bi, Y.; Chen, B.; Liu, G. 2008. An unprecedent extended architecture constructed from a 2-D interpenetrating cationic coordination framework templated by $SiW_{12}O_{40}^{4-}$ anion. *Journal of Solid State Chemistry*. 181: 556–561.

Ye, S.H.; Wang, X.-S.; Zhao, H.; Xiong, R.-G. 2005. High stable olefin-Cu(I) coordination oligomers and polymers. *Chemical Society Reviews*. 34: 208–225.

CHAPTER 7

Angerstein-Kozlowska, H.; Conway, B.E.; Hamelin, A.; Stoicoviciu, L. 1986. Elementary steps of electrochemical oxidation of single-crystal planes of Au—I. Chemical basis of processes involving geometry of anions and the electrode surfaces. *Electrochimica Acta*. 31: 1051–1061.

Angerstein-Kozlowska, H.; Conway, B.E.; Hamelin, A.; Stoicoviciu, L. 1987. Elementary steps of electrochemical oxidation of single-crystal planes of Au Part II. A chemical and structural basis of oxidation of the (111) plane. *Journal of Electroanalytical Chemistry*. 228: 429–453.

Bisquert, J. 2000. Influence of the boundaries in the impedance of porous film electrodes. *Physical Chemistry Chemical Physics*. 2: 4185–4192.

Burke, L.D.; O'Mullane, A.P.; Lodge, V.E.; Mooney, M.B. 2001. Auto-inhibition of hydrogen gas evolution on gold in aqueous acid solution. *Journal of Solid State Electrochemistry*. 5: 319–327.

Chen, A.; Lipkowski, J. 1999. Electrochemical and spectroscopic studies of hydroxide adsorption at the Au (111) electrode. *Journal of Physical Chemistry B*. 103: 682–691.

Damjanovic, A.; Dey, A.; Bockris, J.O.M. 1996. Kinetics of oxygen evolution and dissolution on platinum electrodes. *Electrochimica Acta*. 11: 791–814.

Dickertmann, D.; Schultze, J.W.; Vetter, K.J. 1974. Electrochemical formation and reduction of monomolecular oxide layers on (111) and (100) planes of gold single crystals. *Journal of Electroanalytical Chemistry and Interfacial Electrochemistry*. 55: 429–443.

Doménech-Carbó, A.; Scholz, F.; Doménech-Carbó, M.T.; Piquero-Cilla, J.; Montoya, N.; Pasíes-Oviedo, T.; Gozalbes, M.; Melchor-Montserrat, J.M.; Oliver, A. 2018. Dating of archaeological gold by means of solid state electrochemistry. *ChemElectroChem*. 5: 2113–2217.

Doménech-Carbó, A.; Sabaté, F.; Sabater, J. 2018. Electrochemical analysis of catalytic and oxygen interfacial transfer effects on MnO_2 deposited on gold electrodes. *Journal of Physical Chemistry C*. 122: 10939–10947.

Fabregat-Santiago, F.; Mora-Seró, I.; Garcia-Belmonte, G.; Bisquert, J. 2003. Cyclic voltammetry studies of nanoporous semiconductors. Capacitive and reactive properties of nanocrystalline TiO_2 electrodes in aqueous electrolyte. *Journal of Physical Chemistry B*. 107: 758–768.

Germain, P.S.; Pell, W.G.; Conway, B.E. 2004. Evaluation and origins of the difference between double-layer capacitance behaviour at Au-metal and oxidized Au surfaces. *Electrochimica Acta*. 49: 1775–1788.

Kim, J.H.; Kim, R.H.; Kwon, H.S. 2008. Preparation of copper foam with 3-dimensionally interconnected spherical pore network. *Electrochemistry Communications*. 10: 1148–1151.

Mattarozzi, L.; Cattarin, S.; Comisso, N.; Gerbasi, R.; Guerriero, P.; Musiani, M.; Vázquez-Gómez, L.;

Verlato, E. 2013. Electrodeposition of Cu-Ni alloy electrodes with bimodal porosity and their use for nitrate reduction. *ECS Letters*. 2: D58–D60.

Mattarozzi, L.; Cattarin, S.; Comisso, N.; Gerbasi, R.; Guerriero, P.; Musiani, M.; Vázquez-Gómez, L.; Verlato, E. 2014. Hydrogen evolution assisted electrodeposition of porous Cu-Ni alloy electrodes and their use for nitrate reduction in alkali. *Electrochimica Acta*. 140: 337–344.

Nørskov, J.K.; Rossmeisl, J.; Logadottir, A.; Lindqvist, L.; Kitchin, J.R.; Bligaard, T.; Jonsson, H. 2004. Origin of the overpotential for oxygen reduction at a fuel-cell cathode. *Journal of Physical Chemistry B*. 108: 17886–17892.

Patermarakis, G.; Nikolopoulos, N. 1999. Catalysis over porous anodic alumina film catalysts with different pore surface concentrations. *Journal of Catalysis*. 187: 311–320.

Patermarakis, G.; Moussoutzanis, K. 2001. Formulation of a criterion predicting the development of uniform regular and non-uniform abnormal porous anodic alumina coatings and revealing the mechanisms of their appearance and progress. *Corrosion Science*. 43: 1433–1464.

Patermarakis, G.; Moussoutzanis, K. 2002. Solid surface and field catalysed interface formation of colloidal $Al_2(SO_4)_3$ during Al anodizing affecting the kinetics and mechanism of development of structure of porous oxides. *Journal of Solid State Electrochemistry*. 6: 475–484.

Scholz, F.; Lopez de Lara-Gonzalez, G.; de Carvalho, L.M.; Hilgemann, L.; Brainina, Kh.Z.; Kahlert, H.; Jack, R.S.; Minh, D.T. 2007. Indirect electrochemical sensing of radicals and radical scavengers in biological matrices. *Angewandte Chemie International Edition*. 46: 8079–8081.

Shin, H.C.; Liu, M. 2004. Copper foam structures with highly porous nanostructured walls. *Chemistry of Materials*. 16: 5460–5464.

Simpson, B.K.; Johnson, D.C. 2004. Electrocatalysis of nitrate reduction at copper-nickel alloy electrodes in acid media. *Electroanalysis*. 16: 532–538.

Skowroński, J.M.; Ważny, A. 2005. Nickel foam-based composite electrodes for electrooxidation of methanol. *Journal of Solid State Electrochemistry*. 9: 890–899.

Song, J.; Chen, Y.; Cao, K.; Lu, Y.; Xin, J.H.; Tao, X. 2018. Fully controllable design and fabrication of three-dimensional lattice supercapacitors. *ACS Applied Materials & Interfaces*. 10: 39839–39850.

Tian, M.; Pell, W.G.; Conway, B.E. 2003. Nanogravimetry study of the initial stages of anodic surface oxide film growth in $HClO_4$ and H_2SO_4 by means of EQCM. *Electrochimica Acta*. 48: 2675–2689.

CHAPTER 8

Aurbach, D. 2000. Review of selected electrode–solution interactions which determine the performance of Li and Li ion batteries. *Journal of Power Sources*. 89: 206–218.

Carapuça, H.M.; Balula, M.S.; Fonseca, A.P.; Cavaleiro, A.M.V. 2006. Electrochemical characterization of glassy carbon electrodes modified with hybrid inorganic-organic single layer of α-Kegging type polyoxotungstates. *Journal of Solid State Electrochemistry*. 10: 10–17.

Cataldi, T.R.I.; Guerrieri, A.; Casella, I.G.; Desimoni, E. 1995. Study of a cobalt-based surface modified glassy carbon electrode: Electrocatalytic oxidation of sugars and alditols. *Electroanalysis*. 7: 305–311.

Grygar, T.; Marken, F.; Schröder, U.; Scholz, F. 2002. Voltammetry of microparticles: A review. *Collection Czechoslovack Chemical Communications*. 67: 163–208.

Keita, B.; Nadjo, L. 1987. New aspects of the electrochemistry of heteropolyacids: Part II. Coupled electron and proton transfers in the reduction of silicoungstic species. *Journal of Electroanalytical Chemistry*. 217: 287–304.

Keita, B.; Nadjo, L. 1988. Surface modifications with heteropoly and isopoly oxometalates: Part I. Qualitative aspects of the activation of electrode surfaces towards the hydrogen evolution reaction. *Journal of Electroanalytical Chemistry*. 243: 87–103.

Khan, A.I.; O'Hare, D. 2002. Intercalation chemistry of layered double hydroxides: Recent developments and applications. *Journal of Materials Chemistry*. 12: 3191–3198.

Mahmoud, A.; Keita, B.; Nadjo, L.; Oung, O.; Contant, R.; Brown, S.; Kochkovski, Y. 1999. Coupled electron and proton transfers: compared behaviour of oxometalates in aqueous solution or after entrapment in polymer matrices. *Journal of Electroanalytical Chemistry*. 463: 129–145.

Malinger, K.A.; Ding, Y.-S.; Sithambaram, S.; Espinal, L.; Gomez, S.; Suib, S.L. 2006. Microwave frequency effects on synthesis of cryptomelane-type manganese oxide and catalytic activity of cryptomelane precursor. *Journal of Catalysis*. 239: 290–298.

Martel, D.; Kuhn, A. 2000. Electrocatalytic reduction of H_2O_2 at $P_2Mo_{18}O_{62}^{6-}$ modified glassy carbon. *Electrochimica Acta*. 45: 1829–1836.

Rossmeisl, J.; Qu, Z.-W.; Zhu, H.; Kroes, G.-J.; Nørskov, J.K. 2007. Electrolysis of water on oxide surfaces. *Journal of Electroanalytical Chemistry*. 607: 83–89.

Su, L.-H.; Zhang, X.-G.; Liu, Y. 2008. Electrochemical performance of Co-Al layered double hydroxide nanosheets mixed with multiwall carbon nanotubes. *Journal of Solid State Electrochemistry*. 12: 1129–1134.

CHAPTER 9

Abdelkader, A.M.; Fray, D.J. 2012. Electrochemical synthesis of hafnium carbide powder in molten chloride bath and its densification. *Journal of the European Ceramic Society*. 32: 4481–4487.

Bard, A.J.; Fox, M.A. 1995. Artificial photosynthesis – Solar splitting of water to hydrogen and oxygen. *Accounts of Chemical Research*. 28: 141–145.

Cheng, Y.; Lu, S.; Liao, F.; Liu, L.; Li, Y.; Shao, M. 2017. Rh-MoS_2 nanocomposite catalysts with Pt-Like activity for hydrogen evolution reaction. *Advanced Functional Materials*. 27: artic. 1700359.

Choi, C.H.; Chung, M.W.; Kwon, H.C.; Park, S.H.; Woo, S.I. 2013. B, N- and P, N-doped graphene as highly active catalysts for oxygen reduction reactions in acidic media. *Journal of Materials Chemistry A*. 1: 3694–3699.

Cui, Z.; Yang, M.; DiSalvo, F.J. 2014. Mesoporous $Ti_{0.5}Cr_{0.5}N$ supported PdAg nanoalloy as highly active and stable catalysts for the electro-oxidation of formic acid and methanol. *ACS Nano*. 8: 6106–6113.

Ding, Q.; Song, B.; Xu, P.; Jin, S. 2016. Efficient electrocatalytic and photoelectrochemical hydrogen generation using MoS_2 and related compounds. *Chem*. 1: 699–726.

Dinh, K.N.; Liang, Q.; Du, C.-F.; Zhao, J.; Tok, A.I.Y.; Mao, H.; Yan, Q. 2019. Nanostructured metallic transition metal carbides, nitrides, phosphides, and borides for energy storage and conversión. *Nano Today*. 25: 99–121.

Eng, A.Y.S.; Ambrosi, A.; Sofer, Z.; Simek, P.; Pumera, M. 2014. Electrochemistry of transition metal dichalcogenides: Strong dependence on the metal-to-chalcogen composition and exfoliation method. *ACS Nano*. 8: 12185–12198.

Gupta, S.; Patel, N.; Miotello, A.; Kothari, D.C. 2015. Cobalt-boride: An efficient and robust electrocatalyst for hydrogen evolution reaction. *Journal of Power Sources*. 279: 620–625.

Hunt, S.T.; Nimmanwudipong, T.; Román-Leshkov, Y. 2014. Engineering non-sintered, metal-terminated tungsten carbide nanoparticles for catalysis. *Angewandte Chemie International Edition*. 53: 5131–5136.

Koitz, R.; Nørskov, J.K.; Studt, F. 2015. A systematic study of metal-supported boron nitride materials for the oxygen reduction reaction. *Physical Chemistry Chemical Physics*. 17: 12722–12727.

Lyalin, A.; Nakayama, A.; Uosaki, K.; Taketsugu, T. 2013. Theoretical predictions for hexagonal BN based nanomaterials as electrocatalysts for the oxygen reduction reaction. *Physical Chemistry Chemical Physics*. 15: 2809–2820.

Masa, J.; Weide, P.; Peeters, D.; Sinev, I.; Xia, W.; Sun, Z.; Somsen, C.; Muhler, M.; Schuhmann, W. 2016. Amorphous cobalt boride (Co_2B) as a highly efficient nonprecious catalyst for electrochemical water splitting: Oxygen and hydrogen evolution. *Advanced Energy Materials*. 6: artic. 1502313.

Merki, D.; Vrubel, H.; Rovelli, L.; Fierro, S.; Hu, X.L. 2012. Fe, Co, and Ni ions promote the catalytic activity of amorphous molybdenum sulfide films for hydrogen evolution. *Chemical Science*. 3: 2515–2525.

Ozerova, M.; Simagina, V.I.; Komova, O.V.; Netskina, O.V.; Odegova, G.V.; Bulavchenko, O.A.; Rudina, N.A. 2012. Cobalt boride catalysts for small-scale energy application. *Journal of Alloys and Compounds*. 513: 266.

Saadi, F.H.; Carim, A.I.; Velazquez, J.M.; Baricuatro, J.H.; McCrory, C.C.L.; Soriaga, M.P.; Lewis, N.S. 2014. Operand synthesis of macroporous molybdenum diselenide films for electrocatalysis of the hydrogen-evolution reaction. *ACS Catalysis*. 4: 2866–2873.

Sheng, T.; Xu, Y.; Jiang, Y.; Huang, L.; Tian, N.; Zhou, Z.; Broadwell, I.; Sun, S. 2016. Structure design and performance tuning of nanomaterials for electrochemical energy conversion and storage. *Accounts of Chemical Research*. 49: 2569–2577.

Shi, Y.; Zhang, B. 2016. Recent advances in transition metal phosphide nanomaterials: Synthesis and applications in hydrogen evolution reaction. *Chemical Society Reviews*. 45: 1529–1541.

Tributsch, H.; Bennett, J.C. 1977. Electrochemistry and photochemistry of MoS_2 layer crystals. *Journal of Electroanalytical Chemistry*. 81: 97–111.

Vrubel, H.; Hu, X.L. 2013. Growth and activation of an amorphous molybdenum sulfide hydrogen evolving catalyst. *ACS Catalysis*. 3: 2002–2011.

Wang, B.; Hu, Y.; Yu, B.; Zhang, X.; Yang, D.; Chen, Y. 2019. Heterogeneous CoFe-Co_8FeS_8 nanoparticles embedded in CNT networks as highly efficient and stable electrocatalysts for oxygen evolution reaction. *Journal of Power Sources*. 434: 126688.

Wang, B.; Wang, Z.; Wang, X.; Zheng, B.; Zhang, W.; Chen, Y. 2018. Scalable synthesis of porous hollow CoSe2-MoSe2/carbon microspheres for highly efficient hydrogen evolution reaction in acidic and alkaline media. *Journal of Materials Chemistry A*. 6: 12701–12707.

Yan, Y.; Ge, X.M.; Liu, Z.L.; Wang, J.Y.; Lee, J.M.; Wang, X. 2013. Facile synthesis of low crystalline MoS2 nanosheet-coated CNTs for enhanced hydrogen evolution reaction. *Nanoscale*. 5: 7768–7771.

Yang, D.; Sandoval, S.J.; Divigalpitiya, W.M.R.; Irwin, J.C.; Frindt, R.F. 1991. Structure of single-molecular-layer MoS_2. *Physical Review B*. 43: 12053–12056.

Ye, G.; Gong, Y.; Lin, J.; Li, B.; He, Y.; Pantelides, S.T.; Zhou, W.; Vajtai, R.; Ajayan, P.M. 2016. Defects engineered monolayer MoS_2 for improved hydrogen evolution reaction. *Nano Letters*. 16: 1097–1103.

CHAPTER 10

Allen, M.J.; Ting, V.C.; Kaner, R.B. 2010. Honeycomb carbon: A review of graphene. *Chemical Reviews*. 110: 132–145.

Alpatova, N.M.; Goldshleger, N.F.; Ovsyannikova, E.V. 2011. Electrochemistry of fullerenes immobilized on electrodes. *Russian Journal of Electrochemistry*. 44: 78–90.

Anderson, M.R.; Dorn, H.C.; Stevenson, S.A. 2000. Making connections between metalofullerenes and fullerenes: Electrochemical investigations. *Carbon*. 38: 1663–1670.

Barisci, J.N.; Wallace, G.G.; Baughman, R.H. 2000. Electrochemical quartz crystal microbalance studies of single-wall carbon nanotubes in aqueous and non-aqueous solutions. *Electrochimica Acta*. 46: 509–517.

Bianco, A.; Cheng, H.-M.; Enoki, T.; Gogotsi, Y.; Hurt, R.H.; Koratkar, N.; Kyotani, T.; Monthioux, M.; Park, C.R.; Tascon, J.M.D.; Zhang, J. 2013. All in the graphene family – A recommended nomenclature for two-dimensional carbon materials. *Carbon*. 65: 1–6.

Bond, A.M.; Feldberg, S.W.; Miao, W.; Oldham, K.B.; Raston, C.L. 2001. Modelling of solid-state, dissolution and solution-phase reactions at adhered solid–electrode–solvent (electrolyte) interfaces: Electrochemistry of microcrystals of C_{60} adhered to an electrode in contact with dichloromethane (Bu_4NClO_4). *Journal of Electroanalytical Chemistry*. 501: 22–32.

Chabot, V.; Higgins, D.; Yu, A.; Xiao, X.; Chen, Z.; Zhang, J. 2014. A review of graphene and graphene oxide sponge: Material synthesis and applications to energy and the environment. *Energy & Environmental Science*. 7: 1564–1596.

Eng, A.Y.S.; Ambrosi, A.; Chua, C.K.; Šaněk, F.; Sofer, Z.; Pumera, M. 2013. Unusual inherent electrochemistry of graphene oxides prepared using permanganate oxidants. *Chemistry A European Journal*. 19: 12673–12683.

Grodzka, E.; Grabowska, J.; Wysocka-Zolopa, M.; Winkler, K. 2008. Electrochemical formation and properties of two-component films of transition metal complexes and C_{60} or C_{70}. *Journal of Solid State Electrochemistry*. 12: 1267–1278.

Hirsch, A.; Li, Q.; Wudl, F. 1991. Globe-trotting hydrogens on the surface of the fullerene compound $C_{60}H_6(N(CH_2CH_2)_2O)_6$. *Angewandte Chemie International Edition*. 30: 1309–1310.

Joo, S.H.; Lee, H.I.; You, D.J.; Kwon, K.; Kim, J.H.; Choi, Y.S.; Kang, M.; Kim, J.M.; Pak, C.; Chang, H.; Seung, D. 2008. Ordered mesoporous carbons with controlled particle sizes as catalyst supports for direct methanol fuel cell cathodes. *Carbon*. 46: 2034–2045.

Leroux, F.; Raymundo-Piñero, E.; Fedelec, J.-M.; Béguin, F. 2006. Textural and electrochemical properties of carbon replica obtained from styryl organo-modified layered double hydroxide. *Journal of Materials Chemistry*. 16: 2074–2081.

Loy, D.A.; Assink, R.A. 1992. Synthesis of a fullerene C60-p-xylylene copolymer. *Journal of the American Chemical Society*. 114: 3977–3978.

Meerholz, K.; Tschuncky, P.; Heinze, J. 1993. Voltammetry of fullerenes C60 and C70 in dimethylamine and methyl chloride. *Journal of Electroanalytical Chemistry*. 347: 425–433.

Moore, R.R.; Banks, C.E.; Compton, R.G. 2004. Basal plane pyrolytic graphite modified electrodes: Comparison of carbon nanotubes and graphite powder as electrocatalysts. *Analytical Chemistry*. 76: 2677–2682.

Neufeld, A.K.; O'Mullane, A.P. 2006. Effect of the mediator in feedback mode-based SECM interrogation of indium tin-oxide and boron-doped diamond electrodes. *Journal of Solid State Electrochemistry*. 10: 808–816.

Pandurangappa, M.; Ramakrishnappa, T. 2008. Derivatization and characterization of functionalized carbon powder via diazonium salt reduction. *Journal of Solid State Electrochemistry*. 12: 1411–1419.

Pope, M.A.; Punckt, C.; Aksay, I.A. 2011. Intrinsic capacitance and redox activity of functionalized graphene sheets. *Journal of Physical Chemistry C*. 115: 20326–20334.

Pumera, M.; Wong, C.H.A. 2013. Graphene and hydrogenated graphene. *Chemical Society Reviews*. 42: 5987–5995.

Punckt, C.; Pope, M.A.; Liu, J.; Lin, Y.H.; Aksay, I.A. 2010. Electrochemical performance of graphene as effected by electrode porosity and graphene functionalization. *Electroanalysis*. 22: 2834–2841.

Raut, A.S.; Parker, C.B.; Stoner, B.R.; Glass, J.T. 2012. Effect of porosity variation on the electrochemical behavior of vertically aligned multi-walled carbon nanotubes. *Electrochemistry Communications*. 19: 138–141.

Ruiz, V.; Blanco, C.; Granda, M.; Menéndez, R.; Santamaría, R. 2008. Effect of the thermal treatment of carbon-based electrodes on the electrochemical performance of supercapacitors. *Journal of Electroanalytical Chemistry*. 618: 17–23.

Ryoo, R.; Joo, S.H.; Jun, S. 1999. Synthesis of highly ordered carbon molecular sieves via template-mediated structural transformation. *Journal of Physical Chemistry B*. 103: 7743–7746.

Shao, Y.Y.; Wang, J.; Wu, H.; Liu, J.; Aksay, I.A.; Lin, Y.H. 2010. Graphene based electrochemical sensors and biosensors: A review. *Electroanalysis*. 22: 1027–1036.

Shao, Y.; Jiang, Z.; Zhang, Q.; Guan, J. 2019. Progress in nonmetal-doped graphene electrocatalysts for the oxygen reduction reaction. *ChemSusChem*. 12: 2133–2146.

Streeter, I.; Wildgoose, G.G.; Shao, L.D.; Compton, R.G. 2008. Cyclic voltammetry on electrode surfaces covered with porous layers: An analysis of electron transfer kinetics at single-walled carbon nanotube modified electrodes. *Sensors and Actuators B*. 133: 462–466.

Vairavapandian, D.; Vichchulada, P.; Lay, M.D. 2008. Preparation and modification of carbon nanotubes: Review of recent advances and applications in catalysis and sensing. *Analytical Chimica Acta*. 626: 119–129.

Wang, J.; Xu, F.; Jin, H.; Chen, Y.; Wang, Y. 2017. Non-noble metal-based carbon composites in hydrogen evolution reaction: Fundamentals to applications. *Advanced Materials*. 29: artic. 1605838.

Yamawaki, H.; Yoshida, M.; Kakadate, Y.; Usuba, S.; Yokoi, H.; Fujiwara, S.; Aoki, K.; Ruoff, R.; Malhotra, R.; Lorents, D.C. 1993. Infrared study of vibrational property and polymerization of fullerene C_{60} and C_{70} under pressure. *Journal of Physical Chemistry*. 97: 11161–11163.

Zhang, L.; Lian, J. 2008. The electrochemical polymerization of *o*-phenylenediamine on L-tyrosine functionalized glassy carbon electrode and its application. *Journal of Solid State Electrochemistry*. 12: 757–763.

Zhang, J.; Qu, L.; Shi, G.; Liu, J.; Chen, J.; Dai, L. 2016. N,P-codoped carbon networks as efficient metal-free bifunctional catalysts for oxygen reduction and hydrogen evolution reactions. *Angewandte Chemie International Edition*. 128: 2270–2274.

CHAPTER 11

Adamczyk, L.; Kulesza, P.J.; Miecnikowski, K.; Palys, B.; Chojak, M.; Krawczyk, D. 2005. Effective charge transport in poly(3,4-ethylenedioxythiophene) based hybrid films containing polyoxometallate redox centers. *Journal of the Electrochemical Society*. 152: E98–E103.

Cuentas-Gallegos, K.; Lira-Cantú, M.; Casañ-Pastor, N.; Gómez-Romero, P. 2004. Electroactive organic-inorganic hybrid materials. From electrochemistry to multifaceted applications. In Brillas, E.; Cabot, P.-L. Eds. *Trends in Electrochemistry and Corrosion at the Beginning of the 21st Century*. Universitat de Barcelona, Barcelona, pp. 243–258.

Denny, R.A.; Sangaranarayan, M.V. 1998. Dynamics of electrón hopping in redox polymer electrodes using kinetic Ising model. *Journal of Solid State Electrochemistry*. 2: 67–72.

Gabrielli, C.; Takenouti, H.; Haas, O.; Tsukada, A. 1991. Impedance investigation of the charge transport in film-modified electrodes. *Journal of Electroanalytical Chemistry*. 302: 59–89.

Gomez-Romero, P. 2001. Hybrid organic–inorganic materials—in search of synergic activity. *Advanced Materials*. 13: 163–174.

Gomez-Romero, P.; Sanchez, C. 2004. *Functional Hybrid Materials*. Wiley-VCH, Weinheim.

Hatchett, D.W.; Josowicz, M. 2008. Composites of intrinsically conducting polymers as sensing nanomaterials. *Chemical Reviews*. 108: 746–769.

He, D.P.; Rong, Y.Y.; Kou, Z.K.; Mu, S.C.; Peng, T.; Malpass-Evans, R.; Carta, M.; McKeown, N.B.;

Marken, F. 2015. Intrinsically microporous polymer slows down fuel cell catalyst corrosion. *Electrochemistry Communications*. 59: 72–76.

He, D.; Rauwel, E.; Malpass-Evans, R.; Carta, M.; McKeown, N.B.; Gorle, D.B.; Kulandainathan, M.A.; Marken, F. 2017. Redox reactivity at silver microparticle—glassy carbon contacts under a coating of polymer of intrinsic microporosity (PIM). *Journal of Solid State Electrochemistry*. 21: 2141–2146.

Heeger, A.J. 2010. Semiconducting polymers: The third generation. *Chemical Society Reviews*. 39: 2354–2371.

Hirschhorn, B.; Orazem, M.E.; Tribollet, B.; Vivier, V.; Frateur, I.; Musiani, M. 2010. Determination of effective capacitance and film thickness from constant-phase-element parameters. *Electrochimica Acta*. 55: 6218–6227.

Hyodo, K. 1994. Electrochromism of conducting polymers. *Electrochimica Acta*. 39: 265–272.

Inzelt, G.; Lang, G. 1994. Model dependence and reliability of the electrochemical quantities derived from the measured impedance spectra of polymer modified electrodes. *Journal of Electroanalytical Chemistry*. 378: 39–49.

Inzelt, G. 2017. Recent advances in the field of conducting polymers. *Journal of Solid State Electrochemistry*. 21: 1965–1975.

Kim, D.W.; Park, J.K.; Rhee, H.W. 1996. Conductivity and thermal studies of solid polymer electrolytes prepared by blending poly(ethylene oxide), poly(oligo[oxyethylene]oxysebacoyl) and lithium perchlorate. *Solid State Ionics*. 83: 49–56.

Kulesza, P.J.; Miecznikowski, K.; Malik, M.A.; Galkowski, M.; Chojak, M.; Caban, K.; Wieckowski, A. 2001. Electrochemical preparation and characterization of hybrid films composed of Prussian blue type metal hexacyanoferrate and conducting polymer. *Electrochimica Acta*. 46: 4065–4073.

Kulesza, P.J.; Chojak, M.; Miecznikowski, K.; Leyera, A.; Malik, M.A.; Kuhn, A. 2002. Polyoxometallates as inorganic templates for monolayers and multilayers of ultrathin polyaniline. *Electrochemistry Communications*. 4: 510–515.

Lacroix J.C.; Kanazawa, D.A. 1989. Polyaniline: A very fast electrochromic material. *Journal of the Electrochemical Society*. 136: 1308–1313.

Lee, G.S.; Lee, Y.-J.; Yoon, K.B. 2001. Layer-by-layer assembly of zeolite crystals on glass with polyelectrolytes as ionic linkers. *Journal of the American Chemical Society*. 123: 9769–9779.

MacCallum, J.R.; Smith, M.J.; Vincent, C.A. 1981. The effect of radiation-induced crosslinking on the conductance of $LiClO_4$·PEO electrolytes. *Solid State Ionics*. 11: 307–312.

Maia, G.; Torresi, R.M.; Ticianelli, E.A.; Nart, F.C. 1996. Charge compensation dynamics in the redox processes of polypyrrole modified electrodes. *Journal of Physical Chemistry*. 100: 15910–15916.

MacCallum, J.R.; Smith, M.J.; Vincent, C.A. 1981. The effect of radiation-induced crosslinking on the conductance of $LiClO_4$·PEO electrolytes. *Solid State Ionics*. 11: 307–312.

Malinauskas, A.; Malinauskiene, J.; Ramanavicius, A. 2005. Conducting polymer-based nanostructurized materials: Electrochemical aspects. *Nanotechnology*. 16: R51–R62.

Marchesi, L.F.; Jacumasso, S.C.; Quintanilla, R.C.; Winnischofer, H.; Vidotti, M. 2015. The electrochemical impedance spectroscopy behavior of poly(aniline) nanocomposite electrodes modified by layer-by-layer deposition. *Electrochimica Acta*. 174: 864–870.

Mijangos, C.; Hernández, R.; Martín, J. 2016. A review on the progress of polymer nanostructures with modulated morphologies and properties, using nanoporous AAO templates. *Progress in Polymer Science*. 54–55: 148–182.

Nagashree, K.L.; Ahmed, M.F. 2010. Electrocatalytic oxidation of methanol on Ni modified polyaniline electrode in alkaline médium. *Journal of Solid State Electrochemistry*. 14: 2307–2320.

Pleus, S.; Schulte, B. 2001. Poly(pyrroles) containing chiral side chains: Effect of substituents on the chiral recognition in the doped as well as in the undoped state of the polymer film. *Journal of Solid State Electrochemistry*. 5: 522–530.

Rajesh; Ahuja, T.; Kumar, D. 2009. Recent progress in the development of nano-structured conducting polymers/nanocomposites for sensor applications. *Sensors and Actuators B*. 36: 275–286.

Rassaei, L.; Sillanpää, M.; Milson, E.V.; Zhang, X.; Marken, F. 2008. Layer-by-layer assembly of Ru^{3+} and $Si_8O_{20}{}^{8-}$ into electrochemically active silicate films. *Journal of Solid State Electrochemistry*. 12: 747–755.

Reddy, Ch.V.S.; Zhu, Q.-Y.; Mai, L.-Q. 2007. Electrochemical studies on PVC/PVdF blend-based polymer electrolytes. *Journal of Solid State Electrochemistry*. 11: 543–548.

Rong, Y.Y.; Malpass-Evans, R.; Carta, M.; McKeown, N.B.; Attard, G.A.; Marken, F. 2014. Intrinsically porous polymer protects catalytic gold particles for enzymeless glucose oxidation. *Electroanalysis*. 26: 904–909.

Salaneck, W.R.; Friend, R.H.; Brédas, J.L. 1999. Electronic structure of conjugated polymers: Consequences of electron-lattice coupling. *Physics Reports*. 319: 231–251.

Scholz, F. 2011. The electrochemistry of particles, droplets, and vesicles–The present situation and future tasks. *Journal of. Solid State Electrochemistry*. 15: 1699–1702.

Scotto, J.; Marmisollé, W.A.; Posadas, D. 2019. About the capacitive currents in conducting polymers: The case of polyaniline. *Journal of Solid State Electrochemistry*. 23: 1947–1965.

Shi, W.; Zeng, H.; Sahoo, Y.; Ohulchanskyy, T.Y.; Ding, Y.; Wang, Z.L.; Swihart, M.; Prasad, P.N. 2006. A general approach to binary and ternary hybrid nanocomposites. *Nano Letters*. 6: 875–881.

Watanabe, M.; Nagano, S.; Sanvi, K.; Ogata, N. 1987. Structure-conductivity relationship in polymer electrolytes formed by network polymers from poly[dimethylsiloxane-g- poly(ethylene oxide)] and litegum perchlorate. *Journal of Power Sources*. 20: 327–332.

Wieckzorek, W.; Stevens, J.R. 1997. Impedance spectroscopy and phase structure of polyether-poly(methyl methacrylaye)-$LiCF_3SO_3$ blend-based electrolytes. *Journal of Physical Chemistry B*. 101: 1529–1534.

Wu, D.; Xu, F.; Sun, B.; Fu, R.; He, H.; Matyjaszewski, K. 2012. Design and preparation of porous polymers. *Chemical Reviews*. 112: 3959–4015.

Yoshida, T.; Zhang, J.B.; Komatsu, D.; Sawatani, S.; Minoura, H.; Pauporte, T.; Lincot, D.; Oekermann, T.; Schlettwein, D.; Tada, H.; Wöhrle, D.; Funabiki, K.; Matsui, M.; Miura, H.; Yanagi, H. 2009. Electrodeposition of inorganic/organic hybrid thin films. *Advanced Functional Materials*. 19: 17–43.

CHAPTER 12

Back, R.P. 1987. Diffusion-migration impedances for finite, one-dimensional transport in thin-layer and membrane cells: Part II. Mixed conduction cases: Os(III)/Os(II)ClO_4 polymer membranes including steady-state *I–V* responses. *Journal of Electroanalytical Chemistry*. 219: 23–48.

Brett, C.M.; Oliveira-Brett, A.M. 2011. Electrochemical sensing in solution—Origins, applications and future perspectives. *Journal of Solid State Electrochemistry*. 15: 1487–1494.

Doménech-Carbó, A.; Doménech-Carbó, M.T.; García, H.; Galletero, M.S. 1999. Electrocatalysis of neurotransmitter catecholamines by 2,4,6-triphenylpyrylium ion immobilized inside zeolite Y supercages. *Journal of the Chemical Society Chemical Communications*. 2173–2174.

Doménech-Carbó, A.; Aucejo, R.; Alarcón, J.; Navarro, P. 2004. Electrocatalysis of the oxidation of methylenedioxyamphetamines at electrodes modified by cerium-doped zirconias. *Electrochemistry Communications*. 6: 719–723.

Dong, L.Q.; Zhang, L.; Duan, X.M.; Mo, D.Z.; Xu, J.K.; Zhu, X.F. 2016. Synthesis and characterization of chiral PEDOT enantiomers bearing chiral moieties in side chains: Chiral recognition and its mechanism using electrochemical sensing technology. *RSC Advances*. 6: 11536–11545.

Fletcher, S. 2011. Electronomics. *Journal of Solid State Electrochemistry*. 15: 1451.

Fog, A.; Buck, R.P. 1984. Electronic semiconducting oxides as pH sensors. *Sensors and Actuators*. 5: 137–146.

Gabel, J.; Vonau, W.; Shuk, P.; Guth, U. 2004. New reference electrodes based on tungsten-substituted molybdenum bronzes. *Solid State Ionics*. 169: 75–80.

Guidelli, R.; Becucci, L. 2011. Ion transport across biomembranes and model membranes. *Journal of Solid State Electrochemistry*. 15: 1459–1470.

Inzelt, G. 2020. Future of electrochemistry in light of history and the present conditions. *Journal of Solid State Electrochemistry*. 24: 1–4.

Lai, C.-Z.; Fierke, M.A.; Stein, A.; Bühlmann, P. 2007. Ion-selective electrodes with three-dimensionally ordered macroporous carbon as the solid contact. *Analytical Chemistry*. 79: 4621–4626.

Ou, J.; Tao, Y.X.; Xue, J.J.; Kong, Y.; Dai, J.Y.; Deng, L.H. 2015. Electrochemical enantiorecognition of tryptophan enantiomers based on graphene quantum dots–chitosan composite film. *Electrochemistry Communications*. 57: 5–9.

Peng, Z.H. 2004. Rational synthesis of covalently bonded organic–inorganic hybrids. *Angewandte Chemie International Edition*. 43: 930–935.

Pichon, V.; Chapuis-Hugon, F. 2008. Role of molecularly imprinted polymers for selective determination of environmental pollutants—A review. *Analytica Chimica Acta*. 622: 48–61.

Prodomidis, M.I. 2010. Impedimetric immunosensors–A review. *Electrochimica Acta*. 55: 4227–4233.

Richards, E.; Bessant, C.; Saini, S. 2002. Multivariate data analysis in electroanalytical chemistry. *Electroanalysis*. 14: 1533–1542.

Seeber, R.; Terzi, F. 2011. The evolution of amperometric sensing from the bare to the modified electrode systems. *Journal of Solid State Electrochemistry*. 15: 1523–1534.

Seeber, R.; Terzi, F.; Zanardi, C. 2014. *Functional Materials in Amperometric Sensing. Polymeric, Inorganic, and Nanocomposite Materials for Modified Electrodes*. In Scholz, F. Ed. Monographs in Electrochemistry Series. Springer, Berlin.

Shi, G.; Qu, Y.; Zhai, Y.; Liu, Y.; Sun, Z.; Yang, J.; Jin, L. 2007. $\{MSU/PDDA\}_n$ LBL assembled modified sensor for electrochemical detection of ultratrace explosive nitroaromatic compounds. *Electrochemistry Communications*. 9: 1719–1724.

Wang, Y.; Zhang, H.; Yao, D.; Pu, J.; Zhang, Y.; Gao, X.; Sun, Y. 2013. Direct electrochemistry of hemoglobin on graphene/Fe_3O_4 nanocomposite-modified glass carbon electrode and its sensitive detection for hydrogen peroxide. *Journal of Solid State Electrochemistry*. 17: 881–887.

Zen, J.-M.; Chen, P.-J. 1997. A selective voltammetric method for uric acid and dopamine detection using clay-modified electrodes. *Analytical Chemistry*. 69: 5087–5093.

Zhang, X.; Li, L.; Peng, X.; Chen, R.; Huo, K.; Chu, P.K. 2013. Non-enzymatic hydrogen peroxide photoelectrochemical sensor based on WO_3 decorated core–shell TiC/C nanofibers electrode. *Electrochimica Acta*. 108: 491–496.

Zhao, W.W.; Xiong, M.; Li, X.R.; Xu, J.J.; Chen, H.Y. 2014. Photoelectrochemical bioanalysis: A mini review. *Electrochemistry Communications*. 38: 40–43.

CHAPTER 13

Anand, K.; Singh, O.; Singh, M.O.; Kaur, J.; Singh, R.C. 2014. Hydrogen sensor based on graphene/ZnO nanocomposite. *Sensors and Actuators B*. 195: 409–415.

Domènech-Gil, G.; Barth, S.; Samà, J.; Pellegrino, P.; Gràcia, I.; Cané, C.; Romano-Rodriguez, A. 2017. Gas sensors based on individual indium oxide nanowire. *Sensors and Actuators B*. 238: 447–454.

Drobek, M.; Kim, J.H.; Bechelany, M.; Vallicari, C.; Julbe, A.; Kim, S.S. 2016. MOF-based membrane encapsulated ZnO nanowires for enhanced gas sensor selectivity. *ACS Applied Materials & Interfaces*. 8: 8323–8328.

Esmaeilzadeh, J.; Marzbanrad, E.; Zamani, C.; Raissi, B. 2012. Fabrication of undoped-TiO_2 nanostructure-based NO_2 high temperature gas sensor using low frequency AC electrophoretic deposition method. *Sensors and Actuators B*. 161: 401–405.

Fergus, J.W. 2007. Materials for high temperature electrochemical NO_x gas sensors. *Sensors and Actuators B*. 121: 652–663.

Girdauskaite, E.; Ullmann, H.; Al Daroukh, M.; Vashook, V.; Bülow, M.; Guth, U. 2006. Oxygen nonstoichiometry and electrical conductivity of $Pr_{2-x}Sr_xNiO_{4\pm\delta}$ with $x = 0–0.5$. *Solid State Ionics*. 167: 1163–1171.

Hyodo, T.; Takamori, M.; Goto, T.; Ueda, T.; Shimizu, Y. 2019. Potentiometric CO sensors using anion-conducting polymer electrolyte: Effects of the kinds of noble metal-loaded metal oxides as sensing-electrode materials on CO-sensing properties. *Sensors and Actuators B*. 287: 42–52.

Korotcenkov, G. 2005. Gas response control through structural and chemical modification of metal oxide films: State of art and approaches. *Sensors and Actuators B*. 107: 209 – 232.

Li, J.R.; Kuppler, R.J.; Zhou, H.C. 2009. Selective gas adsorption and separation in metal–organic frameworks. *Chemical Society Reviews*. 38: 1477–1504.

Nisar, J.; Topalian, Z.; Sarkar, A.D.; Osterlund, L.; Ahuja, R. 2013. TiO_2-based gas sensor: A possible application to SO_2. *ACS Applied Materials & Interfaces*. 5: 8516–8522.

Pandey, P.A.; Wilson, N.R.; Covington, J.A. 2013. Pd-doped reduced graphene oxide sensing films for H_2 detection. *Sensors and Actuators B*. 183: 478–487.

Park, C.O.; Akbar, S.A.; Weppner, W. 2003. Ceramic electrolytes and electrochemical sensors. *Journal of Materials Science*. 38: 4639–4660.

Penner, R.M. 2017. A nose for hydrogen gas: Fast, sensitive H_2 sensors using electrodeposited nanomaterials. *Accounts of Chemical Research*. 50: 1902–1910.

Rheaume, J.M.; Pisano, A. 2012. Investigation of an impedancemetric NO_x sensor with gold wire working electrodes. *Journal of Solid State Electrochemistry*. 16: 3603–3610.

Suematsu, K.; Watanabe, K.; Yuasa, M.; Kida, T.; Shimanoe, K. 2019. Effect of ambient oxygen partial pressure on the hydrogen response of SnO_2 semiconductor gas sensors. *Journal of the Electrochemical Society*. 166: B618–B622.

Sun, L.; Campbell, M.G.; Dincă, M. 2016. Electrically conductive porous metal–organic frameworks. *Angewandte Chemie International Edition*. 55: 3566–3579.

Ullmann, H.; Trofimenko, N.J. 2001. Estimation of effective ionic radii in highly defective perovskite-type oxides from experimental data. *Journal of Alloys and Compounds*. 316: 153–158.

Wang, Y.; Qu, F.; Liu, J.; Wang, Y.; Zhou, J.; Ruan, S. 2015. Enhanced H_2S sensing characteristics of CuO-NiO core-shell microspheres sensors. *Sensors and Actuators B*. 209: 515–523.

Wen, Z.; Shen, Q.; Sun, X. 2017. Nano-generators for self-powered gas sensing. *Nano-Micro Letters*. 9: artic. 45.

Weppner, W. 1987. Solid-state electrochemical gas sensors. *Sensors and Actuators*. 12: 107–119.

Zeng, Y.; Zhang, K.; Wang, X.; Sui, Y.; Zou, B.; Zheng, W.; Zou, G. 2011. Rapid and selective H_2S detection of hierarchical $ZnSnO_3$ nanocages. *Sensors and Actuators B*. 159: 245–250.

CHAPTER 14

Abdel Rahim, M.A.; Hassan, H.B.; Andel Hamid, R.M. 2006. A systematic study of the effect of OH– and Ni^{2+} ions on the electro-catalytic oxidation of methanol at Ni-S-1 electrode. *Journal of Power Sources*. 154: 59–65.

Alcaide, F.; Brillas, E.; Cabot, P.-L.; Casado, J. 1998. Electrogeneration of hydroperoxide ion using an alkaline fuel cell. *Journal of the Electrochemical Society*. 145: 3444–3449.

Alcaide, F.; Brillas, E.; Cabot, P.-L. 2004. Limiting behaviour during the hydroperoxide ion generation in a flow alkaline fuel cell. *Journal of Electroanalytical Chemistry*. 566: 235–240.

Auborn, J.J.; Barberio, Y.L. 1987. Lithium intercalation cells without metallic lithium. Molybdenum dioxide/lithium cobalt dioxide and tungsten dioxide/lithium cobalt dioxide. *Journal of the Electrochemical Society*. 134: 638–647.

Cheng, L.; Li, H.-q.; Xia, Y.-y. 2006. A hybrid nonaqueous electrochemical supercapacitor using nano-sized iron oxyhydroxide and activated carbon. *Journal of Solid State Electrochemistry*. 10: 405–410.

Conway, B.E. 1999. *Electrochemical Supercapacitors*. Springer, Boston.

Dai, X.; Chen, D.; Fan, H.; Zhong, Y.; Chang, L.; Saho, H.; Wang, J.; Zhang, J.; Cao, C.-n. 2015. $Ni(OH)_2$/NiO/Ni composite nanotube arrays for high-performance supercapacitors. *Electrochimica Acta*. 154: 128–135.

Delmas, C.; Cognac-Auradou, H.; Cocciantelli, J.M.; Menetrier, M.; Doumerc, J.P. 1994. The $Li_xV_2O_5$ system: An overview of the structure modifications induced by the lithium intercalation. *Solid State Ionics*. 69: 257–264.

Desilvestro, J.; Haas, O. 1990. Metal oxide cathode materials for electrochemical energy storage: A review. *Journal of the Electrochemical Society*. 137: 5C–22C.

Gasik, M. Ed. 2008. *Materials for Fuel Cells*. CRC-Woodhead Publishing. Cambridge.

Guarnieri, M.; Alotto, P.; Moro, F. 2015. Modeling the performance of hydrogen-oxygen unitized regenerative proton exchange membrane fuel cells for energy storage. *Journal of Power Sources*. 297: 23–32.

Guo, G.H.; Tao, Y.T.; Song, Z.P.; Zhang, K.L. 2007. Zinc tetrathiomolybdate as novel anodes for rechargeable lithium batteries. *Journal of Solid State Electrochemistry*. 11: 90–92.

Higgins, D.; Zamani, P.; Yu, A.; Chen, Z. 2016. The application of graphene and its composites in oxygen reduction electrocatalysis: a perspective and review of recent progress. *Energy & Environmental Science*. 9: 357–390.

Ho, M.Y.; Kiew, P.S.; Isa, D.; Tan, T.K.; Chiu, W.S.; Chia, C.H. 2014. A review of metal oxide composite materials for electrochemical capacitors. *Nano: Brief Reports and Reviews*. 9: artic. 1430002.

Kourasi, M.; Will, R.G.A.; Shah, A.A.; Walsh, F.C. 2014. Heteropolyacids for fuel cell applications. *Electrochimica Acta*. 127: 454–466.

Li, H.; Weng, G.; Li, C.Y.V.; Chan, K.-J. 2011. Three electrolyte high voltage acid-alkaline hybrid rechargeable battery. *Electrochimica Acta*. 56: 9420–9425.

Licht, S.; Wang, B.; Xu, G.; Li, J.; Naschitz, V. 1999. Insoluble Fe(VI) compounds: Effects on the super-iron battery. *Electrochemistry Communications*. 1: 522–526.

Liu, Y.; Gorgutsa, S.; Santato, C.; Skorogobatiy, M. 2012. Flexible, solid electrolyte-based lithium battery composed of $LiFePO_4$ cathode and $Li_4Ti_5O_{12}$ anode for applications in smart textiles. *Journal of the Electrochemical Society*. 159: A349–A356.

Maher, K.; Yazami, R. 2013. Effect of overcharge on entropy and enthalpy of lithium-ion batteries. *Electrochimica Acta*. 101: 71–78.

Maher, K.; Yazami, R. 2014. A study of lithium ion batteries cycle aging by thermodynamics techniques. *Electrochimica Acta*. 247: 527–533.

Manjunatha, H.; Shures, G.S.; Venkatesha, T.V. 2011. Electrode materials for aqueous rechargeable lithium batteries. *Journal of Solid State Electrochemistry*. 15: 431–445.

Martos, M.; Morales, J.; Sánchez, L. 2000. Cation-deficient $Mo_ySn_xO_2$ oxides as anodes for lithium ion batteries. *Electrochimica Acta*. 46: 83–89.

Obeidat, A.M.; Luthra, V.; Rastogi, A.C. 2019. Solid-state graphene-based supercapacitor with high-density energy storage using ionic liquid gel electrolyte: Electrochemical properties and performance in storing solar electricity. *Journal of Solid State Electrochemistry*. 23: 1667–1683.

Paloukis, F.; Elmasides, C.; Farmakis, F.; Selinis, P.; Neophytides, S.G.; Georgoulas, N. 2016. Electrochemical impedance spectroscopy study in micro-grain structured amorphous silicon anodes for lithium-ion batteries. *Journal of Power Sources*. 331: 285–292.

Rice, C.; Ha, S.; Masel, R.I.; Wieckowski, A. 2003. Catalysts for direct formic acid fuel cells. *Journal of Power Sources*. 115: 229–235.

Santos-Peña, J.; Soudan, P.; Otero-Areán, C.; Turnes-Palomino, G.; Franger, S. 2006. Electrochemical properties of mesoporous iron phosphate in lithium batteries. *Journal of Solid State Electrochemistry*. 10: 1–9.

Simmons, W.; Gonnissen, D.; Hubin, A. 1997. Study of the initial stages of silver electrocrystallisation from silver thiosulphate complexes: Part I: Modelling of the silver nuclei formation during the induction period. *Journal of Electroanalytical Chemistry*. 433: 141–151.

Sinha, N.N.; Munichandraiah, N. 2008. Synthesis and characterization of submicron size particles of $LiMn_2O_4$ by microemulsion route. *Journal of Solid State Electrochemistry*. 12: 1619–1627.

Skowroński, J.M.; Ważny, A. 2000. Nickel foam-based composite electrodes for electrooxidation of methanol. *Journal of Solid State Electrochemistry*. 9: 890–899.

Snook, G.A.; Kao, P.; Best, A.S. 2011. Conducting-polymer-based supercapacitor devices and electrode. *Journal of Power Sources*. 196: 1–12.

Soudan, P.; Pereira-Ramos, J.P.; Farcy, J.; Gregoire, G.; Baffier, N. 2000. Sol-gel chromium–vanadium mixed oxides as lithium insertion compounds. *Solid State Ionics*. 135: 291–295.

Staffell, I.; Scamman, D.; Velazquez Abad, A.; Balcombe, P.; Dodds, P.E.; Ekins, P.; Shah, N.; Ward, K.R. 2019. The role of hydrogen and fuel cells in the global energy system. *Energy & Environmental Science*. 12: 463–491.

Sunde, S. 1997. Calculations of impedance of composite anodes for solid oxide fuel cells. *Electrochimica Acta*. 42: 2637–2648.

Taniguchi, I.; Bakenov, Z. 2005. Spray pyrolysis synthesis of nanostructured $LiFe_xMn_{2-x}O_4$ cathode materials for lithium-ion batteries. *Powder Technology*. 159: 55–62.

Tilha, M.; Khaldi, C.; Boussami, S.; Fenineche, N.E.; el Kedim, O.; Mathlouthi, H.; Lamloumi, J. 2013. Kinetics and thermodynamics of hydrogen storage alloys as negative electrode materials for Ni/MH batteries: A review. *Journal of Solid State Electrochemistry*. 18: 577–593.

Wang, G.; Zhang, L.; Zhang, J. 2012. A review of electrode materials for electrochemical supercapacitors. *Chemical Society Reviews*. 41: 797–828.

Wu, J.B.; Li, Z.G.; Lin, Y. 2011. Porous NiO/Ag composite film for electrochemical capacitor application. *Electrochimica Acta*. 56: 2116–2121.

Xia, Y.; Yoshio, M. 1996. An investigation of lithium ion insertion into spinel structure Li-Mn-O compounds. *Journal of the Electrochemical Society*. 143: 825–833.

Xie, X.; Zhao, X.B.; Cao, G.S.; Zhong, Y.D.; Zhao, M.J. 2003. Ex-situ XRD studies of $CoSb_3$ compound as the anode material for lithium ion batteries. *Journal of Electroanalytical Chemistry*. 542: 1–6.

Xu, B.; Wu, F.; Su, Y.; Xao, G.; Chen, S.; Zhou, Z.; Yang, Y. 2008. Competitive effect of KOH activation on the electrochemical performances of carbon nanotubes for EDLC: Balance between porosity and conductivity. *Electrochimica Acta*. 53: 7730–7735.

Yang, S.; Song, Y.; Zavalij, P.Y.; Whittingham, M.S. 2002. Reactivity, stability and electrochemical behavior of lithium iron phosphates. *Electrochemistry Communications*. 4: 239–244.

Ye, S.H.; Lv, J.Y.; Gao, X.P.; Wu, F.; Song, D.Y. 2004. Synthesis and electrochemical properties of $LiMn_2O_4$ spinel phase with nanostructure. *Electrochimica Acta*. 49: 1623–1628.

Yu, H.C.; Fung, K.Z. 2003. $La_{1-x}Sr_xCuO_{2.5-\delta}$ as new cathode materials for intermediate temperature solid oxide fuel cells. *Materials Research Bulletin*. 38: 231–239.

Yuan, C.-z.; Dou, H.; Gao, B.; Su, L.-h.; Zhang, Z.-g. 2008. High-voltage aqueous symmetric electrochemical capacitor based on $Ru_{0.7}Sn_{0.3}O_2{\cdot}nH_2O$ electrodes in 1 M KOH. *Journal of Solid State Electrochemistry*. 12: 1645–1652.

Zhang, L.L.; Zhao, X.S. 2009. Carbon-based materials as supercapacitor electrodes. *Chemical Society Reviews*. 38: 2520–2531.

Zheng, J.P.; Jow, T.R. 1995. A new charge storage mechanism for electrochemical capacitors. *Journal of the Electrochemical Society*. 142: L6–L8.

CHAPTER 15

Alhalasah, W.; Rolze, H. 2007. Electrochemical bandgaps of a series of poly-3-*p*-phenylthiophenes. *Journal of Solid State Electrochemistry*. 11: 1605–1612.

Bouhjar, F.; Mollar, M.; Chourou, M.L.; Marí, B.; Bessaïs, B. 2018. Hydrothermal synthesis of nanostructured Cr-doped hematite with enhanced photoelectrochemical activity. *Electrochimica Acta*. 260: 838–846.

Casades, I.; Constantine, M.S.; Cardin, D.; García, H.; Gilbert, A.; Márquez, F. 2000. 'Ship-in-a-bottle' synthesis and photochromism of spiropyrans encapsulated within zeolite Y supercages. *Tetrahedron*. 56: 6951–6956.

Cummings, C.Y.; Marken, F.; Peter, L.M.; Tahir, A.A.; Wijayantha, K.G.U. 2012. Kinetics and mechanism of light-driven oxygen evolution at thin-film α-Fe_2O_3 electrodes. *Chemical Communications*. 48: 2027–2029.

Currao, A. 2007. Photoelectrochemical water splitting. *Chemia*. 61: 815–819.

Doménech-Carbó, A.; Torres, F.J.; Ruiz de Sola, E.; Alarcón, J. 2006. Electrochemical detection of high oxidation states of chromium (IV and V) in chromium-doped cassiterite and tin-sphene ceramic pigmenting systems. *European Journal of Inorganic Chemistry*. 638–648.

Dutta, P.K.; Severance, M. 2011. Photoelectron transfer in zeolite cages and its relevance to solar energy conversion. *The Journal of Physical Chemistry Letters*. 2: 467–476.

Fu, C.; Luang, Y.; Huang, Z.; Wang, X.; Yin, Y.; Chen, J.; Zhou, H. 2010. Supercapacitor based on graphene and ionic liquid electrolyte. *Journal of Solid State Electrochemistry*. 15: 2581–2585.

Ketir, W.; Saadi, S.; Trari, M. 2012. Physical and photoelectrochemical characterization of $CuCrO_2$ single crystal. *Journal of Solid State Electrochemistry*. 16: 213–218.

Kuznetsov, D.A.; Han, B.; Yu, Y.; Rao, R.R.; Hwang, J.; Román-Leshkov, Y.; Shao-Horn, Y. 2018. Tuning redox transitions via inductive effect in metaloxides and complexes, and implications in oxygen electrocatalysis. *Joule*. 2: 225–244.

Liu, H.D.; Zhang, J.L.; Xu, D.D.; Huang, L.H.; Tan, S.Z.; Mai, W.J. 2015. Easy one-step hydrothermal synthesis of nitrogen-doped reduced graphene oxide/iron oxide hybrid as efficient supercapacitor material. *Journal of Solid State Electrochemistry*. 19: 135–144.

Luo, M.; Dou, Y.; Kang, H.; Ma, Y.; Ding, X.; Liang, B.; Ma, B.; Li, L. 2015. A novel interlocked Prussian blue/reduced graphene oxide nanocomposites as high-performance supercapacitor electrodes. *Journal of Solid State Electrochemistry*. 19: 1621–1631.

Morales-Guio, C.G.; Liardet, L.; Mayer, M.T.; Tilley, S.D.; Gratzel, M.; Hu, X.L. 2015. Photoelectrochemical hydrogen production in alkaline solutions using Cu_2O coated with earth-abundant hydrogen evolution catalysts. *Angewandte Chemie International Edition*. 54: 664–667.

Ohtsuka, T.; Iida, M.; Ueda, M. 2006. Polypyrrole coating doped by molybdo-phosphate anions for corrosion prevention of carbon steels. *Journal of Solid State Electrochemistry*. 10: 714–720.

O'Reilly, C.; Hinds, D.; Coey, J.M.D. 2001. Effect of a magnetic field on electrodeposition: Chronoamperometry of Ag, Cu, Zn, and Bi. *Journal of the Electrochemical Society*. 148: C674–C678.

Orimolade, B.O.; Koiki, B.A.; Peleyeju, G.M.; Arotiba, O.A. 2018. Visible light driven photoelectrocatalysis on a FTO/$BiVO_4$/BiOI anode for water treatment involving emerging pharmaceutical pollutants. *Electrochimica Acta*. 307: 285–292.

Oyarzún, D.P.; Córdova, R.; Linares Pérez, O.E.; Muñoz, E.; Henríquez, R.; López Teijelo, M.; Gómez, H. 2011. Morphological, electrochemical and photoelectrochemical characterization of nanotubular TiO_2 synthetized electrochemically from different electrolytes. *Journal of Solid State Electrochemistry*. 15: 2265–2275.

Ragsdale, S.R.; Lee, J.; Gao, X.; White, H.S. 1996. Magnetic field effects in electrochemistry. Voltammetric reduction of acetophenone at microdisk electrodes. *Journal of Physical Chemistry*. 100: 5913–5922.

Rincón, M.E.; Trujillo, M.E.; Ávalos, J.; Casillas, N. 2007. Photoelectrochemical processes at interfaces of nanostructured TiO_2/carbon black composites studied by scanning photoelectrochemical microscopy. *Journal of Solid State Electrochemistry*. 11: 1287–1294.

Sahaya Anand, T.J.; Shariza, S. 2018. A study on molybdenum sulfoselenide (MoS_xSe_{2-x}, $0 \le x \le 2$) thin films: Growth from solutions and its properties. *Electrochimica Acta*. 81: 64–73.

Tributsch, H. 2009. Nanocomposite solar cells: The requirement and challenge of kinetic charge separation. *Journal of Solid State Electrochemistry*. 13: 1127–1140.

Waskaas, M. 1993. Short-term effects of magnetic fields on diffusion in stirred and unstirred paramagnetic solutions. *Journal of Physical Chemistry*. 97: 6470–6476.

Waskaas, M. 1996. Magnetic field effect on electrode reactions. III. Effects on the anodic polarization of an iron electrode in an iron(III) chloride solution under potentiostatic conditions. *Acta Chemica Scandinavica*. 50: 526–530.

Waskaas, M. 1997. Comment on "Short-term effects of magnetic fields on diffusion in stirred and unstirred paramagnetic solutions". *Journal of Physical Chemistry*. 100: 8612.

CHAPTER 16

Attard, G.A. 2001. Electrochemical studies of enantioselectivity at chiral metal surfaces. *Journal of Physical Chemistry B*. 105: 3158–3167.

Awad, M.I.; Saleh, M.M.; Ohsaka, T. 2008. Oxygen reduction on rotating porous cylinder of modified reticulated vitreous carbon. *Journal of Solid State Electrochemistry*. 12: 251–258.

Beley, M.; Collin, J.-P.; Ruppert, R.; Sauvage, J.-P. 1984. Nickel(II)-cyclam: An extremely selective electrocatalyst for reduction of CO_2 in water. *Journal of the Chemical Society Chemical Communications*. 1315–1316.

Bessel, C.A.; Rolison, D.R. 1997. Electrocatalytic reactivity of zeolite-encapsulated Co(salen) with benzyl chloride. *Journal of the American Chemical Society*. 119: 12673–12674.

Borrás, C.; Laredo, T.; Scharifker, B.R. 2003. Competitive electrochemical oxidatiòn of *p*-chlorophenol and *p*-nitrophenol on Bi-doped PbO_2. *Electrochimica Acta*. 48: 2775–2780.

Burke, L.D.; Ryan, T.G. 1992. The role of incipient hydrous oxides in the oxidation of glucose and some of its derivatives in aqueous media. *Electrochimica Acta*. 37: 1363–1370.

Burke, L.D.; Naser, N.S.; Ahern, B.M. 2007. Use of iridium oxide films as hydrogen gas evolution cathodes in aqueous media. *Journal of Solid State Electrochemistry*. 11: 655–666.

Cao, D.; Wang, Y.; Qiao, M.; Zhao, X. 2018. Enhanced photoelectrocatalytic degradation of norfloxacin by an $Ag_3PO_4/BiVO_4$ electrode with low bias. *Journal of Catalysis*. 360: 240–249.

Chen, N.Y.; Degnan, T.F.Jr.; Smith, C.M. 1994. *Molecular Transport and Reactions in Zeolites*. VCH, New York, p. 110.

Comninellis, C. 1994. Electrocatalysis in the electrochemical conversion/combustion of organic pollutants for wastewater treatment. *Electrochimica Acta*. 39: 1857–1862.

Faria, E.R.; Ribeiro, F.M.; Franco, D.V.; Da Silva, L.M. 2018. Fabrication and characterisation of a mixed oxide-covered mesh electrode composed of $NiCo_2O_4$ and its capability of generating hydroxyl radicals during the oxygen evolution reaction in electrolyte-free water. *Journal of Solid State Electrochemistry*. 22: 1289–1302.

Feng, Y.; Cui, Y.; Logan, B.; Liu, Z. 2008. Performance of Gd-doped Ti-based Sb-SnO2 anodes for electrochemical destruction of phenol. *Chemosphere*. 70: 1629–1636.

Grimm, J.; Bessarabov, D.; Maier, W.; Storck, S.; Sanderson, R.D. 1998. Sol-gel film-preparation of novel electrodes for the electrocatalytic oxidation of organic pollutants in water. *Desalination*. 115: 295–302.

Hayes, E.A.; Stemple, J.Z.; Rolison, D.R. 1994. In *Water Purification by Photoelectrochemical, Photochemical, and Electrochemical Methods*. Rose, T.L.; Conway, B.E.; Murphy, O.J. Eds. The Electrochemical Society, Pennington, vol. 94-19, p. 121.

Jonson, A.M.; Newman, J. 1971. Desalting by means of porous carbon electrodes. *Journal of the Electrochemical Society*. 118: 510–517.

Kaluza, D.; Jöhnsson-Niedziólka, M.; Ahn, S.D.; Owen, R.E.; Jones, M.D.; Marken, F. 2015. Solid-solid EC' TEMPO-electrocatalytic conversion of diphenylcarbinol to benzophenone. *Journal of Solid State Electrochemistry*. 19: 1277–1283.

Komorsky-Lovric, S. 1997. Voltammetry of azobenzene microcrystals. *Journal of Solid State Electrochemistry*. 1: 94–99.

Kondintsev, I.M.; Trasatti, S. 1994. Electrocatalysis of H_2 evolution on $RuO_2 + IrO_2$ mixed oxide electrodes. *Electrochimica Acta*. 39: 1803–1808.

Lee, J.-B.; Park, K.-K.; Yoon, S.-W.; Park, P.-Y.; Park, K.-I.; Lee, C.-W. 2009. Desalination performance of a carbon-based composite electrode. *Desalination*. 237: 155–161.

Li, M.C.; Shen, J.N. 2006. Photoelectrochemical oxidation behaviour of organic substances on TiO_2 thin-film electrodes. *Journal of Solid State Electrochemistry*. 10: 980–986.

Martinez-Huitle, C.A.; Ferro, S. 2006. Electrochemical oxidation of organic pollutants for the wastewater treatment: Direct and indirect processes. *Chemical Society Reviews*. 35: 1324–1340.

Molina, V.M.; Frías, A.; González, J.; Montiel, V.; González, D.; Domínguez, M.; Aldaz, A. 2004. Design and development of filter-press type electrochemical reactors for their application in the resolution of environmental problems. In Brillas, E.; Cabot, P.-L. Eds. *Trends in Electrochemistry and Corrosion at the Beginning of the 21*st *Century*. Universitat de Barcelona, Barcelona, pp. 383–399.

Monteiro, O.C.; Mendonça, M.H.M.; Pereira, M.I.S.; Nogueira, J.M.F. 2006. Electrodegradation of organic pollutants at lead and tin oxide thin films. *Journal of Solid State Electrochemistry*. 10: 41–47.

Nafady, A.; Sabri, Y.M.; Kandjani, A.E.; Alsalme, A.M.; Bond, A.M.; Bhargava, S. 2016. Preferential synthesis of highly conducting Tl(TCNQ) phase II nanorod networks via electrochemically driven TCNQ/Tl(TCNQ) solid-solid phase transformation. *Journal of Solid State Electrochemistry*. 20: 3303–3314.

Panizza, M.; Cerisola, G. 2009. Direct and mediated anodic oxidation of organic pollutants. *Chemical Reviews*. 109: 6541–6569.

Saleh, M.M. 2007. Simulation of oxygen evolution reaction at porous anode from flowing electrolytes. *Journal of Solid State Electrochemistry*. 11: 811–820.

Schaufβ, A.; Wittstock, G. 1999. Oxidation of 2-mercaptobenzoxazole in aqueous solution: Solid phase formation at glassy carbon electrodes. *Journal of Solid State Electrochemistry*. 3: 361–369.

Silva, L.M.D.; De Faria, L.A.; Boodts, J.F.C. 2003. Electrochemical ozone production: Influence of the supporting electrolyte on kinetics and current efficiency. *Electrochimica Acta*. 48: 699–709.

Zhao, Y.; Zahng, X.; Zhai, J.; He, J.; Jiang, L.; Liu, Z.; Nishimoto, S.; Murakami, T.; Fujishima, A.; Zhu, D. 2008. Enhanced photocatalytic activity of hierarchically micro-/nano-porous TiO_2 films. *Applied Catalysis B: Environmental*. 1–2: 24–29.

Index

Note: Page numbers in *italics* represent figures and page numbers in **bold** represent tables.

Milton Keynes UK
Ingram Content Group UK Ltd.
UKHW052026141024
449569UK00016B/719